# An Historical Geography of Europe

# An Historical Geography of Europe

Edited by

R. A. Butlin and R. A. Dodgshon

Clarendon Press · Oxford

1998

Oxford University Press, Great Clarendon Street, Oxford OX2 6DP

Oxford New York
Athens Auckland Bangkok Bogotá Buenos Aires Calcutta
Cape Town Chennai Dar es Salaam Delhi Florence Hong Kong Istanbul
Karachi Kuala Lumpur Madrid Melbourne Mexico City Mumbai
Nairobi Paris São Paulo Singapore Taipei Tokyo Toronto Warsaw
and associated companies in Berlin Ibadan

Oxford is a registered trade mark of Oxford University Press

Published in the United States
by Oxford University Press Inc., New York

British Library Cataloguing in Publication Data
Data available

Library of Congress Cataloging in Publication Data
An historical geography of Europe / R. A. Butlin and R. A. Dodgshon.
    p.   cm.
Includes bibliographical references.
1. Europe–Historical geography.   I. Butlin, R. A. (Robin Alan),
1938–   .   II. Dodgshon, R. A. (Robert A.)
D21.5.H57   1998
911'.4–dc21   98–16712

ISBN 0-19-874179-0
ISBN 0-19-874178-2 (Pbk.)

10 9 8 7 6 5 4 3 2 1

Typeset by J&L Composition Ltd, Filey, North Yorkshire
Printed in Great Britain
on acid-free paper by
Bookcraft (Bath) Ltd, Midsomer Norton, Somerset

# Preface

FOR over a century the historical geography of Europe has been a focus of attention for historians and geographers seeking to understand the broad and detailed configurations through time of the complex interactions between humans and their environments. Thus in 1881 E. A. Freeman, Professor of Modern History at Oxford, produced his book *The Historical Geography of Europe*, whose purpose was to review the changing political and territorial boundaries of the states and territories of Europe through time in relation to their geographical background, a work mirrored in a more limited perspective in J. M. Thompson's *Historical Geography of Europe, 1800–1889*, published in 1929 (Freeman, 1881; Thompson, 1929).

These works were produced by historians, but with the development of Geography in the universities of Europe, especially from the late nineteenth century, more overtly geographical accounts of the historical geography of Europe were produced (Butlin, 1993: 12–16). One early overview of this kind was W. G. East's *An Historical Geography of Europe* (East, 1935), which took a strongly thematic approach, though this type of erudite synthesis was still dependent on interpretation of work by historians and archaeologists, as there was little primary work by historical geographers at this time. Increasingly, though, there were similar studies which offered pan-European syntheses of more select themes of the historical geographies of Europe, such as A. Meitzen's study of rural settlements (Meitzen, 1895), and represented towards the mid-twentieth century by such studies as N. J. G. Pounds' *An Historical and Political Geography of Europe* (Pounds, 1947), and H. C. Darby's study of European woodland and its clearance (Darby, 1956).

By the 1960s, however, the nature of the approach to the understanding of the historical geography of Europe was beginning to change, with increasing emphasis on region- or country-based studies, using primary research in historical geography. Examples include E. A. Wrigley's study of the Belgian coalfield (Wrigley, 1962) and A. Lambert's study of the making of the Dutch landscape (Lambert, 1957). This original work now gave scope for broader syntheses, evident in C. T. Smith's richly woven text *An Historical Geography of Western Europe before 1800* (Smith, 1967). There were further notable developments in the late 1950s and the 1960s. A group of historians and geographers with interests in the historical development of the rural landscapes of Europe, who had first met at Nancy in 1957, convened a meeting at Vadstena in Sweden in 1961. The Vadstena meeting brought together a range of European scholars interested in the morphologies of European rural landscapes, and out of these developments was born the Permanent European Conference for the Study of Rural Landscape, a bi-annual conference which discusses the long-term development of the European countryside. The Permanent Conference has provided a forum at which research findings and methodologies from across Europe can be compared (Baker, 1988), and similar though less frequent meetings have provided opportunity for the comparison of work on European urban historical geography. When combined with the explosion of cognate work in economic, social, and cultural history and in historical demography, this now gives a much fuller and more extensively researched foundation for the historical geography of Europe. Its potential has been skilfully realized in N. J. G. Pounds' two-volume study of the historical geography of Europe, subsequently distilled into a single volume (Pounds: 1979, 1985, 1990), and in more thematic studies such as Carter's urban historical geography (Carter, 1983), national studies such as those by Clout (Clout, 1977, 1980) and X. de Planhol on France (Planhol, 1988), and in large-scale regional studies such as C. Delano Smith's historical geography of the Western Mediterranean (Delano Smith, 1979).

The nature and emphases of historical geography have also been changing over recent times, conspicuously evidenced by the broadening of the range of themes addressed, the development and employment of a greater range of methodologies and sources, and all within increasing ideological and innova- tive contexts (Dodgshon and Butlin, 1990: p. vii). Cross-disciplinary explora- tions and connections have continued, and made for richer, more comprehensive and challenging interpretations of the dynamics of the complex European past, which can also have important bearings on the future.

The present volume is the first major multi-authored text on the historical geography of Europe. It brings together a range of scholars who have worked on different aspects of European historical geography and related fields such as history and archaeology. Its main purpose is to provide an informed review of past and current thinking on the major themes of European historical geo- graphy at a time when the geography of Europe, including its political aspects, is changing at great speed but also at a time when our interpretations of the European past are also changing. Produced on a broad canvas, but with attention to regional character wherever possible, it aspires to provide the reader with a summary of the larger issues and themes around which this historical geography can be structured. It attempts to draw out some of Europe's enduring similarities and differences while, at the same time, high- lighting its remarkable capacity for change and innovation.

Euro-centrism, the writing out of history as a study of Europe's inevitable rise to prominence, is currently under fire, especially in studies of imperialism, and for good reason. Yet in questioning how history has been written around Europe and its perspectives, we must not ignore the vital ways in which the European experience is different. As a continent, it boasts a regionally varied landscape, varied not just in a cultural, political, and economic sense, but also in an environmental sense. Further, these regional differences have been—for much of its history—elaborated and enriched by change. If we add to this the fact that its past is amongst the most closely studied of any continent, then we surely have the potential for a historical geography that can both challenge and inform.

The editors owe debts of gratitude to the contributors to this book, to Professor Hugh Clout who translated Chapters 2 and 5, and to departmental cartographers, particularly Ian Gulley and Anthony Smith at Aberystwyth and

Erica Milwain and Peter Robinson at Loughborough University, for drawing many of the maps and diagrams. We are additionally indebted to the editorial officers at Oxford University Press.

REFERENCES

BAKER, A. R. H. 'Historical Geography and the Study of the European Rural Landscape', *Geografiska Annaler*, 70B (1): 5–16 (1988).

BUTLIN, R. A. *Historical Geography: Through the Gates of Space and Time* (London, 1993).

CLOUT, H. D. (ed.), *Themes in the Historical Geography of France* (London, 1977).

—— *Agriculture in France on the Eve of the Railway Age* (London, 1980).

CARTER, H. *An Introduction to Urban Historical Geography* (London, 1983).

DARBY, H. C. 'The Clearing of the Woodland in Europe', in W. J. Thomas, Jr. (ed.), *Man's Role in Changing the Face of the Earth* (Chicago, 1956): 183–216.

DELANO SMITH, C. *Western Mediterranean World, a Historical Geography of Italy, Spain and Southern France since the Neolithic* (London, New York, 1979).

DODGSHON, R. A. and BUTLIN, R. A. (eds.), *An Historical Geography of England and Wales*, 2nd edn. (London, 1990).

EAST, W. G. *An Historical Geography of Europe* (London, 1935).

FREEMAN, E. A. *The Historical Geography of Europe* (London, 1881).

LAMBERT, A. *The Making of the Dutch Landscape* (London, 1957).

MEITZEN, A. *Siedlung und Agrarwesen der Westgermanen und Ostgermanen, der Kelten, Römer, Finnen und Slawen*, 3 vols. and atlas (Berlin, 1895).

PLANHOL, X. de *Géographie historique de la France* (Paris, 1988).

POUND, N. J. G. *An Historical and Political Geography of Europe* (London, 1947).

—— *An Historical Geography of Europe, the 1500–1840* (Cambridge, 1979).

—— *An Historical Geography of Europe, 1880–1914* (Cambridge, 1985).

—— *An Historical Geography of Europe* (Cambridge, 1990).

SMITH, C. T. *An Historical Geography of Western Europe before 1800* (London, 1967).

THOMPSON, J. M. *An Historical Geography of Europe* (Oxford, 1929).

WRIGLEY, E. A. *Industrial Growth and Population Change: A Regional Study of the Coalfield Areas of North-West Europe in the Later Nineteenth Century* (Cambridge, 1962).

# Contents

## Contents

# List of Contributors

PROFESSOR P. ARNAUD, Laboratoire d'Archeolgie, Universite Nice Sophia Antipolis, Parc Valrose, 28 Avenue Valrose, 06108 Nice Cedex 2, France

PROFESSOR R. A. BUTLIN, Principal, University College of Ripon and St John, Lord Mayor's Walk, York, YO3 7EX

PROFESSOR H. CLOUT, Department of Geography, University College, 26 Bedford Way, London, WC1H 0AP

PROFESSOR R. A. DODGSHON, Institute of Geography & Earth Sciences, University of Wales, Aberystwyth, Dyfed, SY23 3DB

DR P. GLENNIE, Department of Geography, University of Bristol, University Road, Bristol, B58 1SS

DR M. HEFFERNAN, Department of Geography, University of Technology, Loughborough, Leicestershire, LE11 3TU

PROFESSOR W. MEAD, Department of Geography, University College, 26 Bedford Way, London, WC1H 0AP

PROFESSOR P. E. OGDEN, Department of Geography, Queen Mary and Westfield College, University of London, Mile End Road, London, E1 4NS

PROFESSOR S. POLLARD (formerly Universities of Sheffield and Bielefeld), 34 Bents Road, Sheffield, S11 9RJ

PROFESSOR B. K. ROBERTS, Department of Geography, University of Durham, Science Laboratories, South Road, Durham City, DH1 5AS

DR A. SHERRATT, Department of Antiquities, Ashmolean Museum, Oxford, OX1 2PH

PROFESSOR I. SIMMONS, Department of Geography, University of Durham, Science Laboratories, South Road, Durham City, DH1 5AS

PROFESSOR A. SUTCLIFFE, Department of Economic and Social History, University of Leicester, Leicester, LE1 7RH

PROFESSOR DR A. VERHULST, Faculteit van de Letteren en Wijsbegeerte Rijksuniversiteit-Gent, Blandijnberg 2, 9000 Gent, Belgium

DR J. R. WALTON, Institute of Geography & Earth Sciences, University of Wales, Aberystwyth, SY23 3DB

PROFESSOR I. D. WHYTE, Department of Geography, University of Lancaster, Bailrigg, Lancaster, LA1 4YB

# List of Plates

# List of Figures and Maps

# List of Tables

# Chapter 1

# The Human Geography of Europe: A Prehistoric Perspective

## A. Sherratt

The cartographic images that confront us on the pages of a modern atlas give the Earth's surface an impression of permanence; but it is a deceptive one. It is not just the brightly coloured political maps whose boundaries periodically change, but also the more subtly toned and naturally coloured maps of the physical landscape which alter over time. Between the last two editions of the *Oxford Atlas*, for instance, the Caspian Sea has altered its outline, and the Aral Sea all but disappeared—the latter under human influence, to irrigate the cotton crops of Kazakhstan. Maps of settlement-distributions and population density require constant updating; but so, too, should 'natural' vegetation and coastlines, and on a longer time-scale even the reassuring averages of precipitation and annual temperature. Every map represents only a sample snapshot, extracted from patterns which are continuously changing, each at its own characteristic (though variable) rate. A historical geography is therefore an account of processes operating over a defined timescale, themselves encompassing briefer, event-like changes, and all set within longer and (on a human timescale) barely perceptible alterations which set the stage but do not interact in the way that human and natural processes have influenced each other over the last ten thousand years. To place these later phenomena in perspective, it is instructive to begin with the deep-seated rhythms of the Earth itself.

The fragment (or, more accurately, collection of fragments) of the Earth's continental crust which is called Europe, with its sedimentary deposits and shallow seas, achieved its position by a series of accidents over the span of geological time; and impacts with other continental masses created the mountain chains that outline its perimeters (Andrews, 1991). For a long part of its existence the patch of continental crust that was to become Europe was joined to North America rather than Asia, and the Ural mountains mark the collision which brought the future Europe and Asia together in the supercontinent of Pangaea some 225 million years ago. Fifty-five million years previously, Africa had made its impact upon this union of Europe and North America, so creating the range of Hercynian mountains which stretch from Bohemia, through Wales to Alabama; two hundred million years before that, the Caledonian mountains had been created when Europe and North America first collided. During the last sixty million years, North America has moved away from Eurasia as the Atlantic Ocean widened; Africa had separated

1

from it earlier, but now began to return, and its impact with the Eurasian landmass all but obliterated the intervening Tethys Sea and created the latest and still sharply upstanding chain of 'Alpine' mountains that marches from the Atlantic to northern India. Like a sheet of pastry, therefore, successively cut and kneaded together at the edges, Europe was configured by the processes of plate tectonics, driven by convection currents deep in the mantle of the Earth. Moving around the surface of the globe in these different combinations, the future Europe found itself at different points on the Earth's surface: at times in the subtropics, at times in the latitudes it now occupies, within a belt of largely temperate climate.

That climate, however, was itself no constant (Street, 1980). The changing sea levels, and especially the changing disposition of the continental masses in relation to the poles, caused major differences in temperature and hence in precipitation as well. For much of the Earth's history, latitudinal differences in climate and vegetation have been much less marked than those of today, and the Earth's climate has been generally warmer. A long-term secular trend of declining temperatures was initiated when Antarctica, formerly linked to both India and Australia, separated from them and found itself situated over the South Pole, some eighty million years ago. As ice accumulated there, global temperatures began to fall, and other events accelerated the process or introduced fluctuations: changes in oceanic circulation consequent on continental convergence (especially the joining of North and South America), or astronomically-caused alterations in the amount and distribution of sunlight reaching the Earth's surface. These caused a build-up of ice, first at the South Pole by forty million years ago, then on mountains in high altitudes; then, over the last two million years, in a series of extensions and retractions, over large areas of mid-latitude continents where precipitation allowed the accumulation of large masses of ice—the continental icecaps which in the northern hemisphere reached as far south as London and Chicago. With large amounts of water locked up as ice, global sea levels fell, exposing large areas of continental shelf; with a more sluggish circulation system, global precipitation lessened, too, creating enlarged areas of desert, steppe, and tundra. Over these last two million years—the period recognized

as the Quaternary or Pleistocene Ice Age, conditions comparable to those of today have alternated with rather longer phases of such cold, dry conditions: 'interglacials' within longer 'glacials'. This information is of no mere antiquarian interest, for the brief warm period in which we live—and which began, together with the earliest attempts at farming, some 12,000 years ago—is simply the latest interglacial in the cycle. Much of the environmental history of this time, therefore, can be considered as the rebound from glacial conditions, the establishment of a precarious interglacial balance, and its long-term rundown (and accelerated human degradation) towards a condition which in its treelessness, though not in its faunal impoverishment, resembles that of a glacial episode.

## The Shape of Europe

Such, in brief, are the processes which have shaped the European landmass. Before we follow its human story as a chronological narrative, it is worth pondering for a moment those shapes within which that story has been set (Andrews, 1991). Chief amongst these is the large internal ocean—the *mare internum* or Mediterranean. Together with the Black Sea, and the Caspian and Aral Seas, this is essentially the surviving remnant of the great Tethys Sea which once separated Laurasia and Gondwanaland, the two successor-continents of Pangaea, of which Eurasia and Africa were respectively parts. The chain of Alpine mountains which lies largely parallel to it on the Eurasian side, separating the Mediterranean from the Black Sea at the Bosphorus and swinging on down to the Himalayas, was created by the convergence of Eurasia with Africa and the increasingly separate Arabia (divided from it by the African Rift system and the Red Sea), and also by collision with the quite independent continental mass of India, whose impact created the much more massive mountains east of the Pamir Knot. As well as being of topographic significance, the Alpine mountains were to be important as sources of metals, particularly copper. Within the Alpine system, some constituent sub-chains, reflecting different episodes in the collision of Africa and Europe, may be distinguished. The line of the southern Alps is prolonged eastward parallel to the Adriatic coastline as the Dinaric Alps, which form the backbone of the Greek peninsula and terminate in the Aegean island arc, underneath which the oceanic crust in front of the converging African plate slowly sinks. (Turkey, scissored between converging Eurasia and Arabia, exerts a westward pressure.) This is the reason for the vulcanism of this region, most spectacularly expressed in the eruption of Thera in the seventeenth century BC. The line of the northern Alps is prolonged eastward in a much more sinuous course as the Carpathians, continued (like a handle to their great sickle-blade) by the Balkan mountains of Bulgaria, whose line is picked up again in the Yaila mountains and Kertch Strait in the southern Crimea, and more spectactularly in the great range of the Caucasus. Between these two chains are the sinking median masses of the lower and middle Danube (Carpathian Basin), which for long were extensions of the Black Sea (extending up to the Vienna Basin and lower Bavaria) before being filled with flysch deposits from the Alps and Carpathians.

North of the line of Alpine mountains are the older mountain roots of the Hercynian system—important, like the Alps, for their metal resources, but having more deeply-formed minerals they have been particularly significant for rare resources like tin and, later in human history, silver. Hercynian mountains enclose Bohemia, and make up Brittany, Cornwall, and part of Wales. Between them are the Ardennes, broken by the Rhine rift, and the Massif Central. In between these upstanding massifs are broad plains, floored with limestones from the successive shallow seas that infilled them, and sometimes concealing deposits of coal from ancient swamplands at the foot of the Hercynian mountains. Northwards from this region of hills and plains, and stretching from the Rhine mouth to the Dniepr, is the great flat expanse of the North European Plain—its low relief created during the glacial periods and consisting of clayey moraines, outwash sands, and wind-blown deposits of sand and silt. The lightest fraction of this rock-dust— created by glacial scour and intensive frost-action, and carried over long distances by the wind across treeless plains—is called loess, and this wind-blown silt mantles both the southern edge of the North European Plain (in a zone called the *Bördeland*) and also the adjacent limestone-floored basins of the central belt of Europe, to provide one of the most fertile and easily worked soils in the continent. This fertility contrasts starkly with the belt immediately to the north: the heavy clays of the *Jungmorän* landscape, covered by ice in the last glaciation and floored with morainic clays, and the acid sands of the *Altmorän* landscape which were dropped by the meltwaters as they carved their way through the decaying ice-sheets along tunnel-valleys or *Urstromtäle*. The more northerly parts of the Plain, in front of the much more ancient hard rocks of the Scandinavian shield, are occupied by the waters of the Baltic: Europe's other, northerly *mare internum*, which has also contributed to the maritime traditions of the continent. Both the Baltic and the Black Sea had interesting salinity histories in the early postglacial period: the Baltic was alternately linked to the widening North Sea and cut off from it by rising land levels rebounding from their load of ice; the Black Sea was freshwater and only incompletely linked with the Mediterranean by a spillway over the Bosphorus sill until 5500 BC, when slowly rising Mediterranean saltwaters spilled into it and raised its level to that of the present day, so drowning extensive flat coastlands. Both Baltic and Black Sea coastlands supported extensive Mesolithic fishing populations, who long resisted the approach of agriculture.

Compared with other mid-latitude continental masses, therefore, Europe has much more extensive and deeply penetrating internal bodies of water, and it is this feature above all others which has contributed to the rapid development of its southern shores. Nevertheless it is remarkable quite how much of Europe's cultural history in the postglacial period has been determined from without. Throughout the last 10,000 years, in a situation only decisively reversed after AD 1600, Europe has been the recipient of innovations which had their origins in the Near East: agriculture, both primary and secondary; urbanism and its arts; iron metallurgy; writing and the alphabet; Christianity (and also Judaism and Islam); the astronomy, chemistry, and medicine of the Islamic world. There is a consistent pattern here; and without its older and more sophisticated east Mediterranean neighbours, Europe could scarcely have achieved what it likes to describe as civilization. Why is that?

## Old and New Geography

To answer the question requires a return to a kind of thinking that has been unfashionable, both in geography and in the social sciences in general, in the period since the Second World War. The principal post-war paradigm—culminating in 1960s 'New Geography'—has been an abstract search for regularities, reflecting a dominant concern with planning and spatial management, rather than an attempt at reading the record of what had happened and learning from it. This 'top–down' approach has been epitomized in economic models of spatial behaviour on isotropic surfaces, from which individual locational particularities had been removed. Such modernist thinking sought universal causalities, for instance in the relationship between population levels and the way resources are used—as in the classic study by Esther Boserup (1965) on agricultural intensification and the introduction of the plough, which relates the appearance of these features to increasing demographic pressure. Models of this kind give the impression that such changes could occur anywhere, whenever conditions were right to enter the next stage of a universal evolutionary succession. But on the even and featureless surface with which many such models begin, it is most unlikely that change of an important kind would have occurred at all: for it is precisely in the interstices and irregularities of the Earth's surface that major innovations have had their origins—the beginnings of farming and urbanism in the Near East and Mesoamerica, or the industrial revolution on Europe's Atlantic coast. Without the prior appearance of such phenomena in these unusual focal areas, it is unlikely that they would have made their appearance more widely in the more extensive areas whose landscapes are typical of the continents as a whole. The explanation of the more general appearance of these features over larger portions of the Earth's surface thus involves historical patterns of dispersal and spread which reflect processes of contact and social interaction in the transmission of new features, and which involve cultural geography as well as arguments from economics (Sherratt, 1995a). These processes of spatial transmission were formerly described, in a rather inappropriate physical analogy, as 'diffusion'. This name suggests an inevitability (and passivity on the part of the receivers) which is very misleading; and the excesses of the 'diffusionist' school led to a major reaction against such ideas. But whilst it is easy to caricature the view that 'everything came from

Egypt', it would be a fundamental mistake to ignore the fact that many things indeed did so. Europe cannot be considered in isolation.

To appreciate these causalities requires a return to the tradition of geographical interpretation of earlier in the century, associated for example with the names of Paul Vidal de la Blache, Jean Brunhes, or Lucien Febvre in France, or such different figures as Ellen Semple, Carl Sauer, and Derwent Whittlesey in the US, and Halford Mackinder, J. L. Myres, and H. J. Fleure in Britain. These writers began with the real configurations of landscape, and made tactical comparisons and generalizations about recurrent phenomena, rather than homogenizing spatial experience in terms of a few abstract principles. It is this sensitivity to the specific character of different regions which can provide an answer to the question raised above.

In this perspective, the Near East appears as one of the most unusual regions on Earth. As the point of contact between the Eurasian and Afro-Arabian plates, it forms an isthmus separating these two continental masses, and also a bridge (via the Red Sea and the Gulf) between the internal sea of the Mediterranean and the much larger Indian Ocean. But it is its environmental peculiarities, as well as its positional characteristics, which make it particularly distinctive. From its latitudinal position, it ought to be a desert; but the arid zone which occupies northern Africa and swings up into central Asia is interrupted by the chain of Tertiary fold-mountains created by the junction of Africa and Eurasia (Fig. 1.1). Occupying the basin between arid north Africa and the mountain chain is the truncated Tethys remnant, the Mediterranean, which permits westerly winds to penetrate from the Atlantic far into the continental interior. Rising as they meet the mountain chain where it swings southwards, they drop their moisture and so create the zonally atypical conditions of the Fertile Crescent—a kind of macro-oasis with run-off from the hills supporting springs, wadis, and the great perennial rivers of Mesopotamia. In a further complication, Mediterranean climates themselves are created by seasonal shifts between tropical desert and temperate rainfall regimes; and Egypt, otherwise completely desert, is supported by the waters of the Nile, fed by tropical rainfall further south. It would be hard to imagine a more unusual conjunction of conditions; and it is this highly

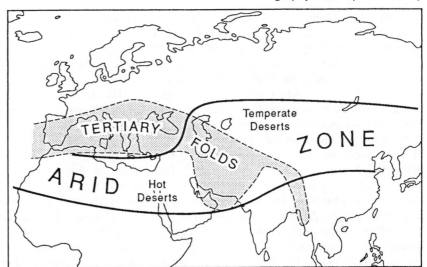

**Fig. 1.1.** Europe in relation to the unusual conjunction of environments in south-west Asia, where the arid zone crosses the chain of Tertiary fold-mountains.

unusual mixture which explains the unique historical role of the Near East (Sherratt, 1996).

The key to its creativity lies in its diversity, and in the intimate juxtaposition of highly contrasting environments. It was this which allowed, for instance, the large-seeded annual grasses that are adapted to summer aridity on the thin soils of the hill country to be 'switched' to the more fertile environments of the plains, where they would have been unable to grow without human aid. This is what created the possibility of farming. In similar fashion, the great exotic rivers of Mesopotamia and Egypt offered at the same time transport arteries, linking environments with highly contrasting resources, and the agrarian productivity that could be unlocked by irrigation. This created the possibility of the first cities. The large steppe and desert interfluves encouraged pastoralism, bringing a third way of life into existence, symbiotic with farming and urbanism. Europe, by contrast, offered a much more homogeneous set of environments, each gently grading into another: terrain where these new lifeways could be adopted, but with its divisions not so sharply drawn as to bring about their rapid genesis. Even if, in the fullness of time, farming and cities might have emerged indigenously in Europe, that time was not allowed: for the spread of farming and urban life to Europe—and, later, the other innovations listed above—provided ready-made patterns for the more temperate continent to follow.

Routes of entry from the North East into Europe lay either overland, across Anatolia (Turkey-in-Asia) or the Pontic steppes, or by water, along the Mediterranean and thence penetrating northwards by the Black Sea and Danube, or over Alpine passes to the Rhine, or else up the Rhône and Aude. Each of these routes was to achieve prominence, successively, as agricultural and then trade and urban activity moved from east to west. As these arteries were opened, and the European landmass slowly became linked in patterns of common intercourse and trade during the metal ages, so the cultural continent of Europe came into being. As urban life spread round the Mediterranean, so these northward links strengthened, bringing the Baltic and the Mediterranean into economic and cultural articulation. At the same time, farming and pastoralism spread eastwards, and domestication of the horse opened up the great expanses of steppe stretching eastwards to the Altai. The mobile populations of this region made a continuing contribution to European development: a 'wild east' comparable to America's 'wild west', where cowboy culture offered Europeans an alternative model to city life (Rolle, 1989).

What, then, is 'Europe'? Geologically it is a continental frontier zone, knocked into shape and tossed between larger landmasses; geographically it is a westward-tapering peninsula bordered by shallow continental seas, well watered on its western edge and increasingly dry to the east; vegetationally

5

it is a block of temperate, broad-leaved forest sandwiched between coniferous woodland to the north and desert to the south, grading eastwards into the more open landscapes of the steppes. Culturally it may be characterized as the temperate hinterland to a Near Eastern focal complex, opened up primarily by expansion along its inland seas and rivers, and exposed to the influence of steppe neighbours; its growing unity came about through the articulation of its northern and southern internal seas across the peninsula, first directly SE–NW or N–S (the 'amber routes') in later prehistory, and then in historical times both via the steppe rivers (Varangian route) and the Atlantic coastal route (Fig. 1.2). The rich metal resources of its central mountains encouraged internal exploration, and their topographic barrier effect, by preventing the creation of a north–south interior water-route (like China's Grand Canal), ensured that bulk transport between North and South would have to travel via the Atlantic. Exploration of that Ocean ended the isolation of the Old World and brought Atlantic Europe a new centrality within global maritime routes, which provided the conditions in which the Industrial Revolution could occur. Such, in brief, is the personality of the continent.

## Glacial Rebound and First Farmers, to 3500 BC

The first human ancestors to appear in Europe entered the continent under interglacial conditions between one million and half a million years ago. Successive glacial advances largely expelled these peripheral populations from most parts of the continent during these recurrent cold phases until in the last glaciation, beginning some 75,000 years ago, a certain very hardy sub-population managed to remain behind and conquer the cold. These were the Neanderthals, probably hairy (like their contemporaries the hairy mammoth and the woolly rhinoceros) and in other ways physically adapted to sub-zero temperatures. This achievement was nevertheless something of a side-show in human evolution, in that fully modern populations of *Homo sapiens*, with a far greater capacity for cultural adaptation (and therefore no need for specialization of physique) were during this time evolving

in Africa, and from 40,000 years ago onwards these incoming populations successively displaced or absorbed the Neanderthal populations of Europe, which were extinct by shortly after 30,000 years ago (Gamble, 1995).

The last glaciation reached its peak around 20,000 years ago.[1] An ice-sheet extended as far south as Szczecin, and tundra and steppe occupied the rest of the European landmass with the exception of a few forest refuges on the western side of the Mediterranean peninsulas (Street, 1980). Sea levels were some 120 m. lower than today; because of reduced evaporation, however, lakes occupied some lowland basins which are now largely dry. Glacial meltwaters occupied a chain of lakes on the margins of the ice-sheet, and a huge area of Siberia beyond the Urals; the overspill found its way south to an enlarged Caspian, which itself discharged along the Manych depression into the then freshwater Black Sea, cut off from the Mediterranean by the Bosphorus sill. Human populations ranged, if only seasonally in pursuit of reindeer, practically to the margins of the ice-sheet; but all the lands further south were inhabited, and in the more temperate parts of southern France and Spain, where the cold was moderated by Atlantic influences, culturally sophisticated communities created the famous cave-art. This pattern began to change from 14,000 BC onwards as global climate began to get somewhat warmer; more decisive warming during the Allerød period was then interrupted by a renewed cold millennium of the Younger Dryas period beginning around 11,000 BC; and after 10,000 BC there was rapid warming to the range of interglacial temperatures which prevail at the present time.

The warming caused a shift from glacial to interglacial modes: sea levels rose as the ice melted, drowning areas of previously exposed continental shelf; and forest vegetation began slowly to spread from its glacial-period ghettos—pioneer species like birch and pine travelling fastest, followed by hazel and oak, lime and elm. It took about 3,000 years for an area such as the North Sea basin to go from periglacial tundra to mixed oak forest, and the

[1] All dates are calibrated radiocarbon dates, using either Holocene dendro-sequences or late Pleistocene marine carbonate data; they thus correspond to true calendar years.

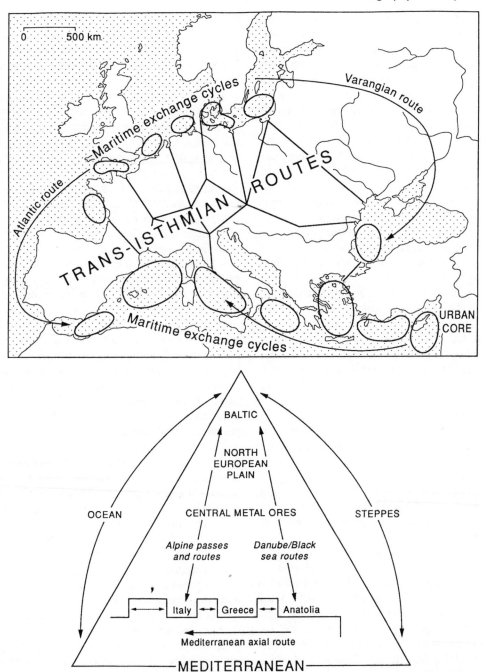

**Fig. 1.2.** Europe only became a unity when articulated by the development of long-distance routes, linking the Mediterranean with the Baltic (*above: map; below: schematic representation*). This occurred during the Bronze Age, when metals were exchanged e.g. for amber. These 'trans-isthmian' routes across the continent, shifting in individual importance, continued to provide the principal channels of north–south contact within Europe, supplemented during the medieval period first by the Russian riverine routes of the Varangian Vikings and then by the Atlantic maritime route. (Compare Fig. 1.6.)

ice did not disappear from Scandinavia until 7000 BC. Temperate animal species moved northwards, and fish recolonized the northern rivers. Animals which formed large herds, like reindeer and horse, became confined to the open areas further north and east, respectively; in the forests, red deer and elk, which could be stalked individually, became the major game. Vegetable foods became more abundant—especially hazelnuts, before hazel was shaded out by incoming oaks. Population became more dispersed, though concentrating in areas of particular abundance along coasts and rivers, especially at estuaries (see also Chapter 16). Dugout canoes, and probably also skin boats, made possible fishing and sea-travel, for instance to islands like Melos in the Aegean to collect the sharp volcanic glass called obsidian, used for making the finest stone blades.

During this period, as ice still covered the northern highlands, Neolithic farming villages appeared in the Near East. The sudden cold snap around 11,000 BC, following easier interstadial conditions, may have been a precipitating factor in necessitating cereal cultivation to support larger populations. Permanent settlements, marked by the first mudbrick architecture, soon came into existence. In the following millennia, animals such as goats and sheep, and then cattle, were domesticated, and their husbandry replaced the hunting of gazelle and deer as a source of protein. New arts, such as pottery-making and weaving, were elaborated. After a period of experimentation, characterized by large sites with major ceremonial installations like Jericho and Çatalhöyük, smaller farming settlements integrating all these innovations began to appear in southern Turkey and across the sea in Greece, from 7000 BC onwards (Scarre, 1988: 78–87). It is certain that an element of migration was involved, since these early farmers were the first human populations to reach Crete and Cyprus; on the mainland, smaller indigenous populations in the areas of fertile soil like Thessaly which were sought out by farming groups were probably integrated into these farming communities; elsewhere, along coasts where native populations were dense, farming was resisted or elements of the total package (like pottery and domestic livestock) selectively incorporated into indigenous ways of life. The two main routes by which farming spread through Europe—the Balkan/central European axis, and the Mediterranean—

form an instructive contrast in this respect: along the Mediterranean, indigenous groups adopted pottery, but continued to live in caves; in central Europe, where indigenous populations were sparse, the introduced pattern of farming villages proliferated, largely by population advance. This occurred very rapidly, from 5500 BC onwards, as village settlements of the *Bandkeramik* culture, now with timber-built rather than mudbrick houses, spread rapidly within the loess zone, both westwards as far as the Low Countries and eastwards to Moldavia. The process picked out the settlement-cells which were to be the heartlands of pre-industrial European demography. This marks the maximum extension of a pattern of village-based horticulture on the Near Eastern model. Beyond these limits, indigenous (Mesolithic) hunting, fishing, and collecting groups continued their existence, and adapted to the changing conditions of an increasingly densely forested continent, often by concentrating on lacustrine and marine resources. In the west Baltic, coastal populations associated with the Ertebølle culture fished from skin boats (and may have kept small quantities of livestock, acquired from nearby farmers) and accumulated great heaps of oyster-shells, sometimes accompanied by small cemeteries, which proudly marked their places of permanent settlement.

These indigenous populations, on three sides of the salient of incoming farming populations in south-east and central Europe, were increasingly drawn into new ways of life made possible by the innovations of farming communities; and from 4500 BC onwards the fusion of these two populations, the natives and the newcomers, created a second generation of European Neolithic cultures with greater regional diversity (Fig. 1.3). At the hub of this process, in south-east Europe (the Balkans and the Carpathian Basin), the predominantly immigrant cultures retained their undiluted distinctiveness, and followed their Near Eastern contemporaries into a Copper Age characterized by a range of largely prestige artefacts (such as shafthole battle-axes based on stone prototypes) made of unalloyed metal by relatively simple methods. Although based on local resources (the copper deposits of the Balkan and Carpathian mountains), and in their distinctive typologies showing no sign of contemporary Near Eastern designs, these objects are nevertheless symptomatic of wider

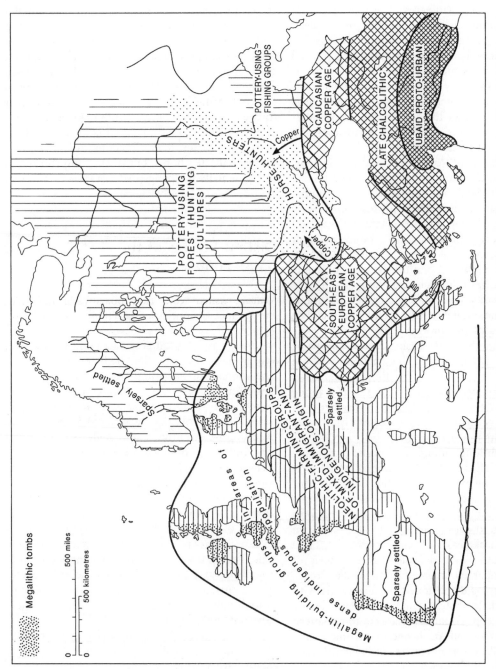

**Fig. 1.3.** Europe between 4500 and 3500 BC: as the societies of the Fertile Crescent of south-west Asia moved towards urban life in the Ubaid period, Europe showed a zonal arrangement of cultures. In the south-east, copper-using groups were the descendants of largely immigrant Neolithic populations of Anatolian origin; in central and western Europe a variety of late Neolithic groups represented a fusion of small numbers of incomers with indigenous (Mesolithic) populations. Areas of particularly dense indigenous population among the Atlantic façade were marked by the appearance of megalithic monuments. Beyond the limits of farming were hunting and fishing cultures which had adopted the use of pottery.

9

networks of contact and exchange. The rich cemetery of Varna on the Bulgarian Black Sea coast, with its spectacular sheet goldwork as well as copper tools and weapons, had access to a web of coastal contact reaching as far south as the Aegean islands, and small quantities of copperwork from the Carpathians was occasionally exported as far afield as Scandinavia and the Ukraine. It was perhaps this new scale of inter-regional contacts, and circulation of desirable new materials, which helped to draw the indigenous communities of outer Europe into more intimate relationship with the central parts.

Over a large arc of western and northern Europe, from Portugal to Poland, the populations and social structures of Neolithic and adjacent Mesolithic societies began to interbreed and fuse into a new synthesis based not on permanent villages but instead on burial and ceremonial monuments built of massive stones: the megaliths (Sherratt, 1990). The builders of these communal tombs probably combined cereal horticulture and small-scale livestock-keeping with a wider range of food-getting pursuits, notably fishing and hunting. The earliest stages of this process were marked by the emergence, in the Paris Basin, central Germany, and Kujavia, of earth and timber long mounds whose design echoed the contemporary timber longhouses of the adjacent loesslands. This shift to a new mode of spatial organization, in which tombs rather than villages were permanent features, seems to have been associated with the first moves beyond the loess (Fig. 1.4), and such long mounds also occur in Brittany and Jutland adjacent to Mesolithic coastal heartlands, where the main concentrations of megalithic tombs were to appear. This process seems to have initiated a continuing fusion of the two populations, which produced cultural groupings—the Chasséen and Funnel-neck beaker (Trichterrandbecher, TRB) culture groups—that occupied large blocks of territory comparable in area to present-day France and Germany, and integrated the loesslands with areas beyond. In eastern Europe, however, at the edge of the steppe country, a quite contrary pattern took shape. Instead of fusing, newcomers and natives—whilst fundamentally influencing each other—starkly differentiated in their ways of life. Increasingly large and well-defended villages of mud-and-timber longhouses, occupied by populations with a provincial version of Balkan Copper Age

lifestyle, the Tripole or Cucuteni culture, pushed out along the forest–steppe margin towards Kiev. Beyond them, in the river valleys running through the steppes, indigenous hunting and fishing populations (using acquired techniques of pottery-making and cattle-raising, and using small quantities of imported copper) gradually elaborated their own culture and economy, hunting the horses which roamed across the extensive interfluves. Groups adjacent to the incoming villages emphasized their distinctiveness by small, round burial-mounds (tumuli). Their so far unassuming material culture disguised the explosive potential of their new domesticate, the horse; but in the long run it was fundamentally to change the character first of Old World, then of New World cultures.

# Distant Echoes of the First Cities 3500–2000 BC

Whilst there were many continuities between this Europe and the one which gradually emerged after 3500 BC, there were also new elements for which the Near Eastern focal area had a continuing relevance, since it was here that the first cities emerged. The immediate sustaining-area of the first urban communities in Mesopotamia, although vast (reaching eastward to Afghanistan to import ritual or luxury items like lapis lazuli), did not encompass even south-east Europe, and its effects were at first very indirect—so much so, indeed, that not all archaeologists recognize western Asia as their source. Nevertheless there is good reason to believe that many of what might be called the secondary features of farming came into widespread use as a result of the enlarged scale of economic activity consequent upon urbanization. This has a bearing on the phenomenon once described (in a phrase which has achieved widespread currency) as the 'secondary products revolution', and it is worth setting this term in context (Sherratt, 1981: 1997).

Animals were first domesticated primarily for meat; but it gradually became evident that they might have a variety of other uses. The practice of milking was perhaps slow to evolve because of adult human intolerance of lactose (milk sugar), but it became advantageous in a variety of environments:

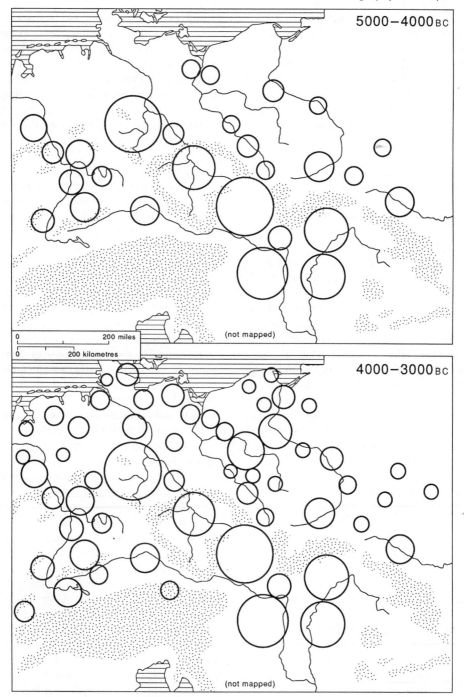

**Fig. 1.4.** Settlement-cells of early farming populations (schematically represented as circles proportional to the occupied area) in central Europe north of the Alps: *above*, in the pioneer Neolithic, fifth millennium BC; *below*, in the later Neolithic (and Copper Age, in the Carpathian Basin), fourth millennium BC. These fertile lowland-areas (initially coinciding largely with patches of loess) remained the focal areas of agrarian population down to the Industrial Revolution.

in northern Europe where low sunlight reduced vitamin-D production and so inhibited calcium absorption (where ability to use fresh milk thus has a positive selective advantage biologically), and in dry environments with low plant productivity, where people needed to live off their herds but without slaughtering them (and where using secondary products like blood and milk has an ecological advantage). Cheese also provided a storable (and tradeable) product. Other secondary products include wool from sheep—again, not a feature of initial exploitation patterns, since wild sheep are hairy rather than woolly, and woolly breeds only appeared after some millennia of domestication. Secondary applications of animals, particularly bovids, included their use as traction animals: initially for ploughing and pulling sledges, then for the specially designed technology of wheeled vehicles, which are known from the mid-fourth millennium onwards in Meosopotamia. Other transport uses, such as riding (or pulling lighter forms of wheeled vehicles) involved a range of animals specially domesticated for these purposes, in a 'second round' of domestication: the horse, ass, and onager, and the two species of camel. New plant crops, too, came into use: especially tree-crops such as vine, olive, fig, date, and pomegranate. Such innovations were appearing in various parts of the western Old World farming area—though most commonly in the areas adjacent to the original beginnings of farming, where experience was longest—from around 6000 BC, in parallel with other innovations such as simple irrigation. Some of these innovations made possible the manufacture of new commodities—as with wine from tree-crops, or woollen clothing from new types of sheep. There was a quickening of pace in the fifth and earlier fourth millennia, however, as the various parts of the Near East interacted and became articulated by exchange networks; and the interest in new forms of animal transport was paralleled by the development of the sail on the great exotic rivers. These new uses of both animate and inanimate energy sources represent the first harnessing of sources beyond the power of human muscles. All these elements, in the context of the ecological differentiation of the Near East, made possible the emergence of urban communities there (Sherratt, 1995a: 17–20).

These early urban communities—initially in Mesopotamia, then soon also in the rather similar ecological conditions (and within the same network of interaction) in Egypt and the Indus valley—drew in raw materials and circulated their manufactured products over a vast area, and stimulated the spread of ideas and resources (like wool-bearing breeds of sheep) even to areas beyond their immediate sphere of active exchange (Algaze, 1993). In this way a number of features were more widely spread, and some innovations 'escaped' from their contexts of origin, and made their appearance in temperate Europe. The rapid successive appearance of carts, ploughs, horses, woolly sheep, and bivalve-mould alloy metallurgy—as well as a prominent role in ceramic assemblages for drinking vessels, which has been taken as evidence for the spread of alcohol production—constitute a 'revolutionary' set of changes which transformed the potential of that area. They should not, however, be seen simply as contributing to productive efficiency in the satisfaction of universal needs, as new sources of energy or material; equally important was their contribution to the stimulation of *demand*—new forms of satisfaction, whether literally consumable (as with alcoholic drinks), metaphorically so (as with clothing of bright colours, made possible by dye-absorbing wool), or through new transport technologies (always sources of prestige) or simply through the new range of artefacts—especially weapons—which could be produced by more sophisticated metallurgy. All these gave scope for some measure of ecological specialization in production, and opportunities for elite monopolization; European society was transformed because its cultural base was altered. Thus the gold in the graves of Varna had a different value from the gold of the Mycenae Shaft Graves, three millennia later: there were simply so many more things for which the latter might be exchanged.

The new features appeared first in south-east Europe, arriving either via Anatolia or around the northern side of the Black Sea. The northern Caucasus became an important contact-area between Near Eastern and steppe traditions, and it was by this route that both woolly sheep and a knowledge of the wheel arrived in the Pontic zone. The addition of wheeled vehicles to the existing technology and cultural pattern on the steppes, manifest in the Pit-grave culture group, produced a great expansion of cattle-herding communities, now able to move supplies as well as personnel over the flat interfluves, both eastwards and westwards. Some groups, building their characteristic tumuli, penetrated to the lower Danube and beyond the Carpathians into eastern Hungary; others, in greater numbers, expanded in the other direction, to the lower Volga and the steppes beyond the Urals. Caucasian metallurgy exercised a formative influence, for instance in the spread of the bivalve mould (Chernykh, 1992). Balkan and Carpathian Europe, closely linked to northern Anatolia, reflect the influence of new, metal drinking-vessels (like those recovered by Schliemann from the destruction levels of Troy) in the design of their pottery: other pottery drinking-cups took the form of model wagons with solid wheels, pulled by oxen—thus celebrating simultaneously two of the novelties associated with emerging rural elites. Their neighbours in the North European Plain, still building megaliths, show by the ploughmarks sometimes preserved underneath their covering earth mounds that ox-traction was known there, too. It seems likely that it became known also in other parts of the megalith-building area, where these monuments became increasingly elaborate (Fig. 1.5).

Of the three cultural models which Europe provided—southern sophistication, eastern mobility, western monumentality—it was a combination of the two former with which the future lay: the last was to disappear without finding an heir to its advanced architectural skills devoted to the realm of the dead. A cosmology set in stone could not encompass the new mobility both of a livestock-raising economy and a broader range of material goods. Instead, the pattern of development in outer Europe was provided by a sequence of major inter-regional exchange networks, occupying successively larger areas, from east to west, providing material symbols and social practices which conveyed the new social values of the times. Because burials survive better than the everyday settings of their use, most of the material evidence for these practices comes from graves and their contents: significantly they are individual burials, often under tumuli, accompanied by drinking equipment and weaponry. The spread of these new ideologies was associated with an increased scale of landscape clearance and livestock raising, often on lighter soils than those selected by the first farmers with their more restricted forms of horticulture. The burial mounds, and the cord-impressions with which the pottery was ornamented, are (like the 'cowboy' ideology) of eastern origin; the drinking emphasis (and the metal-influenced battle-axes or copper daggers) are southern. This pattern spread across the North European Plain and into the Atlantic heartlands of megalithism over the period during the early third millennium BC, eventually reaching as far as western Sicily, and thus largely surrounding the bridgehead of more direct Near Eastern influences in south-east Europe.

During this time the maritime networks of the eastern Mediterranean had developed in scale, from an inter-island canoe traffic best known in the Cyclades (concerned especially with the supply of silver), to a route along the southern coast of Anatolia, from Syria to Crete, that involved plank-built vessels with sails. This seaborne activity explains the prominence at this time of the city of Troy, where boats wanting to enter the Dardanelles had to wait for a favourable wind to take them through. As in some sense the predecessor of Constantinople/Istanbul, Troy can claim to be the first urban foundation on the threshold of Europe; and it is likely that by 2500 BC traffic from there up and down the Black Sea coast had made contact with riverine exchange networks extending up the Danube—and, indeed, called such trading activity into existence by the new and attractive commodities it could supply: bright woollen textiles, bronze ornaments, and trinkets of blue-glazed faience (glass-frit). This channel of concentrated traffic differed in its intensity and directionality from the more diffuse exchange networks mentioned above, and together with the maritime routes of the Aegean it represents a new phase in European interaction with the Near Eastern heartlands.

These differences in connectivity go a long way towards explaining the diversity of settlement patterns over this period. In the south-east of Europe, the Balkans and the Carpathian Basin had for generations formed in many respects a rustic extension of the Near East, with a tradition of mud-built, tell-forming (settlement-mound) villages reaching back to the nucleated pattern of the first Neolithic communities there. As livestock became increasingly important, and areas away from the river valleys were opened up by use of the plough, new

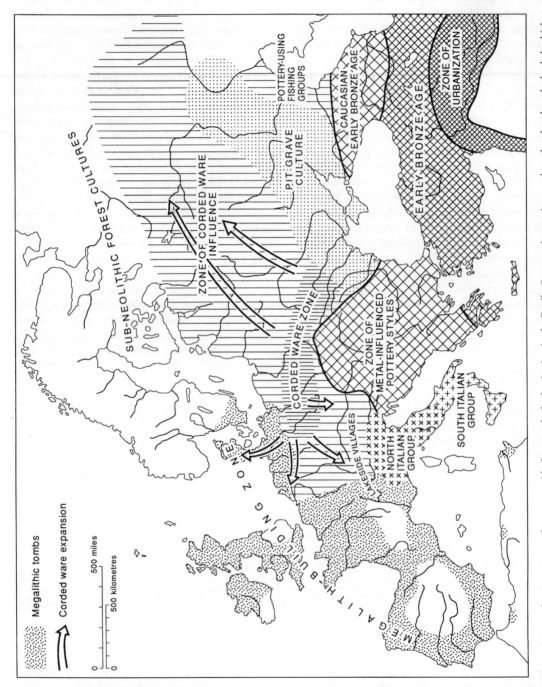

**Fig. 1.5.** Europe between 3500 and 2500 BC: with the spread of urbanization in the Fertile Crescent, innovations such as the plough and wheeled vehicles spread to both temperate Europe and the steppes. Interaction between these two areas created the complex of Corded Ware cultures in north-central Europe, with a new emphasis on livestock rearing. This affected a large part of northern Europe, impinging on the increasingly elaborate megalith-building cultures of the far west.

manifestations of this settlement-type came into being: in the plains (as in central Bulgaria) a few, larger tell sites were fortified with stone walls; but more common were promontory sites on spurs at the edges of the valleys or some way into the uplands. The village community was still a concentrated, visible unit.

Outside this area, however, even the term 'settlement', in the image of a regular lattice of village communities, is potentially misleading. In the areas immediately adjacent to the Carpathians—in Moldavia, Little Poland, Bohemia/Moravia, and to some extent in central Germany—nucleated, fortified sites might be found, but often as exceptions to the general form of dwelling-site, and perhaps serving ceremonial purposes as much as secular ones. In the west and north of Europe, this is quite evident, for the focal points in the landscape were still the communal tombs, perhaps combined with enclosed sites on hilltops whose pattern of interrupted ditches implies a ritual purpose rather than military planning—even though on occasions such sites came to be occupied by large numbers of people, and were sometimes subjected to attack. Where detailed field-surveys have been undertaken, the traces of insubstantial domestic settlements may be detected nearby in the form of surface-scatters of flint and sometimes pottery. The very different appearance of such settlement systems from those in south-east Europe may, however, overdraw the contrast: for both large 'ceremonial' sites (like stone circles) and 'defensive' sites (like fortified tells or promontories) may have served as focal points for inter-regional trade—whether in stone axes or copper ones—and may in this respect be functionally more comparable than a first glance would suggest. One culture couched its wider contacts in the language of religion, another expressed its attitudes more directly, mindful of the chances of attack; but both managed to accomplish the transfer of materials. What both of these major patterns—whether based on villages or monumental ritual sites—also have in common was the need to make permanent inscriptions on the landscape, marking points of face-to-face interaction by fairly large communities of people.

The alternative to this model was something very much more mobile, in which both the dwellings (rectangular on the continent, still round in the British Isles) were relatively insubstantial, and the burial-mounds less monumental and no longer built as if to last through eternity. Nor were there elaborate central ceremonial structures, except in the few places (as at Stonehenge—a rare exception) where older circular monuments retained their power and were still used as meeting-points for local politics or long-distance trafficking. This makes the whole pattern much harder to reconstruct, but is probably an accurate reflection of the rapidly changing landscape which came about as the combination of the plough and livestock-keeping on a larger scale (now, for instance, with horses and wool-bearing sheep as well as draft-oxen to support) gave renewed impetus to forest clearance on lighter soils, and produced a more rapid turnover of occupation sites than in previous millennia. Because many of these soils (like the outwash-sands of the North European Plain) were unable to withstand prolonged cultivation, they often rapidly degraded and today survive as downland, heathland, or light forest, dotted with the typical round burial-mounds or tumuli. The burial-mounds themselves, however, with their objects of bronze, gold, and amber, testify to their builders' ability to acquire such materials through trade.

This pattern, which came to be characteristic of much of Europe outside the Carpathians, did not fundamentally alter until around 1300 BC (as the Bronze Age urban societies of the Mediterranean were collapsing), when the south-east European pattern of nucleated and often fortified sites began to spread over much of the continent. It was at this time that labour was invested in a detailed division of the land, marked by the systems of field boundaries which survive as relics ('Celtic' fields) in areas where cultivation later came to be replaced by extensive grazing. These were accompanied by a pattern of hillforts, villages, and small farmsteads—whose cemeteries no longer had the monumental character even of the tumulus-builders. Effort was now devoted to cultivation and to protecting the prosperity which came from keeping flocks and herds, and making goods such as textiles from their products. It was a consumer revolution, as well as an agricultural revolution—not just an increase either in population or in calories produced, but in how people made use of agricultural products. Temperate Europe thus slowly echoed the kinds of changes which had taken place in the east Mediterranean with the emergence of palace-centred manufacturing towns, where wine, oil, and fine cloth were produced for the benefit of urban populations. It is to these new, mobile forms of material wealth that we must now turn our attention.

## The Articulation of a Continent 2000 BC–AD 1

### Old and New Theories

The ability of *Homo sapiens* to use material culture in social interaction appears to be unlimited (Douglas and Isherwood, 1978): even Pleistocene hunters collected special stones and shell necklaces, while Neolithic sedentism allowed new scope for the accumulation of wealth objects. Since then, humankind has not ceased to accumulate possessions. Because Neolithic communities did not themselves move about the landscape, materials now had to be moved to them; and they were passed through a series of intermediaries to their point of consumption, forming an extensive network of exchanges. As social interactions became more complex, a wider range of materials was required to suit new degrees of difference; and particularly attractive materials, like amber or lapis lazuli, moved very long distances in small quantities. Metals were important as a generally desirable commodity which could easily be transformed into very different forms of artefact, and being very generally exchangeable could therefore provide liquidity—acting as a kind of proto-currency. In the millennia following the beginning of agriculture, the volumes of material transferred between human communities increased exponentially; and the patterns of these flows had a critical influence on demographic growth and agricultural intensification (Schneider, 1977). This formulation of the dynamic of change reverses the commonly assumed formula by which the production of more calories leads to a disposable surplus and a growth of trade—the 'agrarian' model which assumes that 'trade' is secondary. Such a model underlies many theoretical descriptions of economic development, and it has been particularly popular in the post-war period when local autonomy has been stressed in opposition to 'diffusionism'. Such production-oriented models are now being challenged by consumption-orientated descriptions like the one

attempted above; and the following account is written from that perspective (cf. also Sherratt and Sherratt, 1991). This point of view can be described as *interactionist*, by contrast with the *autonomist* perspective which has dominated in the last half-century.

Because of the differential distribution of resources and of transport costs, the flows of traded materials—like those of water in the physical landscape—are soon constrained into channels; indeed, because of ease of access and relative cheapness, trade routes often follow watercourses. So it was with the locations of the first cities, and expansion of the urban network followed the supply routes: up-river and then partly overland, to important resources (Algaze, 1993), but where possible following the easier maritime routes such as those along the northern coasts of the Mediterranean—often going from island to island or peninsula to peninsula. Difficulties of navigation round the latter often led to trans-isthmian short cuts across them which promoted town growth at key transshipment points (i.e. Corinth on the Peloponnese or Hedeby on Jutland. At nodal positions where routes converged, or at trans-shipment points (as well described by the older school of human geography), settlements had an economic advantage and could develop into proto-urban or—with sufficient volumes of flow, depending on their position within the overall system—even urban status. Although industrial economies are dominated by movements of food and fuel, these commodities played a much less significant role in earlier distribution systems; most communities were largely self-sufficient in bulk items, and it was the rarer and less bulky items which moved furthest. Routes created to supply high-value materials gave the occupants of sites along them the opportunity to add other commodities, including those with a higher proportion of added value, such as craft goods, which could support urban populations (Sherratt and Sherratt, 1991). So urban networks grew by accretion, expanding along the principal axes of movement. As the network grew, so its topology changed; and routes or whole areas which had previously been critical within the supply network might be bypassed, and undergo recession. Within a generally dynamic system, individual areas experienced fluctuating fortunes and changing degrees of centrality.

## The Amber Routes

Long-distance 'trade' in prehistoric Europe was mostly the movement of objects for the value of their raw material rather than their form; objects (like pots) whose value lay chiefly in the labour and skill expended on them were mainly consumed within limited areas of shared taste. Urbanized areas, on the other hand, saw long-distance movements of manufactured goods in bulk; economic success lay in importing high-value materials in exchange for (to them) cheap manufactures, thus creating a cultural hegemony of common taste. Raw materials for manufacturing processes were drawn from a periphery, often at a lower level of economic and social organization, which was able to import some manufactured goods in exchange for these primary products; beyond this lay the margin of purely prehistoric cultures. Within this vast outer zone, altered by the spread of new domesticates, technologies, and consumption patterns but not systematically articulated with urban trading systems, the diffuse European exchange networks of the third millennium tended to align themselves into north/south chains (Beck and Bouzek, 1993; Sherratt, 1994).

The first of these, which can be traced already at the end of the third millennium, lay from the Black Sea along the lower and middle Danube (though actually taking a short cut through Transylvania) and thence up to Bohemia and central Germany. No single commodity travelled along its entire length, but a series of complementary resources—including gold, tin, and copper—were linked into a chain of economic activity and exchanges, along which new technological ideas (including tin-alloying) could spread. Central Germany itself maintained contacts with the lower Elbe and the North Sea basin, including Britain and Brittany, as well as the Rhône valley and north Italy, which are indicated by the distribution of solid-hilted bronze daggers associated with the Únětice metallurgical complex. By 1800 BC, three new elements had been added to this route: in Anatolia, an inland route used by Old Assyrian merchants (who had colonies in local Anatolian towns) crossed the plateau and extended as far north as the southern coast of the Black Sea, where it could tap into coastal trade up to the Danube; in Transylvania, a massive exploitation of new copper sources in the

eastern Carpathians began, as well as eastward links across the Pontic steppes; and in the north, via the Oder, the first links were made with Scandinavia, which took amber for the first time to the Carpathian Basin. These strengthened and extended contacts brought about rapid technological innovation: the appearance, both in the Carpathian Basin, Anatolia, and the steppe lands, of the spoked-wheel chariot; and new forms of bronze-casting with analogies far to the east, in the Altai and beyond (Chernykh, 1992). (Chariots themselves were to appear in China as a result of contacts with the steppe-lands, by 1400 BC.) The scale of inter-regional contacts had appreciably enlarged (Fig. 1.6).

The growing volume of trade in central Anatolia was paralleled by a great expansion of seaborne trade in the east Mediterranean, linking the newly emerged civilization of Crete with Syria via Cyprus and Rhodes; and from Crete a more exploratory set of contacts reached out to tap into local exchange cycles both in the Aegean and across the Ionian Sea to southern Italy and the Tyrrhenian. The development of these routes led to the emergence of mainland centres on the route to Italy, at Pylos on the south-west Peloponnesian coast and at Mycenae (the latter commanding a pass to the Corinthian Gulf), where a Greek-speaking civilization first developed. Mycenaean Greeks maintained links—notoriously ambivalent—with Troy and the local Black Sea routes to the Danube mouth, and also with central and northern Italy, whence contacts had already been established over the Alps to central Europe. The chain of contacts from Troy to Scandinavia via the Carpathian Basin was thus to some extent bypassed by a new set from Italy to the Alpine foreland and so to Scandinavia across the lower Elbe (Fig. 1.5, right). It was by this more direct north–south route that amber reached Mycenae by 1600 BC (and even Egypt, in the tomb of Tutankhamun, by 1325 BC)—the first identifiable Baltic import to the Mediterranean (Beck and Bouzek, 1993). This durable, high-value material must have been accompanied on part of its journey by other commodities such as Scandinavian furs and fine skins, exchanged for central European copper and tin.

These two routes (which should be envisaged as multiple-stranded chains of contacts between many different settlements and small polities in fairly

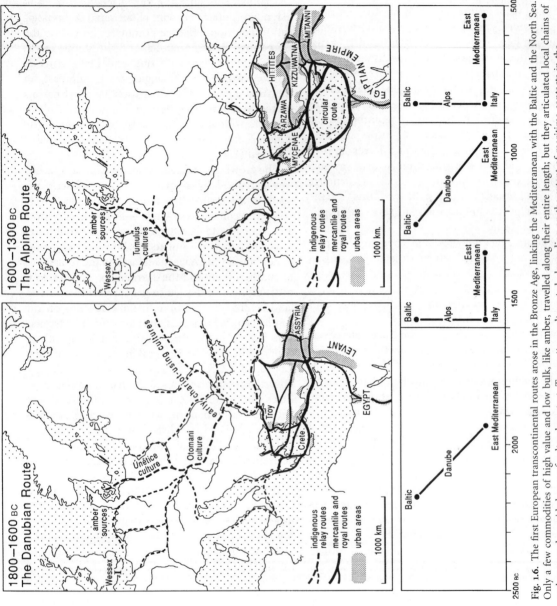

**Fig. 1.6.** The first European transcontinental routes arose in the Bronze Age, linking the Mediterranean with the Baltic and the North Sea. Only a few commodities of high value and low bulk, like amber, travelled along their entire length; but they articulated local chains of exchange and provided axes of cultural contact. Two patterns alternated, depending on the extent of maritime contacts in the Mediterranean: a Danube route and a trans-Alpine one (compare Fig. 1.2.). The same structure of contacts underlay the later division of Christian Europe into Latin and Greek (Orthodox) areas.

broad corridors, based on fluctuating alliances rather than on any imposed authority) must be reckoned as to some degree in competition with each other, or at least as largely incompatible alternatives, for later European prehistory shows an alternation between one and the other: after the floruit of the Danube route from 2300 to 1700, and of the trans-Alpine route from 1700 to 1300, the period of recession or partial collapse in the east Mediterranean at the transition from the Bronze to Iron Ages there caused a retraction in Mediterranean trade, with the fall of Mycenaean and Hittite command economies; some links via the Black Sea were probably maintained after 1300, and the route from the Carpathian Basin to Scandinavia was re-established around 1000 BC; then, after 900 BC (as east Mediterranean contacts with the central and west Mediterranean were re-made), the trans-Alpine route linking Italy with Scandinavia came again into prominence. There is a repeated pattern here which is worth making explicit. As Chapter 2 will explore at greater length, urban civilization expanded from the Levant westwards along the Mediterranean: it extended as far as Greece (with contacts to Italy) in the Bronze Age during the second millennium BC; then, after a retraction, to Greece again and then to Italy (and, at the same time, North Africa and Spain) in the early first-millennium Iron Age (see also, Sherratt and Sherratt, 1991; Sherratt and Sherratt, 1993; Sherratt, 1994). When the centre of gravity has been in the east Mediterranean, the Danube route has been important; as the focus moves westwards the trans-Alpine routes take over. Two crucial urban centres came to occupy nodal points where the east–west maritime trade routes of the Mediterranean articulate with north–south feeder routes from the interior of the continent: Rome and Constantinople. Greece, without a European hinterland of its own, achieved a temporary prominence (first in the Bronze Age, then again in the Iron Age) only in the transition from a centre of economic activity (and political hegemony) in the east Mediterranean to one in Italy. These phenomena are not the result of small local developments; they are located within structural settings of continental dimensions.

This growing articulation of the continent, in terms of increases in the movement of goods of various kinds (articulated by exchanges of bronze artefacts and scrap metal, which acted like a kind of proto-currency Fig. 1.7) was also manifested on the ground. In the lands along the Danube, the pattern of nucleated village settlements had never been abandoned, and since the fourth millennium such villages had often required some form of fortification. With the emergence of transcontinental trade routes, more elaborate systems of fortification came into existence at certain choke-points along the route—especially where traffic was constrained to the valleys going through the Carpathians themselves. Thus the first European 'hillforts'—more than just defended hill-villages, but centres of metalworking and weaving, defended by timber-laced box ramparts—came into existence in Romania and Slovakia in the first centuries of the second millennium, with a florescence of material wealth and craftsmanship associated with the Otomani and related cultures. This has traditionally been seen as due to Mycenaean influence; but the new accuracy of calibrated radiocarbon dating shows that it was rather associated with the importance of Troy, and it took place in the early second millennium whilst the centre of Greek civilization was in Crete rather than on the mainland. With the rise of mainland Greece and the shift to a central north–south European axis of contacts, however, very different mechanisms of trade seem to have been involved, for no fortified sites can be traced along this route (from Denmark through Germany to the Alpine passes) before the first millennium BC, when Italy was already in the Iron Age. Instead, the route can be traced by the pattern of exotic ornaments worn by women buried in the tumuli—marking the pattern of exchanges of marriage partners by which the chain of trading alliances was negotiated. A different and perhaps less sophisticated form of trading relationship is implied. In the absence of defended centres, it is likely that manufacturing techniques were less developed—for instance the plentiful Danish textile remains from waterlogged tumuli show only plain-weaves, while twill was already made in the Carpathians. Thus trading opportunity and cultural or social sophistication did not necessarily coincide: centrality in a trading network no doubt came as much as a surprise to the tumulus-builders of the Lüneburger Heide as it had earlier in the second millennium to the hill-folk of the Slovakian passes.

This perspective makes sense of later settlement patterns, too. It is misleading to analyse Iron Age settlement patterns with the central place models of mature, bulk-production economies: the movement of small quantities of relatively high-value items–crucial as it was to local prosperity—was highly directional, and major sites did not supply their hinterlands with quantities of finished goods in the manner of truly urban settlements with (for instance) facilities for making pottery on the wheel. This condition arrived in temperate Europe only at the end of the prehistoric period, in the century before the imposition of Roman rule, with the appearance of the very large defended centres known to archaeologists as *oppida* (the Latin word used by Julius Caesar). Before this, the pattern of defended centres, whether hillforts, defended marsh-settlements, or the more elaborate 'princely residences' (*Fürstensitze*) of the circum-Alpine region in the seventh and sixth centuries BC, are better treated not as lattices covering extensive areas but as chains along the principal routes of traffic. Rather than the multidirectional links evident in a mature, reticulate, urban settlement pattern, they conform better to a dendritic model—often created by the growth of urban demand in the Mediterranean (Fig. 1.8).

Since the ancient ('classical') world is the subject of the next chapter, there is no need here to rehearse its history; but in tracing the transformation of the still prehistoric trans-Alpine world during this time, the growing articulation of these two components of Europe must be borne in mind. In conformity with the agrarian and demographic enthusiasm of the post-war period, current interpretations of Greek colonization usually describe it as a search for new lands in response to overpopulation. So it was, in a way; but this is to emphasize the negative or 'push' aspects, rather than the positive or 'pull' aspects. Phoenician colonies have always been seen as trading enterprises, and the location of Greek colonies can equally be interpreted in terms of tapping into flows of high-value products in demand at home—not least, metals (Sherratt and Sherratt, 1993; Sherratt, 1995*b*). These two colonizing movements in the central and western Mediterranean (and only slightly later on the northern coastlands of the Black Sea) first brought the Iron Age to these areas, initiated urbanization of their coastal regions, and created a peripheral region of raw material supplying societies in their immediate hinterlands: south-west Iberia, circum-Alpine central Europe, and the lower Dniepr region, yielding metals, hides, salt, and salted preserves. Each of these, moreover, was also the focal area for longer supply routes bringing precious materials over much longer distances across the surrounding margin: gold, amber, and tin (from Bactria, the Baltic, and Brittany), and the more perishable exotic organic materials that travelled with them—often including slaves (Fig. 1.8).

The continuing element of continuity between second-millennium and first-millennium east–west Mediterranean trading systems was the route between Cyprus and Sardinia via Crete. It came into use at the end of the Bronze Age, from the thirteenth century BC onwards, especially following the collapse of the palatial command economies of Mycenaean Greece. Its use was interrupted briefly at the turn of the millennium, but it came to prominence again when the Phoenician cities resumed commercial activity, both in the Aegean and in Sardinia, where they found silver (and perhaps tin). Greeks, too, joined the westward trail in the eighth century, and Phoenicians penetrated further, to the coasts of north Africa and Iberia, while the Tyrrhenian Sea developed as an indigenous economic region focused on Etruria. Around 700 BC there developed a major link between Etruscan and northern Italy and the eastern Baltic, by way of the eastern Alpine fringe, Moravia, Kujavia, and the lower Vistula. This was to be the Iron Age 'amber route'—sometimes drying up, and then reappearing—known to the classical world, and along which a Roman knight was to travel in the 1st century AD in order to procure prodigious quantities of amber to decorate Nero's circus (Pliny, *Nat. Hist.* 37. 45). From the 7th century BC also, routes to the Atlantic seaboard began to tap the tin, gold, and silver sources of the Hercynian ranges in Portugal, Brittany, and Britain: first along the Rhine from Italy and across the Alps, then through the Phokaian colonies of southern France along the Seine, Loire, and Garonne, and ultimately from Iberia and along the Atlantic coast. The role of the original north–south route had been taken over by several transcontinental passages (Brun, 1987).

During the Hellenistic period, the role of these western routes was temporarily overtaken in

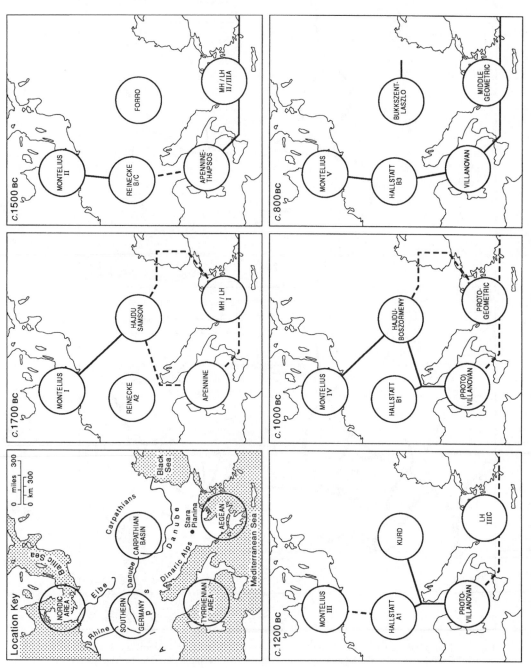

**Fig. 1.7.** Alternative configurations of metal-producers and consumers in the Bronze Age of central Europe, at selected times between 1700 and 800 BC. Copper from the Carpathians and Alps supplied the metal-less North European Plain. The shifts between 'Danubian' and 'trans-Alpine' linkages can be clearly seen. (Names refer to local typological phases, established from bronze implement and pottery styles.)

21

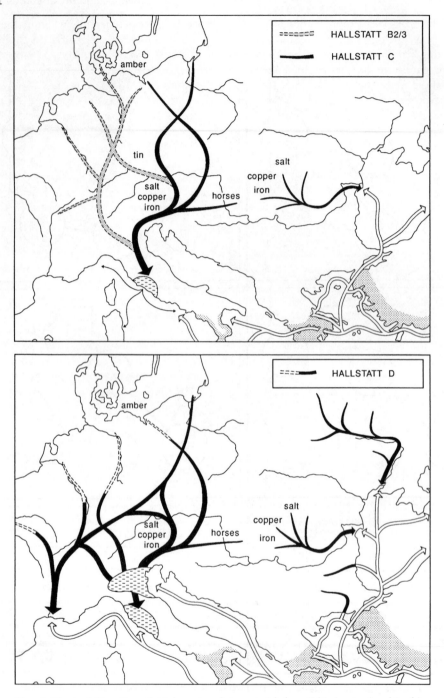

**Fig. 1.8.** The dendritic pattern of supply-routes of raw materials from Celtic (western) and Scythian (eastern) areas of Europe to the urban communities of the Mediterranean, 900–500 BC. *Above*, Hallstatt B2/3 and C (final Bronze Age of central Europe and earliest Iron Age), ninth to seventh centuries BC; *below*, Hallstatt D, sixth century BC. Extensions of the pattern mark the foundation of Greek colonies in the south of France and on the Black Sea. These trade routes determined the prosperity and political development of the areas through which they passed.

importance by Black Sea supply routes, when the economic focus swung eastwards (thus causing Celtic-speaking tribes to move south and east to plunder the goods no longer reaching them by trade); but these western arteries of contact determined the pattern by which the Roman Empire expanded when the focus of economic activity swung back again to the west, and Rome managed to conquer its Punic rivals on the opposite shore of the Mediterranean. The first Roman province was the eponymous *Provincia* (Provence), consisting essentially of the old Phokaian Greek colonies with their crucial access to Gaul and the Atlantic seaboard; then the old Phoenician colonial shores of North Africa and Spain were acquired. Eastward campaigns unified the Mediterranean under Roman control, before Caesar added the fast-developing Gaulish hinterland to its west Mediterranean possessions, which his successors extended to include Britain and the territories up to the Rhine and Danube. There formal control ended, but a lively commerce continued to connect the Rhineland with Scandinavia, the Danube with the eastern Baltic (Hedeager, 1992).

## Coda

Looking ahead, these patterns were both to repeat themselves and to develop further. In the early first millennium AD the economic focus swung eastwards again, as trade in the Indian Ocean expanded; the western end of the urban world was relatively peripheralized, and Germanic tribes from beyond the formal frontier of civilization again moved south to seize the luxuries to which

trade had accustomed them. Soon the east–west traffic across Eurasia, both via the Silk Route and by the Indian Ocean, was to reconfigure the pattern of linkages between Europe and the rest of the Old World: the narrow isthmus of the Near East, separating the Mediterranean from the Indian Ocean, entered upon a second period of prosperity, under Islam (Lombard, 1975); and Europe was encircled by an 'outer ring-road' as long-distance routes developed across the deserts of North Africa, and Viking seamen penetrated both down the Russian rivers to the Caspian and the Black Sea and also across the North Atlantic to Greenland and North America. Partly isolated from its former Mediterranean mentors, the enclosed body of water formed by the North Sea and the Baltic became itself a second Mediterranean in the later first millennium AD; then after the turn of the millennium it became linked again with the other inland sea by a trans-Alpine route to the Rhine and a longer haul by galley round Iberia (Abu-Lughod, 1989: 51–134). The inhabitants of the Atlantic coastlands learned their seamanship in both schools, and pooled their experiences. In the fifteenth century they explored the African coast to outflank the desert caravans; then their vessels both crossed the Atlantic and rounded Cape Horn, to reconfigure the continents once again—joining by their maritime traffic lands that had not been united since Pangaea.

To move through prehistory requires a breathless pace, and the prehistorian finds it hard to decelerate even when faced with the complexities of the historical record. Perhaps this fast-forward technique, by juxtaposing repeated patterns in rapid sequence, may bring out regularities which closer scrutiny conceals.

REFERENCES

*NB   A complete bibliography for such a broad survey would be intolerably long: the works cited here are intended as a guide to further reading. In addition, the reference works edited by Cunliffe, Scarre, and Sherratt may be consulted for general accounts with guides to further reading*

ABU-LUGHOD, J., *Before European Hegemony: The World System AD 1250–1350*, (New York, 1989).

ALGAZE, G., *The Uruk World System: The Dynamics of Expansion of Early Mesopotamian Civilization*, (Chicago, 1993).

ANDREWS, M., *The Birth of Europe: Colliding Continents and the Destiny of Nations*, (London, 1991).

BECK, C. W., and BOUZEK, J., *Amber in Archaeology: Proceedings of the Second International Conference on Amber in Archaeology, Liblice 1990*, (Prague, 1993).

BOSERUP, E., *The Conditions of Agricultural Growth*, (London, 1965).

BRUN, P., *Princes et Princesses de la Celtique*, (Paris, 1987).

CHERNYKH, E., *Ancient Metallurgy in the USSR: The Early Metal Age*, (Cambridge, 1992).

CUNLIFFE, B. W. (ed.), *The Oxford Illustrated Prehistory of Europe*, (Oxford, 1994).

DOUGLAS, M., and ISHERWOOD, B., *The World of Goods: Towards an Anthropology of Consumption*, (Harmondsworth, 1978).

GAMBLE, C., *Timewalkers: the History of Global Colonization*, (Harmondsworth, 1995).

HEDEAGER, L., *Iron Age Societies: From Tribe to State in Northern Europe*, (Oxford, 1992).

LOMBARD, M., *The Golden Age of Islam*, (Amsterdam, 1975).

ROLLE, R., *The World of the Scythians*, (London, 1989).

ROWLANDS, M., KRISTIANSEN, K., and LARSEN, M. T., (eds.), *Centre and Periphery in the Ancient World*, (Cambridge).

SCARRE, C. (ed.), *Past Worlds: The Times Atlas of Archaeology*, (London, 1988).

SCHNEIDER, J., 'Was there a Pre-capitalist World System?', *Peasant Studies*, 6: 20–9, (1977).

SHERRATT, A. G. (ed.), *The Cambridge Encyclopaedia of Archaeology*, (Cambridge, 1980).

—— 'Plough and Pastoralism: Aspects of the Secondary Products Revolution', in N. Hammond, I. Hodder, and G. Isaac (eds.), *Pattern of the Past: Studies in Honour of David Clarke* (Cambridge, 1981), 261–305.

—— 'The Genesis of Megaliths: Monumentality, Ethnicity and Social Complexity in Neolithic north-west Europe', *World Archaeology*, 22 (2): 147–67, (1990).

—— 'What would a Bronze Age World System Look Like? Relations between Temperate Europe and the Mediterranean in Later Prehistory', *Journal of European Archaeology*, 1 (2): 1–57, (1994).

—— 'Reviving the Grand Narrative: Archaeology and Long-Term Change' (David L. Clarke Memorial Lecture, 1995), *Journal of European Archaeology*, 3 (1): 1–32, (1995a).

—— 'Fata Morgana: Illusion and Reality in Greek–Barbarian relations', *Cambridge Archaeological Journal*, 3 (1): 139–53, (1995b).

—— 'Plate Tectonics and Imaginary Prehistories: Structure and Contingency in Agricultural Origins', in D. Harris (ed.), *Origins and Spread of Agriculture* (London, 1996).

—— *Economy and Society in Prehistoric Europe: Changing Perspectives* (Edinburgh, 1997).

—— and SHERRATT, E. S., 'From Luxuries to Commodities: The Nature of Mediterranean Bronze Age Trading Systems', in N. Gale (ed.), *Bronze Age Trade in the Mediterranean*, Studies in Mediterranean Archaeology, 90, (Jonsered, Sweden, 1991), 351–86.

SHERRATT, E. S., and SHERRATT, A. G., 'The Growth of the Mediterranean Economy in the Early First Millennium BC', *World Archaeology*, 24 (3): 361–78, (1993).

STREET, F. A., 'Ice Age Environments', in Sherratt (1980), 52–6.

# Chapter 2

# The Classical World

## P. Arnaud

The title 'classical world' usually indicates a situation that is limited in space and time and is characterized by the Greeks and by those who adopted their way of life after the fifth century BC. That is to say it applies in a strict sense to a very defined area, namely the shores of the Mediterranean. According to the terminology used by ancient writers, the classical world stood in contrast with 'barbarian Europe'; however, contacts between these two expressions of Europe represented one of the dynamic features of the continent in antiquity.

## Before Rome

### Peoples, Cultures, Migrations in Europe at the End of the Bronze Age and in the Iron Age

The history of Europe in antiquity is primarily the story of colonization by external ethnic groups and also of subsequent acculturation which few indigenous groups were able to escape. The European continent remained subject to migration and colonization venture throughout the period we are considering, with the 'great invasions' at the end of the Roman period being the final examples

(Chapter 3). Archaeological criteria are not always sufficient to define the characteristics or the chronology of these migratory flows before the end of the sixth century BC, by which time the great linguistic and cultural areas were tending to become stabilized (Fig. 2.1), despite subsequent changes in detail. We can recognize the following cultural groups:

- those that, for want of a better term, may be called 'indigenous' (Iberian, Ligurian, Finno-Ugrian, Sardian),
- the Semitic group (Phoenician, Punic),
- the Indo-European group (Greek, Italic or Latin, Germanic, Illyrian, Thracian, Iranian, Slavic), and
- a group whose origins are contested and are indisputably not Indo-European, for example the Etruscans whom some scholars take to be indigenous to Europe and others believed to have originated in Asia Minor.

### The Compartmentalized World of the City-States

The principal characteristic of the classical world was its city-states. This distinctive form of political organization characterized the Greek or

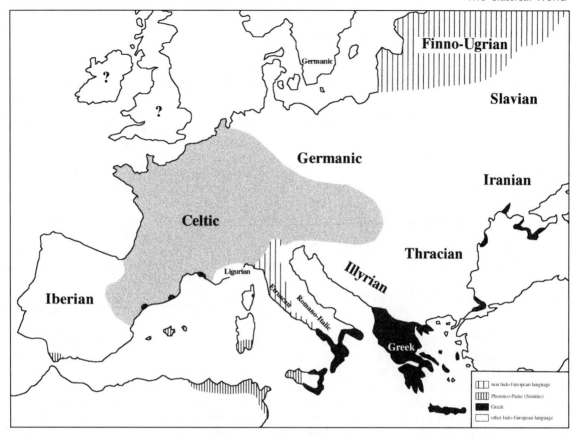

**Fig. 2.1.** Main European Languages, sixth–fifth centuries BC. (P. Arnaud.)

Greek-influenced world, and was different from the strong forms of state which emerged in the irrigated Near East. The city-state was associated with the practice of dry farming. It was to survive the emergence of great states, with the city-state remaining the basic political unit in the Hellenistic kingdoms which appeared after the death of Alexander the Great (323 BC), in the Greek principalities of Sicily or Epirus, and even in the Roman Empire. The city-state is the fundamental feature of the classical world; when city-states appeared in parts of the 'barbarian world' they set the seal on those areas having entered the Graeco-Roman domain; and the disappearance of city-states provided clear evidence of the end of the classical world.

The city-state organized political and economic relations, fashioned the surrounding territory and its landscapes, and provided the basic framework for the human geography of the classical world. It replaced political systems that were founded on the distinction between peoples, tribes or clans, and created a more abstract pattern that was associated with a larger stretch of territory than that related to earlier spatial groupings. The birth of the city-states is usually characterized as a regrouping or fusion of inhabited areas, to produce a cluster of citizens who all contributed to the life of the community according to their means.

Regardless of the constitutional forms that were being devised, the city-state is primarily seen as a kind of closed cell that was self-sufficient and independent, was ruled according to written laws which established relationships between the members of its community, and was characterized by one or several forms of communal worship which sealed the unity of the group and distinguished it from others. Protected by one or several gods, the city-state was above all a territorial unit (*chora* in Greek, *ager* in Latin) that was capable of sustaining the citizens who inhabited it, exploited it, and ensured

27

its military defence. Only citizens were able to own and to work land. They were fundamentally owners of rural property whose position in the class structure of the city-state was determined by the quantity of wheat they were able to produce, in the case of Athens, or the declared value of their land, in the case of Rome. Small and medium-seized property formed the fundamental elements of the landscape of the city-state. Foreigners were not allowed to reside permanently in city-states, where, of course, slavery was widespread.

This type of mono-nuclear organization, although inseparable from urban development, allowed little space for the establishment of real urban networks, or for at least as long as there was no political power at a higher level. The city-state conformed to quite a simple model: in addition to dispersed rural dwellings and villages, its territory included an urban place which was not in itself the 'city', but was its centre of gravity and provided its main image. It was the place where markets were held and all its citizens convened for political meetings. The Greek word *agora* and the Latin word *forum* both indicate the place where these two functions took place. The focal point was also the place where members of the community sought refuge for themselves and their goods by building a citadel. An acropolis or fortified hilltop was the normal response to these requirements.

It is very difficult to set a date for the birth of the city-state, because it was not a single event but rather a slow process whose chronology and complex characteristics must be appreciated. Nor was it a process that was strictly confined to the Greek world. In Italy, more precisely in Latium, similar trends took place at approximately the same time or perhaps even earlier and provide evidence that the same phenomenon may have developed independently in two quite different cultural domains. There is some agreement that the process was completed in Greece during the first half of the sixth century BC. Where it was not completed by that time, the creation of the city-state was sometimes linked to acculturation (as at Carthage) or more often to colonization and conquest.

The birth of the city-state was a highly influential process giving rise to other important features. When viewed in the long term the most influential innovation to date from this time was arguably the invention of coinage. Probably originating in Asia Minor in the early sixth century, the use of coins spread rapidly in the city-states, and more generally throughout the Mediterranean basin (Phoenicia, Persia). At the end of the sixth century BC the whole of the Greek world was familiar with coins, but that does not mean the Greek domain was entirely 'monetarized'.

Expressing the autonomy of the city-state and struck by the state to meet its needs, metal coinage—at least to start with—had only a subsidiary role as a means of allowing differential barter to take place. Its initial purpose was fiscal in character, with coinage being used as a means of controlling and simplifying expressions of weight. But it is clear that until about 400 BC, in the absence of small territorial division, the role of coinage was not initially to encourage internal or external trade. Coins made of precious metal seem not to have been considered as products susceptible of being exchanged for the value of their weight in metal. City-states such as Athens had the opportunity of exploiting important deposits of precious metal and acquired a certain commercial dynamism from that fact. The appearance at the end of the fifth century BC of bronze coinage created conditions for the development of an internal market and for trade based on profit, as well as for the development of a banking system. This enabled the exchange of paper money for coins stamped out of precious metal and enabled more elaborate financial instruments to be developed, such as bills of exchange and cheques. In the fourth century BC, for Aristotle (Rhet. 1. 5. 7) money had become the leading expression of wealth, being followed by land and goods, such as slaves and livestock.

Despite the cultural unity of the Greek world and the homogeneity of its economic development, the conception of the city-state as a rather closed and self-sufficient domain imparted some distinctive features to the Greek economy in ancient and classical times. By contrast with the Phoenicians, the Greek domain initially seemed only slightly inclined to seek out external markets on which to sell its surplus products. Even during the classical period, which was a time of great rivalry for leadership among the Greeks and for control of the sea among the Greeks, Carthaginians, and Etruscans, the rural economy of the city-states appears to have been orientated only slightly to the production of surplus goods for export. The main products that were exported were wine, oil, and manufactured goods (works of art, fine dishes); the main import was corn. The city-states were no less vigilant regarding the introduction of foreign products on to their home markets. If each city-state, as we shall see, was susceptible to enter into long-distance trade, the internal market of the Greek world consisted of a series of city-state territories which in theory were closed to the outside world. Maritime freight was subjected to scrupulous control with boycotts and blockades sometimes being used.

## Greek Colonization

The process whereby city-states came into being was accompanied by migration flows which have been described by the ambiguous term 'colonization'. Initially they were to meet demographic objectives rather than commercial or imperialist ones. As they strove toward self-sufficiency the early city-states seem to have sought to use migration as a way of ridding themselves of surplus population in order to ensure that every citizen should have enough land to meet the needs of the family and of the city-state as a whole. Greek colonization seems to have originated from a food crisis associated with dry-farming, which is less productive than irrigated cultivation and is less able to accommodate an abundant population. It is clear that the Greek colonies were not trading posts but were rather new city-states, totally independent from the parent city-state which had taken the initiative to found them and from which their citizens originated. The selection of the site where they were to be created, in the midst of agricultural land, did not predispose them to become industrial centres or trading posts. If trade followed, and especially trade in manufactured goods, this only occurred in a subsidiary fashion.

However, colonization rapidly extended the outer limits of the Greek world. In the eighth century BC, when colonization began, the west was believed to be a frightening place that was peopled by the kinds of monster evoked in the voyages of Ulysses. At the end of the sixth century, the horizons of the Greeks had been broadened to such an extent that the first geographers attempted to describe the known world and to depict it graphically. One Greek, perhaps from Marseilles (Avienus,

Ora Maritima), knew of the existence of the remote British Isles.

Two main waves of colonization may be distinguished: the first spanned the years between 770 (foundation of Phithekussiai, on the island of Ischia in the Gulf of Naples) and 675 BC. It began in the Doric lands of continental Greece and in Euboea, and involved southern Italy and Sicily as well as Thrace. The second wave had Ionian Greece (Miletus, Phocaea, Athens) as its source, together with the colonies that had been founded in the preceding wave of activity (Sybaris, Syracuse, Zankle, Massalia). This movement, which started c.675 and lasted scarcely after 550, affected several zones: the shores of Thrace, the Black Sea, and the Bosphorus; southern Italy and Sicily; the coastline of Spain and southern Gaul (Mainake in Andalusia, Emporion in Catalonia, Alalia in Corsica, without forgetting Massalia (Marseilles) and its colony Theline (Arles)); and the coastline of Epirus along the channel of Otranto.

These settlements gave rise to a very rapid increase in trade between the greatly enlarged Greek world and the rest of the Mediterranean, as well as with people living in the interior of the European continent. They must not be confused with the commercial or military staging posts which were developed later, such as the satellites of Marseilles (Agathe, Tauroeis, Olbia, Antipolis, Nikaia), which remained politically dependent on their parent city-state.

## Commercial Empires and Emporia

Before the completion of Greek colonization there existed areas of free trading in the Mediterranean basin, and Etruscan coasting-vessels practised small-scale barter of goods along the coasts of Liguria (Arcelin, 1986a). After the middle years of the sixth century BC the European coasts of the Mediterranean were divided among three great trading powers, whose frontiers became more stable after the battle of Alalia (between 540 and 530 BC).

The Phoenicians of Tyre, attracted especially by mineral riches, established a whole series of trading posts, then from the end of the ninth century set up genuine colonies along the Mediterranean and Atlantic coasts of Iberia, on Malta, on the Balearic Islands, and in western Sicily. From the middle of the sixth century their foundation of Carthage became an autonomous power, which established privileged links with the Etruscans against the Greeks. At the end of the sixth century, war set the seal on the limits of the Punic zone of influence, represented by the coasts of Spain to the south of Catalonia, by Sardinia, and by western Sicily (but this last area has been disputed). However, in the absence of territorial continuity it is difficult to talk about real colonization before the third century, when Carthage founded a protectorate around Carthago nova (Cartagena) located between Alicante and the left bank of the Guadalquivir.

The Etruscans, whose archaeological remains date from the eighth century, did not represent a political group but rather a collection of independent city-states. At their greatest phase of expansion in the sixth century BC they were masters of western Italy, all the way from the Brenner Pass and Ventimiglia to Capua, and of Corsica. During the fifth century the independence of Rome, the invasion of the Po plain by the Gauls and the descent of the Samnites in Campania reduced their territory to the original areas of Etruria, between the Arno and the Tiber.

The Greeks, who arrived in the Balkans at the start of the second millennium BC, were fragmented into a myriad of independent city-states, and did not make up a coherent political entity any more than the Etruscans did. Despite their sense of belonging to a single people, the differences in dialect between Ionian and Doric Greeks traditionally masked political divisions. It was only from the beginning of the fifth century that Marseilles and Sicily, and Athens and its allies, represented combinations of some importance. On the other hand, the Greek world after Alexander the Great is normally characterized by the regrouping of city-states either into kingdoms, such as Epirus or Macedonia, or into leagues. Conquered in stages from the beginning of the third century, European Greece came under the definitive control of Rome in 146 BC, with the exception of Marseilles and its territory which was none the less an ally of Rome. Rome, in effect, had become master of the Italian peninsula in 271 BC. At the end of the second century it also possessed Sicily, Sardinia, Corsica, eastern and southern Spain, southern Gaul (with the exception of the possessions of Marseilles), and the coasts of Illyria, Greece, and Macedonia.

The fundamental mechanisms of trade were roughly the same for members of all these groups. With the exception of the barbarians, trade was undertaken by private entrepreneurs rather than by tyrants or kings, even where they still existed. Those holding political power did not play a major role in trading activities but they still exercised meticulous control of how it was carried out. In a context of very fragmented exchange systems, foreigners were excluded from trading on most of the territory of the city-state but were allowed to trade on a small area (the *emporion*) which was specified for that purpose and was supervised carefully by the tax authorities. The loss of autonomy of the city-states and the political rise of the Hellenistic empires reduced the number of places of exchange, from which the distribution of products into their hinterlands might take place. Cargoes were subject to official inventories which listed their point of origin, and foreign sailors had to land their goods in a specified port if they wished to sell them. They were not allowed to trade as or where they wished, unless they chose to operate as smugglers, which must have been exceptional.

This system created protected zones but it did not exclude trade between these political blocs, even though official commercial treaties had not been drawn up. Submarine archaeology has proved that boatmen operating from Marseilles transported Etruscan or Phoenician or Punic amphorae. By fiscal and military means each trading power protected its commercial territory, relying on a network of reception ports, garrisons, and secondary trading posts. Pirates were the main danger. Until the fifteenth century BC piracy was much more commonplace than one might imagine, and often the term 'pirates' simply related to rival powers; for example, the Greeks described their Etruscan rivals as 'Tyrrhenian pirates'. However, the development of maritime .powers reduced this scourge. The decline of the Greek kingdoms after the Peace of Apamea in 188 BC led whole populations (Crete, Cilicia, Liguria) to engage in piracy which consisted not only of inspecting ships, seizing their cargoes, and holding their passengers to ransom, but also launching raids on the nearby shore, especially in Italy. Pompey the Great experienced his hour of glory when he eradicated this problem in 67 BC.

The geographer Strabo, who wrote at the end of the pre-Christian era and assembled information

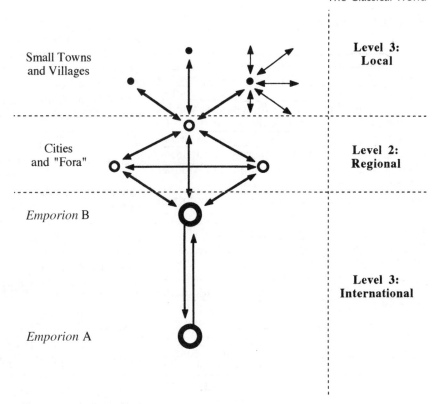

Small Towns
and Villages

**Level 3:
Local**

Cities
and "Fora"

**Level 2:
Regional**

*Emporion* B

**Level 3:
International**

*Emporion* A

Fig. 2.2. The basic pattern of
ancient commerce.
(P. Arnaud.)

from several historic accounts of the world, listed only thirteen emporia of any importance in Europe, including six which he considered to be of prime importance, five of which were in western Europe. Their economic importance encouraged demographic growth and stimulated regional trading activities. In this way a trading system emerged which we have tried to indicate diagrammatically (Fig. 2.2). This would not change greatly until the appearance of the Roman Empire.

Another kind of emporium existed which brought Mediterranean goods into contact with trade in the barbarian world to the north. In this way the whole of Europe was put in contact with the Mediterranean and one might even talk about a kind of colonial trade relationship. In this way, as the previous chapter has described, the Mediterranean world acquired 'raw materials' (ores, metals, foodstuffs, exotic products, even slaves) in exchange for manufactured goods, wine, and other products processed from agricultural goods.

## The Roman Conquest

Europe did not form a particularly coherent entity during the Roman period. According to the limits recognized by the Romans, it extended as far as the mouth of the Don (Tanais), but virtually nothing was known by them about what lay to the north of the Baltic or to the east of the Vistula. By contrast there was a sharp division within the empire between the eastern territories, with their Greek language and culture, and those to the west where Latin was dominant. This dividing line cut Europe into two parts, and linked the Balkans (Achaea, Epirus, Macedonia, and the coasts of Thrace) to the Aegean world and to Asia Minor rather than to western Europe. The major part of Roman Europe belonged in cultural and economic terms to what has been called the 'Roman West', whose limits also included north Africa from the Atlantic to the Grande Syrte.

31

P. Arnaud

## Political Geography

The province was the basic administrative feature set in place by the Romans (Fig. 2.3). This administrative and fiscal unit did not really represent a territory that could be contained within a continuous line as would be the case for a modern administrative unit. It was not until AD 136 that one encounters the only example of boundary stones defining a linear frontier between two provinces. These boundary posts, which delimited the frontier between the provinces of Moesia and Thrace, were very much the exception and related only to that particular region and period. A province was defined as the collection of city-states that were subject to the authority of a particular governor. For this reason, Nikaia (Nice), which was situated to the east of the River Var, should logically have been part of Italy, but was actually part of the territory of Marseilles and hence belonged to the Provincia Narbonensis of Gaul. Italy was the only exception to this rule. As the first territory to be conquered by Rome, the peninsula was considered after the first century BC to be the homeland of its people and thus could never be a province, at least not until the end of the third century AD.

## Development of the Provinces

The presence of Rome has left very varied legacies in the countries of Europe. These differences related initially to the several phases in which the numerous provinces entered the Roman world (*orbis romanus*). The chronology of this conquest reveals considerable variations (Fig. 2.4). Territories that became provinces at an early date enjoyed substantial economic, social, and political integration with Rome. Thus, the earliest provinces approximated most closely to the 'Italian model'; these were the first to receive colonial settlements, colonists, capital, and investments, and the first to be integrated into the Roman economic system. Thus during the reign of Vespasian (AD 69–79), the Provincia Narbonensis could be considered to be like a second Italy, a title that Baetica would surely also have earned.

These differences also relate to the status of the territories that Rome acquired. The decision to create a province was not taken automatically. Very often the Roman state, whether it was Republic or Empire, chose to establish a protectorate composed of client-kingdoms. These remained nominally independent but were subject to Rome in both military and diplomatic terms. In Europe, only the Kingdom of Noricum and the small protected kingdom of King Cottius, in the Alps, were for a time part of this regime. Roman investments were less in these areas than in the provinces, and the impact of passing into the orbit of Rome was less clear. However, as in all frontier territories, Roman businessmen were very active in the protectorates. There was also a very clear hierarchy of provinces in the Roman Empire: so-called senatorial provinces were the earliest to be Romanized and were the best developed (Narbonensis, Baetica, Sicily, Achaia, Macedonia). 'Imperial' provinces involved poorer areas that were less integrated into the Roman system, being governed by procurators. These provinces included a number of relatively isolated mountainous areas such as Epirus, Corsica, the Alps, Rhaetia, and Noricum.

However, we can observe a general trend of centrifugal development. The more one advances in time, the greater the shift of economic dynamism from Italy towards the periphery of the Roman Empire, where investments were less costly. But that is not to say that the economic interests of Italy were not safeguarded by a rather haughty form of protectionism. We know that cultivation of vines and olives was restricted in the provinces, and even reduced as a result of authoritarian decisions by the Roman state. But, little by little, the Romanization of the economy led to the development of local production which replaced imported goods in the local market and brought a sharp shock to provinces which had previously exported their surpluses. At the same time, the importance of market demands on the fringes of the empire, where the army represented a colossal focus of consumption, served to stimulate the economy of regions adjacent to the frontiers of the Roman Empire.

The integration of new territories into the orbit of Rome led to many changes in the landscape and in local economic and social structures. One of the most important undoubtedly was the demographic cost of conquest, which claimed very many lives, especially among those who were beaten, and devastated many landscapes. The total population of the Roman Empire has been discussed many times and

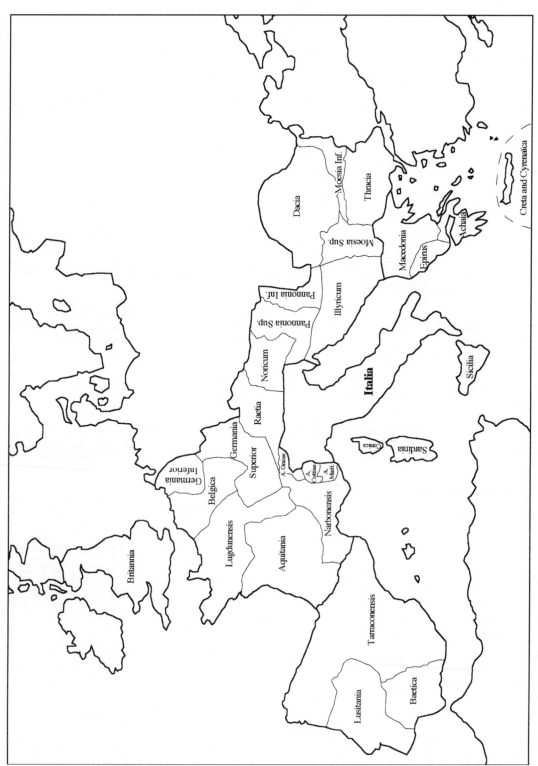

Fig. 2.3. The Roman Empire, AD 117. (P. Arnaud.)

| Province | Conquest started | Conquered | Organized |
|---|---|---|---|
| Achaia | 198 BC | 148 BC | 27 BC (separated from Macedonia) |
| Alpes Cottiae, Maritimae, Poeninae | 15 BC | 14 BC | 14 BC; AD 63–4 |
| Britannia | AD 43 | C.AD 140 | AD 43; divided into two provinces C.AD 200 |
| Creta (and Cyrenaica) | 67 BC | 67 BC | Cyrenaica bequeathed to Rome 95 BC; became a province 75 BC; Creta united to Cyrenaica 67 BC. |
| Dacia | AD 102 | AD 107 | AD 107 |
| Epirus | AD 168 | AD 168 | 148 BC (as part of Macedonia); became a province AD 125 |
| Gallia Aquitania | 51 BC | 19 BC | 51 BC (as Gallia Comata); became a province 24 BC |
| Gallia Belgica | 57 BC | 51 BC | 51 BC (as Gallia Comata); became a province 24 BC |
| Gallia Lugdunensis | 59 BC | 51 BC | 51 BC (as Gallia Comata); became a province 24 BC |
| Gallia Cisalpina | 226 BC | 175 BC | 180 BC; became part of Italy 42 BC |
| Germaniae Superior and Inferior | 56 BC | 51 BC | First, parts of Gallia Belgica, then identified as military districts, they were not organized as two distinct provinces until AD 90 |
| Hispania Citerior Tarraconensis | 202 BC (eastern part) | 19 BC | 197 BC; reorganized 19 BC |
| Hispania Ulterior Baetica | 206 BC | 192 BC | 197 BC; reorganized 19 BC |
| Illyricum | 177 BC | 12 BC | 27 BC; reorganized 11–9 BC and AD 10 |
| Lusitania | 194 BC | 19 BC | 197 BC (as part of Hispania Ulterior); 19 BC |
| Macedonia | 197 BC | 148 BC | 148 BC |
| Moesia | 75 BC | 29 BC | AD 3–14 (?); divided into two provinces AD 86 |
| Narbonensis | 122 BC | 49 BA | 100 BC (?); reorganized by Augustus, 22 BC |
| Noricum | 15 BC | 13 BC | 10 BC; called 'reign' until AD 170 |
| Pannonia | 34 BC | 9 BC | AD 9; later divided into two provinces (C.AD 100–10) |
| Rhaetia | 15 BC | 15 BC | C.AD 100, included Agri Decumates |
| Sardinia and Corsica | 128 BC | 163 BC (Cordica) 115 BC (Sardinia) | 227 BC |
| Sicilia | 248 BC | 210 BC | 210 BC |
| Thracia | AD 46 | AD 46 | AD 46 |

Fig. 2.4. The Roman provinces: a chronological table. (P. Arnaud.)

there is little chance of consensus ever being reached. Estimates range between fifty and seventy million souls at the end of the reign of Augustus. The few estimates of the demographic consequences of war that were communicated by classical authors must be treated with great caution; for example, Pliny the Elder spoke of 1,192,000 enemies being killed in Caesar's foreign wars. However, it is certain that human and material losses were very great for those who were defeated and hence the notion of conquest by the Romans became inseparable from 'devastation' in the writings of many classical authors. Without reaching the dimensions of genocide, the Roman conquest undoubtedly overturned the demography of the subjugated provinces. It was in areas that experienced massive demographic and ethnic losses that the Roman economic system was introduced most thoroughly. Without these demographic losses such changes would have been neither so rapid nor so radical.

One of the most significant impacts was in the use of money, which had obvious implications for areas where this practice was not yet known. This was expressed initially by a rapid increase in the volume of money in circulation and by the growing use of money in trade. It characterized the development of an economy based on the investment of capital that was evaluated in monetary terms in order to promote the means of production. This radical transformation was the immediate cause of the revolt of Boudicca in Britain. Without warning, the immensely rich Seneca demanded the repayment of loans that had been granted to leading Britons. Without these loans, calculated on the exchange value of their properties, and which were reinvested immediately, they would not have been able to operate in the expensive production systems that Roman investors had developed in the provinces, or in the trading systems of the Roman Empire. None the less, the use of money in the provinces resulted from the private actions of capital holders rather than from the initiative of the state.

## Romanization of the Economy

The Romanization of the economy was expressed initially by the integration of the provinces into an open market, which embraced the trading area of the whole empire and within which commercial

taxes were low; for example, an average tax of only 2 per cent was levied on goods being moved from one province to another. In this type of market, the motor of the economy was not reduced to a kind of colonial system focused on Rome, even if the capital, by virtue of the size of its population, did represent one of the leading markets in the empire. Trading commodities between provinces was a new feature for the recent provinces and tended to specialize in particular goods. One may quote the remarkable distribution of Catalan wine from the evidence derived from the diffusion of Pascual 1 and Dressel 2–4 types of amphora (Fig. 2.5). The first type followed the water route from Narbonne through Toulouse to Bordeaux, and then along the Atlantic coast to the south of England or into the mouth of the Loire. To a lesser degree these amphorae were taken up the Rhône valley to Lyons. By contrast, amphorae of the Dressel 2–4 type were especially destined for the Italian and African wine markets.

The Roman economy was characterized by the juxtaposition of two types of economic system: namely micro- and macro-systems. The macro-system had a regional structure, which did not correspond necessarily with provincial frontiers, but covered large areas that were comparable in size with the 'provinces' of France during the *ancien régime*. Unlike in the Greek world, it would seem that the city-state, which to some extent remained a political centre, and its immediate surrounding territory were distinguished as economic areas in only a marginal way and were more integrated into economic structures at a truly regional scale (Fig. 2.6).

The presence of the army, whose use of money played an essential role in the economy, in the development of urbanization, and in the adoption of a Roman-like style, gave rise to new demands and stimulated the appearance of new forms of production to enable them to be satisfied. The demand for foodstuffs was expressed through the consumption of bread (which required greater quantities of flour than traditional gruels had done), of oil, wine, and garum (a condiment made from fish). After having relied on imports, these demands were now rapidly met by local production. Demands associated with changing lifestyles were expressed by the virtual disappearance of simple, domestically made ceramics and the rise

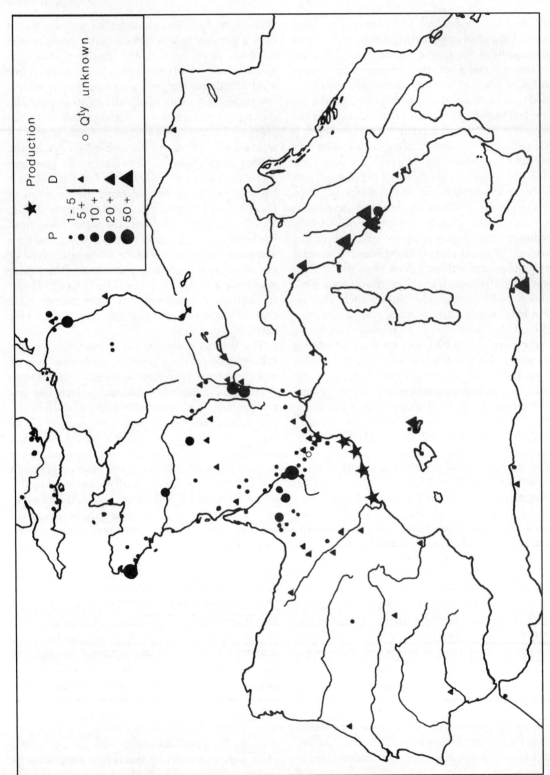

**Fig. 2.5.** Diffusion of Pascual 1 (P) and Dressel 2–4 (D) Catalan wine-amphorae. (P. Arnaud, after P. Leveau, P. Sillières, J.-P. Vallat, *Campagnes de la Méditerranée romaine* (Paris, 1993), 239, fig. 39.)

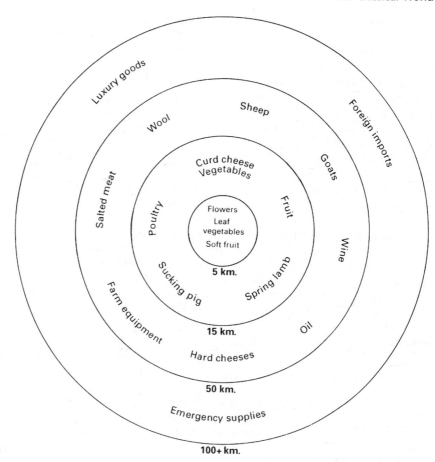

**Fig. 2.6.** Areas of Roman trade (After J. M. Frayn, *Markets and Fairs in Roman Italy.* (Oxford, 1993) 77, fig. 7.)

of independent workshops which distributed their wares locally or on a regional scale. Such workshops were known throughout the Roman west. New forms of construction required the use of new materials, such as hewn stone, lime, bricks, and tiles.

The logic of the macro-system led certain regions to produce with a view to exportation (or to imitate imported products) and gave rise to an inter-regional market. This explains the strikingly cosmopolitan character of some great European market-places, such as Lugdunum (Lyons) and Burdigala (Bordeaux). In this way olive oil from Baetica invaded the market of the Roman west, including Italy. In contrast, Italian viticulture, which initially was geared to exporting wine, especially to Gaul, and generating large profits in the second and first centuries BC, was unable to resist the development of vine-growing in Gaul and the subsequent loss of its former clients. Italian vinegrowers had to turn to the much less lucrative opportunities of their own home market.

Without reaching the level of importance that was achieved in the provinces, Roman businessmen (*negotiatores*) were both active and influential before the conquest of the respective provinces. There is good evidence of their activities in territories that bordered the Roman Empire before those areas were conquered, notably in independent Gaul, where a contemporary of Cicero was able to found an agricultural estate similar in style to a Roman villa. The barbarian world (*barbaricum*) was not cut off from the Roman world. Even when the Roman Empire tended to close itself off beyond military frontiers, those frontier zones remained important areas for trade, which varied according to the seasons and the geographical milieus concerned. The only mention of a frontier in the famous Table of Peutinger, which was a medieval copy of a Roman *mappa mundi*, described it as 'a zone of contact and exchange with the barbarians' (*commercium barbarorum*).

In general terms the presence of Italian interests immediately encouraged the development of a Romanized economic system. It was also a temporary feature in provincial economies which disappeared rapidly once local investments from the province itself or from other provinces made their impact. In Moesia and in Thrace, where these phenomena have been analysed thoroughly, Italians are well represented in the first century AD, but during the second and third centuries they gave way to Romanized citizens of Thrace and to immigrants coming from the eastern parts of the Roman Empire. Whatever the source, such investments allowed the systematic exploitation of resources in the provinces, which was expressed through the quest for mineral deposits and potential quarries and through the development of agricultural and industrial resources as well as routeways.

## Transformation of Rural Landscapes

The presence of Roman power profoundly changed the appearance of the rural environment. One of the most striking aspects of Roman conquest consisted in the creation of centuriated landscapes. By virtue of conquest, the territory of the conquered group became the property of the Roman state, which exerted the right to lease or assign that land to colonists. Cadastral records would be required to manage such land and to raise taxes from it. Such records were often based on the establishment of regularly arranged, rectangular areas of land which usually embraced blocks of 200 *iugera* (*c.*50 ha.) which were bordered by tracks, or by channels in some places, and were called centuriations.

There is good evidence of this system of land division in Italy, Gaul, the Provincia Narbonensis, Spain, and the Balkans, where it was used to allocate land to colonists and to assignees. It was not employed in a systematic fashion, even if it seems much more widespread than its initial function might suggest, nor is its chronology always well understood. Where the confiscated lands were not allocated to Roman citizens, they could be allocated to the local community on condition that rent was paid, or to an individual who leased them out and

drew profit from the transaction. Use of these regular systems of land registration allowed fertilizers to be applied rationally on rented land. Cadastration represented a major reallocation of land holding that was facilitated by the loss of population associated with conquest, and translated the subsequent changes in the political and economic order into the landscape. Quite apart from any consideration of the size of the land parcels, it is undeniable that this type of reorganization changed the rural landscape in a very profound way (Fig. 2.7). Landholding may well have been reorganized and parcels reorientated quite frequently, but caution must be exercised when discussing this topic.

The landscape was also transformed by new methods of production. If one excluded the *vici*, which were small towns rather than villages in the strict sense, and remote or mountainous areas, which were scarcely suitable, the most distinctive feature in the Romanization of the countryside was the appearance of the villa. This was the origin of many forms of dispersed settlement and, following the advice of agricultural writers in classical times, was located in areas close to major centres of food consumption and along routeways (Figs. 2.8, 2.9, and 2.10). The word 'villa' is used to describe Roman buildings that were monumental in character as well as particular types of farm unit. These agricultural holdings were large or medium-sized, and varied according to their geographical location and their position in time, but they had the common objective of rational economic production.

The villa is probably one of the most distinctive features of the Romanized landscape between the second (and especially the first) century BC and the middle of the third century AD, serving as the residence (at least on a temporary basis) of its owner and as a farm unit with housing for its workers. It originated in Italy and Spain in order to meet three objectives: to organize and accommodate a servile and self-sufficient labour force for working an extensive agricultural area; to process goods that were produced and to store a sufficient quantity of them to be able to profit from any price fluctuations; and to provide an appealing residential environment for the owner so that he might manage the holding more effectively. Much conventional wisdom about Roman villas must be treated with considerable caution.

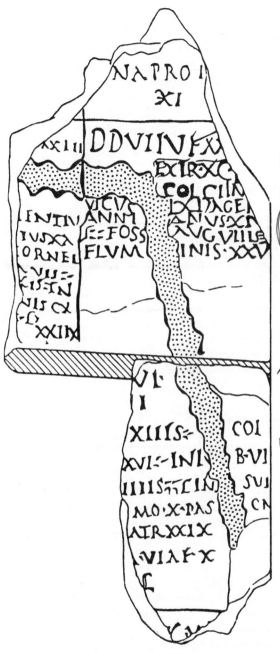

Mediterranean zone. By contrast, a servile labour force could be used on land lying beyond the villas.

2. Villas did not represent a single system of land management.

3. They displayed immense variations in terms of size, wealth, function, and other features.

4. The establishment of villas did not eradicate small landholdings and certainly did not erase small farms.

5. Based on the quest for self-sufficiency, villas could also give rise to a variety of craft activities (basket-making, weaving, woodworking) or industrial functions (ceramics, metal goods).

6. Most villas were owned by members of the urban élite.

7. Villas were not necessarily associated with monoculture.

8. Above all else, villas do not seem to have been equated with vast landholdings (latifundia). The evidence that is currently available suggests that villas were associated with units of land ranging from 10 to 300 ha. For various reasons which related to human safety (for example, fear of uprisings) as much as to the use of finances, there was a tendency to disperse investment in land rather than to concentrate it in a small number of very large units. For this reason the number of production units was high, and contiguous parcels of land were not always placed under the same manager.

Villas operated agricultural systems that stood midway between extensive cultivation and the kind of intensive farming that classical agricultural writers would have us believe took place. The owner initially sought to obtain the maximum return from his rents while making the minimum investment in his property. This explains the relative fragility of the villa system, placed as it was on a precarious financial footing which might encourage the investor to put his money in more viable forms of economic activity. For these reasons many villas underwent frequent reorganization, while others functioned for only a short time, with some simply surviving as places of residence for their owners. Thus villas tended to disappear from the Italian landscape during the second century and were retreating in Baetica; by contrast, they were increasing in other regions of Spain, in Gaul, in Germany, and in Britain, before encountering a critical phase in the third century, doubtless because of economic

**Fig. 2.7.** Orange Cadastre C, fragment 351: the Fossa Augusta (After A. Piganiol and F. Sautel, *Les Documents cadastraux de la colonie romaine d'Orange* suppl. Gallia, XVI. (Paris, 1962), 296.)

1. Villas did not systematically employ chain-gangs or have barracks full of workers. Archaeological evidence shows that barracks for farm workers were rare and were never found beyond the

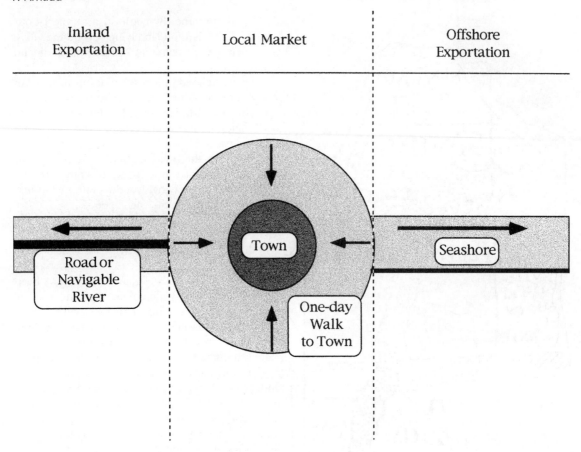

| Inland Exportation | Local Market | Offshore Exportation |
|---|---|---|

Road or Navigable River

Town

Seashore

One-day Walk to Town

**Fig. 2.8.** Theoretical distribution scheme of rural villas according to ancient writers. (P. Arnaud.)

factors as well as the upheavals associated with invasion.

Villas did not disappear from the Romanized landscape until the creation of the first castral mottes, but they did decline in number. This was accompanied by quite a radical transformation of agricultural structures which was expressed through a concentration of land ownership, a shift toward sharecropping and tenancy, the installation of families of slaves on small farms within large landed estates, and the growth of village communities. Villas tended to survive only as the focal points for enlarged estates (functioning as luxurious residences for their owners or places where rents and taxes would be collected), or as locations where farm products would be processed and dues that farmers had paid in kind to their landlord would be stored. With respect to agrarian structures it would be an oversimplification to draw a contrast between enclosed fields (bocage) plus woodlands in north-west Europe, and open fields in other parts of the continent; however, it is clear that the rural landscape of the Romanized world was substantially different from what would have been found in areas that did not come under the influence of Rome.

The engineering abilities of Rome made their contribution to changing the rural landscape and enabled large investments to be made to improve new areas of land. The cadastral records of Orbane provide evidence of land drainage and improvement of islands in the lower course of the Rhône, probably in association with a navigable channel. Lakes and marshes were drained in Italy (Lake Fucino, Lake Velinus, the Pontine marshes, Maremma in Etruria, the Po valley), at the foot of the Alpilles in southern Gaul, along the lower course of the Rhine in Germany, and in Britain, where the

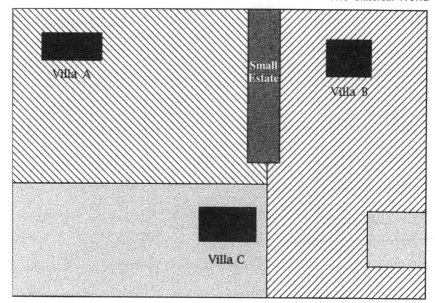

**Fig. 2.9.** Some schemes of land distribution around villas. Each pattern shows pieces of land linked to a single villa (or little estate). (P. Arnaud.)

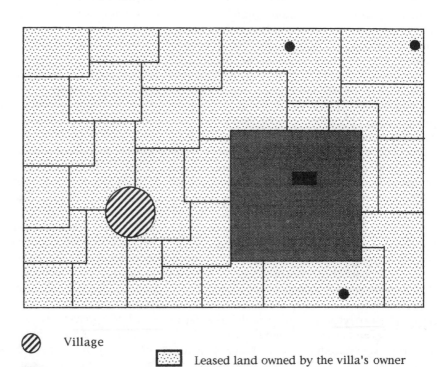

**Fig. 2.10.** Landscape patterns for leased land around villas. (P. Arnaud.)

Village

Villa

Small estate

Leased land owned by the villa's owner

Part of the land cultivated through the villa

cutting of channels linking the Cam to the Ouse, and the Nene to the Witham and then to the Trent enabled parts of the Fenlands to produce cereals. According to ideas recently put forward by Leveau, the improvement of areas of low-lying unhealthy land was not foreign to the policy for creating colonies (for example around Arles in Narbonensis, Valentia and Ilici in Spain, and doubtless in the colonization of the valley of the Po). Finally, although the practice of dry-farming was widespread in Mediterranean Europe, irrigation was developed in Italy (or at least in Latium) with the help of aqueducts and was practised more extensively in southern Spain, where networks of dams and channels were installed.

The use of barrels was a technical innovation that had specific repercussions. Its adoption was rapid, with the number and volume of amphorae in Rome declining rapidly after the first century. At the start of the second century the Roman army adopted this new form of container, and at the end of that century tombs in Lusitania were being fashioned in the shape of barrels. Despite efforts to improve the relationship between the empty weight and the volume of liquid carried in amphorae, these vessels could never compete with barrels. Use of the new containers enabled the cost of carrying freight to be reduced, tended to favour vineyards that had easy access to oak forests rather than those close to deposits of clay (as had been the case in the past), and modified the tastes of wine consumers. The vineyards of northern Italy (notably Aquileia), Bordeaux, and the fringes of the Massif Central benefited greatly from this innovation to the detriment of vine-growing areas in central and southern Italy.

The Romanization of the countryside also had important effects on plants and animals. Roman domination was expressed through spectacular northward extension of vine-growing, which expanded from the shores of the Mediterranean as far as Britain and Germany. New varieties of wine were developed, such as Biturica around Bordeaux and Alba in the Ardèche, and new strains of cereal were perfected in relation to differing farming practices and economic conditions. Completely new crops were introduced by the Romans, including peaches, apricots (acclimatized in Italy after the middle of the first century AD), and cherries, which were brought by Lucullus from the shores of the Black Sea in 70 BC and acclimatized in the villas of the second century. Stock-rearing also saw considerable development in Romanized Europe whilst, in Italy, fish farming and the rearing of shellfish were developed. Rabbits were introduced to the European continent from their native Balearic Islands but their devastating effects were being recorded as early as the third quarter of the first century AD.

## Crafts, Industry and Exploitation of Mineral Resources

Craft activities were located mainly in villages, where they often formed the main occupation; hence one can identify genuine industrial settlement sites in clay vales and on other deposits of raw material. To a lesser extent craftsmen operated in towns or were dispersed on farms from which they served purely local markets. Many activities which were in existence before the Roman conquest experienced great growth subsequently as new markets opened up: the bronze workers and glass-makers of Gaul formed a notable example. However, it was the production of pottery on the wheel that broke completely with earlier traditions and experienced the most spectacular growth. The use of ceramic ware had been unknown in some distant parts of Britain. Pottery workshops produced a range of goods but displayed some specialization. Workshops that made building materials (bricks, tiles) were often dispersed; for example, it was not rare to see a tiler's kiln located where a new villa was being built. Amphorae were also heavy objects that were difficult to transport and were sometimes produced in the same kilns as tiles. Potteries were often complementary or subsidiary activities on an agricultural holding or else fitted into the business activities of a landowner. Workshops that produced commonware for local or regional use were sometimes grouped in villages but might also be dispersed close to centres of consumption which, for these basic forms of pottery, were usually rural rather than urban in character.

More spectacular was the production of high-quality ceramics in much more commercial enterprises which served a large supply area. These goods required a very skilled workforce and were produced in large craft centres. This was the case for the terra sigillata which was made initially in workshops in the Arezzo region (following the earlier production of so-called Campania pottery with black glaze) and whose owners opened branches in Gaul, first at Lyons and then in the Massif Central, with workshops being grouped in several villages of which the best known are Lezoux and Graufesenque. Production of these goods gradually declined during the second century to give way to ceramics made in the Rhône valley, and then in North Africa. The growth of high-quality ceramic production was generally linked to the ability to export it to the four corners of the Roman Empire. Changing fashion accounted for both the rise and subsequent decline of terra sigillata. Pottery was also used as an accompanying cargo in trading vessels or was used as ballast in international shipments that were short in volume or weight. Changes in the geography of pottery workshops was also linked to modifications in maritime trade.

In addition, the exploration and development of mineral deposits was a distinctive feature of the Roman period. Some deposits had been known for a very long time, such as those in the Balkans, Transylvania, Britain, and Spain. Their exploitation was organized rapidly and was well known. The emperor owned them and leased them to entrepreneurs who arranged for the metal to be made into ingots. But technology remained simple, even in these regions. Veins were worked in succession using the minimum number of tunnels. New reserves were sought eagerly, as in the Alps and in the Maures massif, where intense mining activities followed discovery. The development of new construction techniques led to exploration for new materials, including building stone, chalk, limestone, and marble, but also minerals that could be transformed into pigments, such as cinnabar and galenite. Iron deposits were the only type to attract permanent settlement which was linked to the local development of iron-making that displayed considerable geographical dispersion.

## Urban Development and Urban Networks

Contrary to what has long been believed, the Roman conquest did not lead to hilltop settlements being abandoned in a systematic way and replaced by towns on the plains. There were, of course, some examples of this kind of change, such as Augustodunum (Autun) replacing Bibracte, and Augustonometum (Clermont-Ferrand) in place of Gergovia, but these were the exception. Old-established settlements were radically transformed, even if they retained their original names as well as their ancient sites. The real innovation was not a radical change in the geography of urban places in the conquered lands but their transformation into city-states that were Graeco-Roman in style, each being organized around a real urban centre. The most important legacy of Rome was in changing the appearance, character, function, and hierarchy of urban settlements, rather than in creating new towns. Urban changes were none the less rapid and brutal. In some instances it may be possible that old settlements coexisted with new urban centres, but most evidence indicates that the old site was destroyed in order to refashion it according to radically different principles, including organized ground plans (though not as regular as was once thought) which often may still be read in the urban morphology of modern towns. By contrast it is clear that the Romans sometimes operated a policy of relocating population away from small hilltop fortifications, where terrain conditions were unfavourable, and concentrating them in lowland settlements, as was the case after the Roman conquest of Galicia.

The shift from a tribe or a 'people' to a city-state was often accompanied by the foundation of an urban central place, which gave rise to new urban networks and urban functions. Some sites which had been important before the Roman conquest were reduced to the rank of a small town when the conquerors decided not to use that location as the focus of a city-state. Literary sources tell us that this was the fate of fifteen out of twenty-six pre-Roman sites. The Roman conquest gave rise to new urban hierarchies, which might result from the political will of the conqueror or might be due to socio-economic changes and rural depopulation, as was the case in Italy. The Roman administration established a pyramid of settlements with differing kinds of legal status which was expressed in the title of each settlement and determined its place in the settlement hierarchy (*colonia*, followed by *municipium* and ordinary city, with *pagus* at the base). The components in these hierarchies were reorganized frequently as new provincial capitals were created and other settlements (even former cities) were reduced to being dependencies of others. By contrast, some small towns were elevated to city status. Furthermore, the development of towns, which broadly shaped the imperial economy, was linked inseparably to the establishment of lines of communication.

## Development of Means of Communication in the Roman Empire: Road Links

Roman control was expressed almost immediately by the establishment of Roman roads (Fig. 2.11), which formed some of the most distinctive features in the Romanized landscape. From classical times they were perceived as symbols of order and domination by Rome. They were often straight and generally paved, but sometimes were just stabilized routeways. Many fitted into patterns of landholding that had been redistributed and landscapes that had been reorganized by the conqueror. They were equipped with bridges and other architectural features which symbolized the victory of Rome over the natural world. Of course the conquered regions had not been without ancient trackways, and Rome did not necessarily destroy them. This point was to be proved by the speed with which Roman roads were to fall into disrepair at the end of the Roman occupation, and the ancient trackways would come to be used once again. Roads played a major role in the military, administrative, and also conceptual domination of the Roman Empire. They were created and equipped when the conquest took place, and were abandoned when Roman occupation collapsed.

It is traditional to argue that 'all roads led to Rome' but this was only partly true. It is undeniable that the first routes established after the conquest tended to link the nerve centre of a province to Italy and particularly to Rome, but it is also clear that roads were soon built which favoured connections between regions and between provincial urban centres that had been created by Rome. The subsequent development of land-based communications tended to favour inter-regional links. A good example was the establishment of a coherent system of roads from the Rhône valley across the Massif Central to south-west Gaul, during the reign of Antoninus in the middle of the second century AD.

Roads played a determining role in the creation of a centralized administrative system. The Emperor Augustus set in place a series of staging posts which formed the basis of the imperial postal service. This system, which was called the *Cursus publicus*, was improved many times and performed the dual function of allowing imperial functionaries to travel around, and dispatches to be delivered with the minimum of delay. The staging posts also played an essential part in allowing ordinary travellers to move around the empire, in which personal mobility was becoming more commonplace. Inns, hostels, and even villas provided accommodation for travellers, albeit with differing degrees of security. By contrast, roads played a rather more limited role in the movement of freight. They seem to have been used for moving goods over relatively short distances and hence were important in the functioning of secondary markets.

## Importance of Seaborne Freight

The great majority of cargo was moved by water which was faster and cheaper than transporting goods by road. Classical sources inform us that an ordinary merchant vessel would travel at 4.5 knots, under normal conditions of wind and load. The busiest period for seaborne traffic (*mare apertum*) was defined in two ways: the broad definition ran from early March to 11 November while the strict definition extended from 27 May to 14 September. Maritime trade was slight outside this period but continued none the less and was encouraged by fiscal incentives.

Despite the hazards of seaborne travel, moving goods by boat was three times faster than overland, and allowed much greater volumes of freight to be transported. Very early on, from the sixth century BC, appropriate containers were devised to optimize the loading and unloading of boats, including special amphorae that could be loaded in layers and, later on, veritable tankers that carried wine in bulk. No form of land-based transport could offer such advantages with respect to volume, speed, and hence price of return. Using figures provided by the 'maximum edict' of Emperor Diocletian at the end of AD 301 we may calculate that to transport a cargo of corn over 100 miles cost 1 per cent of its overall value if carried by sea, 4.9 per cent if shipped by river, and 28 per cent (or even 56 per cent depending on how the document is interpreted) if moved overland. These proportions were very close to those noted for eighteenth-century

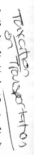

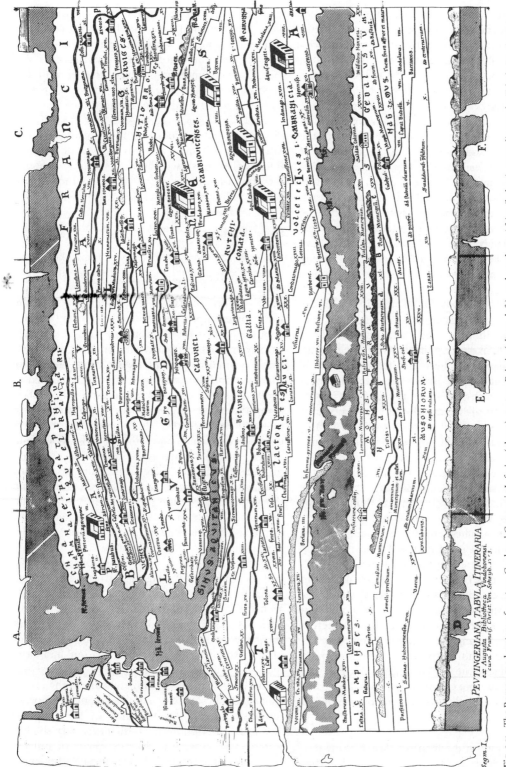

Fig. 2.11. The Roman road system of western Gaul and Germania Inferior (above), eastern Britain (above, left), and northern Africa (below), according to the Tabula Peutingeriana (thirteenth century). (Copy of a Roman original of the mid-fourth century AD), after F. C. von Scheyb (Vienna, 1753).)

45

England. In other words, a journey of 300 miles would double the price of a cartload of corn; hence it was more economical to transport cereals from remote parts of the Mediterranean than from a distance of more than seventy-five miles overland.

Only the savings in time and money that were possible by moving goods at sea can explain the remarkable security measures that were taken to improve river-based and maritime trade in the Mediterranean and along the Atlantic coasts of the empire. As well as the two great praetorian fleets of Messina and Ravenna, which respectively guarded the Tyrrhenian and Adriatic coasts, each stretch of coastline had its own fleet. With the exception of major periods of crisis when they were remarkably ineffective, they seem to have managed to keep piracy under control following the general eradication of this scourge by Pompey in 67 BC.

Major works also left abundant evidence of the interest directed to navigation on the high seas, including lighthouses (Fig. 2.12) and harbour improvements. Following the example of Claudius, Trajan called on the best architects to improve docking facilities and means of trans-shipment between the seaport of Ostia and its river port at the mouth of the Tiber, thereby affording very large vessels the closest possible access to Rome. Trajan also had a harbour built further north at Centumcellae (Civitavecchia) which would remain the port of Rome in medieval times. Finally, in spite of what has been written, the scheme to cut through the isthmus of Corinth, on which several thousand Jewish prisoners were employed, should not figure in the list of Nero's follies. The project could have been completed and it should be noted that the present canal used the route that had been abandoned following the suicide of Nero. Other equally ambitious projects were completed; for example, Marius linked Arles to the Mediterranean by means of a canal, the Fossa Mariana (or Fossae Marianae), and thereby enabled Arles to function as an estuarine port.

It is very difficult to chart the pattern of maritime trade during the Roman period. However, it is clear that there were two types of navigation based on two types of vessel and two scales of commercial activity.

1. Navigation on the high seas lasted many days and linked major trading centres without putting into port. It made use of the latest vessels, which were long, streamlined, and had three or more mainmasts, and by virtue of their great size and displacement could not enter every harbour.

2. Coastal shipping involved smaller and less prestigious vessels, which were more pot-bellied and were only suited to local trade.

Finally, ballast cargoes, such as quality pottery, must be incorporated into any analysis of seaborne trade. Ballast was used above all else to ensure that the ship was buoyant, but was also included to increase cargoes that were deficient in weight.

## River Traffic

Rivers were used as intensively as maritime routes. Three fleets guarded the empire's two river frontiers, namely the Rhine and the Danube. Seagoing vessels sailed considerable distances upstream, with, for example, Seville and Arles being considered as seaports by the Romans. Navigable waterways were used by vessels of many sizes, known to us only through descriptions or images that have survived from Roman times. Sails were hoisted in wide rivers, and use of oars and especially haulage along the banks enabled boats to be moved both up- and downstream. The importance of waterborne freight encouraged the imperial powers to undertake major schemes, including the excavation of canals. Schemes were drawn up for cutting channels to link the Saône and the Rhine and the seaport of Pozzuoli (Puteoli) in Campania to the Tiber. Such great schemes were technically impossible at this time but more modest examples were achieved.

Navigation channels were dug alongside the Po and its tributaries during the first century BC and along the course of the Rhône. We have already seen the example of the canal opened by Marius (the Fossa Mariana) and the cadastres of Orange show the existence of a canal cut through a sector of the Rhône valley where navigation had been made difficult because of the presence of numerous islands. The name of this channel, the Fossa Augusta, indicates that it derived from an imperial initiative probably at the time of the Emperor Augustus. Inscriptions tell us that even very shallow watercourses were used by rafts made of inflated goatskins, for example in the Provincia Narbonensis,

**Fig. 2.12.** Places of commerce (emporia) according to Strabo (d. *c.* AD 20) and Roman lighthouses (P. Arnaud): ■ first-class emporia; □ other emporia; △ lighthouses; ● other important places.—1. Gades; 2. Carthago nova; 3. Narbo Martius; 4. Arelate; 5. Lugdunum; 6. Burdigala; 7. Corbilo; 8. Cenabum; 9. Gesoriacum; 10. Dubris; 11. Massilia; 12. Genua; 13. Ravenna; 14. Aquileia; 15. Centumcellae; 16. Ostia; 17. Misenum; 18. Dikearcheia—Puteoli; 19. Caprae ins.; 20. Corinthus; 21. Aegina ins.; 22. Delos ins.; 23. Olbia; 24. Panticapaea; 25. Tanaïs.

near Lyons, and on the Danube. Under Roman domination the development and deliberate equipment of routeways by land, river and sea played a determining role in the expansion of both long- and short-distance trade, of urban networks and of agrarian systems.

The ancient world drew to a close over the third and fourth centuries AD. The second century AD was the golden age of the European classical economy; by contrast, the third century AD was a period of severe crisis in Europe and more so than in other regions. This was certainly the time of the first invasions, but they do not explain everything, and current interpretations tend to scale down their effects. In this period of military anarchy, an abundance of troops stationed along its frontiers turned Europe into a battlefield where pretenders to the imperial throne confronted one another, and towns were besieged and pillaged. The anarchy that characterized this time gave rise to the development of bands of brigands in each of the western provinces: the most notorious band was known as the Bagaudes. Saxon piracy occurred along the Channel between Britain and Gaul. Systems of communication were disorganized or destroyed and needed to be entirely remade. In addition to this insecurity and political isolation in the western provinces, increasing costs, resulting from higher levels of taxation and an enlarged state apparatus, gave rise to galloping inflation which largely contributed to the disorganization of the economy.

Changes in population at this time are not clearly understood and are difficult to quantify. Some scholars identify a manpower shortage which would have serious implications for the military potential of the empire. However, it is certain that, following a period of demographic growth, a pandemic affected the empire after 165 AD. In Europe its effects were reinforced by more isolated epidemics which resulted from general insecurity and poor health conditions, for example associated with the fracturing or poor maintenance of aqueducts. These epidemics particularly affected urban populations and gave rise to urban depopulation. In spite of movements which affected rural areas along the most exposed frontiers (which had to be repopulated by barbarians, the Laeti) and a shortage of labour resulting from employment conditions, the countryside suffered less than the towns. One might even suggest that during the fourth century the villages were taking their revenge on urban places.

It is undeniable that western cities experienced a crisis which began in the second half of the second century AD. Most cities declined considerably, for example in the fourth century Saines occupied only one-fifth of the surface it had covered in the first century, and then stabilized at the size at which they would remain down to the Middle Ages. In some cases urban decline was directly the result of social upheaval, for example in northern and eastern Gaul and in the Danubian provinces, but contraction was much more widespread and was particularly intense in southern Spain. It was certainly linked in a major way to social and cultural causes which were found throughout the empire, notably the trend that was increasingly common after the end of the second century AD for urban élites to leave the cities whose functions and lifestyle they had maintained, in order to live on their rural properties with which they identified more and more. This movement shifted into the countryside numerous functions that were associated with these formerly urban élites. The parallel shrinkage of cities has been inferred from evidence on the dilapidation of public monuments, but it is not exactly clear how this related to the collapse of urban populations.

This movement fluctuated through time and space and had varying effects but it certainly contributed to the complicated logic which produced the decline of the city, the cornerstone of the classical world. New taxation introduced in AD 212 weakened the position of Roman citizens and free men throughout the empire, and brought to an end the urban hierarchy that had been based on differences in the legal status of settlements. Urban networks were reassembled according to new criteria, becoming more complicated and founded on more elaborate hierarchies. These were reinforced by the reforms of Diocletan which, after AD 284, reduced the size of the provinces, increased their number (and the number of their capitals), and regrouped them into larger units or dioceses (Fig. 2.13). These were placed under the authority of four emperors or co-emperors, three of whose capitals were located in Europe, at Trier, Milan, and Thessaloniki.

The development of a more burdensome taxation system proved to be more complicated to

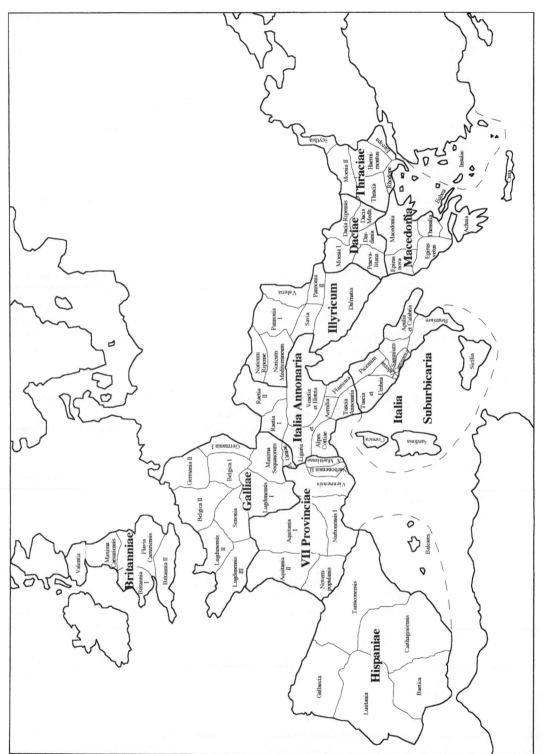

Fig. 2.13. The later Roman Empire.

operate in the countryside than in cities and required a close surveillance of taxpayers. This, in turn, led imperial subjects to associate themselves with specific geographic areas and with specific trades. In some respects these spatial and social changes anticipated conditions in medieval times. Despite some recovery in the fourth century (the so-called renaissance at the time of Constantine), the third century saw increasing disparities between the eastern and western sections of the empire, and certainly witnessed an eastward shift in the imperial centre of gravity. Centres of economic activity within Europe were displaced towards the periphery, along an axis extending from Trier through Milan, Aquileia, and Thessaloniki, to Byzantium. Mediterranean Europe, which had been the economic lung of the whole empire, experienced a loss of vitality from which it would never recover.

## References

AERTS, E., ANDREAU, J., and ORSTED, P., *Models of Regional Economies of Antiquity and the Middle Ages to the 11th Century*, (Louvain, 1990).

ALCOCK, S. E., *Graecia Capta: The Landscapes of Roman Greece*, (Cambridge, 1993).

AMPOLO, C., *La città antica*, (Bari, 1980).

ANDREAU, J., 'La cité antique et la vie économique', *La Cité antique? A partir de l'œuvre de M. Finley*, (= *Opus* 6–8 1987–89), 175–85.

ARCELIN, P., 'Activités maritimes des sociétés protohistoriques du midi de la Gaule', *L'Exploitation de la mer de l'antiquité à nos jours: la mer, moyen d'échange et de communication, VIe rencontres internationales d'archéologie et d'histoire d'Antibes*, (Valbonne, 1986), 11–30.

—— 'Le Territoire de Marseille grecque dans son contexte indigène', *Études Massliètes*, 1: *Le territoire de Marseille grecque* (Aix-en-Provence, 1986), 43–104.

AVIENUS *Ora Maritima (or Description of the Seacoast [from Brittany round to Massilia])*, J. P. Murphy (ed.), (Chicago, 1977).

AUDOUZE, F., and BUSCHENSCHÜTZ, O., *Villes, villages et campagnes de l'Europe celtique*, (Paris, 1989).

BARKER, G., and LLOYD, J. (eds.), *Roman Landscapes: Archaeological Survey in the Mediterranean Region*, Archaeological monographs of the British School at Rome, 2 (London, 1991).

BASS, G., et al., *History of Seafaring Based on Underwater Archaeology*, (London, 1972).

BATS, M., *Définition et évolution du profil maritime de Marseille grecque (VIe–Ier s. av. J.-C.)*, in *L'Exploitation de la mer de l'antiquité à nos jours: la mer, moyen d'échange et de communication, VIe rencontres internationales d'archéologie et d'histoire d'Antibes*, (Valbonne, 1986), 31–53.

BAYARD, D., and COLLART, J.-L. (eds.), *De la ferme indigène à la villa romaine– La Romanisation des campagnes de la Gaule. Actes du 2e colloque de l'association AGER tenu à Amiens (Somme) du 23 au 25 septembre 1993* (*Revue Archéologique de Picardie*, (Amiens, 1994), 11).

BEKKER-NIELSEN, T., '*Terra incognita:* The Subjective Geography of the Roman Empire', *Studies in Ancient History and Numismatics Presented to Rudi Thomsen*, (Aarhus, 1988), 148–61.

BENDER, H., and WOLFF, H. (eds.), *Ländliche Besiedlung und Landwirtschaft in der Rhein-Donau-Provinzen des römischen Reiches*, Passauer Universitäts-Schriften zur Archäologie, 2 (Espelkamp, 1994).

BELOCH, K. J., *Die Bevölkerung der griechich-römischen Welt*, (Leipzig, 1886).

BÉRARD, J., *La Colonisation grecque*, (Paris, 1957).

BERTUCCHI, G., *Les Amphores et le vin de Marseille, VIe s. avant J.-C.–IIe s. après J.-C.*, (Paris, 1992).

BOAK, A. E. R., *Manpower Shortage and the Fall of the Roman Empire in the West*, (Ann Arbor, Mich., 1955).

BOARDMAN, J., *The Greeks Overseas: Their Early Colonies and Trade*, (London, 1980).

BOWEN, E. G., *Britain and the Western Seaways*, (London, 1972).

BRESSON, A., 'Les Cités grecques et leurs emporia', in Bresson and Rouillard, (1993), 163–226.

—— and ROUILLARD, P. (eds.), *L'Emporion*, Publications du Centre Pièrre Paris, 26, (Paris, 1993).

CAMPOREALE, G., *I commerci di Vetulonia in età orientalizzante*, (Florence, 1969).

CAPOGROSSI, L., *La struttura della proprietà*, (Milano, 1969).

CARANDINI, A., *Schiavi in Italia. Gli strumenti pensanti dei Romani fra tarda repubblica e medio Impero*, (Rome, 1988).

CASSON, L., *Ships and Seamanship in the Ancient World*, (Princeton, 1971).

CHEVALLIER, R., *Les Voies romaines*, (Paris, 1972).

CHOUQUER, F., and FAVORY, F., *Les Paysages de l'antiquité*, (Paris, 1991).

CRAWFORD, M., *La moneta in Grecia e a Roma*, (Bari, 1982).

COUNILLON, P., 'L'Emporion des géographes grecs', in Bresson and Rouillard, (1993), 47–57.

CUNLIFFE, B., *La Gaule et ses voisins: le grand commerce dans l'antiquité*, (Paris, 1993).

CURCHIN, L., 'Vici and Pagi in Roman Spain', *REA* 89: (1985), 319–26.

DE SALVO, L., *Economia privata e pubblici servizi nell'Impero Romano–I Corpora navicularorium*, (Messina, 1992).

DILKE, O. A. W., *The Roman Land Surveyors: An Introduction to the Agrimensores*, (Newton Abbott, 1971).

—— *Greek and Roman Maps*, (London, 1984).

DUBY, G., and WALLON, A., *La Formation des campagnes françaises des origines au XIV$^e$ s.*, (Histoire de la France rurale, 1 (Paris, 1975).

DUMASY, F., 'Les Villes de la Gaule romaine au Haut-Empire', in J.-L. Huot, (ed.), *La Ville neuve: une idée de l'Antiquité?*, (Paris, 1988), 147–83.

DUNBADIN, J., *The Western Greeks*, (Oxford, 1948).

DUNCAN-JONES, R., *The Economy of the Roman Empire: Quantitative Studies*, (Cambridge, 1974).

—— *Structure and Scale in the Roman Economy*, (Cambridge, 1990).

DUVAL, A. (ed.), *Les Alpes à l'Âge du Fer*, (Paris, 1991).

DYER, J., *Ancient Britain*, (London, 1990).

ÉTIENNE, R., 'L'Emporion chez Strabon', in Bresson and Rouillard, (1993), 23–46.

FEBVRE, L., 'Limites et frontières', *Annales ESC*, (1947), 201–7.

FENTON, A., PODOLÀK, J., and RASMUSSEN, H. (eds.), *Land Transport in Europe*, (Copenhagen, 1973).

FERDIÈRE, A., *Les Campagnes de la Gaule romaine*, (2 vols.; Paris, 1988).

FÉVRIER, P.-A., 'Villes et campagnes des Gaules sous l'Empire', *Ktèma*, 6: (1981), 359–72.

FICHES, J.-L., and FAVORY, F. (eds.), *Les Campagnes de la France méditerranéenne dans l'Antiquité et le Haut Moyen Âge. Études microrégionales*, (Paris, 1993).

Finley, M., *The Ancient Economy*, (London, 1973).

—— *Ancient Sicily*, (London, 1978).

Frank, T. (ed.), *Economic Survey of Ancient Rome*, (Baltimore, 1933–40).

Frayn, J. M., *Markets and Fairs in Roman Italy*, (Oxford, 1993).

Garnsey, P., and Whittaker, C. R. (eds.), *Trade and Famine in Classical Antiquity*, (Cambridge, 1983).

Gerov, B., *Beiträge zur Geschichte der römischen Provinzen Moesien und Thrakien*, (Amsterdam, 1980).

Giardina, A. (ed.), *Roma: politica, economia, paesaggio urbano, ii. Società romana e impero tardoantico; 3. Le merci. Gli insediamenti*, (Bari, 1986).

Gras, M., *Trafics Tyrrhéniens archaïques*, (Rome, 1985).

—— Rouillard, P., and Teixidor, J., *L'Univers phénicien*, (Paris, 1989).

Harmand, L., *L'Occident romain*, (Paris, 1969).

Jacob, J.-P., Jacquet, B., and Mangin, M., *Les Agglomérations secondaires en Franche-Comté romaine, Ann. litt. de l'Univ. de Besançon*, (Paris, 1984).

Janot, J.-R., *A la recherche des Étrusques*, (Rennes, 1987).

Jones, A. H. M., *The Later Roman Empire, 284–602: A Social, Economic and Administrative Survey*, (4 vols.; Oxford, 1964).

Köster, A., *Das antike Seewesen*, (Berlin, 1923).

Lancel, S., *Carthage*, (Paris, 1992).

Latouche, R., 'Aspect démographique de la crise des grandes invasions', *Population*, (1947), 681–90.

Laubenheimer, F., *Le Temps des amphores en Gaule. Vins, huiles et sauces*, (Paris, 1990).

Leroux, P., *Romains d'Espagne. Cités et politique dans les provinces, IIe siècle av. J.-C.–IIIe siècle ap. J.-C.*, (Paris, 1995).

Leroy-Ladurie, E., *Histoire du climat depuis l'an Mil*, (Paris, 1967).

Leveau, P., 'La Ville antique et l'organisation de l'espace rural: villa, ville village', *Annales ESC*, (1983–4), 920–42.

—— 'La Ville romaine et son espace territorial', *XIV$^e$ Congrès International d'Archéologie Classique, Tarragone, 5–11 Sept. 1993*. (forthcoming).

—— Sillières, P., and Vallat, J.-P., *Campagnes de la Méditerranée romaine*, (Paris, 1993).

Mangin, M., and Petit, J.-P. (eds.), *Les Agglomérations secondaires. La Gaule Belgique, les germanies et l'Occident romain*, (Paris, 1994).

Margary, I. D., *Roman Roads in Britain*, (London, 1973).

Maurin, L., *Saintes antique*, (Bordeaux, 1978).

Miles, D., *The Romano-British Countryside*, British Archaeological Reports, 103, (Oxford, 1982).

*La Mobilité des paysages méditerranéens; Hommages à P. Birot* (*Revue archéologique des Pyrénées et du Sud-Ouest*, Travaux, 2, (Toulouse, 1984).

Mocsy, A., *Gesellschaft und Romanisation in der römischen Provinz Moesia Superior*, (Budapest, 1970).

Morrison, J. S., and Williams, R. T., *Greek Oared Ships, 900–322 BC*, (Cambridge, 1968).

Moschetti, C. M., 'Aspetti organizzativi dell'attività commerciale marittima nel bacino del Mediterraneo durante l'impero romano', *Studia et documenta historiae et iuris*, 35: (1969), 374–410.

Nicolet, C., *L'Inventaire du monde*, (Paris, 1988).

Pallotino, M., *Genti e culture dell'Italia preromana*, (Rome, 1981).

PERCIVAL, J., *The Roman Villa*, (London, 1976).

POTTER, T. W., *The Changing Landscape of South Etruria*, (London, 1979).

*Les Princes celtes et la Méditerranée* (Paris, 1988).

RIVET, A. F. L., *Gallia Narbonensis, with a Chapter on Alpes maritimae*, (London, 1988).

RODWELL, W., and ROWLEY, T. (eds.), *The Small Towns of Roman Britain*, British Archaeological Reports, 15 (Oxford, 1975).

ROMAIN, D., and ROMAN, Y., *Histoire de la Gaule. VIe siècle av. J.-C.–Ier siècle ap. J.-C.*, (Paris, 1997).

ROUGÉ, J., *Recherches sur l'organisation du commerce maritime en Méditerranée sous l'empire romain*, (Paris, 1966).

ROUILLARD, P., *Les Grecs et la péninsule ibérique*, (Bordeaux and Paris, 1991).

SALMON, P., *Population et dépopulation dans l'Empire romain*, (Brussels, 1974).

SALVIAT, F., 'Cadastre et aménagement: Quinte-Curce, les *insulae furianae*, la *fossa Augusta* et la localisation du cadastre C d'Orange', *Revue Archéologique de Narbonnaise*, 19: (1986), 101–16.

SALVIOLI, G., *Il capitalismo antico*, (Bari, 1985).

SARTRE, M., and TRANOY, A., *La Méditerranée antique*, (Paris, 1990).

SZABO, M., *Sur les traces des Celtes en Hongrie*, (Budapest, 1971).

TCHERNIA, A., *Le Vin de l'Italie romaine*, BEFAR, 261, (Rome, 1986).

THOMAS, C. (ed.), *Rural Settlement in Roman Britain*, (CBA; Oxford, 1966).

THOMSEN, R., *The Italic Regions*, (Copenhagen, 1947).

TRANOY, A., *La Galice romaine*, Publications du centre P. Paris, 7, (Paris, 1981).

VALLET, G., *Rhégion et Zanklè, commerce et civilisation des cités chalcidiennes du détroit de Messine*, (Paris, 1958).

*Le Vicus gallo-romain, coloque, Paris, 1975* Caesarodunum, 11 (Tours, 1976).

*Les Villes de Gaule Belgique au Haut-Empire* (= *Revue Archéologique de Picardie*, NS 3 (4) (1984).

WALBANK, F. W., *The Awful Revolution: The Decline of the Roman Empire in the West*, (Liverpool, 1969).

WALSER, G., *Studien zur Alpengeschichte in antiker Zeit*, (Stuttgart, 1994).

WHITTAKER, C. R., 'The Western Phoenicians', *Proceedings of the Cambridge Philological Society*, NS 20: (1974), 58–79.

WHITE, K. D., *Roman Farming*, (London, 1970).

—— *Greek and Roman Technology*, (London, 1984).

WILL, E., 'La Grèce archaïque', *Deuxième colloque international d'histoire économique, Aix-en-Provence*, 1962, (Paris and The Hague, 1965), 41–96; 97–115.

WOODS, D. E., 'Carteia and Tartessos', *Tartessos y sus problemas, V Symposium Internacional de Prehistoria Peninsular*, (Barcelona, 1969), 251–6.

ZIMMERN, A., *The Greek Commonwealth*, (Oxford, 1967).

# Chapter **3**

# The Socio-Political Map of Europe, 400–1500

R. A. Dodgshon

The political map of Europe was repeatedly redrawn over the period AD 400–1500. The divide between Roman and Barbarian Europe began to dissolve into a more complex political geography from the fourth century onwards as large areas, including most of the western empire, were overwhelmed by the migration of Germanic tribes like the Goths, Sueves, Vandals, and Saxons. Out of these dislocations emerged the outlines of a new ordering of political space, one whose formative concepts and institutional forms are fundamental to any definition of the medieval world. In trying to understand this new ordering, and the processes which shaped its development over the medieval period, we need to look closely at four aspects of the problem. First, we need to grasp the nature and extent of the migrations which redrew the tribal and ethnic map of early medieval Europe; secondly, we need to consider how political space was redefined over the medieval period, notably through the institution of the state; thirdly, we need to review the actual pattern of European states as it unfolded over the medieval period; and, fourthly, we need to consider whether political space was structured in ways other than through the state system.

## The Migrations of Early Medieval Europe

Even before the disintegration of the empire, the Romans had used Germanic soldiers as *foederati*, or hired soldiers, and some German tribes had crossed its frontiers. In some instances, these movements were substantial. An incursion into Dacia by Goths over the third century AD, for example, effectively redrew the frontier along the line of the Danube. By AD 400, the scale of these movements changed dramatically. Driven by the threat of Huns moving westwards out of Asia, tribes living along the northern and eastern edges of the empire becan to migrate in numbers. In some cases, their movements were long-distance and rapid, such as when around AD 400 a group of tribes made up of Vandals, Sueves, and Alans moved swiftly westwards along the northern edge of the empire before crossing into Gaul and then, by AD 409, down into Spain. The Sueves established themselves in Galicia, maintaining political control until AD 585. The Vandals and Alans, meanwhile, took control over the rest of the peninsula, but their control lasted barely twenty years before they crossed to a base in

North Africa (Collins, 1983: 16–19). Contemporary with these movements, another Germanic tribe, the Burgundians, also moved from the north German plain down into Gaul, establishing themselves in the Rhône valley by AD 406.

The movement of the Visigoths was comparable in its scale and range to that of the Vandals, Sueves, and Alans. Reported in AD 376 as occupying lands in the lower Danube valley beyond the Roman frontier, they were allowed to settle south of the frontier in AD 382 when under threat from the Huns. Soon after, they appear to have reacted against Roman authority and to have taken up a campaign of warring and looting. By the close of the century, they had moved through the eastern empire to the Adriatic coast, reaching Italy by AD 402. Within a few years, they had campaigned down through the Italian peninsula, sacking Rome itself in AD 410. By AD 418, though, they had moved westwards into Gaul, establishing a base in Acquitaine. When the Vandals and Alans left Spain for North Africa in AD 429, the Visigoths seized the opportunity to cross the Pyrenees, establishing a hegemony in Spain that lasted until the early eighth century. The Italian peninsula's respite from the Goths was short-lived. By the late fifth century AD Ostrogoths from the Danube valley had also invaded Italy, seizing control of the north and bringing the history of the western empire finally to an end. Control changed yet again when, in AD 568, the Lombards, a Germanic tribe previously settled in Pannonia (Hungary), established a hegemony over both northern and southern Italy, their complete grip on the peninsula being broken only by a narrow central belt left under imperial control.

On the north-western edge of the empire, the British Isles were affected by similar migrations. The Roman withdrawal from Britain was followed by a movement of Saxons, Angles, and Jutes across the North Sea from the mid-fifth century AD onwards, though some were probably present before the Roman withdrawal. Further west, the weakening of Roman authority was followed by groups of Irish moving into western Scotland, into north-west and south-west Wales, and into south-west England, whilst Britons from the latter moved across into Brittany. As in other parts of Europe, these movements had lasting significance for the political and cultural geography of Britain.

The scale and complexity of these cultural migrations lessened after the seventh century, but they did not cease. Slavic tribes living in the area of modern-day Poland began to migrate, some moving eastwards across the north German plain, others moving north-eastwards towards the Volga, and another stream working their way southwards. Amongst the latter were Serbs and Croats who moved deep into the Balkans, eventually reaching the Adriatic Sea by the seventh century. In south-west Europe, meanwhile, the death of the Visigothic king in AD 711 led to a raid into Spain by Arabs from North Africa. What was meant to be exploratory turned into a rapid and large-scale conquest. By AD 715, much of Spain had been turned into a new Muslim-controlled province, al-Andalus, ruled by the Umayyad amirs but ultimately under the caliphs of Damascus, a conquest that involved substantial movements of Berbers as well as Arabs into Spain (see Fig. 3.1). At the other end of Europe, the ninth century saw the start of Viking raids and conquests, initiating one of the most expansive of all the cultural movements over the early medieval period. From bases in Scandinavia, the Norse expanded across the North Sea, settling in Iceland, Greenland, the Faeroes, the northern and western isles of Scotland, Ireland, the Isle of Man, north-west England, and parts of Normandy, as well as establishing a presence in both Portugal and the western Mediterranean, whilst the Danes settled along coastal areas of the European mainland and eastern England. The eastwards movement of mainly Swedish Vikings was no less dramatic. Over the ninth and tenth centuries, Swedes, or Varangians as they became known, crossed the Baltic and penetrated north-flowing rivers like the Dvina and from bases established on the watershed, like Polotsk, Smolensk, and Kiev, began moving down south-flowing rivers like the Dnieper and Volga.

Finally, in western Europe, the period between the tenth and thirteenth centuries saw a new wave of migrations erupting out of the former territories of the Frankish empire. In the west, William the Conqueror's victory at Hastings (1066) paved the way for a flow of Normans, Bretons, and Flemish knights and their followers across the Channel into England. By the end of the century, they had also conquered the more fertile parts of eastern and southern Wales, and by the late twelfth century had secured a strong but far from stable position in central and eastern Ireland and, by more peaceful means, had been invited by David I into southern and eastern Scotland. Further afield, they established themselves by 1100 in southern Italy and Sicily, turning the latter into a great centre of cultural activity. Soon after Norman lords and knights had started laying the foundations for 'an aristocratic diaspora' (Bartlett, 1993: 24) by expanding westwards and southwards, we find German lords and knights from beyond the Rhine also starting a long, more drawn-out process of expansion eastwards around the Baltic and into Silesia, and south-eastwards down the Danube into Bohemia and Hungary.

## The Emergence of the European State System

By the eighth century, the political geography of Europe began to change in other ways. In addition to the movement and mixing of ethnic groups, changes also began to take place in the character of political systems. By the tenth century these changes had started to mature into the first European states. A number of factors helped to shape the character of the state as a political form.

With its pockets of rich riverine lowlands separated by more impoverished uplands, Europe's physical character has been seen as favouring a network of locally centred polities rather than a monolithic and long-standing empire like the Roman Empire or those of the Middle and Far East (Jones, 1981). Significantly, when the first European states emerged, most were initially focused on a fertile core of land, their maturation as states involving an outward extension of territorial control and political alliance into less fertile surrounding areas (Pounds and Ball, 1964: 24–40). Thus, the Île de France provided the nucleus for the growth of what became France, whilst the rich alluvium of the Hungarian plain provided a productive core for Hungary (see Fig. 3.1). Like all such generalizations, though, there are exceptions. Switzerland is an obvious one, yet as a federal state it did have a nucleus. Its integration started in 1291 with the three forest cantons around Lake Lucerne: Walden, Uri, and Schwyz. Most other cantons had joined by the mid-thirteenth century, with the most easterly,

Vaud and Fribourg, joining in 1513. Portugal and Spain, too, were exceptions, their growth erupting from Christian enclaves in the poorer north-west corner of the Iberian peninsula. Germany provided a different kind of exception. Each of its five original stem duchies had pockets of fertile soil, a factor which may have worked against their easy integration. In fact, it was Brandenburg, a late-settled and less fertile duchy which eventually provided the focus for a more integrated German state (Pounds and Ball, 1964: 38–9).

The multi-centred nature of European political space also affected state development by fostering intense political rivalry. In practice, this rivalry was articulated through two quite different forms of competition: coercion-intensive military systems and capital-intensive trading systems (Tilly, 1992). The former generated expansive land-based states, the latter, city-based states which, whilst restricted in size, worked to expand continually through their networks of trade (Rokkan, 1975; Tilly, 1992). Yet whilst the multi-centred focus of political space set states one against the other, providing a context within which local forms continually struggled for advantage through institutional innovation and development, we need to keep in mind the added fact that the spread of Christianity—like trade—provided a unifying capstone emplaced across the system at large. Indeed, for some, this juxtaposition of the regional and extra-regional provided a vital means by which local institutional innovation and advantage could be more widely diffused. As one author put it, the Christian church was 'the ghost of the Holy Roman Empire', providing the ideological power that 'cemented the space previously integrated by the legions' (Hall, 1986: 121).

A basic challenge facing early states was how to project their authority at a distance. This was hardly a new challenge. The Roman Empire faced spectacular problems of integration. With its collapse, political forms became extensively and profoundly disaggregated, with many parts of the empire becoming reconfigured around tribal kingships. The challenge of integrating space was renewed. The political forms that erupted out of the fertile core areas of Europe were the first to respond to it. In France, for instance, political geography became restructured around tribal groupings like the Franks and Burgundians. At first,

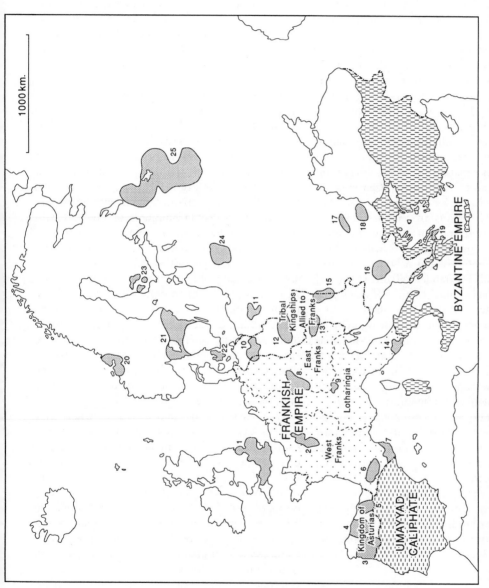

Fig. 3.1. European political geography: empires and states c.800–1250. European political geography, c.800, was dominated by three large-scale political structures or empires: the Byzantine Empire, Umayyad Caliphate, and the Frankish, later the Carolingian, Empires. Even by AD 847, when Charlemagne's empire was divided between his three sons, these large-scale structures started to be slowly replaced by smaller, state-based polities which emerged and expanded out of fertile core areas over the tenth–thirteenth centuries: those areas numbered on the map provided the cores for 1. England; 2. France; 3. Portugal; 4. and 5. Spain (León); 6. and 7. Spain (Aragon); 8. Germany (Saxony); 9. Switzerland; 10. Germany (Brandenburg); 11. Poland; 12. Bohemia; 13. Austria; 14. Italy; 15. Hungary; 16. Serbia; 17. Wallachia; 18. Bulgaria; 19. Greece; 20. Norway; 21. Sweden; 22. Denmark; 23. Finland; 24. Lithuania; 25. Russia. (*Source*: State nuclei based on Pounds and Ball, 1964: 29.)

these tribal groups may have been only loosely confederated, but they soon formed into extended alliances stabilized around the figure of a dominant tribal king, with the more successful establishing control across large territories or provinces, such as Francia, Burgundy, and Gothia. These broad regional provinces provided the building blocks for the French state, but their stable integration as a unified state was a prolonged affair. Though linked for a time as part of the Carolingian Empire and though Hugh Capet asserted an extensive kingship in AD 887, late ninth-century France could still be described as 'a ragbag of old sub-kingdoms and peoples', with marked regional differences of language, tradition, and law still persisting (Dunbabin, 1985: 4). Yet already institutional changes were in progress that would make this wider integration possible. To the north, the early experience of Germany was similar. The fifty or so German tribes identified by Tacitus in the first century AD were thinned and transformed by the great migrations. By the third century AD, tribes in the more fertile areas, both old and new, had grouped themselves into stable, extended alliances, such as the Alemmani, Franks, and Saxons. These major alliances formed the basis for the stem duchies that were to dominate political development in Germany over subsequent centuries. What has been described as 'their rootedness in the landscape' (Fleckenstein, 1978: 12–13) made them the cornerstone for the Holy Roman Empire and any process of future state development in Germany.

It was in these burgeoning core areas of Europe that larger, more integrated political systems based on more intense and routinized forms of government began to emerge by the tenth century. Yet, given the flux of earlier systems, the question we should ask is not why larger, more integrated systems now emerged, but why they endured. The answer lies in understanding how new institutional forms and relationships were used to redefine the nature of political space.

## The Socio-Cultural Context of State Systems

A critical dimension to these institutional changes lies in appreciating how the large-scale migrations outlined earlier affected political space. Few migrations involved the pioneering of wholly new landscapes. In this respect settlements like the Norse land-take in Iceland (Byock, 1988: 55–71), were exceptional. More usually, the tribal armies and hordes that swept across Europe over the early medieval period moved into areas that were already settled. Furthermore, the idea that they invariably cleared or depopulated such landscapes before settling them is too simplistic. In many cases they found settled landscapes and adjusted to them. However, what we cannot rule out is the possibility that the immediate post-Roman world saw a drastic disease-induced 'thinning of the rural world' on a greater scale than the Black Death of the fourteenth century, so that new or fresh opportunities for colonization were widely available and may have been a factor in the early migrations as well as in some of the institutional changes that occurred (Cheyette, 1977: 197). Though the extent to which some tribal armies drew large-scale folk migrations behind them has been questioned, with movements like the Anglo-Saxon conquest of England being reinterpreted by some as involving only small armies led by powerful warrior aristocracies (Sawyer, 1978), most early migrations involved sizeable movements of farmers as well as fighting men. In the case of the migrations which took place in the tenth–thirteenth centuries, when written evidence became available, the extent to which farmers followed fighting men is well documented, the gross effect of all the migrations being to transfer population from the more settled core areas of western Europe out towards the less settled peripheries, in both the west and east (Bartlett, 1993: 111–16).

Whether involving hired soldiers or whole tribes like the Visigoths, such movements were already a feature of Europe's changing cultural geography even by the closing years of the Roman Empire. Events after the empire's collapse simply continued them. When the Visigoths finally reached Spain, estimates put their number at about 30,000. This might be compared with the estimated 30,000–50,000 Arabs and Berbers who led the Muslim conquest and settlement of Spain. At first, this settlement was confined to southern areas around Cordoba, Granada, and Seville, and consisted largely of military settlements, but in time many acquired estates in river valleys like those of the Ebro and Guadalquivir, or in the fertile lands around Toledo. Far from settling or creating an empty landscape, though, elements of the local Spanish, Roman, and Gothic population stayed on to become Mawallads or Muslim converts, whilst many Arabs and Berbers intermarried with them. Similar hybridization followed the Ostrogothic and Lombardic invasion of Italy, with many settling as hospitallers on existing estates (James, 1989: 50–2). By comparison, the effect of the migrations on France was to create a broad regional patchwork of different ethnic groups and languages (Bonnaud, 1977: 42–62).

The impact of these migrations was that even by the eighth century many parts of Europe had acquired a complex cultural character. In the process, the character of political space changed. Few areas could now be described as neatly compartmentalized around tribal kingships and chiefdoms that were bonded together by kinship. Those that still were, like Ireland (Byrne, 1973) or the North German plain east of the Elbe (Jaeger, 1982), had been little affected by either the Roman Empire or the migrations of the early medieval period. Elsewhere, the organization of political space around distinct tribal groups, or *pagi*, had been undermined by the interactions and migrations of the Roman and post-Roman period (Reynolds, 1983: 375–90; James, 1989: 47–9). Many of the bigger tribal groups were now largely synthetic, based as much on assumed as on real kinship ties and bonded through increasingly more complex political alliances.

In character, many parts of Europe were now ethnically diverse. Over and above its pre-Roman population, for instance, Spain had absorbed Romans, Vandals, Suevi, Alans, Visigoths, Franks, Arabs, and Berbers in varying numbers by the eighth century, and even though ethnic groups like the Vandals and Alans had largely moved on to North Africa, what remained was a culturally mixed population. At first, where different tribal or ethnic groups came to coexist within the same political framework, each continued to possess its own law and custom and, therefore, to maintain its own ethnic identity and process. 'Everywhere', wrote Reynolds, 'immigration kept muddling the general rule' (Reynolds, 1984: 19; Reynolds, 1983: 382). In time, though, a critical shift took place. As the more powerful kings extended their rule over new tribal groups and larger areas, they secured their control by imposing a uniform law. Law became territorialized, exercised by kings over all people within a particular territory rather than over specific ethnic groups or kinsmen. Arguably,

this generalization of society was a critical stage in the emergence of state systems. Yet the ideological power of claiming a common origin for society was not lost on would-be state rulers. In spite of their increasingly diverse origin and their increased numbers, and in spite of the changing framework of law and custom, many proto-states simply acquired new enlarged myths about the common origin of their peoples (Reynolds, 1983: 382; James, 1989: 41).

## Feudalism and the Control of Space

The emergence of the states also involved radical changes in the way space was defined and controlled. In part, this had to do with how rulers faced the problem of controlling territories beyond their immediate grasp. The tribal kingships pieced together after the collapse of the Roman Empire integrated space through networks of kinship extended out from ruling families and, on a larger scale, through alliances between ruling families. When seen in the fifth century AD, areas like northern Gaul and Spain were controlled by such alliances, with an over-king being elected from amongst the leaders or nobles of the various constituent tribes. This system of elective leadership served to hold alliances together, giving the various constituent tribes a vested interest in maintaining the whole.

Under such a system, political landscapes would have been in constant flux, due in large measure to the role played by warring in the organization of society. Campaigning and plundering in a search for booty was a regular, annual affair, encouraged by the contrasts in wealth that existed between the barbarian world and what lay beyond the Roman frontier (Reuter, 1991: 23 and 94). For tribal kingships, booty was needed to reward the retinue or immediate followers that formed the core of a tribe's fighting men. By the fifth and sixth centuries AD the character of this retinue had started to change amongst Germanic tribes in a fundamental way. Initially comprized of a chief's network of close kin, they now acquired a more mixed character, with individuals being rewarded with a share of plunder in return for their military service (Thompson, 1965: 55–60). Encapsulated in this shift was a new concept of social relations, one that was to spread rapidly by the eighth century as extended political systems like the Carolingian Empire emerged.

The challenge facing rulers of extended polities like the Carolingian Empire (see Fig. 3.1) was how to turn their assertion of lordship over all men and all land into a system of government. They responded by evolving new political relationships based on the concept of military feudalism. Building on the ties of vassal homage and commendation binding members of their retinue to them, rulers allocated out territories and estates, or benefices, to their leading followers in return for military service, the latter serving to project the king's authority at a distance. In effect, feudal grants projected the organization of the royal household on to the wider sphere of the state, with benefices providing a new system of maintenance for members of the royal retinue. By way of a cautionary note, though, some would argue that feudal land grants could and did have an origin that was independent of the royal household and its needs, whilst others would argue that the precise legal character of the benefice or fief as a dependent tenure and a reward for service, and therefore the whole idea of a feudal society *sensu stricto*, did not develop until the eleventh century (Reynolds, 1994: 48–74). Rights of jurisdiction were also granted out by kings, so that feudalism became the basis for a system of territorial administration based on devolved forms of lordship. In time, the great territorial princes and lords invested lesser lords and knights with land in return for service, thereby creating an elaborate hierarchy of feudal control and administration.

A weakness of military feudalism was that, though designed to project the king's power at a distance, it created provincial centres of legitimate power and jurisdiction that could easily fracture a state. To survive, feudal states had to maintain a balance between their centrifugal and centripetal forces. Many struggled to do so. Over the ninth and tenth centuries, a weakening of central authority in France led to a 'dissociation territoriale' that shifted effective power from the great territorial counts down to dense patterns of local bannan lordships and castellanies, each exercising ban or power to command and punish over the surrounding district (Teunis, 1978: 242). A similar fragmentation of power occurred in twelfth- and thirteenth-century Germany, but there it had more to do with an upsurge of provincial energies

amongst secular dynasties, powerful bishops, and newly-forged urban authorities than any loss of control by the centre, for the Holy Roman Empire was never a strongly integrated state (Barraclough, 1938: 84–6; Arnold, 1991a: 5).

## The Manorial Geography of Europe

As well as projecting the king's authority into the provinces, the granting out of land and jurisdiction provided a means by which the maintenance problems of the king's household and those of his nobles were spread more widely. Maintaining a knight was a costly business. This was the reason why military feudalism, and the holding of land in return for a knight's service, or a knight's fee, was largely a feature of the richer, fertile areas of Europe. It was rarely found in any developed form in livestock areas such as the coastal lowlands beside the North Sea or Alpine areas (Slicher van Bath, 1963: 37; Bartlett, 1993: 52–3). For feudal rulers, the costs of subduing the upland or marginal regions was rarely matched by the local surplus that could be extracted.

Once emplaced on their own estates, lords and knights maintained themselves. They did so through the institution of the manor. The manor represented a local unit of lordship, one which brought the socio-political and economic dimensions of feudalism together. In its classical form, it consisted of a small and relatively compact estate, though in some areas, such as in parts of Italy, Germany, and the eastern Midlands of England, manorial estates tended to be fragmented (Duby, 1968: 53). Yet, whether compact or dispersed, manors comprised two parts: a demesne kept in the lord's hands and used to sustain his household and the land leased out to tenants in return for labour services (see also Chapter 4). The growing use of labour services was underpinned by an erosion of tenant freedom, many tenants becoming servile or unfree. They now had the lord's protection but in return they were tied to him and the manor and burdened by a range of services and dues.

Manorial systems had developed in the more fertile parts of the Carolingian Empire by the late eighth century, especially on Carolingian and ecclesiastical estates, probably in association with a wave of colonization and village formation (Cheyette,

1977: 182–206). In Italy, it developed widely across Lombardy. Outside Lombardy, its development was uneven and largely confined to the north, though slave labour was extensively used in the far south if not within a manorial framework. Across the Alps, classical manors were to be found throughout the central, eastern, and northern parts of France, being especially well developed between the Loire and Rhine (Slicher van Bath, 1963: 46; Duby, 1968: 53). Their development only lightly affected the south-west, but across the Pyrenees manorial structures were present in Rousillon and Catalonia (Smith, 1966: 433). They had also emerged in parts of southern and central Germany by the eighth century but their character was varied and uneven, especially in Saxony. By the ninth century they had spread north-westwards into parts of Flanders, Brabant, and the eastern Netherlands, and soon after into Denmark and England. By the twelfth century, the Anglo-Normans had pushed the manorial system into Atlantic Europe, establishing manors in the fertile parts of Wales and eastern Ireland. Some form of dependent tenure may have existed in these areas before the arrival of Normans, particularly on ecclesiastical estates, but it now became organized around the manor and its court. In both areas, though, the creation of the manors was probably based on existing units of landholding and settlement (Simm, 1988).

The subsequent history of the manorial system shows broad regional divergences. Even by the ninth century, some manors, particularly in core areas of the Carolingian Empire like northern Italy, Lorraine, Burgundy, and Bavaria, were already being broken up and parts sold or leased out for cash rents instead of labour services, as siegneurs withdrew from the direct cultivation of their estates, a switch that amounted to 'la révolution censive' (Herlihy, 1959: 63). The unsettled political conditons of the late ninth and tenth centuries contributed to this physical and economic disintegration of the manor but the growth of local trade and the high rents which could now be exacted for land within reach of growing markets also played a part, especially in areas like Lombardy (Duby, 1968: 46 and 51; Herlihy, 1959: 68–9). The concentration of property into large estates and the use of dependent cultivators to work the land recovered over the eleventh century, but labour services did not regain their former importance in the core areas of the old

Carolingian Empire. Indeed, by the twelfth century, most areas were experiencing a final shift out of labour services into cash rents or at least rents in kind (Jones, 1954: 30; Pounds, 1974: 209–10). However, we must be careful not to see this decline of labour services as marking the end of feudalism. A wide range of other feudal dues and exactions remained in place, ensuring that there were still other ways in which peasants in these areas were burdened.

Elsewhere, other far-reaching changes were starting to transform the socio-political character of the countryside by the late eleventh century. Population growth encouraged the colonization of new areas (Darby, 1956: 194–5). As these new areas were opened up, new and freer forms of tenure, less burdened by the demands of either manorialism or lordship, were established. Along with the more dispersed forms of settlement that existed in many of these areas (see Chapter 4), their effect was to foster a different, freer form of society. Overall, there were two frontiers of colonization for this freer form of society. First, there was an inner frontier, one that surrounded the extensive waste areas still to be found within and around older feudalized areas. Such waste areas included lowland heath and forest areas like those of northern Flanders, parts of Brabant, and the livestock woodland districts of Midland England, together with upland massifs like the Harz mountains, Eifel, Westerwald, Black Forest, and Thuringian Forest in Germany and parts of the Massif Central such as Mâconnais and the Beaujolais mountains (Darby, 1956: 194; Ganshof and Vershulst, 1966: 296; Koebner, 1966: 72–3; Duby, 1968: 83–4; Pounds, 1974: 168).

The second frontier was the extensive wave of colonization that developed along the eastern flanks of the Holy Roman Empire. Between the twelfth and fourtenth centuries, colonists poured across this frontier, creating vast tracts of newly settled countryside across Brandenburg, Mecklenburg, Pomerania, and Prussia. Further south, sizeable flows of colonists moved into Silesia and parts of Bohemia. As with the inner frontiers of feudal Empire, this 'peopling of the border', as Thompson called it (1928, 496) was facilitated by the creation of freer, less feudally-burdened tenures and communities. It was organized by lords from the feudalized western parts of Germany, with offers of greater freedom being used to attract colonists. In areas like the lower Elbe and Weser, those who responded included Dutch and Flemish colonists as well as Germans (Thompson, 1928: 575).

The emergence of freer conditions east of the Elbe served to create a broad contrast with the older still feudalized parts of Germany. By the fourteenth century, though, the contrast had begun to change. Estate owners in newly colonized areas to the east of the Elbe, what Bartlett has called the 'new colonial aristocracies' who sprang up in these new areas of settlement (Bartlett, 1993: 52), began to acquire more rights of lordship from the various ruling princes, and the power to impose dues and services (Aubin, 1966: 470). By c. 1500, many peasants in late colonized areas, but especially along the rivers that flowed northwards into the Baltic, found their conditions worsening, as large estates developed and the estate economy began to shift into grain production using labour provided by an increasingly burdened and servile peasantry (Topolski, 1981: 377–8). In some parts of eastern Germany, Poland, and Lithuania, these changes actually took place within a manorial framework, with tenants working part of the lord's estate in return for land, but elsewhere we find a plantation-like system in which extensive estates were worked by slave or servile labour force. By this time, peasants west of the Elbe enjoyed much freer conditions, so that the contrast apparent back in the eleventh and twelfth centuries had effectively been inverted.

## State Systems and the Territorialization of Space

Together, military feudalism and manorialism constituted a revolution in the definition of political space. Of the essence to this revolution was the way feudalism created a contractual basis to sociopolitical relations. Territories and estates were granted out to lords and knights in return for military service, whilst at the level of the peasant holdings were granted out in return for labour service. This contractual basis introduced the need for definition, with the extent of territories, estates, and holdings being defined on the ground and linked to the amount of service owed. As jurisdictions also became defined and attached to

grants of lordship, the whole pattern of feudal right and obligation became territorialized (Dodgshon, 1987: 145–63, 166–92; Reynolds, 1994: 19). In the process, the nature of political space shifted from being defined through society, or the tribes and kin-groups which occupied it, to being defined, first and foremost, through territory and the network of jurisidictions and rights into which it was partitioned (Sahlins, 1968: 5).

The first signs of law being territorialized are evident by the sixth and seventh centuries (Reynolds, 1983: 282). For instance, beneath the broad regional framework of duchies and provinces, the territory of the Frankish Empire became ordered around the county or comitatus, a unit over which counts or comites acted as both judges and representatives of the king. They were laid out over the earlier form of the civitates in the western or Roman areas of the empire, and over the *gau*, an older Germanic unit found in the eastern empire (Fleckenstein, 1978: 68–73; Reuter, 1991: 27). A still closer partitioning of local political space began to appear by the tenth century, when areas like East Saxony, Thuringia, and Wessex acquired units of local jurisdiction as building blocks for a now routinized system of law (Leyser, 1982: 99; Smith 1984: 166–7). By the eleventh century, such close territorialization was widespread throughout western Europe, supported more and more by an emergent bureaucracy or expert government and by the use of charters. Seen from a geographical point of view, it can be taken as a hallmark of the early state, with a top–down structuring of rights and jurisdictions from the great territorial lordships and fiefs down to the individual peasant holding, all emplaced and bounded in the landscape.

## The Nature of the European State System

The institutional and conceptual changes described so far were associated with the start of far-reaching changes in the political map of Europe as the first states emerged. There was nothing predictable about the way in which this map was drawn. The many problems which would-be state-rulers faced when extending their power from a regional base out over surrounding tribes ensured this. Indeed, on reviewing the history of key states, one is struck by the alternative trajectories that might have been as much as by what actually happened. What survived in the long term was the outcome of a long and competitive selection process.

This is well shown by the history of the Frankish Empire and its political offspring, the French and German states. From an initial base in Austrasia—roughly Belgium, northern France, inland Holland, and the Rhine, Moselle, and Main valleys—Frankish control was extended under Clovis to embrace most of what had been northern Gaul and south-western France, or Neustria and Aquitaine respectively. The Merovingians added Burgundy, the mouth of the Rhône, and part of the Alps. Finally, under Charlemagne, were added the Spanish March, central and northern Italy, Bavaria, Austria, and Saxony, together with a broad march belt to the east stretching from the North Sea down to the Adriatic, over which Charlemagne exercised lordship (see Fig. 3.1). The problem for the Carolingian Empire was that its rulers practised partible inheritance. Following Charlemagne's death, it was divided between his three sons by the Treaty of Verdun in AD 843, a partition which created the kingdom of the West Franks, occupying most of France; Lotharingia, which occupied a broad corridor of land from Holland down through Flanders, across Burgundy, Provence, and Switzerland into Italy; whilst the third portion, occupied by the East Franks, covered the eastern territories of the former empire as far as Saxony and Bavaria. Within a few decades, though, the Treaty of Mersen (AD 888) had redivided the middle kingdom, or Lotharingia, between the West and East Franks, that is, between what largely became France and the Holy Roman Empire (see Fig. 3.1).

Amongst the West Franks, it was the Capetian dynasty based in the Île de France that claimed the kingship. At first, it was an empty claim. Over the tenth and eleventh centuries, effective power shifted away from the Capetians and down to the bannum lordships and castellanies that came to dominate many parts of France. Berry, the pagus de Bourges—to the south of the Île de France—was typical, with over twenty different castellanies established by the twelfth century (Devailly, 1973: 172). To the east, the comparatively large county of Champagne had been divided into twenty-six castellanies by the mid-twelfth century (Evergates,

1975: 61–2). The assertion of a closer, more direct authority by the Capetian kings was a slow, difficult process. By the end of the eleventh century, Normandy was under the direct control of the English kings, whilst Brittany owed more allegiance to them than the Capetian kings. By the mid-twelfth century, the prospect of a united France had receded further, with the whole of Aquitaine and Toulouse owing allegiance to the English kings. Though the French kings established a direct control or fiefdom over most of France except for Gascony by the 1320s, the early fifteenth century saw further reversals, with the English kings resuming temporary control over Normandy and Gascony and the House of Burgundy not only asserting the independence of Burgundy but also gaining control over Franche-Comté and Flanders from the French kings. This was a successful if brief phase for the Dukes of Burgundy for they also acquired control over various Netherland provinces—including Hainault, Brabant, Holland, and Utrecht—from the German kings, politically separating out both parts of the Low Countries or Netherlands from their more immediate neighbours. However, when Duke Charles the Bold was killed in 1477, his control of Flanders and the Dutch provinces passed to the Habsburgs, whilst Burgundy passed back to the French kings. Despite the permanent loss of Flanders, the possibilities of an integrated French state now recovered quickly. By the time the medieval period drew to a close, the French kings had reasserted control over most parts of France, either in the form of direct kingship, fiefs, or appanages.

The early evolution of the East Frankish state was even more beset by centrifugal forces. Following the Treaty of Verdun (AD 843), it comprised the four original stem duchies: Saxony, Franconia, Swabia, and Bavaria, plus a broad belt of marcher areas including Brandenburg, Bohemia, Austria, and Carinthia. The four stem duchies provided the core of the emergent East Frankish state from which the emperors were elected. Later, it expanded to embrace Lower Lorraine, Upper Lorraine, Burgundy and Arles, Brandenburg, Austria, and Carinthia together with central and northern Italy, including the Papal States, the latter enabling it to assume the mantle of the Holy Roman Empire. By the tenth century Lower Lorraine and Bohemia had also become established as stem duchies. Despite their central role within the empire, the duchies formed fairly independent regions and peoples, a fact that did not ease the path to German unity. Nor did the fact that, as the Holy Roman Empire, it drew what cohesion it had as much from the rule of the Pope as from the political assumptions of the Emperor. The Empire's physical character as a sprawling assemblage of peoples laying athwart the continent from the North Sea to the Italian peninsula and, broken by the Alps, also inhibited political cohesion.

With such foundations, it was not surprising that the emergence of a German Reich was not straightforward. The strengthening of provincial power through the creation of new duchies and counties and the further subdivision of lordship into a dense local pattern of castellanies or burgbezirk created 'a complex intricate landscape' (Arnold, 1991*b*: 3) made up of 'diverse legal custom, local economies and small-scale aristocratic, ecclesiastical, and urban jurisdictions' (1991*b*: 59–60). As these local jurisdictions asserted themselves, a process that advanced rapidly during the thirteenth century, but especially under the Hohenstaufen regime, it reduced the would-be German state to a loosely bonded assemblage of regions and communities (Barraclough, 1938: 84–6; Arnold, 1991*b*).

How to control their eastern march was no less of a problem. Like the Carolingian Empire before it, the Holy Roman Empire leant eastwards. States like Austria, or Ostmark, and Bohemia were established initially as march states along the eastern frontier. By the tenth century, though, they had been incorporated into the main body of the empire and what constituted the march had been pushed further eastwards. By the twelfth century, renewed pressure led to the large-scale eastward movement of lords and colonists described earlier. The sheer scale of this movement was such as to reposition the fulcrum of the German state itself. Areas like Mecklenburg, Brandenburg, and Pomerania bore the initial brunt of colonization, but by the late thirteenth century German settlers led by the Teutonic Knights had established settlements in Prussia and around the Gulf of Riga, whilst others had reached beyond Prussia into Poland. Though the waves of colonists who moved eastwards across the North German plain and along the Baltic coast formed the major thrust of this movement, others were also to be found moving east into Bohemia and south-eastwards into the middle and lower Danube valley.

To the north, the political geography of Scandinavia was organized around tribal chiefdoms and petty kingships down to the tenth century. Typically, these chiefdoms and petty kingships were highly unstable forms, competing with each other through alliances and hierarchies that quickly formed and just as quickly collapsed (Jensen, 1982: 267; Odner, 1972: 623–51). As with tribal chiefdoms and kingships elsewhere, the quest for plunder and tribute payments provided a powerful driving force behind their regular raiding and warring. To this the Vikings also added a growing exploitation of trade as a source of wealth. The earliest signs of political integration appear in the ninth century and were focused on central Denmark, or Fyn (Randsborg, 1980: 32–3). This provided the core area for the Danish state. By the tenth century, its hegemony had been extended to embrace peripheral areas. In the process, feudal control over landholding was a key instrument for extending political hegemony. The spread of Christianity, and its administrative needs, was also critical in giving the emergent state an ideological unity. By AD 1000, it embraced Jutland, Fyn, southern Slesvig, and, across the Øresund, the provinces of Skåne, Halland, and Blekinge. Traditional views would have us believe that, by this point, a Swedish state had also taken shape to the north when the Svear conquered the Gotar during the ninth or tenth centuries. More recent views, though, have tended to stress the continued political fragmentation of the area down to at least the eleventh century (Löfving, 1991: 147–56). Likewise, though Svealand has traditionally been seen as its core area, a case has been made out for seeing Gotaland as the original nucleus of the Swedish state. Again, the feudal control over landholding and the spread of Christianity played a critical role in forging its unity. By the thirteenth century, a territorially structured system of administration had taken shape, with nobles exercising Crown authority over local districts, or counties, from a network of newly built castles (i.e. Stockholm, Jönköping) and supporting themselves with land grants from the king and by local taxes. Yet though the pattern of castles and counties was elaborated further over the fourteenth and fifteenth centuries, it did not produce a more integrated state by the end of the fifteenth century, but one in which the Crown and a provincial nobility struggled with each other for power (*Atlas över Sverige*, 1967: 133–4).

Even before state systems had emerged on their own soil, the Vikings had helped to establish the foundations for a state in eastern Europe. By AD 800, Vikings had crossed the Baltic and had penetrated rivers like the Vistula and Dvina, moving deep into Russia. From bases established on the watershed, they developed trade along south-flowing rivers like the Dnieper and Volga, effectively linking the Baltic with the Black and Caspian Seas. By the 860s, a Varangian dynasty had established itself at the centre of a political and mercantile hegemony that stretched from Novgorod, its initial base, to the mouth of the Dniester on the Black Sea. By AD 900, the Varangians had moved their capital to Kiev, strategically located on the edge of the northern forests and the steppe. Though Kievan Russia expanded further for a time, even extracting tribute from lands as far west as the Volga, sustained growth was constrained by the expansion of two rival systems. Growth to the south and east was brought to an abrupt halt in 1223 by the appearance of the Tatars or Mongols, a pastoral, nomadic people whose horse skills exploited the rapid movement allowed by the steppes. By 1240, Kiev itself had been sacked and the Tatar had established a rule over Russia that was to last until 1480, Kiev being displaced by the Tatar capital at Saray on the Volga. Equal pressures on the proto-Russian state were also exerted on the west where a revitalized Grand Duchy of Lithuania, a Slavic state, began to expand dramatically over the fourteenth century. Taking advantage of the fact that the Tatars concentrated their control to the east and were more interested in exacting tribute rather than the close settlement of territory, the Lithuanians seized the extensive catchments of the Dniester and Dnieper, including the old capital at Kiev. In time, though, the real threat to Tatar control was from within. From about 1300 onwards, the Principality of Moscow, protected from the heavy overlordship of the Tatars by the extensive forest environments of the area, began to extend its base. It became the religious and political centre for Russians and over the fourteenth and fifteenth centuries developd a strong rural base, pioneering large areas of new farm land out of the forests. Its growth provided a basis for the eventual overthrow of Mongol rule and the emergence of a new Russian state.

Between the Holy Roman Empire and Russia lay a series of buffer states—Poland, Bohemia, and Hungary—whose early history shared similar problems. Each had open western frontiers which absorbed German colonists, and ill-defined eastern frontiers which encouraged would-be rulers to press their own expansionist claims. As a cohesive political unit based on local tribes, Poland first emerged during the tenth century in the area known as Great Poland. It quickly and successfully expanded southwards into what became known as Little Poland. Its expansion westwards and northwards proved less secure. In the shadow of the Holy Roman Emperors, to whom its kings paid tribute, its northern and western edges were heavily infiltrated by German settlers over the twelfth century. Areas like Silesia, Pomerania, and Prussia acquired a mixed identity, having German as well as Polish communities by the fifteenth century. Dynastic settlements created links with both Hungary and, more critically, Lithuania, by the fifteenth century.

Bohemia to the south was closely involved with the Holy Roman Empire being, for a time, one of its constituent states. Though inhabited by a Slavic people, its ruling dynasty became Germanic under the Luxembourgs and even candidates for the title of Emperor at one stage. Further south, the Magyars had already begun to establish a settled and territorialized polity across the fertile soils of the Great Hungarian Plain by the eleventh century. Its complex dynastic relationships did not make for stability. Though successful in extending its bounds into peripheral areas like Croatia and Transylvania and in establishing vassalage over Bosnia, Serbia, Wallachia, and Moldavia around its southern and eastern borders, its growth was brought to a halt when, over the second half of the fifteenth century, these peripheral areas were absorbed into the Ottoman Empire.

Though the western part of the Roman Empire collapsed under pressure from the barbarian invasions, the eastern half survived. Justinian succeeded in reimposing imperial authority over Dalmatia, Italy, and parts of southern Spain in the sixth century, but, with the exception of small sections of central Italy, these were lost again by the seventh century. In time, it also lost Thracia and Moesia, and the lands of its most easterly province, Oriens, including Egypt, but it maintained control over Turkey and most of the coastal areas of Greece.

Byzantium, as this residue of the eastern Empire was called, provided an area of continuity between the Roman and post-Roman worlds though what was left was little more than a Greek Empire. A culturally diverse and politically fragmented empire, it managed to survive and even to re-expand northwards at the expense of the Bulgars during the early eleventh century and into Serbia and Croatia by the twelfth century. By the early fourteenth century, though, the Ottomans began to expand out of their core area in northern Turkey. By the end of the century they had secured control over most of Turkey and, across the isthmus, had pushed northwards across the Rhodope and Balkan mountains as far north as the Danube and westwards as far as Albania. Over the late fifteenth century, they extended their control still further, absorbing Moldavia and Wallachia to the north, Serbia, Bosnia, Herzegovina, and Albania to the west, and Achaea, Morea, and Athens to the south-west.

The Ottoman conquest of Byzantium balanced the Muslim expulsion from south-western Europe. The Muslim control of Spain and Portugal lasted three centuries. During that time, Christian societies survived only in the difficult terrain of north-west Spain, centred first in the Asturias but later—by the tenth century—in León. Despite their limited resource base, these societies led the successful re-expansion of Christian Spain from the eleventh century onwards, a re-expansion that led to the formation of both the Portuguese and Spanish states. Part of the reason for their success lies in the crusading zeal with which they fought to expel the Moors and part in the fragmented political state of al-Andulus after the tenth century. The Reconquista started in the mid-eleventh century and progressed along a number of different fronts. From an initial base in León, it had carved out an expanded territory—New Castile—as far as Toledo by the end of the century. To the west, the Portuguese progressed southwards along the coast, whilst in the east, the Aragonese pushed more slowly south towards the Ebro. By the mid-thirteenth century, the Portuguese had already carved out what was to be their kingdom, whilst the Aragonese had also pushed further south, creating a kingdom that stretched from the Pyrenees to the fertile huerta of Valencia and inland to beyond Saragossa. In the centre, the Castilians had pushed the Moors south

beyond the Guadalquivir, confining them to the largely mountainous province of Granada by 1264, though local pockets of Moorish population also survived around Jaen and to the north of Seville. The Moors held on in Granada until 1492, finally losing control just as the Spanish conquistadores were about to find new frontiers elsewhere and just as the Muslim grasp reached its maximum extent in south-eastern Europe.

Though reviewed last, the changing political geography of the Italian peninsula during the medieval period provides some of the most interesting themes, with old and new political forms competing one beside the other to fragment the political landscape. With the collapse of the western half of the Roman Empire, the grip of the eastern empire on the peninsula was reduced to the extreme south and Sicily. By c. AD 500, the north had become the core territory, first for the Ostragoths and then for the Lombards. In time, the Lombards also secured control over much of southern Italy, or what had been the Duchy of Benevento. In between these two portions of the Lombardic kingdom lay the Papal States, around Rome, Naples, and Ravenna. By the eighth century, though, the Lombardic kingdom had been incorporated into Charlemagne's empire, and with the Treaty of Mersen (AD 888) had passed under the control of the Holy Roman Empire, though it was not formally joined with the German kingdoms until AD 951. Over the late eleventh and twelfth centuries, the southern part of the peninsula passed into the hands of the Two Sicilies, by then a Norman kingdom.

By this point, though, the political geography of the peninsula was being affected by a more significant development. The towns of central and northern Italy had begun to assert their independence to the extent that those of the Lombard League were seen as a threat to the authority of the Holy Roman Empire. The outcome of their growing independence was the Italian city-states, powerful independent cities that controlled not only what lay within their walls but also large surrounding hinterlands, or contadi. Many were dominated politically by signori, or powerful families, with parts of some cities, like Genoa, sectorized between their kin. Seen in its wider context, what empowered the Italian city-states was the wealth generated by trade and commerce, providing them with a different constitution to the coercion-intensive feudal states. In strongly centralized systems, such wealth would have been taxed and access to it regulated by state rulers or territorial lords. In the fragmented political system of central and north Italy, the initiative was more easily seized by towns themselves. A direct comparison can be drawn here with other trading centres that developed on the trade routes across Europe. Some of those which emerged on the edge of political systems, such as the towns of the Hanseatic League on the Baltic, also succeeded in acquiring independence from state rulers though not to the same extent as Italian city-states.

## Beyond the European State System

The character of European political space as it developed over the medieval period was not geographically uniform or neatly segmented. However structured and territorialized the *pays legal* of the early state might have been, the *pays réel* comprised 'varying conglomerations of multiple, overlapping, intersecting networks of power' not 'simple totalities' (Mann, 1986: 521). In part, this reflected the diverse ways in which states had taken shape as well as the diverse ways in which power and jurisdiction became territorialized, with some territories involving wholly new systems of order and others reinvigorating old ones. In part, it also reflected the way in which many states developed out of assertive core areas, a source of growth which almost inevitably created a mosaic of local core-periphery systems across which power and authority lay unevenly. Whereas the Roman Empire divided itself into provinces and a single but extended frontier region, the Europe which emerged over the medieval period had a more complex frontier geography. In addition to the frontiers of late colonization developed around physically marginal areas, both within and without the older cores of feudal settlement, there were the political frontier zones, such as those between Muslim and Christian Spain, or which fronted the outer reach of the Anglo-Normans in Wales, Scotland, and Ireland. Whether constituted in farming or political terms, the settlement of these frontier areas gave rise to a different kind of society and political space compared with core areas with state rulers and lords offering freer, less burdensome conditions of tenure

as an inducement to would-be settlers. The social and political consequences were apparent everywhere (Bartlett, 1993).

When German, Dutch, and Flemish colonists poured eastwards across the Northern and Eastern March into Mecklenburg, Brandenburg, and Pomerania, it enabled Albrecht the Bear to 'build not just a state' but also 'a society de novo' (Thompson, 1928: 519), one that contrasted with older and heavily feudalized areas of the empire. Similar differences were produced by late colonization throughout Europe, creating local as well as regional contrasts in social character that had far-reaching political implications (Dodgshon, 1987: 251–9). Military frontiers, like that created by the Christian reconquest in Spain over the eleventh–fourteenth centuries, generated similar conditions. The Spanish Reconquista was not a sudden military event but the gradual southwards shift of a frontier. The conditions created by this frontier 'moulded Spain's historical development . . . when there was no longer a frontier, the formative period of Spanish history ended'; the key to its impact was 'the transforming disciplines—that is, they had to forgo luxuries and adapt to those habits and institutions which the reconquest demanded' (MacKay, 1977: 2). Exactly when an area was part of the frontier also mattered, so that both tenures as well as the character of society in Old Castile differed from those of New Castile (MacKay, 1977: 3; Smith, 1966: 434).

The political geography of medieval Europe was differentiated in other equally fundamental ways. Particularly relevant are the ideas which Fox developed around the basic distinction between, on the one hand, land-based feudal societies, and on the other, town-based commercial societies, a distinction that can be aligned with Tilly's contrast between coercion-intensive and capital-intensive societies. Fox saw each type as organized initially around linear networks of interaction that could overlap or criss-cross (Fox, 1971: 37–8). In time, though, capital-intensive societies became associated with a burgeoning pattern of trade that developed along a corridor stretching from the city-states of northern Italy, via the Alps, to the North Sea and Baltic. For Fox, as for others, this was a fundamental line of cleavage running through European political geography. At its heart lay a network of flourishing towns and communities whose growth depended on trade. Many

**Fig. 3.2.** The structure of European political space *c.*1500. This diagram maps the broad patterning of European political space as proposed by Fox (1971: 37–8) and Rokkan (1975: 576), with a central 'town-based trade corridor' dominated by independent and semi-independent city-states and trading towns, bordered on the western side by emergent sea-based 'empires' and on the eastern side by emergent land-based 'empires', each organized around a powerful and well-established state system.

towns sought to convert the liberties and privileges which they exercised over trade into greater political power. Some did so by forging trading alliances like the Rhenish League, an alliance based on towns like Cologne in the middle reaches of the Rhine. The most extensive and longest-lasting was undoubtedly the Hanseatic League, a vast conglomerate of trading interests that embraced ports around the North and Baltic Seas, as well as centres inland along the rivers that flowed northwards across the German plain. Others worked for political as well as economic independence by turning themselves into city-states. Though towns like Florence, Venice, Milan, and Berne succeeded, others which tried, like the Flemish towns of Bruges and Ghent, failed. As the more successful land-based feudal societies expanded, they found it extremely difficult to absorb these trading towns and city-states (Fox, 1971; Rokkan, 1975: 576). The outcome was a politically and culturally distinct society based on cities that 'were located between, rather than within, the rising monarchies' (Fox, 1971: 33).

In fact, the trade route belt had a distinct history that stretched back beyond its emergence as a city-state band, to use Rokkan's term for it (1975: 576). Large and significant parts of it formed part of the middle kingdom that was created out of the Carolingian Empire by the Treaty of Verdun, AD 843. Even then, it 'was an unmanageable unit' (Pounds, 1947: 90). By the partition of Mersen in AD 888, it had acquired a more broken structure, with Burgundian and Italian portions. In fact, it long remained a fragmented march belt, a fact which merchants may have exploited. Even by 1500, it still functioned as a march belt between what Rokkan described as the seaward peripheries and empires that were now emerging on one side and landward empires and buffers that now flourished on the other (see Fig. 3.2).

The political geography of Europe changed in two critical ways over the period AD 400–1500. First, there was the gradual re-emergence of larger, more integrated political systems. Following the collapse of the Roman Empire, the political map of Europe had become disaggregated into numerous small, local polities, many based on tribal kingships and chiefdoms. Though there were still parts of Europe where these small, local polities survived, the political map was altogether simpler by 1500, with far fewer but larger polities. Second, there were fundamental changes in the character of political space. New political forms emerged, with political space becoming structured around the territorialized jurisdictions and administrative routines of states and their rulers. Initially, the latter coped with their expanding polities by building a hierarchy of feudal ties and obligations around them. As a solution to controlling the provinces, though, this carried risks, not least because it empowered a network of territorial lords whose ambitions could easily fracture nascent states. By the end of the medieval period, state rulers had begun to assert a more centralized and absolute power, diminishing the power of their nobles and barons whilst raising large and powerful armies under their direct command and seeking alliances with new interest groups such as merchants. Together, these twin pathways of change, one changing the scale and the other the character of political integration, created a qualitative as well as a quantitative change in the nature of political forms.

REFERENCES

ARNOLD, B., *Count and Bishop in Medieval Germany: A Study of Regional Power 1100–1350* (Philadelphia, 1991a).
—— *Princes and Territories in Medieval Germany*, (Cambridge, 1991b).
*Atlas över Sverige* (Stockholm, 1967).
AUBIN, H., 'The Lands of the Elbe and German Colonization Eastwards', in M. M. Postan (ed.), *The Cambridge Economic History of Europe*, i. *The Agrarian Life of the Middle Ages*, (Cambridge, 1966).
BARRACLOUGH, G., *Medieval Germany 911–1250: Essays by German Historians*, (2 vols.; Oxford, 1938).

BARTLETT, R., *The Making of Europe: Conquest, Colonization and Cultural Change 950–1350* (London, 1993).

BONNAUD, P., 'Peopling and the Origins of Settlement', in H. C. Clout (ed.), *Themes in the Historical Geography of France*, (London, 1977), 21–72.

BYOCK, J. L., *Medieval Iceland. Society, Sagas and Power*, (Berkeley, 1988).

BYRNE, F. J., *Irish Kings and High Kings*, (London, 1973).

CHEYETTE, F., 'The Origins of European Villages and the First European Expansion', *Journal of Economic History*, 37 (1977), 182–206.

COLLINS, R., *Early Medieval Spain: Unity in Diversity 400–1000* (London, 1983).

DARBY, H. C., 'The Clearing of the Woodland in Europe', in W. L. Thomas (ed.), *Man's Role in Changing the Face of the Earth*, (Chicago, 1956), 183–216.

DEVAILLY, G., *Le Berry du Xe siècle au milieu du XIIIe*, (Paris, 1973).

DODGSHON, R. A., *The European Past: Social Evolution and Spatial Order*, (London, 1987).

DUBY, G., *Rural Economy and Country Life in the Medieval West*, (London, 1968).

DUNBABIN, J., *France in the Making 843–1180* (Oxford, 1985).

EVERGATES, T., *Feudal Society in the Bailliage of Troyes under the Counts of Champagne, 1152–1284* (Baltimore, 1975).

FLECKENSTEIN, J., *Early Medieval Germany*, 1976, trans. B. S. Smith (Amsterdam, 1978).

FOX, E. W., *History in Geographic Perspective: The Other France*, (New York, 1971).

FUHRMANN, H., *Germany in the High Middle Ages* c.*1050–1200*, trans. T. Reuter (Cambridge, 1986).

GANSHOF, F. L., and VERHULST, A., 'France, the Low Countries and Western Germany', in M. M. Postan (ed.), *The Cambridge Economic History of Europe*, i. *The Agrarian Life of the Middle Ages*, (Cambridge, 1966), 290–339.

HALL, J. A., *Powers and Liberties: The Causes and Consequences of the Rise of the West*, (Harmondsworth, 1986).

HAVERKAMP, A., *Medieval Germany 1056–1273*, trans. H. Braun and R. Mortimer (Oxford, 1988).

HERLIHY, D., 'The Agrarian Revolution in Southern France and Italy, 801–1150', *Speculum*, 33 (1958), 23–37.

—— 'The History of the Rural Seigneury in Italy, 751–1200', *Agricultural History* (1959), 58–71.

JAEGER, H., 'Reconstructing Old Prussian Landscapes, with Special Reference to Spatial Organization', in A. R. H. Baker and M. Billinge (eds.), *Period and Place: Research Methods in Historical Geography*, (Cambridge, 1982), 44–50.

JAMES, E., 'The Origins of Barbarian Kingdoms: The Continental Evidence', in S. Bassett (ed.), *The Origins of Anglo-Saxon Kingdoms*, (Leicester, 1989).

JENSEN, J., *The Prehistory of Denmark*, (London, 1982).

JONES, E. L., *The European Miracle*, (Cambridge, 1981).

JONES, P. J., 'An Italian Estate, 900–1200', *Economic History Review*, 2nd ser. 7 (1954), 18–32.

KOEBNER, R., 'The Settlement and Colonization of Europe', in M. M. Postan (ed.), *The Cambridge Economic History of Europe*, i. *The Agrarian Life of the Middle Ages*, (Cambridge, 1966), 1–91.

LEYSER, K. J., *Rule and Conflict in an Early Medieval Society: Ottonian Saxony*, (London, 1979).

—— *Medieval Germany and its Neighbours 900–1250* (London, 1982).

Löfving, C., 'Who Ruled the Region East of the Skaggerak in the Eleventh Century?', in R. Sampson (ed.), *Social Approaches to Viking Studies*, (Glasgow, 1991), 147–56.

MacKay, A., *Spain in the Middle Ages: From Frontier to Empire 1000–1500* (London, 1977).

Mann, M., *The Sources of Social Power*, i., *A History of Power from the Beginning to AD 1760* (Cambridge, 1986).

Mathews, D., *Atlas of Medieval Europe*, (Oxford, 1983).

Nitz, H.-J., 'Feudal Woodland Colonization as a Strategy of the Carolingian Empire in the Conquest of Saxony', in B. K. Roberts and R. E. Glasscock (eds.), *Villages, Fields and Frontiers: Studies in European Rural Settlement in the Medieval and Early Modern Periods*, (BAR International Series, 185; Oxford, 1983), 171–83.

—— 'Settlement Structures and Settlement Systems of the Frankish Central State in Carolingian and Ottonian Times', in D. Hooke (ed.), *Anglo-Saxon Settlement*, (Oxford, 1988).

Odner, K., 'Ethno-History and Ecological Settings for Economic and Social Models of an Iron Age Society: Valldalen, Norway', in D. L. Clarke (ed.), *Models in Archaeology*, (London, 1972), 623–51.

Poly, J.-P. and Bournazel, E., *La Mutation feodale Xe–XIIe siècle*, (Paris, 1980).

Pounds, N. J. G., *An Historical and Political Geography of Europe*, (London, 1947).

—— *An Economic History of Medieval Europe*, (London, 1974).

—— and Ball, S. S., 'Core-Areas and the Development of the European State System', *Annals of the Association of American Geographers*, 54 (1964), 24–40.

Randsborg, K., *The Viking Age in Denmark*, (London, 1980).

Reuter, T., *Germany in the Early Middle Ages 800–1056* (London, 1991).

Reynolds, S., 'Medieval *Origines Gentium* and the Community of the Realm', *History*, 68 (1983), 375–90.

—— *Kingdoms and Communities in Western Europe 900–1300* (Oxford, 1984).

—— *Fiefs and Vassals: The Medieval Evidence Reinterpreted*, (Oxford, 1994).

Rokkan, S., 'Dimensions of State Formation and Nation Building: A Possible Paradigm for Research on Variations within Europe', in C. Tilly (ed.), *The Formation of National States in Western Europe*, (Princeton, 1975), 562–600.

Sahlins, M., *Tribesmen*, (Englewood Cliffs, NJ, 1968).

Sawyer, P. H., *From Roman Britain to Norman England*, (London, 1978).

Simms, A., 'The Geography of Irish Manors: The Example of the Llanthony Cells of Duleek and Colp in County Meath', in J. Bradley (ed.), *Settlement and Society in Medieval Ireland*, (Kilkenny, 1988), 291–326.

Slicher van Bath, B. H., *The Agrarian History of Western Europe AD 500–1850* (London, 1963).

Smith, R. M., 'Modernization and the Medieval Village Community', in A. R. H. Baker and D. Gregory (eds.), *Explorations in Historical Geography*, (Cambridge, 1984), 140–79.

Smith, R. S., 'Spain', in M. M. Postan (ed.), *The Cambridge Economic History of Europe*, i. *The Agrarian Life of the Middle Ages*, (Cambridge, 1966), 432–48.

Teunis, H. B., 'The Early State in France', in H. J. M. Claesson and P. Skalnick (eds.), *The Early State*, (The Hague, 1978), 235–55.

THOMPSON, E. A., *The Early Germans*, (Oxford, 1965).

THOMPSON, J. W., *Feudal Germany*, (Chicago, 1928).

TILLY, C., *Coercion, Capital and European States AD 990–1992* (Oxford, 1992).

TOPOLSKI, J., 'Economic Decline in Poland from the Sixteenth to the Eighteenth Centuries', in P. Earle (ed.), *Essays in European Economic History*, (Oxford, 1974), 127–42.

—— 'Continuity and Discontinuity in the Development of the Feudal System in Eastern Europe (12th–17th Centuries)', *Journal of European Economic History*, 10 (1981), 373–400.

WICKHAM, C., *Early Medieval Italy*, (London, 1981).

# Chapter 4

# Rural Settlement in Europe, 400–1500

## B. K. Roberts

Within the complex demographic, historical, and environmental forces which have swept the continent since the collapse of Roman power in the west four principal themes can be defined, namely: continuity, the sustaining of elements derived from older orders into new arrangements; colonization, involving folk movements, the intaking of new lands, and state-building; cataclysm, the impact of rapid and sometimes devastating changes; and finally the emergence of new economic systems as population increased and as trade and other linkages expanded and strengthened, processes involving the transition from economies heavily based upon subsistence to others incorporating a large amount of industry and trade. Within these immense contexts, all European settlement patterns and forms have been subject to long-sustained mechanisms of generation, usage, change, and destruction. Across the broad face of a whole continent and throughout the thousand years before 1500 we are faced with almost unimaginable complexity resulting from interactions between these four themes. They underlie the elements of the distribution map appearing as Fig. 4.1, a map which will be used to bind the discussion together.

## The Settlement Base: A Summary of Patterns and Forms

Fig. 4.1 builds around a map by Schröder and Schwarz (1978) summarizing northern European settlement forms. This has been extended in a more summary fashion into France, Great Britain, and Scandinavia. Throughout the continent clustered settlements predominate. Between the Seine and the Elbe extending southwards to the Alpine foreland these tend to be irregular hamlets and villages, which also appear in the Danish peninsula and islands and, curiously, as an outlier in the scarplands of central England; settlement patterns based upon regular street and street-green plans appear in regions roughly peripheral to this core, notably east of the Elbe and extending into Czechoslovakia, Hungary, and Poland, but also in eastern France, particularly Lorraine, and in northern England, the Danish islands, and to a degree in Sweden. However, to the west and north, western France, Cornwall, Wales, Ireland, and Scotland, together with much of Scandinavia, more difficult marginal environments, and in the Alpine forelands and middle Danube valley, mixtures of dispersed hamlets and single farmsteads are found. These mixtures also appear as isolated pockets amid zones more generally dominated by nucleation, for example, in the valley of the upper Ems, in a zone extending westwards to the Lippe and the Rhine and in south-eastern England. Mediterranean Europe is not included in the map, but Italy possesses a settlement pattern based upon nucleated villages, often irregular in character and set upon hilltops, interspersed with areas dominated by hamlets and dispersed farmsteads. This generalization appears valid for the whole zone, extending from Spain to Greece.

For trans-Alpine Europe this broad description suggests a zonation, involving core, periphery, and outer rim, a zonation which transcends the boundaries of the historic nation states (Simms in Smyth and Whelan, 1988: 22–44). This poses many questions about the mechanisms which have created what can be seen and the chronology of its development. If crude environmental controls are eschewed, that is, villages on fertile plains, hamlets in wooded interfluves and single farmsteads in the more marginal pastoral zones—a generalization which does indeed have a limited validity—then the map establishes no more than a framework for basic descriptions (Smith, 1967: 268–95). However, in any synoptic view a time dimension is unavoidable; at its most simple, each component of the overall distribution must represent at least one genetic layer, a settlement genotype, created within a definable time period—however short or long—and existing until it mutates beyond recognition or is wholly destroyed. Each definable element must have either been established within virgin land or, more probably, superimposed upon yet earlier landscapes, which were in turn partly adapted and partly destroyed. To give one example, discussed more fully below, the street-green villages, street villages, and other linear forms of eastern Europe are intimately associated with German colonization, the *Drang nach Osten*, 'drive to the east', of the medieval period, but it is hardly to be doubted that the large, regular, planned colonizing villages of this phase were superimposed over an older generation of Slavic settlements, while the regular village plans of northern England, created in the later eleventh and twelfth centuries, must overlie as yet indistinct and ill-defined Anglo-Saxon antecedents, which limited excavation shows ranged from high-status single farmsteads to hamlets and sprawling agglomerations. This is not to embrace the old chestnut of 'racial' types of settlement, but to point out that in a map which compresses time into a single plane any 'windows', or local regions, can be important, preserving settlement characteristics which may antedate those of a dominant provincial type. These may range in size from a single individual settlement to the patterns and forms of substantial regions bypassed by the historical processes which brought dominant new forms to surrounding landscapes. In general the evidence suggests that in many parts of Europe the large, rather irregular villages and highly planned formal village layouts are features broadly characteristic of the centuries after the first millennium AD, and replaced older generations of irregular, rather smaller hamlets, whose essential features, a form of timbered longhouse, sometimes arranged radially, sometimes in parallel lines, and sometimes rather more randomly, are of an ancestral type found throughout the whole of trans-Alpine Europe.

These large-scale transformations must be associated with others occurring at a more local scale. Successful small places increase their populations and grow: a single farmstead can expand to become a hamlet, a hamlet a village, while a village, with the right locational and political connections, can become a market town. Such transformations have constantly affected the individual settlements making up a pattern. Conversely, other places can decay, decline, and shrink, for change can operate in two directions, and while post-classical deserted towns are not proportionally numerous in the total European pattern, they do occur, as for instance at Hedeby in Schleswig, where the sole visible remains of a thriving Viking market centre is its great rampart; on the other hand depopulated villages, hamlets, and single farmsteads are common throughout the entire continent (Beresford and Hurst, 1971; Chapelot and Fossier, 1985: 161–6; Abel, 1980: 16–91). This dynamic aspect cannot easily be incorporated in any map, but is of importance when assessing specific places.

## Security and Insecurity: The Collapse of the Roman World

The cultural migrations and folk movements which accompanied the 'cataclysm' of the decay and collapse of the Roman Empire (see Chapter 3) had profound effects on settlement: first, those parts of Europe within the empire's frontiers have antecedent settlement forms which to a greater or lesser degree reflect Roman influence, including both traditional native patterns and colonial components derived from Mediterranean roots. Second, the settlements beyond the limes, while not uninfluenced by Rome, derive directly from prehistoric antecedents. The folk movements brought these barbarian traditions westwards and superimposed them upon the more civilized order, creating the twin foundations from which the settlement systems of the great western European lowlands eventually developed. The complexity of this blending is most clearly illustrated by the hard physical structures of churches: in Aachen, in the early 790s Charlemagne began the construction of a great palace, including a round chapel, modelled on the sixth-century Byzantine church of San Vitale, Ravenna (Dixon, 1976: 113), a conscious and deliberate linking of the old order with the new. More fundamentally, and much more tenuously, the aisled *horrea*, granaries, once common throughout the Roman world in forts, towns, and adjacent to villas, appear to have links with the great aisled halls which became so much part of the new barbarian kingdoms, a form manifest in churches, aristocratic dwellings, and, completing the circle, the great barns of medieval Europe (Kirk, 1994: 61; Arnold, 1984: 62; J. Hadman in Todd, 1978: 187–94).

An example of a 'barbarian' settlement can be illustrated by the case of Vorbasse, in mid-west Jylland, Denmark (Randsborg, 1980: 61–9), in existence by about AD 400: it comprised a series of large, well-built longhouses, 44 metres in length, and divided into several rooms, including a living-room, a cattle-stall, and possibly a barn. These were set in a series of approximately rectangular enclosures, house plots or tofts, arranged in blocks or compartments. Wide streets, indeed almost narrow greens, lay between the compartments, and the whole excavated complex extended over 120,000 square metres, the village tofts enclosing about 2,500 square metres and including between twenty and thirty farmsteads. Vorbasse represents only one phase of occupation within a limited area, the main focus of settlement shifting over the centuries from the earlier Iron Age to the Viking period. The site lay upon poor soil but with access to large meadow areas, and this, together with larger than usual cattle bones from the site, indicates a form of mixed farming economy. Evidence for iron production, weaving, potting, and the making of grindstones appears, associated with an absence of any imported luxuries. The settlement was deserted in the fifth century but refounded soon after AD 800, when a new village was established, whose form appears to have been less regular. By the late tenth century the village was at least in part replaced by three gigantic magnate farms, which were then deserted, for they do not continue into the Middle Ages. Nevertheless, here, in some detail, we can observe a complex process of transformation and the archaeological evidence throws light upon several general mechanisms. Two in particular are important: first, the fifth-century settlement is of a type whose general plan and dimensions can be found in many thousands of hamlets and smaller villages during the medieval period. It had clearly been planned, and suggests that the ideas which eventually became manifest in the regulated villages of the twelfth and thirteenth centuries, and to be discussed below, were already to some degree latent in these early societies. How far these originate in wholly barbarian contexts and how far in the order of the classical world cannot be determined, neither can the lines of transmission from one culture zone and region to another. The deliberation involved is important, for as Éliade emphasizes in this archaic context every construction or fabrication has cosmology as a paradigmatic model (1959: 49), but the application of skills and attitudes derived from Romanized areas, conceivably by links as tenuous as service with the Roman army, cannot be ignored. Carefully ordered layouts are found in other settlements, for example, at Saedding, near Esbjerg on the west coast of Jylland, where a rectangular open green, 150 m. by 30 m., remains in the same location throughout the history of the site, while the orientation of the longhouses changes from east–west to north–south. These important and complex themes in European settlement must be reconsidered in a later section.

As a second point, Vorbasse had a nucleus that migrated within an area of approximately one square kilometre between the first century BC and the eleventh century AD, but was finally located on a site which was then occupied from the eleventh century to the present. This is a process of local site shifts preceding site-fixation, now well attested in Denmark, in the Netherlands, and in Anglo-Saxon England (Hedeager, 1992: 193–201; Chapelot and Fossier, 1985: 97–100; Heslerton *ex inf.* Domenic Powlesland, the excavator). If there was indeed continuous occupation, and also social and economic continuity, then this suggests that ownership was focused upon the territory rather than the settlement. Movements may have been stimulated by the limited life of timber buildings and the recognition of the degree to which the soil under the settlement was eventually enriched by the inflow of urine from byres and privies, as well as actual ordure, both animal and human (Heidinga, 1987: 37–8, fig. 19). Former toftland—that land immediately adjacent to or actually containing the farmsteads—could, in a poor soil area, be notably improved, and this could occasion a site shift, with former toftland becoming new arable. Further, even on well-drained soils surface compaction within the well-used parts of the settlement could have led to much ordure-impregnated mud in wetter seasons making a site move attractive, while episodic devastation from war or storm, a recurrent feature of early medieval life, must have been other powerful incentives for moves. On the other hand, it is as likely that the change from migratory to fixed-site settlement was associated with crucial changes in landownership, tenure, and perhaps even documentation. That this was taking place is shown by the appearance of very large farmsteads occupying an area once utilized by many smaller ones—as at Vorbasse—suggesting that within the community wealth was becoming less evenly distributed, and was consolidated in fewer hands. In this matter a major deficiency of the archaeological record reveals itself, for neither tenure nor inheritance practices can be excavated, and this is a fundamental barrier to reconstructing the early history of all European settlement; further, successful villages are not easily available for any excavation, still less the total excavation which is essential to reveal the full history of site occupation. Hence, all excavated examples are in some sense atypical, being of sites which were deserted at some stage in their development. Jepperson's work on the Danish island of Fyn shows that Viking Age settlements undoubtedly existed beneath surviving, living hamlets (Jeppersen, 1981), while work at Wharram Percy (Beresford

and Hurst, 1990) in England attest persistence of site occupation for long periods.

The impact of barbarian culture upon the rural settlement forms of Romanized Europe varied greatly: no doubt in the more remote areas, the mountains of the Iberian peninsula or the highlands of Britain, where civilization was thinly spread, the tenor of life was sustained, with the barbarian incursions causing no more than the usual disruption brought by the tax-gatherer, the passage of armies and raiders, and the demands of lordship. In contrast, in more prosperous farming areas a classic indicator of Romanization is the rural villa (see Chapter 2).

In sharp contrast to England, France and Belgium are particularly rich in the survival of place-names which derive from the names of villa estates—indicated by local variations in the suffix *-acus* or *-acum*. Percival cites one area just to the south of Paris where within an area of 30 × 25 km. some thirty such names survive. Of these no fewer than twenty-one appear in documents earlier than the twelfth century. While only two are known to contain villa remains, the balance of probabilities is that Roman sites are indeed present beneath many surviving settlements. However it be defined, qualified continuity is present, and Percival paints a convincing picture of the way in which aggregation of lesser tenants round a villa during the economic and social upheaval of the barbarian invasions established a nucleated settlement from the large house and its farmstead. Church and later manor house might occupy the Roman site, at first utilizing the buildings but, as these decayed, replacing them with other structures. Of course, by definition, the sites which are not disturbed by later settlement are the ones which did not persist beyond the Roman period; the longer and the more successfully they persisted, the less easy they are to discover and explore (Percival, 1976: 145–99). Examples of villas in close association with churches are widespread in France (Gaul), Germany, and England, in increasing if not large numbers. Percival concludes 'whatever the degree of continuity, it was sufficient in hundreds, perhaps thousands, of cases to ensure the survival of the original estate name' (1976: 176–80). To the south of Amiens, in the valleys of the Somme, Luce, Avre, and Noye, Agache (Percival, 1976: *77*) noted that, while aerial photography allows many villa sites to be detected, many others are probably concealed by surviving villages; indeed it is almost normal for deserted villages to have villas under them. There are general lessons in this: in the civilized, Romanized parts of Gaul, the barbarian invasions stimulated a major settlement transformation, from one settlement system to another. It is an open question how many small, nucleated tenant settlements existed alongside the villas, but aggregation to nucleations meant a redistribution rather than a rise of population and, as Percival makes clear, villas could experience varied fortunes. At the two extremes some could have fallen victim to the invaders and become depopulated ruins, the archaeological sites we now find, while others may have survived more or less intact into later centuries. The vast majority will have suffered varied fortunes, some being abandoned and reoccupied, others struggling on at a lower economic level, some becoming nucleations, while others changed character completely, emerging into the medieval centuries as cemeteries, chapels, or monasteries.

One important caveat must be entered here: the villas are, by definition, normally represented by substantial remains but numerically speaking they must have been vastly outnumbered by humbler steadings. These tend to be most clearly visible in upland areas, where clearance stone was available for construction purposes; but in lowlands, where building materials were more ephemeral timber and turf, wattle, daub, and thatch, the remains of the steadings of humbler cultivators have been more susceptible to wholesale destruction by later cultivation. Nevertheless, these humbler people, bound to the soil metaphorically if not literally, must have had a cumulative impact upon new colonists. The soils of their tilled, manured, warmed fields (*ex inf.* Axel Steensberg) must have been attractive, while even their farming skills—the 'proper' way to do this task or that—must have often been passed on; a thread of continuity, always present but rarely visible.

## The Rise of Feudalism

The migration period falling between the late fourth and the seventh centuries saw great folk-movements in which threads of continuity from the Roman world, although never wholly distinguished, became diluted and tenuous. In this cauldron of change, two parallel tendencies provide important contexts for the genesis of rural settlement: first, the social changes associated with the rise of feudalism; and secondly, the development of regional polities, states, from nuclear cores, areas from which or about which a state originates and spreads, by either conquest or accretion. Socially the feudal state was, as is commonly recognized, made up of those who fought, those who prayed, and those who worked to keep the rest alive. Traders, craftsmen, and town-dwellers were also present, for trade, local, regional, and international, offered opportunities, but the vast majority of the population were peasants, tied to the demands of agricultural production. In contrast to older societies, where it represented sacred roots, land in feudal societies was wealth, and in this hierarchical society the manor, estate, and barony were the building blocks from which states emerged.

However, as Dodgshon (1987: 178) points out, for the great mass of society, feudalism was not about the exercise of lordship, but about the burdens and obligations, about labour services, feudal dues, and personal servitude incumbent upon peasant farmers. Lords and vassals who performed military service or official duties and clerical functions had to be fed and clothed, and in communities where money was scarce—for one of the effects of the conquests of Islam was to cut western Europe off from gold supplies in the Near East—this was achieved with the help of the *manor*. This was an agrarian estate, usually divided into two parts, a large share being set aside for the lord, the demesne, lordship, or manor farm, either worked under the lord's own direction or managed by a steward, with the remainder being distributed as farms to tenants of varied status, part of whose obligation was to provide labour and other services for lord's portion (Slicher van Bath 1963: 40–53). In this socio-economic development there is much which is crucial for the development of settlement, for the rise of the manor undoubtedly parallels a crucial social transformation from small hamlets based upon the obligations of kinship to hamlets and villages based upon tenancies, integrating the labour and productive capacities of families who were not necessarily kin. In practical terms, there is a world of difference between a settlement cluster based upon kinship, in which partible inheritance was practised, dividing the lands of a father between sons and perhaps even daughters, and a settlement based upon tenants, where the lord or overlord had a deep interest in the integrity of the holding, an integrity sustainable by impartible inheritance—primogeniture—normally inheritance by the oldest son (G. Pfeifer in Thomas, 1956: fig. 78; Thirsk in Goody, Thirsk, and Thompson, 1976: 179).

Several points must be kept in mind about the organization of manors: first, they could range in size from vast estates, consisting of hundreds of farms spread over a very large territory so as to embrace a diversity of terrain types, each with their varied productive capacities, to the properties of small landlords embracing a dozen or so farms and a more limited range of economic potential. Secondly, the concordance between the manor and the vill or township, the basic unit of practical production, could vary greatly, for while in some instances manor and vill could be coincident, a single vill could contain many whole manors, sub-manors, and portions of manors, whose centres were based in other localities. For example, in 1280 the Warwickshire village of Harbury was 'held' by some six substantive manorial lords, each possessing separate demesne lands and tenants, without the three other lords of lesser status and a parish church sufficiently well-endowed to possess itself a demesne of 100 acres and eight free tenants (Hilton, 1966: 125–6). Such complexity appears to derive from two sources: on the one hand, manors as administrative territories, given by lords to vassals and tenants, were artificial creations, involving the severance of portions of existing settlements, but on the other hand, this complexity could reflect an older underlying settlement frame, in which what eventually emerges in the period of written records as a single village could have originated as two, three, or even more separate settlements. Finally, there were areas, both large and small, where small holdings or great single enterprises prevailed. This background has been recounted because it is quite essential for understanding the evolution of settlement in central and western Europe. How did settlement respond to these developments? The core–periphery dimension is present in many ways: in contrasts between the nuclear-core and the peripheries of the state, where external political frontiers are to be found; between the more fertile areas of intensive settlement and cultivation in the nuclear-core and the internal frontiers provided by environmental contrasts; between the cores of great manors and their peripheries and appendices; between upland and lowland; between forest and fen and between upland pasture and coastal marshes; in the gradients between well-populated zones and thinly populated zones, and above all in the economic contrasts which result from the presence of both the deep-seated environmental variations and the existence of internal and external frontiers. All of these provided varied contexts for the development of local settlement.

## Antecedents of the Mature Forms and Patterns: Settlement Before 1000

The German heartlands, essentially the inland areas between the Rhine and the Elbe, have been historically dominated by irregular agglomerations of hamlet and village size, the largest appearing as dense, irregular nucleations known as *Haufendörfer* (Fig. 4.1). The balance of opinion is that these developed from smaller hamlets by accretion of new steadings and by the splitting of established new ones (Mayhew, 1973: 26; H. Grees in Dussart 1971: 196). The excavated settlement of Kootwijk in the Netherlands may give a glimpse of the process: the settlement developed from a hamlet of six to eight farmsteads present in about AD 750 which a few decades later was a village of some twenty farmsteads with an unknown number of scattered steadings. On the very sandy soils of the Veluwe, the lower Rhine, water supply was crucial, so that siting was dependent upon the presence of a pond, and the settlement's final desertion, about AD 1000, was associated with the drying up of this water source. The marginality of the site, leading to desertion, allowed the settlement's history to be reconstructed. The plan is best described as an 'approximate grid'—itself an interesting observation hinting at careful ordering such as that seen at Vorbasse—with the great longhouse farmsteads occupying enclosures, tofts, related to the main north–south village road and lateral axes. There were frequent rebuildings and a peculiarity was that as new additions were made the expansion always took place in the side bordering the heath or woodland, with the well-fertilized sites of abandoned steadings gradually being absorbed into the arable area, a logical enough process in the light of points made earlier in this chapter. The buildings were large, generally somewhat over twenty metres in length, with accommodation for cattle at one end, a main central hall containing side benches and a near-central fire, and often a small private room, for the womenfolk or storage, at the opposite end. There are indications that these were not random structures, but were carefully laid out using two ellipses, which results in gently curving sides. The excavators saw this as having an aesthetic or symbolic rather than a functional significance. Within the enclosed precincts of these great buildings there were subsidiary structures, barns, stores, and workshops (Heidinga, 1987).

The land around the village seems to have been divided into small enclosures of about 0.2 ha., sometimes smaller, and a form of horticulture is postulated, the small closes—block fields or *Blockfluren*—virtually gardens, which received fertilizer from both byre manure and the folding of stock. The wealth of the settlement, the cattle, were probably moved seasonally to more distant grazings, a form of transhumance. It is worth emphasizing that the excavators did not feel that any form of communal agriculture based upon strip systems was present during the life of Kootwijk: however, *overlying* the deserted village is a layer of improved *plaggen* soils, fields which have been treated annually with a mixture of animal dung and heather turves, together with grass sods, forest litter, peat, clay, or sand. While these manufactured soils feather out sideways to a depth of a few centimetres, at their thickest they can achieve a half a metre, or even in the most dramatic cases, a metre or more of accumulation. It appears that in this area the earliest of the *plaggen* soils are no older than the tenth century and have been linked with the introduction of rye cultivation in situations where there was no other opportunity of expanding the arable land, and as a way of minimalizing periods of fallow and sustaining year-by-year grain production. Heidinga, however, suggests that the value of turves as manure was limited and that their addition was more a way of protecting soils from wind-blow during periods when long, cold, dry winters delayed spring growth and exposed the soil to the dangers of erosion (Heidinga, 1987: 125–7).

Such cultural soils are common throughout the sandy morainic tracts of north-western Europe where they are associated with loose groupings of farmsteads which possess the large arable strips on the *plaggen* soil area or *Esch* (Mayhew, 1973: 18–22; Werkgroep Brinken, 1981: 20). Sometimes the farmsteads are rather scattered, but in others they are concentrated into hamlets, *Drubble* settlements, ranging between three and fifteen farmsteads, having strips on the arable and common grazing rights in the surrounding wastes. It is possible that from early times the community was socially divided between full-status farmers, *Vollerben*, but partible inheritance, division between heirs, resulting in a group with only partial rights, *Halberben* and *Drittelerben*; but eventually a class appeared with no rights in the core arable at all, who were allowed to settle as cottagers (*Kötter*). Similar forms, associated with intensive manuring rather than *plaggen*, appear in other settlement regions, not least in the marginal rimlands of western Europe, in Ireland, Scotland, and northern England. It was once thought that the *Esch* with its associated *Langstreifenflur* (with strips in excess of 500 m. in length) was a type antecedent to the large agglomerated villages with their associated strip systems, and in some areas this may be so, for example, Krenzlin cites work from Lower Saxony (A. Krenzlin in Helmfrid, 1961: 201–3) which suggests that a single-field system could result from the amalgamation of at least three independent field systems built of massive furlongs based upon sets of long rather narrow strips between two and six hundred metres in length and down to ten or a dozen metres in width; each set was once associated with a separate settlement, one of which has been depopulated, one of which remains as a hamlet, and one of which has grown to become a *Haufendorf*. In Lower Franconia, central Germany, Krenzlin has reached two important conclusions: first, the complex field systems based upon thousands of strips of between one hundred and two hundred metres in length in some villages evolved as a result of subdivision from a framework of rather rectangular blocks or *Blockfluren*, and in others from large surviving strips, several hundreds of metres in length, which were subdivided into narrower units; secondly, the complex tangles of plots in the *Haufendorf* result from the subdivision of earlier rectangular plots (Krenzlin, 1959). This discussion raises many practical questions: such developments can be very difficult to date, and there is always a danger of projecting to early periods situations only securely documented in post-medieval centuries. An example from Lower Saxony discussed by Krenzlin was given by two brothers to the abbey of Fulde in 804, and specifically noted as being an old village (*in antiqua villa*) suggesting continuity through a very long time trajectory. Nevertheless, fully fledged *Haufendörfer* are probably a post-medieval development as a result of partible inheritance, breaking farms into many smallholdings, and leading to the appearance of large numbers of cottagers (Krenzlin and Reusch, 1961: 17–26, map 3; H. Grees in Dussart, 1971: 179–203). There are wide indications throughout Europe to suggest that the division of the antecedent field systems into two types, block-enclosures on one hand and systems based upon long strips on the other, may represent a division between areas of ancient and sustained settlement and more marginal areas where organized colonization took place (W. Matzat in Höppe, 1988: 146; H.-J. Nitz in Höppe, 1988). In both zones, population increases, associated with partible inheritance, probably represent key mechanisms generating fragmentation. A glimpse of another type of antecedent system in environments where crude physical deterministic factors play a powerful role is obtained from examples seen on seventeenth- and eighteenth-century maps from Sweden, where hamlets based upon small numbers of farms may be divided into two or more separate systems (Sporrong, 1985: 106–15). The Swedish landscape still bears more signs than most of the raw rock-strewn terrains from which agricultural lands have been torn, and the fine maps capture field systems which are simpler than those of more favoured environments, with small arable areas divided into both blocky and strip shares amongst no more than two, three, or half a dozen farms, yet which show many and subtle local variations; these are tangible reminders of the complexity which formed the antecedents of the complex systems in favoured agricultural zones. Where agricultural potential was present the early systems were gradually overridden by later field developments and the settlements were gradually expanded, as at Kootwijk, by adding new units and subdividing existing house precincts. Nevertheless, one caveat must always be entered: unless archaeological evidence is used,

most of these arguments are creating models based upon evidence which is not only after AD 1000, but is often largely post-medieval, while there are also profound questions concerning the migration of agricultural practices and ideas from one area to another.

One of the most remarkable early medieval settlements to come to light in England is at West Heslerton, East Riding, Yorkshire (*ex inf.* Domenic Powlesland). The site is distinctive, on the south side of the Vale of Pickering, below the chalk scarp, but well above the lacustrine flats of the main basin. This zone of mixed soils has been subject to sand-blows and hill-wash for millennia, in fact, since the ice departed, processes exaggerated by cultural activity, clearing, grazing, and ploughing. On this sloping bench, Powlesland has excavated an early Anglian settlement extending north–south along a shallow spring-fed stream for about 500 m., a linear dimension comparable to a moderately large medieval village. In this case, however, the settlement has been interpreted as divided into five distinct functional areas, a higher-status zone, a complex multifunctional zone, a zone for agricultural processing, a zone where craftwork was practised, and a zone of hall-houses (i.e. separate dwellings, in contradistinction to the aisled longhouses of the Continent), presumably associated with tenants. The relationships to underlying Romano-British and prehistoric settlements are unbelievably complex, but no sharp break need be postulated, while the core of the medieval village lies some distance away along the bench. The settlement reached some 20 ha. in extent in the sixth century, but occupation began earlier and appears to have continued until the late eighth or early ninth centuries. This settlement was well organized rather than obviously planned, although the seventy-five post-hole structures of the 'housing zone' show a clear tendency for an east to west orientation. This complex site seems to point towards two broader tendencies: first, an 'early Saxon shuffle', which Powlesland thinks may even be late Roman, with marked site changes taking place, leaving older sites unoccupied, while secondly, there appears to be a similar major shift, the 'late Saxon shuffle' between the generation of settlements represented by the present small village of West Heslerton and those occupied during the documented period of the twelfth century and later.

Excavations represent a window into the past, yet may give a wholly atypical picture. In this, which may seem like special pleading, it is important to set the few excavated sites against the tens of thousands of occupied places—'documented' by place-name evidence or by written records—which will never be more than sampled (Jepperson, 1981). Nevertheless, the question remains: if these settlement shifts did take place, and are widespread, what caused them and over what time-periods were the shifts occurring? Practical clues have already been provided: the importance of settlement land as *manured* land and the presence of sufficient space into which a move could occur. Increased populations, increased lordly power, and the advent of written records would appear to be powerful fixative forces, binding both settlement and tenant to the land beneath, no matter what practical problems manifested themselves. In general, it is probable that, within the framework provided by the fundamental themes listed at the beginning of this chapter, four further powerful factors were at work affecting settlement: first, improving agricultural technology, particularly crops, manuring, and rotations but perhaps also ploughs, which gradually raised yield ratios (W. Matzat in Höppe, 1988). Secondly, this in turn supported higher populations, causing changes in the character of the settlement by creating both expansion and further subdivision. Thirdly, a complex transition from kinship to tenancy relationships was also associated with differentiation caused by the maintenance of partible inheritance in some regions on one hand and the more feudally acceptable primogeniture imposed on other regions (J. Thirsk in Goody, Thirsk, and Thompson, 1978: 179) although local practices brought variation to the overall pattern. Finally, what is clearly documented is the presence of hierarchical systems of status variations, which are well documented in Carolingian surveys in France (Duby, 1968: 361–78) and in Anglo-Saxon England, where not only are they present but it is clear that a system of standard holdings or fiscal tenements was emerging. Standard peasant holdings were the basis of existence for the family and the unit from which the lord took rent and tax (Finberg, 1972: 430–48, 507–25). It is lordship which discouraged the more natural peasant tendency to subdivide amongst heirs, particularly as the land in old settled areas filled up, so that residual areas of waste and woodland became less viable for supporting separate entities and became more economically significant to existing communities.

The settlements discussed are, so far as can be determined, ordinary farming communities, for the archaeological evidence reveals few variations in social status although individual variations in wealth must have been present. The examples chosen here have been selected because extensive excavations have revealed substantial portions of their plans, and in spite of the ultimate judgement of desertion, these were for a time generally successful communities. In all of excavated examples of settlements falling before about AD 1000, and where timber was a primary building material, i.e. in the more agriculturally favoured lowlands, two features are noteworthy: first, the buildings needed frequent reconstruction, sometimes encouraging slight shifts of location, and this presents the excavator with vastly complex problems of defining contemporaneity and sequence (Losco-Bradley and Wheeler in Faull, 1984: fig. 2); while secondly, even when it is appreciated that fences are particularly difficult to detect, there is a strong feeling when studying the partial plans of excavated hamlets that each farmstead is sited without close reference to the others of the group and as a corollary that no overall authority was imposing a preconceived plan. However, the more completely excavated Vorbasse is a reminder of the dangers of assuming that such formalization was never present, and while it is tempting to suggest that social relations were normally kin-based, the sheer size of Kootwijk warns against this assumption. It is clear, however, that the roots of later, deliberately conceived settlement plans were already latent before AD 1000.

## Polities and Estates

In the early 790s, when at the height of his power, Charlemagne began the construction of his great palace at Aachen, comprising a massive stone hall with other residential accommodation linked by an elaborate two-storey corridor to a great two-storey round chapel where his throne stood, looking across a central octagon towards the altar. The model for the chapel was—as was noted earlier—the sixth-century Byzantine church of San Vitale,

Ravenna, aglow with splendid mosaics, and from Theodoric's palace at Ravenna came the marble columns, whose capitals, despite their purely classical Corinthian style, are of local stone. Charlemagne's intent was political, and came as a result of his strong contacts south of the Alps. This renewal—his biographer uses the Latin *renovatio*—had political, ecclesiastical, architectural, and literary implications for the whole of western Europe, in a measure creating a bridge between the cultural achievements of the classical world and the successor barbarian states, and the court style of his realm formed the basis of developments in western Europe for generations to come. The adoption of a fixed political capital was itself a Mediterranean feature because, while rural hamlets and villages formed the base of the economy, chieftains and even kings, including Charlemagne, spent their time in a progress from one part of their lands, from one centre, to another, literally consuming the surplus. This tradition persisted into the medieval period proper, so that the Norman kings of England literally progressed with their household and officers of state around their kingdom, and was necessary because of the difficulty of transporting the food rents—grain and meat, honey and milk, cheese and ale—over great distances. This system persisted in Wales to be documented in laws written down in the tenth century (Jones in Finberg, 1972: 299–308). Whatever the roots, rural estates tended to be organized hierarchically long before the full development of feudalism, with farmsteads, hamlets, and villages being administered from localities of higher status—royal vills and estate centres. *Appendicia*, appurtenances, dependencies, 'berewicks', as seen in Domesday Book, or graphically in one document *geburatunas*—the 'farmer's hamlets' (Craster, 1954)—all words which emphasize this status difference (Barrow, 1973: 7–68; Dussart, 1971: 251–64). It follows that there were classes of central places to which rents were paid: in England at both Yeavering and Cheddar great complexes of timbered royal halls have been excavated, the former being what Bede termed the *villa regalis ad Gefrin*, used in about AD 600 by the Northumbrian kings, and the latter a rural mansion of the kings of Wessex and England, a timber *palatio regis* of King Alfred's successors. Outside Italy the great stone palace at Aachen remained unique, although at Tilleda, Harz, a Saxon royal citadel

contained a small fortified tower (Dixon, 1976: 114). In contrast the craftsmen of the outer bailey occupied a series of sunken huts containing evidence of iron smelting and working, bronze working, leatherworking, pottery manufacture, the working of goat horn and even of hippopotamus ivory, with stoneworking and weaving.

Moreover, many other lesser sites began to show traces of the wealth available to other lords, with great hall, tower, and often a private church (*Eigenkirke*), along with the essential farm buildings, forming a defended or defensible nucleus, parts of the growing framework of manorial centres and the focuses of humbler sub-infeudated holdings. With the major estate centres there were also associated two other elements characteristic of medieval landscapes, the castle and the monastery. Both are ancient ideas, but both were renewed and revitalized during the early part of the half-millennium after AD 1000. High-status dwellings, with towers or other defences, were known long before this date, but in England, Wales, Scotland, and Ireland the advent of the earthen castle mount with associated enclosure—motte and bailey—brought fortifications to many rural sites by the twelfth century. Throughout trans-Alpine Europe the twelfth and thirteenth centuries saw the appearance of hundreds of semi-fortified manor houses, some formally crenellated, with defending wall, towers, and a water-filled ditch, but others, at lower levels on the social scale, comprised no more than the 'protection of insurance', enclosing the valuable goods and stores of the seigneurial household, and protecting them from casual theft, occasional brigandage, and—perhaps the greatest threat of all—from fire. Where terrain forbade wet ditches, walls served, while from Spain to Italy the tenth and eleventh centuries saw the grouping of the rural population into villages, many of which were fortified by their lords—a process of *incastellamento* (Hodges, 1991: 32–4). It has been argued that this was less the result of persuasive insecurity than representing deliberate acts of colonization which involved taking a firm hold over the local inhabitants, what may be termed a 'communality of enforcement' as opposed to communality of agreement—where kin are involved—or communality to economize, designed to increase agricultural production by pooling limited technical resources. *Incastellamento* is a widespread and much debated process (Hodges, 1991: 32–4). Of course, there were in addition large castles associated with the powerful rulers: thus in England a series of great royal castles attest the power and wealth of the Crown, while the status of sites such as Prague castle, with its great Romanesque basilica of St George still standing behind the medieval cathedral of St Vitus, and both still adjacent to the seat of government, attests the importance and continuity of the royal sites, and their frequent separation from the river-based merchant power.

Nevertheless, the Church must be considered in two ways, as landowners and as an ingredient of rural settlements: on the one hand are the seats of the great bishoprics, and in England between 1000 and 1100 one must think of men such as Lanfranc of Canterbury, Maurice of London, Wulfstan of Worcester, Robert of Hereford, Osmund of Salisbury, Remigius of Lincoln, Gundulf of Rochester, and William of St Calais, Bishop of Durham—statesmen and architects, scholars, justiciars, and administrators, men of business, energy, and success, in whose capable hands rested the strings and, often, the trappings of government. Most of these centres had more than great cathedral churches: with them were associated landed monasteries, which lands and sub-infeudations were so extensive that the Crown demanded that they return knights for the royal host (H. R. Loyn in Holt, 1987: 7). Often of ancient foundation, during the eleventh and twelfth centuries these were intellectual powerhouses, and the far-travelled and literate clerics associated with them helped mould Europe quite as much as did the mailed knights. Thus, William of St Calais, Bishop of Durham, may well have been in charge of that most remarkable of great surveys, Domesday Book, which gives an unparalleled view of England in the year 1085–6 (P. Chaplais in Holt, 1987: 65–77). Later on in the twelfth century a new order, the Cistercians, was at the forefront of colonizing ventures, rapidly acquiring lands in frontier zones as the mailed hands of the knights carved new territories. At the other end of the scale, from the middle of the tenth century parish churches, at first of timber but then of stone, were constructed throughout Christendom; indeed it has been estimated that by 1100 there were between six and seven thousand churches in England alone (Morris, 1989: 167). Unlike the great abbey churches, these were simple and unpretentious, and their appearance, often alongside the hall of the lord, brings together the three ingredients of the classic medieval village, (1) the hall or manor house, the seigneurial or magnate dwelling, (2) the church, and finally (3) the farmsteads and cottages of the varied classes of tenant. Each of these bears a distinctive burden of associations, in contained artefacts such as weapons and tools, objects of use, objects of value, as economic entities, and as social expressions, and all are parts of that complex unity within diversity which is characteristic of mature cultural landscapes.

This is a context from which the nucleated village emerged throughout Europe: the residence of tenants of the local lord and itself divided between land in lordship—the demesne farmstead of the lord—and the tenant acres, whose occupants owed services in work, kind, and cash, tithes to the lord's church, and above all practising a form of mixed farming, in which cattle were the prime movers, drawing the ploughs vital to the tillage of the extensive acres on which wheat, barley, rye, mixed grains, and oats were produced. Bread grain (wheat and rye), drink grain (barley), and finally, horse grain (oats), horses being both symbolic and vital for the aristocratic knightly elements of the social order. At the peasant level there is a communality of effort, for work shared was at least made more bearable, and in years of better harvests, population growth, and local prosperity, defined social and economic obligations did allow new clearance, new intaking, and the extension of tilled acres towards the limits of hamlet and village territories. This brought problems, not least those of soil exhaustion and pasture shortages to the more populous, closely settled, and long-tilled regions.

## Physical Environmental Change

No discussion of rural settlement can fail to take account of the physical environment: along with population, this is a vital element. In no way does it wholly control the course of development, but it has always been a powerful factor, framing what was possible. There are three broad interlocking ingredients: climate, vegetation, and soils. The European continent extends from the distinctive Mediterranean zone, with its evergreen forest, the outcome of a climatic regime with summer drought and winter rains and a foliage adapted to a period of enforced rest during the summer heat, to the treeless arctic tundra found in the northlands and on the highest mountain areas of the Alps, the Pyrenees, and the uplands of Scotland and Wales. To the south of this arctic fringe lies an area naturally dominated by coniferous forest, but a dominant zone of temperate deciduous forest occupies a broad tract extending north from the Alpine forelands to southern Norway and Sweden, and inland from the changeable, often wet and windy conditions of the Atlantic edge towards the continental interior of the eastern marches of Poland, where hot summers alternate with savagely cold winters dominated by Siberian air masses. Earlier scholars divided Europe into two contrasting types of region, areas of difficulty, having high-relief and/or high rainfall, and lowland areas with richer soils, where effort is rewarded. In fact, each and every environment offers a livelihood to those culturally adapted to its vicissitudes, and this adaptation is a product of the possession of certain basic skills, be these in hunting and gathering, fishing and cultivation, pasturing and woodcraft, linked to experience achieved over many lifetimes. Of course, throughout the whole continent, travel has always been possible, for those with courage, wealth, determination, and adaptability; bishops, missionaries, and other churchmen, moving from the provinces to Rome and back on business, merchants carrying trade-goods, together with warrior adventurers and pilgrims. Along such channels, as has been repeatedly stressed in this chapter, flow crops, animals, and manufactured goods, skills, diseases, words and ideas, attitudes and questions.

Throughout the period between AD 500 and 1500 there are signs that the climate was not static, and this had effects upon both ecology and cultural life. The evidence often appears to be flatly contradictory: this may be so, but contradictions may arise from the sheer complexity of the changes, which create subtly different effects in different areas. There are indications of a colder and more disturbed climate in the period AD 500–1000, with increased wetness in the Mediterranean and a tendency to more extreme winters and glacial advances in the mountains. Lamb summarizes this by noting that the 'reconstituted western empire of Charles the Great (Charlemagne) did not coincide with a particularly favourable climatic

period' (Lamb, 1982: 163 ff.). However, during the three centuries after about AD 800, the climate seems to have been warming up. This had the double effect of allowing both trees and cultivation to climb higher up mountain slopes. It is likely that the hotter summers were balanced by colder winters, in effect a more continental type of regime. This was the context in which Vikings, and indeed Irish monks, taking advantage of ice-free sea routes, sailed northwards towards the limits of oceanic arctic Europe, reaching Spitzbergen, Greenland, and the North American continent. More importantly, the warmth allowed the cultivation of grain and the grape to be pushed towards new latitudinal and altitudinal limits, in Iceland and in upland Europe (Parry, 1978: 95–134). The warm phase seems to have already passed its peak in Greenland in the twelfth century, for it was by no means coterminous everywhere, yet continued in Europe until the first decade of the fourteenth century. In the Mediterranean zone, this was a time of greater moisture.

The climatic deterioration affecting Greenland by the middle years of the thirteenth century—perhaps earlier—struck continental Europe in the early decades of the fourteenth century, taking the form of severe storm floods and sand-blows in coastal areas and a general cooling of the climate associated with increasing wetness. This change, in temperate Europe, was devastatingly sudden, with an extraordinary run of wet summers, and mostly wet springs and autumns between 1313–14 and 1317, continuing without intermission until 1321 and bringing in their wake a succession of harvest failures. These changes ushered in a phase of wild and long-lasting weather variations in western and central Europe, with a gradual run into more severe winters. Eventually, but not necessarily immediately, the changes became associated with the desertion of farms and villages, most notably in those areas where the warmer conditions had allowed cultivation to be pushed to the limits, sites such as Hound Tor on Dartmoor, but also within the agricultural heartlands of Germany and England, where settlements founded upon strong clays or poor thin soils were always susceptible. To link these in a direct cause–effect relationship would be wrong, for matters are much more complex. Nevertheless, amid a matrix of demographic, social, and economic changes, the variability of climate has a legitimate place.

Sufficient has been said already to draw a picture of sustained grain cultivation throughout all of the lowland areas of temperate Europe, and underlying this lay two tendencies: first, a push towards those limits of cultivation imposed by nature, high land, wet land, and dry land; while secondly, sustained grain cultivation will always tend to depress yields, the ratio of grain sown to that produced as crop. The addition of manure, animal dung, seaweed, and/or shell sand or turf, readily improves fertility and brings increased yields, but the problem for the medieval peasant in areas of extensive grain production was always to find a sufficient quantity of manure for the arable acres. In short, while fallowing, resting the soil for a short period, ameliorated the problems, medieval farming tended towards soil exhaustion, and farmers were engaged in a constant battle to sustain and improve this. In fact, the real opportunity for change in the arable areas—those zones in Fig. 4.1 dominated by nucleated villages and hamlets—did not appear until the middle decades of the fourteenth century, when plagues, by severely reducing population numbers, allowed the survivors to experiment and adapt in ways which were not possible under the pressures caused by large numbers pressing against available agrarian resources. Of course, in regions where this type of economic system had never been present, alternative possibilities had always existed.

## The Rural Base: Core, Periphery and Rim Characteristics after about 1000

There are profound methodological problems associated with treating medieval settlement, not least because the maps which allow large-scale mapping and the analysis of detail are all post-medieval, so that much must be built upon inference from later sources. While local or even regional studies can lead, by retrogressive analysis, to reconstructions of medieval conditions, this exercise is never available at a national scale, still less a continental scale. The map appearing as Fig. 4.1 provides a necessary framework, but three things must be established: first, the terms *nucleated* and *dispersed* may be thought of as representing the two ends of a spectrum of settlement types. At one end, all of the farmsteads within a settlement's territory may be concentrated together, so closely in fact that in the Mediterranean zone many of the dwellings may be conjoined, but more normally they will be separated by short distances; at the other end of the scale, all farmsteads may lie scattered, each wholly and distinctively separated from its neighbour, by at least the 'hailing distance' of about 150 m. and, as an English sixteenth-century topographer put it, 'eache in the midst of his owne occupieng'. Between these two extremes there lies a gradation of types, with some patterns, for instance, comprising congregations of tightly clustered hamlets, others with nucleations breaking up into loosely textured ribbons or skeins (Uhlig and Lienau, 1972; Roberts, 1987). Secondly, when these types are multiplied through large areas of countryside, while areas of chaotic mixtures can appear, zones can be detected, with each dominated by broadly similar settlement patterns, some almost wholly nucleated, others largely dispersed, and others containing more complex mixtures. Thirdly, where nucleations appear, they are seen to have different shapes, plans which are sufficiently varied to have attracted the attention of scholars since the late nineteenth century. Finally, any discussion of this complexity must adopt an order, a sequence, and this is dangerous because, while this may generate no more than a catalogue, a more sophisticated discussion needs to explore a possible chronological sequence. What follows adopts an approach which moves outwards from core Europe to peripheral Europe via series of short discussions, a framework which also allows economic systems to be described and chronological implications considered.

### Nucleated Settlements

Much of western Germany, northern France, eastern Denmark, and central England were by the nineteenth century dominated by a form of nucleated settlement which is to be concisely described as irregular agglomerations of hamlet and village size. As has already been suggested, they probably originated in accretive growth around one or both of the cultural nuclei of church and a high-status farmstead or hall. Paradoxically, settlements of this type are distinguishable because they *lack* the notable regularities of other categories of nucleation (see below): nucleated settlements without a regular street plan, with variable degrees

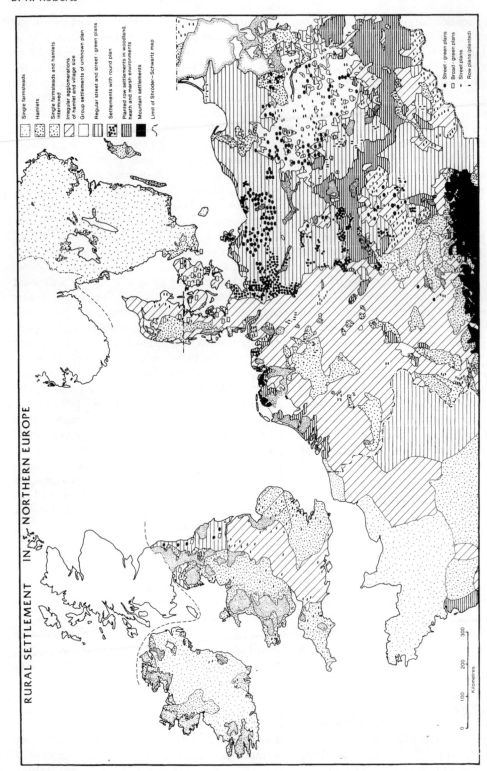

Fig. 4.1. Rural settlement in northern Europe

of actual concentration of dwellings, are found throughout the whole zone, some with, others without a central open space or green, and are normally supported by a communally organized field system. It is probable that these nucleations were already present by the thirteenth century. Sidetracking the troubled question of how these are defined, as open, common subdivided systems, the English word 'townfield' remains a useful neutral generic term, being equally applicable to small primitive systems found amid the woodlands of Sweden or the complex, large, highly fragmented systems of the great fertile plains of Germanic Europe. In fact, a crucial element of definition, in addition to the structural morphology and functional dimensions, is the actual area of the settlement's territory which is under communally organized cultivation, and this must be an important diagnostic feature in any classification. Within the zones under discussion, such fields normally represent over 60 per cent of a settlement's territory, and it is possible that under the population pressures of the later thirteenth century even larger proportions were present, bringing fundamental problems with the shortages of grazing lands, woodlands, fuel supplies, and land for expansion. Such pressures led to the fragmentation of holdings and great pressure on resources and in extreme cases, led to soil exhaustion.

Work in East Swabia by Grees (1975: H. Grees in Dussart, 1971: 179–203) illuminates the general problems: he sees the origins of these *Haufendörfer* in small clusters of original farmsteads: sometimes one, sometimes several, of these small nuclei form the kernel from which expansion occurs. Fragmentation of the original farms and the influx of some smallholders' dwellings had already enhanced the original cluster by the end of the thirteenth century, by which date fragmentation ceased because of the introduction of undivided inheritance. Succeeding centuries saw growth through the addition of smallholders' dwellings, and it was through this growth that the original small hamlets became substantive villages, with the smallholders earning a living by day labour on the larger farms or as craftsmen. On this, several points need emphasis: first, the large villages of later centuries originated in smaller hamlet clusters; secondly, this took place in a context of agricultural potential, where the arable fields could be extended as population

increased; thirdly, the presence of a cloth industry allowed smallholders to supplement their incomes; fourthly, a distinction existed between 'open' villages, with many smallholders from earliest times, and 'closed' villages, with fewer, and landholders who discouraged their appearance—this contrast radically affecting a settlement's potential for growth or stability; finally, the effect of inheritance is important. It is easy to overemphasize the effects of undivided inheritance, for as Howells has shown in Kibworth Harcourt, England, legacies for non-inheriting children could take the form of cash provided before the death of the parents (C. Howells in Goody, Thirsk, and Thompson, 1978: 154–5). Nevertheless, it is likely that this model is a useful one: Mayhew, using much German work, has drawn attention to the presence of status variations amongst tenants, distinctions between *Vollbauern*—full-status farmers, *Halbbauern*—half-status farmers, and *Erbkötter* and *Seldner*—cottagers and smallholders (Mayhew, 1973: 41), divisions reflected in the distinction between *villani*, *bordarii*, and *cotarii* found in eleventh-century England (Finberg, 1972: 507–25).

Similar developments have been documented in Denmark (V. Hansen in Dussart, 1971: 205–18) and they represent a useful generalization. Such agglomerated settlements reveal many subtle variations: some are tight clusters while others are loose; some have one centre, others are composite or polyfocal; while others possess a clear outer boundary which possibly had a defensive purpose (Roberts, 1987: 63–86). However, it is particularly important to note three things about this type of settlement. First, the argument about their origins presented above bridges the gap between two apparently distinct types of settlement pattern, one dominated by truly nucleated villages, the other, an earlier pattern, dominated by scattered hamlets, perhaps kin-based. Secondly, these settlement forms are associated with the great fertile plains and scarplands of western Europe, the heartlands of European states, and we must presume that most have ancient roots. It is notable that these areas, dominated by irregular nucleated villages and hamlets, are intermixed with local regions where mixtures of single farmsteads and hamlets appear, seen on the heathlands of north Germany, parts of Scandinavia (rather more in fact than the map suggests), and in south-eastern and western England.

It is possible that these latter are areas which have followed different trajectories to the most richly endowed zones, and thus preserved a semblance of older characteristics. To this already complex set of ingredients must be added a third component: if the type trajectories postulated here can be seen as *organic*, then a second type of trajectory was present from the earliest times, *planned* developments.

Planning takes many forms: the most irregular formal layout can be socially or economically organized, that is, a settlement's space can be perceived as organized by its inhabitants even if the outsider can only see chaos. While it is perhaps possible to detect economic zonation within a settlement by archaeological methods, with areas of living space producing different finds from areas where stock was kept or metalworking or weaving pursued, it is wholly impossible to dig up the more subtle social contrasts. Settlement planning is most tangible where it results in structured, formalized layouts, where settlement morphologies are clearly influenced by the principles of Euclidian geometry, notably the triangle (or wedge), the square, and parallel lines, imposed upon a settlement's framing structures, its property boundaries and roads. What complicates the issue, and makes it more fascinating, are two tendencies: on one hand it is possible to achieve planning without perfect accuracy, so that the end-product is an approximation, particularly if the new plan is modelled upon older underlying field patterns, which will impart a softer, curved or sinuous pattern to lines that pure geometry would make straight (Roberts, 1987: 196–203). On the other hand, no 'medieval' settlements are translated to the present century wholly unaltered: plots are subdivided and combined, new streets, lanes, and paths are added or extracted, open spaces infilled or created, and above all farmsteads and buildings are added, subdivided, or deserted and removed. This kaleidoscope of change, brought about by post-medieval social and economic forces, acts as a powerful filter.

An intermediary in this is the *street plan*, a universal form, ranging in length, degree of compaction, and regularity: in fact all settlements possess streets, and true *linear* settlements are distinguishable from ribbons by the degree of organization present in the layout of their property boundaries and open spaces. The distinction is made clear in the north of England, where two-row plans, built of two compartments of house-plots or tofts, are often *not* orientated along the road on which it lies, which actually changes direction as it enters and leaves the main street of the village, emphasizing the discord present. This, by suggesting that the orientation of the village matters more than that imposed by adopting the line of the street, introduces another concept closely linked to regular village plans, that of regulation. This implies the existence of organizational arrangements which link together a settlement's plan, its field system, the social hierarchy present, and the farming system applied, and goes far beyond mere physical planning. It reveals the existence of a 'village idea', a conceptualized framework, used as an exemplar for an on-ground layout. An added layer of complexity is found in the fact that not only were these ideas used fully, or partially, in the planting of wholly new plantations, the ideas were also applied to existing, older settlements, which could be replanned or reorganized not once, but at several stages in their history. It is a thread whose roots and chronology have to be convincingly traced, but slight hints are seen in the plans recovered from sites such as Vorbasse and Heidinga. It is possible that the transmuted and translated ideas of the *agrimensores*—the Roman land surveyors—are ultimately involved, although the need for all families, or all tenants of equal status, to an aliquot share of limited agrarian resources is one common to many human societies (Homans, 1960: 83–106; S. Göransson in Helmfrid, 1961).

Using data from the seventeenth century, Hansen examined settlements in Grinsted parish on the glacial outwash plains of Denmark: there were no fewer than nineteen small hamlets and scattered single farmsteads in 1683, the largest housing five farmsteads and ten cultivators, the smallest merely one of each. Grinsted itself contained four farms: one of these, the glebe, tilled half of the arable land of the hamlet, while the rest of the land lay in strips, scattered among numerous furlongs. While in regulated settlements it is customary to find the same sequences of strips from one open field furlong to the next, often falling in an order determined by the sequence of dwellings and tofts within a geometrically regular village plan, in this case the most frequent successions of strips were 4-4-2-1 or 4-4-3-3-4-4-2-1 or 4-3-3-4-2-1, where the figures stand for farm numbers and 4 corresponds to the glebe or parson's farm. V. Hansen (in Dussart, 1971: 212–13) sums this up with admirable clarity:

this is in fact sun-division [*solskifte*—a form of regulation, see Homans, 1960: 96–102], but with the one deviation that number 3 is absent in some furlongs—bundles of strips tending to be ploughed as a unit—and wherever it appears, it has two adjoining strips in between strips belonging to number 4. This farm always has two strips for each one belonging to farms number 2 and number 1 . . . The most obvious explanation is that farm number 3 is a later addition to the hamlet and that it came into existence by a division of farm number 4 . . . there is no possible way to tell when this division took place but there is no doubt as to what the surveyor had in mind . . . the creation of a fourth farm equal in size to the other two small ones.

In fact, as Göransson has shown, the roots of subdivision in both England and Scandinavia are indisputably medieval, appearing in the centuries before AD 1000. This complex concept was an ingredient of no small importance and, along with the gradual addition of new furlongs brought into cultivation from the waste and the intricacies generated by inheritance practices together with a market in land, brought into being the great organized field systems comprized of dozens of furlongs and hundreds, even thousands, of strips. The specific case of Grinsted, cited here because of the useful intricacy of a simple case, is neither wholly anachronistic nor over-complex! In fact there is fundamental debate over this matter. In England, a context is seen to be provided by emergent feudalism, even before the Norman conquest, and scholars also stress the role of lordship in organizing the planning process. In contrast, in Scandinavia, the role of the farmers themselves is perceived as dominant, and the planned hamlets and villages are seen more as a response to the imposition of royal taxation. Cutting across this question is the role of inter-regional contacts, and not only were there Viking conquests in England between the 830s and the 950s, Canute and his sons (1016–42) ruled both England and Denmark (S. Göransson in Helmfrid, 1961: 98–101).

This discussion has carried the argument from the accretive complexity found in the older village lands of Europe, to developments in areas such as Scandinavia and the north of England, peripheral regions, where the reorganization of the older settlement system based upon kinship hamlets can

just be glimpsed in the earliest local documents. The physical expansion of Europe noted earlier, by conquest and by colonization, generated distinctive settlement forms, notably the great formal street-green villages of eastern Germany and the Polish plains (*Angerdörfer*), supported by massive and highly organized grain-producing strip field systems, but also the settlements of the German woodlands, extending into the Czech Republic and Hungary, where clearances of woodland or heath based upon broad but very long strips (*Waldhufendörfer*, *Heidenhufendörfer*), with farmsteads strung along a roadway, so that the subdivision and addition of new cottages eventually resulted in elongated street villages (*Strassendörfer*) that stretched for several kilometres. The term *Hufe* embedded in these compound German terms implies the presence of fiscal farms, aliquot units for taxation purposes, deep structures which lie within and beneath the practical on-ground details of settlement and field system (G. Pfeifer in Thomas, 1956: 254–8). In wetland areas, settlements based upon similar broad-strips appeared—*Marshufendörfer* (Mayhew, 1973: 66–84; Uhlig and Lienau, 1967: E54–9). German evidence shows that these structured plans could involve the active participation of a land agent or *Lokator*, whose task it was, following a grant of land for the purpose by a lord, to initiate the settlement, to lay out the initial plots, and to find the colonists. Often he was able to obtain for himself a large share in the lands of the new enterprise (Duby, 1968: 79, 393–403; Simms in Smyth and Whelan, 1988: 25–8; Mayhew, 1973: 47, 55–9). The impact of these activities upon eastern Europe can be appreciated from Fig. 4.1, where the 'regular street and street-green plans' and 'planted row settlements' appear to reflect its prevalence.

In sharp contrast to this plantation of new settlements, it is widely recognized throughout Europe that many villages became wholly deserted and depopulated during the later medieval centuries, the result of climatic variation, plague, and economic change during the fourteenth century or more episodically as a result of warfare, landlord action, and economic instability (Chapelot and Fossier, 1985: 161–6; Abel, 1980: 80–91; Slicker van Bath, 1963: 162–6; Beresford and Hurst, 1971). Just as increases in population, the fragmentation of holdings, and the addition of new farmsteads and cottages altered the character of both long-established and new nucleations, falling populations and economic difficulties, by reducing the numbers of farmsteads in many but by no means all successful villages, brought changes. These took the form of larger holdings for those who could work them, expanded demesne farms and more elaborate dwellings with surrounding parkland for those lords who sought them, and even settlement shuffles as a by-product of shrinkage and expansion. It is probable that the nature of the tenancies was crucial in affecting trajectories: a village comprising tenants rendering wholly customary dues to a single lord—a closed village—would be much more susceptible to seigneurially induced change, while one under the control of many lords, or with a strong body of freeholders, tenants with a long-term legal interest in their land—an open village— would be far more resistant to catastrophic change, and more likely to accrete new smallholder and cottage tenants and sustain itself; indeed, this very labour reserve was itself a potential catalyst for further change, allowing diversification into industrial activities. In summary, the village, in origin an assembly of separate elements with different economic and social agendas—lord's hall and church, tenants, sub-tenants, and humbler folk—proved to be an adaptable, durable, and long-lived settlement type. Hamlets and villages were never wholly rural in character: in seventeenth-century Northumberland each coastal village tended to have attached to it a small coastal hamlet, a fishing village, and there are no reasons for assuming that these had been recent arrivals. The logic of the link and the pattern of exploitation is an ancient one. Other settlements specialized in industrial production: Potter Hanworth, Toynton All Saints, and St Peter and Boarstall in England possessed vigorous pottery production (McCarthy and Banks, 1988: 70–1). Nevertheless, it is too easy to present a medieval world in which the village is the only settlement element, and we must now turn to dispersion.

## Dispersed Settlements

It is one of the peculiarities of medieval England that some of the most wealthy, most populous regions are to be found in the south and east, areas where a rather thin scatter of nucleations are associated with a settlement pattern which is made up of dispersed elements (B. K. Roberts, S. Wrathmell, and D. Stocker in Fridrich *et al.*, 1996: 74; Darby, 1973: 139, 191). Throughout the landscapes of the south-east ancient hall centres—manorial focuses and other isolated farmsteads and even isolated churches, where there are sometimes traces of a degree of nucleation in centuries before AD 1000 (P. Wade-Martins in Fowler, 1975: 137–57), intermingle with large numbers of small farms and cottages scattered through each township's territory; the latter are often loosely arranged around pieces of common grazing or 'greens' (Smith, 1967: 272; Warner, 1987). These, it cannot be sufficiently emphasized, are *not* the same as the 'green village' layouts associated with formally planned and regulated nucleations; indeed the fact that the word 'green' often appears as part of their place-name, as in Potters Green, Saxstead Green, reveals their presence. They are more loosely structured than formal villages, and often lie towards the peripheries of townships and are dominated by individuality rather than communality. Many 'greens' are first documented in the thirteenth century, but this could hardly be otherwise, for earlier documentation tends to record the presence of the central place, the township centre, be this a hall, church, or small nucleation. Wade-Martins has shown that these loose settlement structures are to be linked with movements of population away from older hamlet cores, or to be the result of population increases, taking advantage of the extensive common grazings. Shared arable fields, with strips, only appear near the small hamlets which bear parish or township names, the oldest settlement focuses, and in general the surrounding field systems incorporate a complex mixture of enclosures and communally organized blocks, with communities tending to be separated by tracts of common grazing and sometimes woodland. Such features have been recognized for many years (Davenport, 1906), but they have not been as closely observed or studied as they may warrant, for their very lack of substance—location of farmsteads and other buildings are probably rather ephemeral—may conceal both their antiquity and flexibility as the physical framework for a productive system.

Given the fact that cases are known where Roman roads appear to cut across this type of countryside (Williamson, 1993: 26–7), a reasonable interpretation is that some of these are 'ancient landscapes', continuously occupied and adapted, where settlement focuses have tended to be less stable than is seen as 'normal' for the landscapes dominated by nucleated villages. If correct, this is an exciting vision and in England is leading towards the argument that the villages are relative latecomers to the settlement scene appearing in contexts where limited rather than high population levels are concentrated at central places with communally organized grain-producing field systems. Disentangling the development of these long-sustained landscapes is in fact much more difficult than treating the relatively catastrophic changes associated with aggregation to villages. To argue that this is universally true would be to underrate the vast complexity of the European settlement scene, but throughout this chapter the argument has been that places either *become* villages through accretive expansion and the concentration of a territory's population at one locus, or they are *made* villages by the plantation of colonists. It remains a fact that dispersed settlement—perhaps older—is a difficult will-o'-the-wisp to grasp and take hold of for investigation, and this subtly affects what is studied. Chapelot and Fossier have noted that in northern France place-names ending in -*court* or -*y* suggest that ancient forms of isolated settlements survived into the medieval period (1985: 134). As Steensberg showed in Denmark, the excavation of a medieval village needs large-scale area stripping to even begin to touch the real complexity, but dispersed entities, though often functionally, socially, and economically linked, sprawl across archaeologically vast areas (Steensberg and Østergaard Christensen, 1974).

On a broader scale, as the map (Fig. 4.1) shows, the distribution of areas with single farmsteads or mixtures of single farmsteads and hamlets throughout Atlantic Europe, from France to Ireland, in scattered regions throughout the Low Countries, Germany, along the northern Alpine forelands, in Scandinavia, and further south throughout central Italy, parts of southern France, northern Spain, and much of south-eastern Spain, points to a vast diversity of origins. Of course, in spite of the arguments presented above, it is certain that in many cases the establishment of this dispersion was post-medieval or even quite recent, following land reclamation (Uhlig and Lienau, 1972: 146–9), the consolidation and/or enclosure of once fragmented fields or simply the result of the abandonment of older hilltop

defensive sites with the arrival of more peaceable conditions. Thus, the role of slave-seeking pirates in the Mediterranean, for example, Saracen piracy in the eighth to the tenth centuries, in encouraging avoidance of the coast and/or strong nucleation in its vicinity, should never be underrated. Periods of anarchy and unrest, linked with episodic devastation, were normal components of everyday life. For example, at Rougiers, Mediterranean France, a twelfth-century stone-planted hilltop village developed around an early twelfth-century castle foundation—a case of *incastellamento*—but classic Gallo-Roman villa-type settlement had been present on the plain below and there is clear evidence that the medieval site lay within an area which had been long utilized. Writing in 1980 Chapelot and Fossier make the point that Rougier was then a structural model without precedent, save in northern Europe, but much recent French work has tended to focus upon these tangible sites (J.-M. Bazzana and A. Poisson in Fridrich *et al.*, 1995: 176–202). There was a gradual movement of dwellings downhill, and at Rougiers a new nucleation had become the core of the village by the seventeenth century.

Dispersed farmsteads are important in the landscapes of Scandinavia, indeed while large villages are generally to be found in eastern Jutland and the Danish islands, the southern province of Sweden, Skåne, and in the island of Öland, evidence from about 1500 shows that in the more favoured portions of central Sweden, Norway, and Finland permanent settlement comprised single farms and small 'villages' of up to six farmsteads (hamlets in English terminology), with the remoter areas generally having over 50 per cent of all habitations as single farms. Work which concentrated upon settlement developments between 1300 and 1540 suggests that farm desertion during this period was very extensive, over 40 per cent in Norway and even in parts of Denmark, 25–40 per cent in Sweden (Bohuslän and Östergotland), but falling to below 15 per cent in most other areas. These figures are based upon taxation records and place-name studies, and stress not the colonizing processes and population rises before 1300 which created the settled landscapes, but the processes associated with retreat, decolonization, and desertion after 1300 (Gissel *et al.*, 1981). This was not a uniform process: it is possible that the abandonment of older farms could also have been accompanied by the creation of new ones, while in areas with some nucleation scattered farms may have accreted to these, so that settlement may contract as a result of reorganization rather than a more obvious physical factor such as climatic deterioration (see below). In fact, Scandinavian scholars see several possible causes: first, the impact of plague in the middle years of the fourteenth century was a key demographic factor, for even if a population was not wholly eliminated—and it rarely was—the reduction of the labour force linked with the viability of other holdings in more favoured regions was a powerful force for change; secondly, phases of climatic deterioration leading to a run of long winters and/or bad harvests can render a steading nonviable, while more cataclysmically examples are known from Norway of farmlands being swamped by advancing war and its side-effects, trade policies and economic policies can all play a part. The omission from this list is landownership; in Scandinavia, while many farmers were rent-payers and all were taxed by the Crown, the formal elements of feudalism appear to have been generally absent, and scholars have been less inclined to explain farmstead depopulation as being the result of the policies of lords or the location of a farmstead, hamlet, or village within the economic context of a great estate. Nevertheless, it remains a fact that in all but Finland, of those Scandinavian areas where farm histories have been summarized, in most areas over 50 per cent of the farms belonged to the nobility, gentry, or the Church, suggesting that even in these northern regions wider estate policies cannot be ignored.

Nevertheless, while it is possible to generalize about farm desertion in the later Middle Ages, local regional tendencies vary greatly: thus in Denmark economic reasons are pre-eminent, and a peak seems to have been reached after crises developing in the first half of the fourteenth century, with maximum desertions after this, perhaps even between 1375 and 1400; in Sweden the evidence is poorer, but the peak may have been later, in the fifteenth century, while in Norway, perhaps because of both climatic and economic marginality, the whole process was protracted throughout the two centuries after 1300. On balance it is the fifteenth century which is the period when deserted farms are registered, pointing towards the later fourteenth

century and the first half of the fifteenth century as the key period. Recolonization appears to have been a general phenomenon of the sixteenth century. This rather abstract picture fails to present real images of the traditional farmsteads of Scandinavia; half-timbered, with dwelling, barns, byres, and stables ranged around a single great square of polygonal courtyard in Denmark and southern Sweden, with two courtyards in much of south central Sweden, with the stable between the two, because the 'horse is a creature between man and beast', breaking into loose conglomerations of separate buildings further north and in Norway. Most of the surviving buildings, except for a few distinctively Scandinavian Romanesque churches, are recent. Time, decay, and fire wreak steady havoc with timber buildings, but archaeological work shows the medieval antiquity of these building traditions (Hauglid, 1970; Steensberg and Østergaard Christensen, 1974).

Farmsteads and steadings which were isolated normally practised some cultivation. In remoter, more constrained environments this involved the cultivation of grain (bread grain and drink grain) virtually as a garden crop on heavily cultivated land, together with legumes and vegetables which could not be provided by nature, while even today the natural berry crop of the woodlands and heaths remains an item in Scandinavian and central European diet. Precious cattle, pigs, and fowl were produced and sustained by vast labour, but the greater the marginality the more that a subsistence economy emerged which included hunting as well as gathering. This is one dimension: there was, however, another; upland and mountain pastures, marshlands, and sea-coasts were exploited by lord and peasant alike. The variability of pasture for sheep and goats, cattle and horses, stimulated two developments: first, transhumance, the movement of stock, generally over substantial distances, which Braudel has documented from late medieval sources in Spain, France, Italy, and the Balkans (Braudel, 1972: 85–102); in northern Europe, where the distances were often less, the upland summer pastures were characterized by specific types of place-name, *shieling, airigh, ruigh, booley, saeter* (Adams, 1976: 134–4), where milking and cheese production were pursued and where temporary shelters were established (Smith, 1967: 288–9). These, as work on Scandinavia shows most clearly, could, as a result of manuring by dung and urine, prove attractive for the establishment of more permanent settlements (Smith, 1967: 236). Nevertheless, in England there appears to be little evidence to suggest that such names are generally embedded in areas where mature settlement has appeared, implying that the more specialist land use is subject to environmental control although the outlying portions of Anglo-Saxon estates in the Midlands of England suggests that there may be deep-rooted causes.

Secondly, there were many specialist dependent enterprises, the vaccaries and bercaries of secular and ecclesiastical landowners, cattle and sheep stations, stud farms, enterprises for the extraction of minerals, lead and gold, iron, tin, and copper, enterprises which were more than mere subsistence production, and looked towards the wider markets. In this, the remoter wilderness areas of both insular and continental Europe were procurement zones, where specialist gatherers, from habitat and terrain, collected or extracted natural products, furs and resins, ivory and skins, precious and semi-precious stones and gold and other minerals, for consumption in richer areas. Networks of administration and war, pilgrimage and trade, literacy, intellect, and enquiry bound all areas of rural Europe into a series of functional systems.

## Demographic Change

Estimates of the population of Europe between AD 500 and 1500 are problematic. A limited number of documentary sources provide records of households or houses, the numbers taxed or the numbers paying rent but, illuminating as these may be, these are, in practical terms, mere fragments, and synthesizing them into local, regional, national pictures is near-impossible (Slicher van Bath, 1963: 77–97; Pounds, 1994: 125–63; Abel, 1980: 21–3, 34–43, 92–5; Dodgshon and Butlin, 1978: 87–9, 123–6, 199–237; Dodgshon and Butlin, 1990: 70–3, 93–5). Even Domesday Book, giving some information for over 13,000 settlements in England in 1086, with about 300,000 landholders, provides no accurate assessment of population. Multiplying this crude estimate by 3.5 (J. C. Russell's assessment of household size) gives a gross total of somewhat over 1 million, but household sizes of 4.5 or even 5.0

(P. D. A. Harvey in Hallam, 1988: 48) lifts this to 1.5 million. However, Postan notes that the numbers of households actually present in 1086 may be 50 per cent higher than the number of tenants actually listed (Postan, 1972: 28), which would suggest a gross population of 2.25 million (cf. Harvey in Hallam, 1988: 46–9).

An estimate in excess of 2 million is reasonable, and an even higher figure is not excluded. Campbell suggests that rates of population growth after this date could be between 0.5 and 1.3 per cent per annum (Dodgshon and Butlin, 1990: 93): a rate of 0.5 per annum applied to a base population of 1.5 million and sustained to 1300 would give 3.78 million, and to a base population of 2.25 million would have given a population of some 6.54 million. These appear to be realistic estimates, but growth rates in excess of 0.5 need only have been sustained for short periods or within particular regions, and were probably balanced by slower rates or even declining local populations in other areas. However, assuming a rate of 1 per cent per annum before 1250 and 0.5 per cent between 1250 and 1300, the 2.25 million in 1086 would have given 11.5 million by 1250, and 12.75 million by 1300— both improbably high figures. Postan's estimate of the peak in about 1300 as 'nearer 7 millions than [Russell's] 3.7 million' must be compared with Hallam's estimates, which begin with a base of two million in 1086, are based upon a close examination of dozens of surveys falling between 1086 and 1328, and take account of regional variations, to suggest a range between 4.5 and 7 million by 1300 (Hallam in Hallam, 1988: 512–13, 536–7).

To attempt to cut through the many contradictions involved in estimating medieval populations, and the citation of a few examples of particular cases, which may or may not be typical, two pieces of information are needed: first, some gross measures are undoubtedly required, base points, to provide indications of how many folk were actually present; this is why the figures for England in 1086 have been addressed in some detail, but, secondly, the likely pattern of broad trends between AD 500 and 1500 is more important for this is particularly useful in the context of settlement evolution and encourages focus upon rather different issues. Figures in Table 4.1, mostly derived from Abel, for the three largest west European countries provide both some useful gross indicators and a beginning.

B. K. Roberts

Table 4.1. Population trends 1200–1620 (millions)

| Year | France | Germany | England | Total |
|------|--------|---------|---------|-------|
| 1200 | 12 | 8 | 2.2 | 22.2 |
| 1340 | 21 | 14 | 4.5 | 39.5 |
| 1470 | 14 | 10 | 3 | 27 |
| 1620 | 21 | 16 | 5 | 42 |
| 1990 | 56 | 79.5 | 57[a] | 192.5 |

[a] UK.

*Source*: After Slicher van Bath, 1963: 80.

Van Bath summarizes what is a generally accepted picture:

a period of decline occurred at the time of the late Roman Empire and the barbarian invasions. The low point . . . probably came . . . between AD 543 and 600, when serious epidemics were raging. The upswing which follows AD 700 (the Carolingian era) sinks again in about AD 900. After the spectacular rise from about AD 1150 to 1300 there follows the decline of the fourteenth century; this is partly due to the Black Death which took such toll between 1347 and 1350. But this epidemic was only one of a long series which had already begun in the second decade of the fourteenth century. From 1400 to 1450 a fairly quick recovery took place, followed by a steady rise. (Slicher van Bath, 1963: 78)

It will be observed that the figures for England are in general accord with those discussed above, and that after 1200 the populations of France and Germany follow this broad model; such figures, applied to the general trend, provide the necessary fixation points. Further, by altering no more than one phrase within this description—replacing 'after the spectacular rise from about 1150 to 1300' with 'after a period of growth and consolidation between 1150 and 1300'—crucial questions are raised about the balance between this growth and consolidation, and the trends between AD 500 and 1000.

Sawyer formulated an important argument when he suggested that 'the rural resources of England were almost as fully exploited in the seventh century as they were in the eleventh and that although some settlements were established, or moved, in the late Anglo-Saxon period, the settlement pattern is, in general, much older than most scholars have been prepared to recognize' (1976: 2). He demonstrated that Domesday Book omits many places; in

Kent while some 347 localities are listed by name, another 159 places are known to have had baptismal churches which fail to appear in the record. While the Weald of Kent with its peripheral estate centres may be exceptional for the high number of omissions, this is a sobering view, and even if we assume that Domesday Book underestimates actual settlements by only one fifth and not one third, there are profound implications because for England the exceptional record of 1086, compiled twenty years after the crucial Norman conquest of 1066, represents a pivotal point. Sawyer's point appears substantively correct, and it is likely that Domesday Book either conceals or omits much evidence for the dispersed elements of settlement. His view is the basis for using the population estimate of 2.5 million instead of 1.5 in the calculations above, and must raise questions about population trends before 1086 as well as the nature of the increase after that date. For England, Smith's imaginative presentation of the data, the estimates, and the problems reveals the qualifications and caveats inherent in all such figures (Dodgshon and Butlin, 1978: 199–237, particularly table 8.1 and fig. 8.1), and a figure in excess of 4.5 million but below 5 million appears most probable for the last decade of the seventeenth century (and note here the contradiction with Abel's estimate in Table 4.1). Looking back from this point, post-Civil War, post-Restoration England, with improved and improving husbandry practices and increasing yields and powerful incipient industrialization, with the beginning of empire, must we indeed estimate a gross population for 1340 substantially in excess of this? In fact, given the extension of arable cultivation known in England, to heights between 350 and 400 metres in the north Pennines, a figure between 6.5 and 7 million is not an unbelievable total.

Further, while in England much is written of the *documented* reclamations after 1086, their areas are rarely vast, particularly if one excludes such distinctive forward movements as the reclamation of the Fenlands (see Miller and Hatcher, 1978: 30–1) and perhaps some areas in the uplands. On balance, in England it is all too easy to overestimate the contribution of colonizing movements after 1000 to the picture which emerges from post-medieval sources, and Oliver Rackham's map of Anglo-Saxon place-names containing the suffixes -*leah* (*ley*), *-hurst*, *-thwaite* and *-feld* provides clear evidence that in England the central settlement province, extending from the vale of York through the valleys, scarplands, and low plains of the Midlands to Somerset and Dorset, was, as Domesday Book confirms, already substantively cleared of woodland by the middle Saxon period, a conclusion in accord with Sawyer's postulation (O. Rackham in Woodell, 1985: fig. 3.23; Darby, 1986: fig. 64). The fact that these developments left firm imprints upon the settlement landscapes of the nineteenth century (B. K. Roberts, S. Wrathmell, and Stocker in Fridrich *et al.*, 1996: fig. 2) emphasizes their importance.

Slicher van Bath (1963: 77–90), Abel (1980: 21–3), and Pounds (1994: 125–63) include discussions of material having a bearing on the medieval population of Europe, but only Slicher van Bath, using older sources, notably Bennett and Russell, attempts an overall assessment of long temporal trends for a number of modern nation-states. In part this is fashion, for historical scholarship, aware of the fundamental deficiencies of the basic sources, eschews such overviews, and relies upon local studies where these can be more effectively taken into account. As Pounds notes, 'a familiar study of the medieval population of Europe, excluding "Slavia" (whatever that means) gives 23.2 million in AD 800 and 28.7 two centuries later. For neither of these totals, nor for the regional figures upon which they are based, is there a shred of acceptable evidence' (Pounds, 1994: 144). If we accept the common view that in temperate Europe population, in general, increased during the centuries before the famine years beginning in 1314, and then declined, to begin to recover at some stage during the fifteenth century, then each country presents crucial questions concerning rates of increase and decrease, the precise chronology of expansion, peak, and decline, the actual densities achieved at the peak and the presence—or not—of a Malthusian pressure on available resources. Medieval taxation records show a deplorable tendency to become a fixed annual render, while even weather changes which induced famines were by no means always uniform in their effects from region to region (Pounds, 1994: 146–7). Abel puts another perspective on the problems: assume that in the year 1300 the area contained within the 1930 boundaries of Germany had 13 million inhabitants, that is 24 to the square

kilometre. This may well be a minimal figure. The estimated population density of France at the same time was 35 to the square kilometre, that of Flanders at least 60 per square kilometre (Abel, 1980: 41). Significantly Pounds cites comparable figures for two seigneuries in Bas Languedoc: Vézenobres between 1295 and 1321 rose from 33.9 to 36.6 per square kilometre, while Nogaret between 1304–6 and 1321–2 rose from 26.1 to 32.2 per square kilometre, figures in accord with the figure Abel cites. Table 4.2 summarizes in forty-year jumps the changes in the population densities of the Low Countries and Burgundy between 1360 and 1540 and is based on listings of hearths. The way this table is presented exaggerates the gaps in the records, but the fact remains, only for Hainault is the curve near-continuous. Nevertheless, we are left with a further problem: if these are indeed households, then by what figure or figures—for each region may vary—must each total be multiplied in order to arrive at an estimate of the population per square kilometre? How many people were there to each household? If we adopt four, a conservative estimate, then the trend for Hainault will be 26.8, 22.0, 24.0, 18.0, and 24.0, so that overall these figures for west Belgium are significantly below those for Flanders cited by Abel. Common sense suggests that all such figures must be treated with scepticism rather than caution. A glance at Laslett's extensive analysis of the *Household and Family in Past Time* (1972) can only increase doubts, but this is not to dismiss the efforts of those scholars who seek to estimate both local and national populations for the medieval centuries. As Slicher van

Table 4.2. Hearths per square kilometre in the Low Countries and Burgundy, 1360–1540

|          | 1360 | 1400 | 1440 | 1480 | 1520 |
|----------|------|------|------|------|------|
| Brabant  |      |      | 8.8  | 8.5  | 9.0  |
| Hainault | 6.7  | 5.5  | 6.0  | 4.5  | 6.0  |
| Dijon    |      | 3.7  | 4.0  | 6.0  |      |
| Beaune   |      | 3.0  | 3.2  |      |      |
| Auxois   |      | 2.9  | 1.5  |      |      |
| Châtillon|      | 1.5  |      |      |      |

*Source*: Pounds, 1994: fig. 4.9.

Bath notes, 'Better a few figures—which although not entirely reliable, give at least *some* guidance—than no figures at all' (1963, 78).

If the general European trends sketched above do indeed contain grains of truth, then they must be related to settlement tendencies: village desertion and shrinkage, and some measure of retreat from marginal land all appear to follow the down-swing of climate, the associated famines, and the advent of plague cycles, which are chronologically located in the fourteenth century. The world which reassembles itself after these cataclysms lays new foundations, for which one icon is the discovery of the New World of 1492. Less well-known but no less important was the publication in 1477 in Bologna, Italy, of an atlas containing the maps of Claudius Ptolemy, compiled from a manuscript prepared in Alexandria as early as AD 120. Various other editions quickly followed. Not only did this atlas imply that the world was a globe—a fact long known—but the places included were given co-ordinates, and the imposition of a grid upon the known world represented an important change in perception—a crucial renaissance in thought. The expansion of established nucleations, the multiplication of villages and hamlets, and the spread of single farmsteads are without doubt associated with the two or three centuries before 1300, though we should be aware of the crudity of our models, for the operation of the four processes which form the theme of this discussion, continuity and cataclysm, colonization and fundamental economic change, have always operated very unevenly in space and time, so that any synoptic view is doomed to failure. This argument has been presented because it raises important questions of perception: whatever the rate of growth between 1100 and 1300, it was absorbed in two ways, by the intensification of occupation in areas of old settlement, and by the expansion of surplus into the frontiers, both internal and external. Nevertheless, arguments are locked into models which are generated in spite of severe limitations upon the quantity of evidence (see R. H. C. Davis in Holt, 1986: 15–29). Silence is proof of nothing; the point was—perhaps innocently—made by a student many years ago: 'little is known of the historical geography of the Dark Ages, but much can happen in the dark'!

## Conclusions

To summarize is not easy: to write a history of medieval settlement in Europe

means including and setting side by side the religious, political, feudal, and social history of several nations. It is necessary to tabulate the various influences which resulted in the attainment of different standards in individual countries, to analyse their interaction and define the results; to take into account local traditions, tastes, customs, rules imposed by the use of materials, trade links, and the genius of individuals who influence events either by hurrying them on or by causing them to turn aside from their natural courses. Finally it is necessary to keep under observation the ceaseless investigations of a civilization in process of formation, and to immerse oneself in the religious and philosophic trends of the Middle Ages.

These words, by Eugène-Emmanuel Viollet-le-Duc, were written by a craftsman-scholar who had got closer than most to the medieval spirit, for he was, between 1844 and 1864, engaged in the restoration of the most prestigious and venerable of French cathedrals—that of Notre-Dame. He was writing not of settlement, but of medieval architecture, but his opinions remain as true today as when they were written. Any view of the rural settlement landscapes will inevitably be partial, fragmented, localized, but three concluding points can be made: first, the Mediterranean–Temperate division of Europe remains as a persistent thread throughout all discussion, regions with contrasting *genres de vie*; secondly, within each of these broad zones, while subtle, local, and often important differences are always manifest, radically affecting lifestyles, there are underlying similarities present in social and economic structures and in landscapes which, at least at a superficial level, present a remarkably uniform picture; thirdly, there are some grounds for suspecting that these underlying similarities were already present by the Carolingian period (to take but one example, the rise of Romanesque architecture in the centuries between 900 and 1100). From the continental interior of Russia to the Atlantic coasts of Ireland, from the splendours of Rome and Ravenna to Kirkwall in Orkney and the intricate wood-carvings of Norway, this reveals stylistic developments and mixtures which speak of intellectual cohesion wrought by sustained contacts and the movement of ideas.

REFERENCES

ABEL, W., *Agricultural Fluctuations in Europe from the Thirteenth to the Twentieth Centuries*, trans. Olive Ordish (London, Hamburg, and Berlin, 1966, 3rd edn. 1978, 1980).

ADAMS, I., *Agrarian Landscape Terms: A Glossary for Historical Geography*, Institute of British Geographers, Special Publication 9 (London, 1976).

ARNOLD, C. J., *Roman Britain to Saxon England*, (London and Sydney, 1984).

BARROW, G. W. S., *The Kingdom of the Scots*, (London, 1973).

BERESFORD, M. and HURST, J., *Deserted Medieval Villages*, (London, 1971).

—— *Wharram Percy: Deserted Medieval Village*, (London, 1990).

BOARDMAN, J., GRIFFIN, J., and MURRAY, O., *The Oxford History of the Classical World*, (Oxford, 1986).

BRAUDEL, F., *The Mediterranean and the Mediterranean World in the Age of Philip II*, being a translation of the 2nd rev. edn. of 1966 (London, 1972).

CHAPELOT, J., and FOSSIER, R., *The Village and House in the Middle Ages*, (London; tr. from 1980 French edn., 1985).

CORNELL, T., and MATTHEWS, J., *Atlas of the Roman World*, (Oxford, 1982).

CRASTER, H. E., 'The Patrimony of St. Cuthbert', *English Historical Review*, 69 (1954), 177–99.

CUNLIFFE, B., *The Oxford Illustrated Prehistory of Europe*, (London, 1994).

DARBY, H. C., *A New Historical Geography of England*, (Cambridge, 1973).

—— *Domesday England* (Cambridge; paperback of 1977 hardback, 1986).

DAVENPORT, F. G., *The Economic Development of a Norfolk Manor, 1086–1565* (Cambridge, 1906).

DAVIES, W., *An Early Welsh Microcosm: Studies in the Llandaff Charters*, Royal Historical Society (London, 1978).

DIXON, P., *The Making of the Past: Barbarian Europe*, (Oxford, 1976).

DODGSHON, R. A., *The European Past: Social Evolution and Spatial Order*, (London, 1987).

—— and BUTLIN, R. A., *An Historical Geography of England and Wales*, (London, 1978).

—— —— *An Historical Geography of England and Wales*, 2nd edn. (London, 1990).

DUBY, G., *Rural Economy and Country Life in the Medieval West*, (London, 1968), trans. Cynthia Postan from the French edn. (1962).

DUSSART, F. (ed.), *L'Habitat et les paysages ruraux d'Europe*, Les Congrès et Colloques de l'Université de Liège, 58 (Liège, 1971).

FAULL, M. L., *Studies in Late Anglo-Saxon Settlement*, Oxford University Department of External Studies (Oxford, 1984).

FINBERG, H. P. R. (ed.), *The Agrarian History of England and Wales*, (Cambridge, 1972).

FOWLER, P. J., *Recent Work in Rural Archaeology*, (Bradford on Avon, 1975).

FRIDRICH, J., KLÁPŠTE, J., SMETÁNKA, Z., and SOMMER, P. (eds.), *Ruralia I* (Conference Ruralia I, Prague, Pamatky Archaeologické—Suppl. 5, 1996).

GISSEL, S., JUTIAKKALA, E., Österberg, E., SANDES, J., and TEITSSON, B., *Desertion and Land Colonization in the Nordic Countries c. 1300–1600* (Stockholm, 1981).

GOODY, J., THIRSK, J., and THOMPSON, E. G., *Family and Inheritance: Rural Society in Europe 1200–1800* (1976; Cambridge, 1978).

HALLAM, H. E. (ed.), *The Agrarian History of England and Wales*, ii. *1042–1350* (Cambridge, 1988).

HAUGLID, R., *Norwegian Stave Churches*, (Oslo, 1970).

HEDEAGER, L., *Iron-Age Societies*, (Oxford, 1992).

HEIDINGA, H. A., *Medieval Settlement and Economy North of the Lower Rhine: Archaeology and History of Kootwijk and the Veluwe (the Netherlands)* (Assen, 1987).

HELMFRID, S. (ed.), 'Morphogenesis of the Agrarian Cultural Landscape', *Geografiska Annaler*, 43 (1961), 1–2.

HILTON, R. H., *A Medieval Society*, (London, 1966).

HODGES, R., *Early Medieval Archaeology*, Headstart History Papers (Bangor, 1991).

HOLT, J. C. (ed.), *Domesday Studies*, (Woodbridge, Suff., 1986).

HOMANS, G. C., *English Villagers of the Thirteenth Century*, (New York, 1960).

HÖPPE, G. (ed.), 'Landscape History', *Geografiska Annaler*, Ser. 70 B (1) (1988).

JEPPERSEN, T. G., 'Middelalder-landsbyens Opståen', Fynske Studier XI, Odense Bys Museer (Odense, 1981).

KIRK, M., *The Barn: Silent Spaces*, (London, 1994).

KRENZLIN, A., 'Blockflur, Langstreifenflur und Gewannflur als Ausdruck agrarischer Wirtschaftesformen in Deutschland', in *Géographie et histoire agraires*, Actes du Colloque International (Nancy, 1959), 353–69.

—— and REUSCH, L., *Die Entstehung der Gewannflur nach Untersuchungen im nördlichen Unterfranken*, Frankfurter Geographische Hefte, 35 (Frankfurt, 1961).

LAMB, H. H., *Climate History and the Modern World*, (London and New York, 1982).

LASSUS, J., *The Early Christian and Byzantine World*, (London, 1967).

LEWIS, A. R., 'The Closing of the Medieval Frontier', *Speculum*, 33 (1958), 475–83.

McCARTHY, M. R., and BANKS, C. M., *Medieval Pottery in Britain AD 900–1600* (Leicester, 1988).

McEVEDY, C., *Penguin Atlas of Medieval History*, (Harmondsworth, 1961).

MAYHEW, A., *Rural Settlement and Farming in Germany*, (London, 1973).

MILLER, E., and HATCHER, J., *Medieval England: Rural Society and Economic Change 1086–1348* (London, 1978).

MORRIS, R., *Churches in the Landscape*, (London, 1989).

PARRY, M. L., *Climatic Change, Agriculture and Settlement*, (Folkestone, 1978).

PERCIVAL, J., *The Roman Villa*, (London, 1976).

POSTAN, M. M. (ed.), *The Cambridge Economic History of Europe*, i. *The Agrarian History of the Middle Ages*, (Cambridge, 1966).

—— *The Medieval Economy and Society: An Economic History of Britain 1100–1500* (London, 1972).

POUNDS, N. J. G., *An Economic History of Medieval Europe*, (London and New York, 1994).

—— and BALL, S. S., 'Core-Areas and the Development of the States System', *Annals of the Association of American Geographers*, 54 (1) (1964), 24–40.

RANDSBORG, K., *The Viking Age in Denmark*, (London, 1980).

ROBERTS, B. K., *The Making of the English Village*, (London, 1987).

RUSSELL, J. C., *Medieval Regions and their Cities*, (Newton Abbot, 1972).

SAWYER, P. (ed.), *Medieval Settlement*, (London, 1976).

SCHRÖDER, K. H., and SCHWARZ, G., *Die ländlichen Siedlungsformen in Mittel-europa: Grundzüge und Probleme ihrer Entwicklung*, Forschungen zur Deutschen Landeskunde, 175 (Trier, 1978).

SEEBOHM, F., *Tribal Custom in Anglo-Saxon Law*, (London, 1911).

SLICHER VAN BATH, B. H., *The Agrarian History of Western Europe*, trans. Oliver Ordish; original publication in the Netherlands (London, 1963).

SMITH, C. T., *An Historical Geography of Western Europe*, (London, 1967).

SMYTH, W., and WHELAN, K., *Common Ground: Essays on the Historical Geography of Ireland*, (Cork, 1988).

SPORRONG, U., *Mälarbygd: Agrar bebyggelse och odling ur ett historisk-geografiskt perspektive*, Meddelanden series B 61, Kulturgeografiska Institutionen Stockholms Universitet (Stockholm, 1985).

STEENSBERG, A., and ØSTERGAARD CHRISTENSEN, J. L., *Store Valby* Det Kongelige Danske Videnskabernes Selskab, Historisk-Filosofisk Skrifter 8, 1, Part 1 (Copenhagen, 1974).

THOMAS, W. L., *Man's Role in Changing the Face of the Earth*, (Chicago, 1956).

TODD, M. (ed.), *Studies in the Romano-British Villa*, (Leicester, 1978).

UHLIG, H., and LIENAU, K., *Flur und Flurformen, Materalien zur Terminologie der Agrarlandschaft*, with English translation (Giessen, 1967).

—— —— *Die Siedlungen des Ländlichen Raumes*, with English translation (Giessen, 1972).

WARNER, P., *Greens, Commons and Clayland Colonization*, Department of English Local History, Occasional Papers, 4th ser. 2 (Leicester, 1987).

WILLIAMSON, T., *The Origins of Norfolk*, (Manchester and New York, 1993).

WOODELL, S. R. J., *The English Landscape, Past, Present and Future* (Oxford, 1985).

Werkgroep Brinken, *Brinkenboek: on verkenning van der brinken in Drenthe*, (Assen, 1981).

# Chapter 5

# Towns and Trade, 400–1500

## A. Verhulst

## The Legacy from Earlier Times (AD 300–700)

As Chapter 2 has shown, the city, both as a geographical phenomenon and as a way of life, was introduced into Mediterranean Europe by the Greeks and spread through western Europe by the Romans. Its diffusion through the latter paralleled the expansion of the Roman Empire. That is particularly true for north-west Europe, whose geographical position between the Rhine and Britain was particularly favourable for the implantation and development of towns like Paris, Trier, Rheims, Amiens, Cambrai, Arras, and several others. Towns and military camps were also established along the Rhine and the Danube, notably at Xanten, Neuss, Cologne, Bonn, and Mainz, and at Augsburg, Regensburg, and Passau. The same was true in England, where the coloniae of York, Lincoln, Gloucester, Colchester, and London were founded, as well as a large number of chief towns of *civitates* (for example Carlisle, Wroxeter, Leicester, Winchester, Exeter, Canterbury) and other important towns (Bath, Ilchester, Chelmsford). Their foundation goes back to the first and second centuries AD. In the third century AD, particularly at the end of the century, and during the fourth

century AD, the crisis of the Roman Empire brought this trend of urbanization to a halt. Oppressive taxation and fragmentation of the economic system led to towns being abandoned by their dominant classes. The dangers associated with the barbarian invasion caused cities to become military strongholds, notably by the construction of defensive walls around inhabited centres which were often reduced to just a few hectares apiece. Within the city, administration and political power increasingly came into the hands of the local bishop, who during the sixth and seventh centuries AD would play an essential role in ensuring continuity from the town of classical times to that of the early Middle Ages.

In Italy and on the northern shores of the Mediterranean in general, the urban crisis did not start seriously until after the middle of the sixth century AD. It was expressed in other ways from what happened in north-west Europe. Certainly urban population declined dramatically, as at Rome which, it has been argued, lost 90 per cent of its inhabitants. The inhabited areas inside cities also shrank, for example at Bologna from 70 ha. to 25 ha. But urban life continued and most Roman towns survived. Administrative functions, for both the towns and their surrounding countryside, remained firmly established in urban places, even under the Lombard kings, and involved not only the bishop and his clerical entourage but also lay people. Members of the aristocracy did not leave the cities and continued to draw in all the resources they needed from low-lying surrounding areas where their estates were located. These two factors, the administrative function of the city and the fact that it was the residence of the aristocracy, assured the continuity of urban life from an economic point of view, especially since the town remained as the dominant point for the consumption of resources that originated in the surrounding countryside. This is why some scholars have remarked that it was the surrounding hinterland which assured the continuity of urban life in Italy. In many cities, especially those in northern Italy, this continuity is expressed in the survival of the Roman grid in the urban plan even to the present day.

In southern Gaul, at Toulouse, Bordeaux, Bourges, Poitiers, and other towns south of the Loire, the elements of continuity in the urban plan (including walls) that were observed in Italy were equally present, but to a lesser degree than in Italy. The nadir of urban decline in southern Gaul was in the eighth century AD, but at this time the first signs of urban renewal were already to be seen in Italy and to the north of the Loire. The real problem of continuity from classical times to the early Middle Ages is expressed most clearly, at least with respect to urban phenomena, in the territory between the Loire and the Rhine, where the presence of the Romans from the first to the fourth centuries was certainly real but not as complete as in Mediterranean Europe, and where the urban crisis was both longer and more serious (from about AD 300 to 700). With respect to both time and space, the issue of urban continuity was expressed in differing ways from region to region. There was rarely functional continuity but spatial continuity often occurred, since the Roman city or its ruins, or more often just a simple fortified centre (*castrum, castellum*) or non-fortified place (*vicus*) served as a point of attraction for a new population. This occurred in the case of small centres (Namur, Huy, Koblenz) and larger towns (Trier, Metz, Cologne, Bonn, Maastricht), or with respect to just a fragment of the former urban settlement or even a point on its periphery where suburban churches or cemeteries formed the subsequent point of attraction. In this way continuity was purely spatial since the material remains of Roman occupation simply served as anchoring points for entirely new settlements which were created in the ninth century. This was the case in areas far from the middle sections of the Rhine and Meuse valleys, notably along the Danube (Augsburg, Regensburg, Passau, Vienna) and the Scheldt (Ghent, Antwerp), and in England (London, York, Canterbury).

## New Forms of Urban Development (AD 700–850)

At the end of the seventh century, new centres for long-distance trade appeared around the shores of the North Sea. These were often described as emporia in the Latin texts of the time or had the Germanic word *wik*(*wiic, wich*) as a suffix. The most important and certainly the best-known by virtue of its archaeological excavations is Dorestad, to the south-east of Utrecht, at the point where the Lek separates from the Rhine. Others include Quentovic, on the south bank of the Canche estuary to the south of Boulogne; Hamwic, which was to be the site of a medieval district in Southampton on the south coast of England; the Strand area of London; the settlement discovered on Domburg beach, on the north-west coast of Walcheren island in Zeeland; and Ipswich in East Anglia. These towns had several points in common, namely their location at the protected estuarine site not far from a coastline which also served as a territorial frontier for a kingdom; their temporary character, since most disappeared between AD 850 and 950 without giving birth to a medieval town; and the fact that they would be succeeded during the tenth century by one or several towns in their immediate surroundings, often less important or of quite a different type. They were important collection points for tolls on water-borne goods and their traders, who included many foreigners, were protected by the Crown, whose local representative was a prefect or count (*prefectus* or *comes*) and who was also the customs controller along the local section of coast. These places did not function as market or administrative centres for the surrounding countryside, with which they were very poorly linked, even though some important merchants, who were often larger farmers from the surrounding countryside, had a more or less temporary residence there. The function of these places was rather as a 'port of trade' or as a 'gateway', to which luxury products (such as wine, textiles, arms, slaves, amber, skins, furs) were brought over great distances. Such goods were destined for members of important families and for the aristocracy. Some of these products originated in workshops located on large rural estates in the interior, others in these emporia themselves, although they were not primarily industrial centres. Their urban structure is not clearly known, although a street along a quay, either natural in origin or constructed of wood, seems often to have been their principal axis. This street would have wooden houses, which also served as warehouses, along the waterfront. Further inland were to be found farms that were surrounded by craftsmen's workshops. Such a settlement would not have a market-place, since this was not part of its function. It was not fortified nor was it the site of an abbey or the seat of a bishop, except on a temporary basis at Dorestad. However, abbeys were careful to acquire some buildings in these emporia from which their trading agents could acquire the luxury goods brought from overseas and purchased by the abbeys in the same way as the king or members of the aristocracy. With a surface of only several tens of hectares, the population of each of these towns would not have exceeded one thousand people. The decline and disappearance of these emporia, whose changing fortunes may be traced from the number of coins they minted and which found their way to places as far apart as Sweden, the lower Rhône, Northumbria, and Liguria, has been explained by a number of specific factors, ranging from invasion by the Normans to changes in local hydrology. Their strong association with the changing fortunes of the Frankish kings and their very special and almost exclusive trading function leads one to believe that more general causes, linked to economic and political trends, explain both their success and their decline. Certainly, it is clear that these emporia did not play a direct role in the creation of the medieval city.

## The Formation of the Medieval City (AD 850–1000)

The fact that there is no direct continuity between the Carolingian emporia around the North Sea and genuinely medieval towns does not indicate that these latter settlements dated only from the tenth and eleventh centuries, as Henri Pirenne believed, and which he explained by reference to a revival of international trade at that time. The history of the Carolingian emporia shows that commercial revival occurred much earlier, although on a more modest scale, towards the end of the seventh century. This trading renaissance was simply the replacement of a Mediterranean trading system by an international trading circuit organized around the North Sea. This shift to the north was not, as the bold theory of Pirenne proposed, the negative effect of the Arab conquest of the western Mediterranean basin. It seems that it was part of a long economic and political process which extended over several centuries. After the instability of the barbarian invasions and the plague of the sixth and seventh centuries, north-west Europe and Italy experienced a period of economic and demographic growth from the eighth to the tenth centuries, whose effects were first experienced in the countryside: land was cleared for cultivation, large estates with numerous peasant farms were set up, and these also supported many rural craftsmen. A new political structure, whose economic impact was underestimated by Pirenne, was set up initially in north-west Europe. This was the Kingdom of the Franks, which before the end of the eighth century was also to make its appearance on the northern shores of the Mediterranean.

Alongside the Carolingian emporia, whose very special characteristics and indirect role in medieval urbanization must not be overlooked, came a number of new urban centres which were smaller but more durable. These originated principally during the ninth century, that is before the Norman invasions whose end (*c*.900) was considered by Pirenne to be one of the essential conditions for the urban renaissance of the tenth and eleventh centuries. Modest river ports (*portus*) appeared along major rivers, notably the Meuse (Maastricht, Huy), the

Scheldt (Valenciennes, Tournai, Ghent), the Rhine (Xanten, Duisburg, Cologne, Bonn, Mainz). Some of these, such as Maastricht and several centres on the Rhine, perhaps owed their commercial role to the influence of emporia located closer to the sea, such as Dorestad. It would be an exaggeration to consider that they were all involved in international trade, at least from the beginning. Unlike the emporia, these *portus* seem to owe their commercial character to their trading links with the surrounding countryside. The same explanation is probably valid for many towns on the Po plain in Italy, to which cereals, wine, and oil were brought from their rural hinterlands. These products were exchanged in these northern Italian towns for salt, brought initially from the Po delta by the knights (*milites*) of Comacchio, and later for glassware and spices from Venice. During the ninth and tenth centuries the towns of the Po plain progressively became involved in international trade which, from the ninth century onwards, extended through Venice to Byzantium and the Islamic Near East. Venice exchanged wood, arms, and slaves for spices and precious fabrics. Other Italian ports, of which Amalfi was the most dynamic, also served as bridgeheads for trade with the East during the ninth and tenth centuries. In western Europe this connection between the regional trade of river ports and flows of international trade seems to have been held back for half a century or even a century, probably by the Norman invasions. Those invasions were not over until the tenth century and it was not until the end of that century that merchants from Flanders, the Meuse valley, and the Rhinelands made their appearance in London. The settlement of the Vikings in Normandy and in eastern England contributed to urban foundations, in the form of York and Dublin during the early tenth century in the western part of the Viking realm, and at Staraya Ladoga, Novgorod, and Kiev to the east. Between these two extremes, a great phase of urban development occurred on the northern and southern shores of the Baltic, and on the island of Gotland. This process renewed some of the Scandinavian merchant settlements of the ninth century, such as Haithabu (soon to be replaced by Schleswig) on the Schlei isthmus on the Baltic coast of Jutland, and Birka, which was soon to be overtaken by Sigtuna, on Lake Mälar close to Stockholm. In central Europe similarly the tenth century was the period when a particular type of urban development proliferated, which was characterized by the emergence, near an existing fortification (*Burg*, Latin *civitas*, *urbs*, *castrum*), of a quarter (*suburbium*) of merchants and/or craftsmen. Hamburg, Würzburg, and Magdeburg are examples between the Rhine and the Elbe, with further instances being Mikulčice and Stare Mesto in Moravia, and Prague in Bohemia. A similar type of semi-seigneurial and semi-urban settlement developed in Pomerania and Poland (Wollin, Stettin) as well as in Russia (Smolensk, Pskov).

In western Europe, between the Loire and the Rhine, as well as in England, a kind of feudalization took place which recalls, albeit with some differences, the kind of development occurring in central and eastern Europe. Fortifications were established in most of the river ports (*portus*) that had been created during the ninth and tenth centuries, or more precisely adjacent to their central cores which were settled by merchants. These fortifications usually consisted of a fortified residence in the form of a tower or keep, initially built of wood but later of stone, and a district of craftsmen and other people who were in the service of the lord of the castle. Such lords were often territorial princes, or, in the case of England, the king. These urban castles were, in effect, the material expression of the hold that the feudal lords exerted over pre-existing merchants' settlements. At the same time, and especially where mercantile settlements had not existed before, these castles encouraged local commercial activity, since the agricultural products of the seigneurial estate were not entirely consumed by the seigneur and his entourage and hence were available as a surplus that could be sold by the merchants who lived in the *suburbium* beyond the castle.

The great difference between the seigneurial type of town, which was often founded by a deliberate act of the sovereign in central and eastern Europe, and the kind of town encountered in the west, related to the emancipation of the inhabitants of a merchant quarter (*suburbium*), many of whom were not initially free or were servants of a church or an abbey. Emancipation was achieved earlier in western Europe, sometimes as early as the tenth century. In any case traces of a certain collective identity among these inhabitants with regard to the seigneur of the urban castle were in evidence from this time onwards, that is a century or sometimes two centuries before the merchant town was completely emancipated and endowed with its own institutions.

# The Medieval Town in its Prime (Eleventh–Twelfth Centuries)

## Typology and Topography

There was no single type of medieval town throughout Europe, hence urban typologies have been prepared to enable one to grasp the diverse reality of urban development and to better explain phenomena by identifying specific groups.

The typology proposed by Edith Ennen takes into account both the date and the origin of towns, as well as the geographical diffusion of the various types. The first type of town originated in classical times and is characterized by a certain unity and continuity into the Middle Ages. Such towns are typically found in Mediterranean Europe, with Italy offering the best examples; the Roman towns in Spain and Sicily that were conquered by Islam displayed too many major changes to belong to this first type.

The second type is composed of towns that were created in the early Middle Ages, without a preconceived plan, and grew up around a variety of different elements of which one or even several (such as fortifications, walls, Roman grid plan) would date from classical times. This type of town was characteristic of a good part of western Europe.

New towns, lacking features that go back to classical times or even prior to the year 1000, represent the third type, with two variants which are encountered in central and eastern Europe and in Scandinavia. These new towns might grow up spontaneously, without the intervention of higher authority, or they might be founded, planned and entitled to certain privileges from the moment of their creation.

Topographic factors, while not being a determinant in this typology, are none the less always present, most frequently in influencing the form of the ground plan of the town or the location of a stronghold somewhere in its territory. It is, however, impossible to establish a historical typology simply on the basis of the layout of towns, since regular urban plans were encountered both in Roman towns and in the planned towns of the twelfth century.

By contrast, the presence of some kind of fortified point on the territory of the town was at the heart of the typology drawn up at the start of the twentieth century by Henri Pirenne. This famous Belgian historian recognized the spatial duality of a seigneurial fortification and a commercial agglomeration as the characteristic of most medieval towns in Europe. This duality seemed to be more important and acceptable to historians when it was linked, following Pirenne's argument, to the origin of the town and to its functions. According to Pirenne, the military and protective function of a fortification—whether it was Roman in origin, as in the case of the *civitas*, or medieval, as in the form of a feudal *castrum* formed by a keep, would have attracted merchants to establish a settlement nearby. This quite simple argument, in which military functions are replaced by a more complicated explanation, is no longer accepted. Many medieval towns were preceded by the establishment of several nodes, which were not necessarily fortified, which served as points of attraction not because of their military character but because of their economic importance. Spontaneous towns, that is to say towns that were not specifically founded by charter, grew up around these points of attraction. These were the 'dominant centres' around which towns came into being.

In towns originating in Roman times and corresponding to the first type in the typology proposed by Edith Ennen, the classical forum maintained its role and the medieval community (commune) of the twelfth–thirteenth centuries often established its focus close to the old forum. However, new centres, which were generally religious in character, were created outside the walls during the earlier Middle Ages, for example by newcomers such as the Lombards in Italy. These centres became points of attraction which favoured fresh urban growth and were integrated into the urban area when new walls were built between the eleventh and thirteenth centuries. This was the case at Milan, Pavia, and Bologna.

In towns of the second type, which developed spontaneously from the ninth and tenth centuries onwards, especially between the Loire and the

Rhine, the 'dominant centre' was either the *civitas* or the keep. The *civitas* was the site inherited from classical times which the master of the town, typically the king or more often the bishop, had fortified or where he had established a sanctuary and used as his residence (for example, London, Cologne, Mainz, Utrecht, Arras). Feudal keeps were built from the tenth century onwards and were surrounded by a ditch (fosse) and a rampart equipped with palisades (Ghent, Bruges). The keep has sometimes, as at Ghent, been preceded by another dominant point such as an abbey within the urban perimeter. Elsewhere, as at Bonn, a sanctuary built in early medieval times outside the Roman walls attracted population and thereby determined the structure of the medieval town. These processes of attraction and expansion in medieval times were somewhat disorganized, and had to take the pattern of local watercourses into account which could be quite a complex matter on the north European plain.

The new towns of the third type in Ennen's typology, namely those which originated spontaneously, for example in Russia in the ninth and tenth centuries (Kiev, Suzdal, Vladimir), involved the coalescence of several dominant points (including a fortified kremlin) and to some extent remind us of the feudal type described in western Europe. Although being established at the foot of a kremlin situated on an opposite bank of the River Volchov from where foreign merchants had settled, the town of Novgorod was endowed with a regular street plan during its expansion phase in the tenth century. This plan comprised two major axes parallel to the river which were crossed at right angles by other streets. This street plan recalls the layout of Lübeck, a new town that was founded towards the middle of the twelfth century on a hill separating two rivers just before their confluence. Two long parallel streets along which perpendicular streets descended to the waterfronts linked a series of churches and other important buildings that were located at both ends and midway along the two main streets. Other towns founded on the southern shores of the Baltic Sea (Riga 1201, Rostock 1218, Stralsund and Danzig c.1230) also have a regular layout which is characteristic of a type of urban settlement that proliferated in western and northern Europe during the twelfth and thirteenth centuries. These were the *problaciones* in Christian Spain, the *bastides* of Gascony, and the colonial towns of Ireland, Wales, eastern Germany, northern Poland, and elsewhere.

## Functions

### Crafts and Industry

What, in functional terms, differentiated the medieval town, strictly speaking, from expressions of urbanism prior to the year 1000 was not the fact that the town was the headquarters of regional or international trade. Instead it was the transfer into these trading settlements of craft production (even on an industrial scale) which had flourished previously on great rural estates, either on the seigneurial farm or on peasant holdings. This transfer occurred around the year 1000 or during the eleventh century and represented the birth of urban manufacturing, which was long dominated by textile production. We believe that it may be explained by the disintegration of the estate system which began during the tenth century and speeded up in the eleventh century. This transfer was accelerated by the temporary increase in wool production in certain regions, such as the salt marshes on the coastlands of Flanders. Increased output reached such a volume that the wool could no longer be worked solely in the countryside. The urban merchants, who purchased wool directly from the producers and brought it to the towns, were illiterate and consequently were unable—at this stage—to co-ordinate the processing of wool in the countryside by means of an integrated putting-out system. Instead, and for a long time, the preparatory activities associated with working wool remained dispersed and partially located in the countryside, where they were undertaken by women. By contrast, the main operations, such as weaving, fulling, and dyeing were undertaken elsewhere, creating a division of labour which increased the quality of the wool. These latter stages were associated with urban craft guilds which were being organized from 1120–40 onwards. Production of wool, like other agricultural goods, was subject to feudal dues which had to be paid to local seigneurs. The wool could be transported to sites of seigneurial power, notably the urban *castra*, where surplus quantities were sold and thus contributed to the birth of a market.

## Trade

The market became one of the fundamental elements of the medieval town, strictly defined, and a feature which distinguished it from earlier forms of urban settlement, notably the Carolingian emporia in which markets do not seem to have been present. Urban markets served especially for trade between the town and its region. By comparison, the stagnation of international trade, as well as trade linked to emporia, during the late ninth and tenth centuries was followed by a recovery after the year 1000, especially in the great medieval cities. This resurgent international trade served international fairs. Some of these, like the fairs of Saint-Denis near Paris, went back to the early Middle Ages, but most were set up in the eleventh and twelfth centuries by territorial princes as protected meeting points for merchants operating over long distances. They were often located beyond towns, on their periphery or in places that could scarcely be described as urban, at least in their early phases.

Echoing the ideas of Henri Pirenne, what has been called the renaissance of long-distance trade after the year 1000, principally in the Mediterranean, was not only or simply the restoration of an earlier situation which had been temporarily interrupted by the Arab conquest of the western Mediterranean basin. This commercial expansion had been preceded by a small amount of international trade in the ninth century to the profit of the aristocracy and occurred especially in the North Sea; further expansion developed there and in the Mediterranean after the start of the tenth century.

In the Mediterranean this trade profited from old-established links which towns such as Venice, Amalfi, and Bari had with Byzantium. Other maritime towns around the Tyrrhenian Sea, such as Pisa and Genoa, managed to take military action to dispel the Muslim threat in the western Mediterranean during the eleventh century, and, following the example of Venice, established commercial privileges and the right to set up permanent trading posts (*fondacchi*) in numerous ports in the Orient, including Antioch, Alexandria, Caesarea, Beirut, and naturally Byzantium. These privileges were granted after the end of the eleventh century in recognition of assistance from these towns in the Crusades.

Italian cities imported expensive goods such as oriental silks, spices, ivory, and perfumes, and exported slaves and raw materials. With the exception of arms and woollen fabrics which they, in turn, had imported from western Europe, they did not export great quantities of manufactured products. Their export trade was not primarily supplied from their own urban manufacturing, as would be the case one or two centuries later. However, Lucca and Venice were exceptions to this rule since from the beginning their craftsmen had exported luxury goods, respectively golden thread and glass.

By contrast, in north-western Europe a great export trade destined for Italy and the shores of the Baltic developed from the high-quality textiles and metallurgical goods that were manufactured in the exporting towns themselves. Customs returns for London and for Koblenz show that this trade started towards the end of the tenth century. At this stage, it was principally limited to regions around the southern shores of the North Sea, involving on the one hand Flanders, the Meuse valley, and southern and eastern England and on the other the Rhinelands. It had begun at commercial sites which had emerged during the ninth century as regional ports (*portus*) on the coast and along the valleys of the Scheldt, Meuse, and Rhine, but which during the course of the tenth century had developed regional trade, and then towards the year 1000 acquired supra-regional trading functions. This commerce acquired its international character as high-quality goods which had been manufactured in urban centres, especially in Flanders (Ghent, Bruges, Arras, Saint-Omer, Douai, Ypres) and in the Meuse valley (Dinant, Huy, Liège) reached Italy by means of the Rhine and especially by the land routes which crossed eastern France.

The northern extension of international trade is only known because coins that had been struck in Flanders or at Rouen have been discovered in parts of Scandinavia. These coins were possibly related to the slave trade, because Rouen had been the distribution point for slaves since the Norman invasions. But practically nothing is known about trading activities with northern Europe, nor about the origin of numerous coins that had been struck in the east and were found in Sweden and Denmark. These discoveries suggest that these areas played an intermediate role between western and eastern Europe at this time. Frisian trade seems to have held up in northern Germany and in the Baltic until the eleventh century, but its relative decline in the twelfth century and in later years was to the advantage of Hanseatic traders.

## Social, Commercial, and Industrial Organization

Traders came from a variety of social origins in the towns of north-west Europe, with some having been former merchants for the abbeys and enjoying their fiscal privileges; others having been free merchants in Carolingian times; and yet others having been landowners who indulged in commerce. In the first half of the eleventh century, they began to group themselves in guilds. Often known as charities or fraternities, these guilds had a purely social function to begin with, notably to encourage mutual self-help and solidarity. This objective declined during the twelfth century as guilds, which at first had only been found in Valenciennes, Saint-Omer, Arras, and Ghent, also appeared in other towns in this part of Europe. Guilds had initially admitted artisans but gradually excluded them and only welcomed great merchants. They became closed associations, which tended to restrict trade in cloth and other important products for the benefit of their own members, and also to control the production of cloth by dominating and regulating the artisans' looms. In some towns these merchant guilds merged with associations of merchants known as 'hanses', which developed during the twelfth century especially at Ghent and Saint-Omer but also in other towns. Some of these local hanses entered into federations, and had the objective of reserving certain aspects of long-distance trade for the benefit of their members. They not only restricted membership to the richest merchants but also created a bond of solidarity abroad which was invaluable when their members were arrested and their merchandise confiscated because of unpaid debts, political pressure, or some other reason.

During the thirteenth century the hanses lost their leading function in international trade which, by this point, was expanding rapidly in response to new trading techniques. They reacted by becoming political organizations in the hands of the great merchants, rather like the former merchant guilds, and acquired the monopoly of power in some cities. A certain democratization occurred when artisans were admitted, especially weavers who had risen to the rank of independent entrepreneurs, together with small and medium-scale local merchants. This was paralleled by the progressive eviction of Flemish merchants from the English wool trade by Italians and members of Hanseatic League during the second half of the thirteenth century.

Corporations of artisans were recorded in France, England, Germany, and the Low Countries from 1100 onwards. They originated from the convergence of voluntary associations of artisans for religious, fraternal, and mutual assurance (hence the old name 'fraternity' for some corporations in the twelfth century) and the wish of the great merchants, veritable city fathers, to make use of these associations to control production and guarantee its quality. This wish was shared by the craftsmen, who restricted production to the benefit of their own members so as to exercise a kind of monopoly.

The social organization of Italian towns was different from what has just been outlined because of the specific demands of their widespread network of trading relations, notably in the Near East and also around the Black Sea. Initially, in the eleventh and twelfth centuries, temporary association (*commenda* and then *societas maris* at Genoa and *colleganza* at Venice) were established between sedentary partners who provided capital and merchant adventurers who had responsibility for selling the goods. These associations ensured that risks were shared and profits were divided in the ratio of 1:3 or 1:2.

Very different were the family 'companies' that were set up in the towns of the Italian interior, such as Florence, Piacenza, Milan, Lucca, and Siena, which reached their high point during the thirteenth century. These companies were formulated for two to twelve years but arrangements could be renewed many times across a whole century. They owed their existence and their success to the advances made in Italy in the use of writing, which permitted branches and correspondents to be set up in all the great towns of Italy and the East. Writing was of particular importance in trading activities with respect to the drawing up of bills of exchange. This originated from the fact that there were many different currencies in circulation in Europe which were converted into cash by money changers in each location. Written contracts of exchange that were approved by lawyers were in existence in Genoa as early as the mid-twelfth century. Their rapid diffusion was favoured by the fact that Genoese merchants needed access to cash in the market-places of the Near East and especially at the fairs of Champagne.

The different kinds of association that have just been discussed had a political as well as an economic and social role to play, in as much as they were instruments for acquiring power and exercising power. During the eleventh and twelfth centuries they contributed to the formation of urban laws which were distinct from the customary laws of the surrounding countryside and were better suited to a dynamic society based on money transactions. The merchant guilds, and particularly the collective identity that they engendered, were largely responsible for the fact that towns acquired their own statutes and distinctive institutions after 1066 (the date of the charter of freedom for Huy on the Meuse) and especially during the twelfth century. The bench of aldermen, sometimes in isolation and sometimes together with representatives of the legal community, became the political expression of the dominant class in the town. In towns controlled by bishops, a type that was particularly common in north-west Europe, this emancipation of the urban community was often achieved only after a harsh struggle with the ecclesiastical authorities. Soon, as in Flanders in 1127–8, the urban community, and especially the rich merchants who controlled and sustained it, would come to play a political role at the level of the state by allying with or opposing the local prince or sovereign. In Italian towns the situation was very different from the start, not only because of the fragmentation of political power in the peninsula, but above all because of the presence (sometimes under force) of the nobility in the town. This stood in contrast with what was found in the countries of north-western Europe. This nobility had given military support to Italian merchants in their colonial enterprises in the Near East and had seized the opportunity to enter into commercial operations. The surrounding countryside, the *contado* where the nobility had their landed estates, also came under urban control. However, this did not prevent communities of jurymen also developing in Italian towns, especially those in the north. Consuls emerged as urban leaders in most towns after the end of the eleventh century. The captains and armourers of the Italian fleets, upon which Pisa and Genoa depended for their colonial empire, formed the main focus of power. Their emergence was less a matter of obtaining urban privileges and defending liberties against the local seigneur, as was the case in north-west Europe, and more a matter of maintaining peace within the city. Rivalries between clans and noble families often threatened urban tranquility, which led many towns to select a chief magistrate (*podesta*) from among the military powers. The consul system also penetrated the towns of southern France as early as the first half of the twelfth century.

## The Medieval Town at its Peak: The Long Thirteenth Century: 1180–1315

### Demography

Approximate figures for the population of European cities start to be known for the fourteenth century and by virtue of their decline from that time onwards allow one to believe that the demographic optimum of most urban centres was achieved at some stage during the thirteenth century. At a minimum, the great cities of Europe housed 20,000–30,000 inhabitants apiece at this time. They were concentrated in north-west Europe and northern Italy, with the largest being Paris (estimates range from 80,000 to perhaps 200,000 inhabitants), London (50,000 or 100,000), Ghent (64,000), Bruges (42,000), Cologne (35,000), Ypres (30,000), Leuven or Louvain (20,000), and Brussels (20,000). In northern Italy, Venice, Milan, Florence, and Genoa each reached approximately 100,000 inhabitants. Naples, Palermo, Bologna, and Rome housed about 50,000 people apiece, since they experienced rather less commercial prosperity than those in northern Italy. From an economic point of view it was not the population of each individual town that was important but rather the concentration of large and very large towns in the two most urbanized and densely populated regions at that period, when on average 30–40 per cent of the population of Europe lived in towns.

## Topography

This expansion of population led to an expansion of urbanized areas, which was marked by the construction of a succession of longer ramparts, often arranged in concentric rings around the city between the eleventh and fourteenth centuries. In towns where the terrain made that kind of expansion impossible, as at Genoa, population growth led to increasing densities, sometimes reaching more than 500 persons per hectare. These densely packed cities also lacked large open spaces, such as market squares, of which there were often several in an expanding medieval city, or pieces of farmland which were found inside the ramparts of many northern cities. In many cities earthen ramparts were replaced during the thirteenth century by stone walls, being reinforced with towers at regular intervals and by great gateways made of stone and wood which controlled main access routes. This use of stone during the thirteenth century and, in the case of the plains of northern Europe, of brick affected not only private houses but also public buildings, bell towers, and churches.

## Industrial Production: Objects and Directions of Trade

In the regions where urban institutions acquired great power, notably in Flanders and northern Italy, the numerous trades were increasingly placed under regulations that were devised by the municipal authorities during the thirteenth century. These regulations specified strict internal organization, involving a threefold hierarchy of labour (master craftsmen, apprentices, menial workers), the need to produce a 'master piece' to become a master, and above all to pay an entry fee. Very restrictive rules regarding the organization of work, techniques, and raw materials to be employed sought to avoid competition between master craftsmen and the concentration of production in the hands of a few great merchants, but also, and above all else, aimed to ensure that a quality of production was maintained in order to withstand foreign competition.

This competition became particularly challenging for Flemish cloth makers during the second half of the thirteenth century, since Italian merchants, and especially the permanent representatives of Italian companies in north-west Europe, came to England to acquire for themselves quantities of fine English wool, rather than importing cloth that had been woven in Flanders and transported across the land routes of Champagne in order to be dyed and finished in certain Italian cities, especially Florence. Thereafter the Italians carried the precious raw material from England by sea, sailing through the Strait of Gibraltar and using larger and larger vessels which they navigated with the help of compasses and portulan charts. As a consequence every stage in the processing of English wool was undertaken in Italy. The direct sea route between Italy and north-west Europe, which opened in 1277 and operated regularly after 1318–28, was one of the causes of the decline of the Champagne fairs in eastern France, where important contacts between Italian and Flemish merchants had taken place since the end of the twelfth century. Four small towns (Troyes, Bar-sur-Aube, Provins, and Lagny) had hosted six fairs (two at both Troyes and Provins) during the course of each year, which formed a regular cycle of trading activity that was protected by the counts of Champagne, the counts of Flanders and the kings of France, and benefited from a distinctive commercial jurisdiction. Trading activity was accompanied by intense financial activity, which eventually exceeded it in importance after 1230. It was then that the Italians developed their techniques of credit, notably the bill of exchange, from the earlier contract of exchange which had been inaugurated in Genoa during the second half of the twelfth century. Soon the presence of Italians and their companies in Bruges would make the city the great competitor with the fair towns of Champagne in terms of money exchange. In addition, the development of the Italian cloth trade using English wool imported directly by sea, especially at Florence, meant that the Italians were less interested in Flemish cloth, which in earlier times they had purchased at the Champagne fairs.

The widening of the commercial horizons of Italy toward north-west Europe by means of direct navigation was, however, less important than advances in their trading techniques, which enabled them to become more involved in financial dealings in the Near East and along the northern and southern shores of the Black Sea. After the 1270s the commodites that were known collectively as 'spices' (including dyestuffs, sugar, oil, leather) represented a growing trade between the Black Sea and Genoa or Venice. Italian imports from these regions during the period from 1250 to 1350 involved silk produced around the Black Sea and the Caspian Sea and even imported from China following the voyages of Marco Polo at the end of the thirteenth century. But after 1260 the most important import in terms of volume was alum mined in Asia Minor (Phocea and Koloneia), which was indispensable for textile production in western Europe. Genoa acquired the monopoly of this trade from Venice after the fall of the Latin empire of Constantinople (1261) and the subsequent reconquest by Byzantium. The alum trade with north-western Europe also helps explain the embellishment of the direct sea route by Genoese galleys between Italy and Bruges through the Strait of Gibraltar. By the early fourteenth century Bruges had become not only a meeting point for traders from the Mediterranean and northern Europe, but also the principal port on the shores of the North Sea and, above all else, the largest redistributing market in the western world.

To the north and north-east from Bruges lay the German trading area of the North Sea and the Baltic, which was comparable to the Mediterranean trading area controlled by the Italians. This domination was principally due to the foundation of the German Hanse, which had a trading house (the Guildhall) in London from 1157. The German Hanse was soon directed by Lübeck, whose merchants were installed from 1161 on the island of Gotland in the middle of the Baltic and especially in the port of Visby. The merchants of the German Hanse received important trading privileges in Bruges in 1252–3 but they did not control a specific part of the city; by contrast they possessed the Steelyard in London, the Deutsche Brücke in Bergen (Norway), and the Peterhof in Novgorod, which formed a closed settlement with its own churches and protective walls within the Russian town. During the thirteenth century the merchants of the German Hanse founded the towns of Rostock, Riga, Stralsund, and Danzig on the southern shores of the Baltic, while to the north they played a large part in the foundation of Stockholm in the middle years of the century. Together with other cities in northern Germany, such as Hamburg, Bremen, and Cologne, these towns became members of the German Hanse, which changed in the fourteenth century from being an association of merchants, which was the original meaning of the term 'hanse', into a league of cities under the direction of Lübeck, which had its own organization, its special meetings when important decisions were taken, and even its own armed fleet. Victory over the Danes by its armed fleet (1369) gave the German Hanse control of the Sound and hence the monopoly of navigation and trade in the Baltic. Herring fished along the coast of Scania were packed in salt from Lüneburg or imported from Bourgneuf Bay on the Atlantic coast of France, and sent in great quantities to the city of Bruges. The Esterlins, as they were called, brought fur, wood, ashes, tar, iron and amber to Bruges and in exchange took Flemish cloth, Bordeaux wine, salt, and spices that had been brought to the city by the Italians. Importation of grain to the Low Countries from the Baltic lands and Poland was not to become established until the fifteenth century.

## The Later Middle Ages (Fourteenth and Fifteenth Centuries)

### Urban Decline and Newly Urbanized Regions

The great famine of 1315–17 affected several large cities in north-west Europe, including Bruges and Ypres, whose populations declined by 5–10 per cent, but it did not reach the Mediterranean regions. It was the great plague of 1347–52 which spread from Italy to reach the whole of Europe that represented the real starting point for the demographic and economic decline or stagnation of almost all the great cities of Mediterranean and north-west Europe. Some Italian cities, such as Florence, lost about half of their population because of the plague, but the great towns of Flanders, such as Bruges, seem to have been much more resistant. The demographic collapse of Ypres from 30,000 in the fourteenth century to 10,000 inhabitants in the fifteenth century was due to the decline of Flemish cloth.

This negative trend was counterbalanced by the growth of towns in regions which previously had been less affected by international trade. Only a few towns located in the regions of old-established urban development experienced real urban growth, notably Antwerp and Lyons.

## The Atlantic Façade of Europe

Meanwhile ports along the maritime route used by the Italians to reach Bruges were stimulated into new activity. This involved ports on the Atlantic coasts of Spain (Cadiz, Seville), Portugal (Lisbon), and France (Bordeaux, La Rochelle, Rouen). Having experienced a strong urban tradition under the Muslims, then being reconquered by the Christians and undergoing an important growth in trade, the Andalusian city of Seville reached a population of 75,000. It stood in marked contrast with Barcelona, whose population fell from 50,000 c.1340 to 20,000 in 1477 because of epidemics, famines, and above all else because it was not located on the new maritime trade routes which attracted population to southern Spain and to Majorca. Lisbon profited less than Cadiz and Seville from the presence of rather fewer Italian merchants and financiers and owed its expansion above all else to the maritime expeditions of the Portuguese to the coasts of northwestern Africa, Madeira, and the Azores. The coastlands of western France did not benefit from the growth of trade encouraged by the Italians and suffered severely from the political difficulties between France and England during the Hundred Years War. Only after the catastrophic crisis of 1453–70 did Bordeaux manage to regain its position as a great trading city exporting wine to England and Flanders, and Rouen recovered thanks to the presence of Spaniards and Italians who brought wool from Castille, alum, and spices from Lisbon.

## Southern Germany

In the southern half of Germany, towns such as Augsburg and Nuremburg were located on the new trading route across the continent which extended from Antwerp and Cologne, with one branch reaching Venice and another running through central and eastern Europe from Lvov and Prague to Kiev. As well as their favourable position on great international trading routes these cities benefited from a real growth in their industries. The proximity of mines in Bohemia and Hungary (especially silver mines), whose production was controlled by the merchants of Nuremberg, contributed to the brilliant fortunes of this economic capital of southern Germany. The metal trade was in the hands of several great dynasties whose members personified genuine urban capitalism. In the sixteenth century the Fuggers of Augsburg would become the best known of these families. Merchants from Nuremberg were grouped in Venice in the famous *Fondaco dei Tedeschi* where they purchased large quantities of cotton brought from Syria and Egypt. It would be used to make fustian, a new textile that would be made alongside cotton cloth in the area around Lake Constance, especially in the small town of Ravensburg. In 1380 this place became the headquarters of a company, the *Grosse Gesellschaft*, which in the fifteenth century would become more important than the great companies of Nuremberg and Augsburg. It would come to monopolize the purchase of cotton in Venice and would control the production of cotton cloth and fustian in the whole region, enabling its trading network to extend to all large towns of western Europe by 1530.

## The Fairs of Lyons and Geneva

The success of towns in southern Germany contributed to the growth on the other side of the Alps of the manufacturing and trading cities of Lombardy and of the fairs of Geneva and Lyons. In Lombardy, Milan specialized in the production of silk and velvet and possessed several famous arms manufacturers. Both Lyons and Geneva owed their growth during the fifteenth century to the success of their silk fairs. The population of Lyons doubled from 20,000 to 40,000 between the mid-fifteenth century and 1500, especially because of the decision of King Louis XI in 1462 to forbid his subjects frequenting trade fairs in Geneva which reached their high point of activity at about that time. Fairs retained an essential role in commercial activities in the interior of the continent at the end of the Middle Ages.

## The Rise of Antwerp

The growing importance of central Europe during the fifteenth century was to the benefit of Antwerp on the western side of the continent. The city saw its population increase from 20,000 to 50,000 in the second half of the century, largely due to the fact that it overtook Bruges as a major trading centre, with increasing numbers of Portuguese and English traders visiting Antwerp. Commercial activities in

Antwerp included metal products, especially armaments, from Germany's growing industries, that were being exchanged for spices, slaves, and gold brought by the Portuguese from the western coasts of Africa which they had progressively explored following the encouragement of King Henry the Navigator (1433–60). The fairs of Brabant that were held at Bergen-op-Zoom as well as at Antwerp further contributed to the success of the latter city. They attracted fustian and other products from southern Germany and were frequented not only by Portuguese merchants but also by the English who brought quantities of semi-finished cloth which then finished in Antwerp by merchants from Hanseatic towns (principally Cologne), and by the Dutch.

## Holland

Holland underwent very rapid urban development during the fifteenth century with the rise of industrial towns such as Leiden, Delft, Gouda, and Haarlem. Several of these soon exceeded 10,000 inhabitants apiece, and among the islands of Zeeland a merchant fleet was being assembled which would soon challenge the German Hanse in the Baltic and a century later would serve as the base for Dutch colonial enterprises in the Far East and in America.

## Italy

Italy still retained its strong trading position and its competence in commercial activities at the end of the fifteenth century, but developments and changes in the direction of international traffic made it more difficult for Italian coastal towns to participate in the new commercial operations that were developing along the coasts of the Atlantic from north Africa to Antwerp. Following the taking of Constantinople by the Turks in 1453, the Genoese retreated from their colonies on the shores of the Black Sea and in Asia Minor, concentrating initially on the island of Chios in the Aegean Sea. In 1462 alum mines were opened in the pontifical states of central Italy and four years later the monopoly for their exploitation was granted by the Papacy to the Florentine company of the Medicis. These developments definitely reorientated Genoese activities toward the west, where they reinforced

their presence along the Mediterranean coasts of Spain and northern Africa, but rather less beyond Gibraltar except for certain points in Andalusia. The stagnation and subsequent decline from 1560 of the company of the Medicis, which closed its branches in London, Bruges, and Avignon in the 1570s before finally collapsing in 1594, symbolized the decline of the Mediterranean in the world economy, which would soon turn towards the New World.

## Northern Europe

Trade in northern Europe also experienced considerable upsets after the late fourteenth century, which were characterized by the difficulties which affected the German Hanse and by the appearance of English traders on major international commercial routes. Following the emergence of the English cloth industry, especially involving small towns in the Midlands, growing quantities of English cloth had been exported during the fourteenth and especially the fifteenth centuries, notably by the Merchant Adventurers. English merchants, with the protection of their king, introduced a tougher policy with respect to foreign merchants and more severe control of the activities of Hanseatic and Italian traders. The monopoly of the Hanse came under serious threat following the appearance of the English in the Baltic ports after the beginning of the fifteenth century, especially Thorn and Danzig. This was intensified by the Dutch who carried their textile products not only by land to the fairs of Leipzig and Poland but also in the vessels of their rapidly growing merchant fleet to the ports of the Baltic and Norway. This offensive by the English and Dutch coincided with a serious weakening of the internal cohesion of the Hanse, because of the independent initiatives taken by Cologne regarding the trade fairs of Frankfurt and the strong competition coming from the trading cities of southern Germany.

## The Establishment of Urban Hierarchies and the Political Role of Cities

Changes in the urban geography of Europe were accompanied through functional changes by the growth of an urban hierarchy and the establishment of veritable urban networks in which the economies and functions of specific cities became increasingly complementary. At the same time the political role of large cities increased, even if they were stagnant or declining demographically. The power of these large urban centres expressed itself not only over small and medium-sized towns but also in the surrounding countryside. Europe's largest cities increasingly exercised political, economic, and cultural dominance over their tributary regions. This trend gave rise to problems in terms of power relationships with the central authorities of the state which also became stronger at this time, notably by the establishment of a state apparatus with its own system of taxation, civil service, and bureaucracy. Some large towns, especially those in small political units such as the principalities of the Low Countries (Flanders, Brabant, Holland), acquired a major influence on the trading policy of their states, but the general tendency was for the centralizing power of kings and princes to overtake the great cities, often with the support and connivance of small and medium-sized towns.

## Changes in Industrial Organization and Social Effects

Within the textile cities, the workshop remained the usual unit of production even though many entrepreneurs controlled several looms, for example Francesco Datini in Florence and Jehan Boinebroke in Douai. This phenomenon may be explained by the growing intervention of merchants in the wider production process, as suppliers of raw materials or as sellers of the finished product, from the late thirteenth century onwards. In some cities, notably in southern Germany where the development of the textile industry was recent, large merchants now controlled supplies of raw material and established contracts with weavers through the putting-out system to deliver products to them and to be paid on a piece-work basis. This trend towards increasing the scale of manufacturing activity was even more pronounced in industries which experienced technical progress and required large enterprises and which, in turn, demanded great amounts of capital and a complex organization of labour. Examples might be drawn from the great arsenal at Venice, from canon foundries, and from printing works. This trend would run up against restrictive craft regulations which were made even more rigorous during the fourteenth century in towns that were undergoing economic decline. An example would be found in Ypres where attempts were made to withstand competition by producing increasingly exclusive cloth. In towns where the establishment of particular industries was more recent, as in Holland and southern Germany, and in places where craft workers made up only a minority of the population (as in Bruges), manufacturing displayed a greater degree of flexibility in adapting to new technological and economic conditions. This enabled these towns to overcome the crisis which affected traditional industry towards the end of the Middle Ages and to avoid the social unrest which occurred in many old-established industrial centres including Ypres, Ghent, and Florence.

References

Beresford, M. W., *New Towns of the Middle Ages*, (London, 1967).

Bocchi, F., *Attraverso le città italiane nel Medioevo*, (Bologna, 1987).

Bookmann, H., *Die Stadt im späten Mittelalter*, (Munich 1986).

Christie, N., and Loseby, S. T. (eds.), *Towns in Transition: Urban Evolution in Late Antiquity and the Early Middle Ages*, (Aldershot, 1996).

*La Città nell'alto Medioevo*, Settimane di studio del Centro Italiano di studi sull'alto medioevo, 6 (Spoleto, 1959).

Clarke, H. B., and Ambrosiani, B., *Towns in the Viking Age*, rev. edn., (London, 1995).

—— and Simms, A., *The Comparative History of Urban Origins in Non-Roman Europe*, BAR International Series, 225 (Oxford, 1985).

Dollinger, P., *La Hanse (XIIe–XVIIe siècle)* (Paris, 1989).

DUBY, G., 'Les Villes du Sud-Est de la Gaule du VIIIe au XIe siècle', *La Città nell'alto Medioevo*, Settimane di studio del Centro Italiano di studi sull'alto medioevo, 6 (Spoleto, 1959), 231–58.

—— *Histoire de la France urbanie*, ii. *La ville médièvale* (Paris, 1980).

ENNEN, E., *Frühgeschichte der europäischen Stadt*, (Bonn, 1981).

—— *The Medieval Town*, trans. N. Fryde (Amsterdam, 1979).

—— *Die europäische Stadt des Mittelalters*, 4th rev. edn. (Göttingen, 1987).

FASOLI, G., and BOCCHI, F., *La città medievale italiana*, (Florence, 1973).

GANSHOF, F. L., *Études sur le développement des villes entre Loire et Rhin au Moyen Âge*, (Paris, 1943).

GAUTIER, DALCHE, J., *Historia urbana de Léon y Castilla en la Edad Media* (siglos IX–XIII) (Madrid, 1979).

*La Genése et les premiers siécles des villes médiévales dans les Pays-Bas méridionaux. Un problème archéologique et historique*, Actes du 14e Colloque International (6–8 Sept, 1988) du Crédit Communal, Crédit Communal, Coll. Histoire in 8°, no. 83 (Brussels).

HEERS, J., *La ville au moyen âge*, (Paris, 1989).

HODGES, R., *Dark Age Economics: The Origins of Towns and Trade AD 600–1000* (London, 1982).

—— and HOBLEY, B. (eds.), *The Rebirth of Towns in the West AD 700–1050*, CBA research report 68 (London, 1988).

JANKUHN, H., SCHLESINGER., W., and STELLER, H., *Vor- und Frühformen der europäischen Stadt in Mittelalter*, Abhandlungen der Akademie der Wissenschaften in Gottingen, Phil.-Hist. Klasse, 3rd ser. 83–4 (2 vols.; Göttingen, 1974–5).

MITTERAUER, M., *Markt und Stadt im Mittelalter*, Beiträge zue historischen Zentralitatsforschung (Stuttgart, 1980).

PIRENNE, H., *Medieval Cities: Their Origins and the Revival of Trade*, 1986 edition (Norwood, 1925).

PLANITZ, H., *Die deutsche Stadt des Mittelalters*, (Cologne, 1965).

PLATT, C., *The English Medieval Town*, (London, 1976).

RENOUARD, Y., *Les Villes d'Italie de la fin du IXe siècle au début du XIVe siècle*, (Paris, 1969).

REYNOLDS, S., *Introduction to the History of English Medieval Towns*, (Oxford, 1977).

RÖRIG, F., *The Medieval Town*, (Berkeley, 1967).

*Topografia urbana e vita cittadina nell'alto medioevo in Occidente*, Settimane di studi sull'alto medioevo 21 (2 vols.; Spoleto, 1974).

VAN WERVEKE, H., 'The Rise of Towns', in *The Cambridge Economic History of Europe*, ii (Cambridge, 1963), 3–41.

VERHULST, A., 'The Origins of Towns in the Low Countries and the Pirenne Thesis', *Past and Present*, 122 (1989), 1–35.

# Chapter 6

# Geographical Knowledge and the Expansion of the European World after 1490

R. A. Butlin

## Introduction

The historical geography of Europe has never been self-contained: in prehistoric times and for most of its history it has had links with Africa, Asia, and North and South America, and in later periods with the more distant lands of Australasia and the remoter Arctic and Antarctic wastes. The processes whereby these links were initiated and modified were exceedingly complex, comprising kaleidoscopic mixes of cultures, racial attitudes, geographical knowledge and broader belief systems or cosmographies, geopolitical ambitions and failures, scientific understandings through experiment and ambition, and changing processes of resource evaluation and realization.

Our understanding of these processes and changes has not been static, and in the course of the twentieth century in particular we have seen major changes in the historiography and interpretation of the imperialist and colonialist processes which were characteristic of the late nineteenth and early twentieth centuries, not simply in terms of moral judgements about the attempted appropriation of labour forces and resources of overseas lands, but also in terms of the environmental

impacts of European economic activity. Political ambitions for overseas influence and settlement waxed and waned over most of the period from the late fifteenth century onwards, reflecting the geopolitical intelligence and interpretations available to and collected for visionary and ambitious state rulers. The extent to which the populations of the countries and regions of Europe generally supported such endeavours is, however, difficult to discern. Even at the peak of European imperialism there was significant opposition to colonial and imperial policies from within such countries. In addition a recent historiographic trend has been the attempt more emphatically to examine the experiences of peoples who were subject to the various imperial and colonial impulses: an important part of current revisionist critiques.

## Voyages and Perspectives

During the course of the fifteenth century a combination of new ideas and events led to an acceleration of interest by several European powers in the acquisition and application of geographical knowledge of territories outside Europe, and to the intensification of maritime and land-based exploration of those territories in what generally has become known as the Age of Discovery. The motivations comprised a complex mixture of desire to promote and proselytize the main branches of the Christian religion, the advancement of trade, scientific experiment (albeit of a basic kind), the testing of theories about the nature of the earth, the investigation of those parts and their inhabitants farthest away from Europe, and the acquisition of territories whose resources might help secure the economic well-being of the European countries involved.

The broader background was that of the Renaissance—a rebirth or reawakening—of Western learning after the 'dark ages' and also what has been described as a crisis of feudalism, that is the perceived need of European countries, as a result of epidemics, climatic change, low productivity in agriculture, and war, to enhance economic conditions by

locating, seizing, and distributing resources available beyond the European frontiers. The movement to the New World, the establishment of forts and trading posts along the coasts of Africa, the entry into the Indian Ocean and the China seas, and the spread of the fur trade through the boreal forests of America and Asia all represent ways in which these goals were sought and fulfilled. (Wolf, 1982: 109)

The acquisition of new wealth in this way also required the development of strong mercantile states, new navigational skills, strong ships, and specialist merchant companies that would share with the state the risk capital required.

Broc (1986) has suggested that the prime factors linking the Renaissance to the famous voyages of geographical discovery were: a general curiosity for the discovery of new lands; the political consequences of the Treaty of Tordesillas (1494); religious interest in finding new territories and peoples for evangelization; and the opportunities for trade. The actual process of the revolution in geographical knowledge in the fifteenth and sixteenth centuries was, of course, more complex than this, and Livingstone (1992: 34–5) has characterized this time of transition in geographical knowledge and scientific discovery as a period in which both science (navigation, geographical exploration, and mapping) and myth (both as imaginative and fantastic accounts of exotic peoples, flora, fauna, and places,

and as 'a narrative in which some aspect of the cosmic order is manifest, and accordingly is an expression of the collective mentality of any given age') were closely linked together.

Geographical knowledge of the regions which were to form the focus of European imperial and commercial ambitions expanded rapidly from the late fifteenth century, but was initially related to older sources. Thus European ideas about southeast Asia derived in part from the records of the exploits of Alexander the Great in India, from the classical geographies of Strabo (c.64 BC–AD 20) and also from Ptolemy (c. AD 90–168). Ptolemy's writing was one of the sources of origin of a range of late medieval and early modern concepts and myths of South-East Asia as a region with large sources of gold, notably in the region described as the Golden Khersonesus or Khersonese (Aurea Chersonesus— the Golden Peninsula), variously located by later scholars as perhaps Cambodia or peninsular Malaya and Sumatra, notably the region of the two Mounts Ophir either side of the Malacca Straits (Savage, 1984; Wheatley, 1961).

Early European perception of these remoter parts of the globe, deemed to be ripe for trade and ultimately for settlement and conquest, was a mixture of myth, frequently based on ancient geographical sources and the early reports of contemporary travellers, and increasingly accurate geographical knowledge, based on exploration and trade contacts, this knowledge now being more frequently incorporated into modifications of Ptolemy's world map in his Geography (translated into Latin in 1410 and republished with its maps in Bologna in 1477), Waldseemüller's globe of 1507, and world maps such as those of Fine (1531) and Mercator (1538).

In the fifteenth century the European geographical centre for international maritime (and to a degree overland) trade was the Mediterranean together with the Atlantic seaboard of Iberia. Within the Mediterranean the major city-state ports in the fifteenth century were Venice, Genoa and Florence. Venice had developed as the most powerful of the three by the second quarter of the century, this position being strongly influenced by changes in the trading patterns within and outside Europe and also by geographical advantage, including market advantages consequent on the earlier medieval expansion of Germany eastwards, connections with bankers and entrepreneurs of south Germany and with the north German Hanse. Venice had a greater degree of political and social stability than the rival cities Genoa and Florence, and adopted a strong and protectionist kind of mercantilist policy allied to a sophisticated commercial and diplomatic intelligence network (Van der Wee, 1981). While the extent of Venetian mercantile hegemony, primarily related to the trade in spices and luxury commodities, was largely confined to the eastern Mediterranean, the Levant, and Asia Minor, such activities connected in these regions with overland and sea routes from Asia, and thus identified possibilities for long-distance trade that were more successfully developed by Portugal, Spain, the Netherlands, and Britain at later dates. Additionally, the navigational knowledge of Venetian and Genoan sea-captains was a major contribution to the late fifteenth-century voyages of discovery to the New World, notably the Columban voyages from 1492.

In the early and mid-fifteenth century the Portuguese, using astronomical forms of navigation, explored and charted the islands of the eastern Atlantic (Canaries, Azores, Cape Verde, and the Madeiras) and established a series of stations along the north-western and western coasts of Africa and in the Atlantic islands, but the major acceleration came in the later fifteenth century, with the rounding of the Cape of Good Hope and the beginning of the voyage in 1487 by Vasco da Gama to East Africa and south-western India. According to Scammell (1981: 227), 'The roots of Portuguese expansion lay in those Christian assaults on a weakening Moslem power in the early Middle Ages, and the later and largely Italian penetration of Africa and the North Atlantic'. Although the early voyages to north-west and western Africa were both environmentally daunting and at first commercially unrewarding, the persistence of the ambitious visions of Prince Henry, ultimately, after his death in 1460, known as 'the Navigator' led to successful exploitation of the resources of the Far East.

The voyages of discovery initiated by the Portuguese in the aftermath of Prince Henry's death, begun during the reign of King John (1481–95), included: a voyage under Bartholomeu Dias in 1487 down the coast of West Africa, round the southern tip of Africa and up the east coast; and the expedition under Vasco da Gama which began in 1497 and which reached the coast of south-west India in May 1498.

Thence Africa, India, and the East became a major target of Portuguese mercantile expansion, offering a vast range of goods for trade. The scale of activity was approximately seven ships per year from Lisbon to India in the period 1500–1635, each of about 400 tons in 1500, the size however rising to 2,000 tons by the end of the sixteenth century. The main focus of trade with India was the spice trade, but the number of commodities produced and traded was complex, following a similar production and trading pattern to that established by the Portuguese in West Africa, with trade in such exotic commodities as gold and spices being supported by agricultural products grown for local markets by Portuguese settlers (Phillips, 1990).

Westward, and by a chance deviation of an expedition to India (in 1500), Portugal's major discovery and subsequent colonial territory was Brazil, which on account of climatic, topographic, and logistical obstacles was largely ignored until the early seventeenth century, though trading with the Amerindians of garments and mirrors for brazilwood (which provided dye for European textiles) had begun. Acceleration of Portuguese interest in the early seventeenth century was stimulated by the parallel interest of Spain, France, England, and Holland in Brazil (Scammell, 1981: 247). Great freedom was given by the Portuguese state to high-born and middle-class settlers, but in spite of this freedom settlement and economic development from the mid-sixteenth century was slow. However, the introduction of sugar plantations, especially on the rich soils of the Atlantic coast, initiated a rapid development of sugar production, initially using indigenous labour, but after mid-century more frequently using slave labour from Africa. Elsewhere Portugal's major interests were in Africa, where short-lived and tenuous interests were developed in Morocco, Ethiopia, Mauritania, and the Gulf of Guinea, and also in East Africa, succeeded by longer-term involvement with the Congo basin and Angola, mainly in respect of the slave trade. Portuguese commercial activity extended to the Far East, where trading settlements were founded in Malacca in 1511 and in Macao in 1557, the latter with permission to trade with the Japanese port of Nagasaki.

R. A. Butlin

# Columbus and the Enterprise of the Indies

In the late fifteenth century the major European 'encounter' westward was that made by the Genoese captain Christopher Columbus in his 'enterprise of the Indies' in 1492. Having sailed from Palos in the south of Spain in August 1492 under the patronage of King Ferdinand and Queen Isabella of Spain and a Spanish financial consortium (having previously been refused support by King João of Portugal), Columbus arrived at an island in the Bahamas on 12 October, and at Cuba on 28 October. Born in Genoa c.1451, Columbus made extensive sea-journeys between c.1470 and c.1485 within the Mediterranean, and further afield to Iceland, the Azores, and the Guinea Coast in the Atlantic, remarkable for the times but none the less consistent with the geographical knowledges, navigational competences, and commercial and political interests and ambitions of fifteenth-century Genoa. As Fernandez-Armesto indicates in his biography of Columbus, over the course of the fifteenth century the Genoese city-state moved from an active mercantile role based on collaboration with other states and foreign princes to one with greater mercantile and minor imperial ambition; there was 'a sense—sometimes a muted sense—of national solidarity, supplemented and often exceeded by family ties' (Fernandez-Armesto, 1992: 7). These families were the great merchant trading families such as the Centuriones, by whom Columbus was employed. Columbus was therefore a participant in the growing trade and colonization linkages between Genoa and the Atlantic. Notable developments were the sugar estates in the Atlantic islands such as Madeira, the western Canaries, the Cape Verde islands and the islands of the Gulf of Guinea. These were partly modelled on Genoese trading and mercantile settlement patterns in the western Mediterranean, particularly Andalusia, with Cadiz and Seville as the major bases, and the Columban encounters in the western Atlantic owe much to the experience of Columbus as a captain and trader for the Centurione family in the western Mediterranean and the eastern Atlantic. An additional influence was the attraction of the Atlantic, with its mythical islands portrayed on contemporary maps, and much contemporary speculation about its use as a route to the East and thus to wealth through trade and the acquisition of precious goods and metals. Thus 'Columbus's design for a new Atlantic voyage clearly belongs in the context of an age of vigorous speculation about the secrets of the Open Sea. Almost every element in the thinking that underlay his enterprise was part of the common currency of the geographical debate of his day' (Fernandez-Armesto, 1992: 21).

The first of the Columban voyages westward across the Atlantic took place in 1492, when with three ships he sailed to the Canaries and thence to the Bahamas, and subsequently to Cuba and Hispaniola. A settlement—La Navidad—was established in Hispaniola (in that part which is modern Haiti) when the *Santa Maria* was grounded on the reefs and a small fort constructed to accommodate the thirty-nine sailors, who were left to trade with the Taino Indians and to look for gold, though by 1493 the sailors were all dead and the fort destroyed (Deagan and Cruxent, 1993). Later voyages by Columbus in 1493 (with seventeen ships and the title of Admiral of the Ocean Sea and Viceroy of the Indies), 1498, and 1502 led to the 'discovery' of islands in the lesser Antilles, of Trinidad, of what would become known as Venezuela, and various parts of the western Caribbean. He reached the Isthmus of Panama on Christmas Day 1502, but was unable to understand the significance of his proximity to the Pacific and the route to Asia (Scammell, 1981: 306).

The impact of these early encounters with the New World has been extensively evaluated, with particular intensity of interest around the time of the 500th anniversary of the Columban rediscovery of the Americas. Butzer (1992), for example, in a study of the Spanish encounter with the New World in the fifteenth and sixteenth centuries, emphasizes the innovative and enthusiastic character of descriptions of the New World by soldiers, sailors, administrators, and missionaries, and contrasts them with the inadequacy of the preceding classical and medieval cosmographic geographies and philosophies. He stresses the multilinearity of Spanish geographical knowledge of this period, as evidenced by the observational skills of Christopher Columbus; the landscape taxonomy schemes of Fernando Colón (Columbus's son, who proposed and executed a survey of the geography of Spain, begun in 1517, and replicated in New Spain in 1547–71); the work in natural history and biotic taxonomy of Gonzalo Fernández de Oviedo (1478–1557), writing of the history of Spanish exploration and conquest of the New World but including an extensive taxonomy of the plants and biota, modelled on the Pliny classification; the recording of cultural landscapes by Las Casas in the 1540s and 1550s and by early Franciscans in Mexico. Additional indicators of Spanish geographical knowledge of the time included: the regional and synthetic accounts of the geography of Peru in 1554 by Cieza de Léon; the role in the urban planning ordinances of 1573 of Juan López de Velasco (c.1530–99); and the new scientific framework for data on the New World produced by Joseph de Acosta (c.1540–1600).

Additionally, further progress has been made in the assessment of the complex role of cartography in the process of exploration and settlement at this time. Harley (1990, 1992) has placed the maps associated with the Spanish colonization of Central and South America in the context of two debates, the first about the meaning of 1492—essentially whether that year signifies a great achievement or a major crime—and the second about the nature, and the meanings, of maps, especially the recognition and acknowledgement of the significance of indigenous mapping traditions in the Americas. Presenting two contrasting perspectives, Harley (1992: 528–9) contended that the early European maps of the Americas, 'usually are strident political documents. Above all they bear the traces of the territorial moves by which the colonial powers of early modern Europe sought to delimit, divide, and assert control over their overseas territories', instruments of both political and religious imperialism, and of more abstract ideologies through the naming of places and features, the creation of the Americas with a European identity, and the unicultural geography of absences. The indigenous maps in contrast, according to Harley (1992: 527), can be seen as part of means of resistance to colonial claims for the appropriation of land, hence 'making a map became a conscious strategy of resistance. From Mesoamerica, where pre-Conquest cartography was a well-established tradition, come numerous instances of maps being used in this way'.

The territorial impact of the Spanish involvement in the New World from the fifteenth century to the eighteenth century is well documented and reasonably well understood. The chronology of conquest and settlement testifies to the swiftness of the process: attempted settlement of Hispaniola in 1493, of the mainland around the Gulf of Urabá and the coast of the Isthmus of Panama from 1509, followed by the conquest of central Mexico by Cortés in the 1520s and of the Inca kingdom in Peru in the 1530s, thence Guatemala (1523–42), New Granada (1536–9), Central Chile (1540–58), and from the late sixteenth century the area of New Mexico in the North American south-west, whose northern limit was the upper Rio Grande. From 1564 the Spaniards also had settlements in the Philippines, and began trade with China. The initial motives for the conquest and colonization of Central and South America by the Spaniards was a mixture of greed and religious fervour, and also took place at a time when the reconquest of Spain had been completed, so that, as Meinig has argued (1986: 45), the spirit of reconquest was exported to the New World. The need for material gain was, it is thought, caused by domestic financial difficulties in Castille, and the search for silver and gold was unremitting. Gold was first exploited from the Antilles, and then more successfully from Colombia, but the principal exported wealth to Spain was silver, discovered in Bolivia in 1545, then in various locations in upland Mexico in the 1540s. There and elsewhere it was mined with a massive and greatly exploited indigenous labour force, and according to Wolf (1982: 139) seven million pounds in value of silver was sent to Spain from the New World in the period 1503–1660. The process inevitably involved the large-scale transfer of land to the Spanish, especially after the initial protection for indigenous cultivated land was removed by the demographic crises and depopulation that followed the Conquest. Prem (1992: 458) has shown that in Central Mexico 'alienation of Indian landholdings began on a large scale after 1580, and by 1620 most Indian properties in the Basin of Mexico and around Puebla had been awarded as land grants to Spaniards'. Control and management of land in these newly conquered territories was initially effected by grants of trusteeship or *encomienda*. In time (mainly in the seventeenth and eighteenth centuries), these *encomienda* were replaced by the landed estates known as *haciendas* and managed by estate owners with an indigenous labour force, over land obtained by territorial disappropriation, conscription, and inducement (Wolf, 1982: 143). This process was complex and varied regionally: Lovell, for example, has shown that in the Cuchumatán Highlands of colonial Guatemala there was significant retention of ancestral lands under Indian control, notwithstanding the development of the *hacienda* (Lovell, 1992a).

The nature of land-use changed through time, initially based on an indigenous cultivation system, to which the Spanish introduced new crops such as indigo, wheat, and barley, the vine, olive, sugar plantations, horticulture, together with sheep, cattle, pigs, horses and mules, and new ploughing techniques (Scammell, 1981; Meinig, 1986). They also cultivated indigenous crops such as cotton, maize, and tobacco, cacao to feed the fashionable habit of drinking chocolate, and cocaine (for indigenous use, in Peru). Major environmental changes were brought about by the extensive development of pastoral farming systems based primarily on sheep and cattle. Oxen, horses, and pigs were also bred in very large numbers. Scammell has emphasized the speed and variety with which Spanish agricultural systems were developed in the Americas:

Feckless the agrarian economy of the Spanish colonies undoubtedly was, yet there had been established, in a surprisingly short time, a relatively complex and specialized agriculture, more varied than that of Portuguese Brazil, sensibly utilizing the wide range of climatic conditions encountered, and shaped by local and regional demand. Though initially based on, and long sustained by, unfree or slave labour it came to employ many who worked for wages, and it was developed and directed by a society considerably more adaptable and less parasitic than its antecedents and original behaviour would have suggested. (Scammell, 1981: 324)

Dutch imperial ambitions and achievements accelerated from the early seventeenth century at a time when the force of Spain and Portugal was weakening; although the main focus of activity was commercial, deriving from a long tradition of maritime trade in the Baltic, none the less Dutch influence extended in due course to North and South America, the Caribbean, Southern Africa (the settlement at the Cape of Good Hope was established in 1652), and South-East Asia, especially Indonesia, with a major trading influence in West Africa. The extent of Dutch settlement was limited, partly owing to the lack of a large population willing to settle overseas, so that in North America, for example, they established only a limited number of settlements in the Hudson valley and the seaboard of Jersey. Their model of colonization has been characterized as one of highly decentralized free-trade capitalism, with a multiplicity of widely scattered ports as the bases of activity (Earle, 1992). In Asia the Dutch actively removed Portuguese influence and England's role in the spice trade, and established a strong regional presence.

England and France were the main rivals of the Dutch in the process of European expansion overseas in the seventeenth century. Voyages of discovery starting from England in the fifteenth, sixteenth, and seventeenth century were mainly concentrated in the North Atlantic and the Arctic Ocean, and included Cabot's rediscovery of Newfoundland in 1497, part of the search for the North-West Passage across the top of North America from the Atlantic to the Pacific, Frobisher's voyages of 1567, 1577, and 1578 to Frobisher Bay, Henry Hudson's voyages between 1607 and 1611 (on the last of which he discovered Hudson Bay), and William Baffin's voyages between 1612 and 1616. The parallel search for a route to Asia to the north of Europe (the North-East Passage) led to expeditions by, among others, Willoughby (1553), Chancellor (1553), Bassendine (1568), and Hudson (1607 and 1608).

Building on the experience of her medieval trade, England began in the sixteenth century to develop overseas settlement and trading and military posts. The earliest ventures were in North America. They occurred against a background of successful privateering of Spanish bullion, notably by Sir Francis Drake (who had completed a circumnavigation of the globe to England in 1580), and were encouraged by the national appetite for investment in the New World, by a weak economy and poor balance of trade, and by new conceptual models of colonization proposed by the Hakluyts (Earle, 1992). The first settlement began in 1585 with Raleigh's short-lived colony at Roanoke on the Outer Banks of North Carolina, and was followed by the settlements of the Virginia Company from 1607. By the end of the seventeenth century much of the eastern seaboard of North America had been colonized by English migrants, the numbers of English settlers who had migrated to North America and the Caribbean during the course of the seventeenth century being estimated at 375,000 (Earle, 1992:

496), considerably greater in number than those of any other European country.

The regional economies of the new colonies varied: the economy of the Virginia Company settlements was closely related to the production of tobacco, that of the later settlements to a mixture of farming, fishing, and the fur trade. Further south, in the West Indies, the tradition of privateering continued and territory was steadily acquired in the early seventeenth century, including St Kitts, Barbados, and parts of the Lesser Antilles from the 1620s and Jamaica in 1655 (Scammell, 1981: 492), at a time of weakening Spanish influence and of growing demand for sugar in Europe. A major consequence was the extensive development of sugar plantations in the British West Indies, associated with the notorious slave trade. A parallel ambition to that of acquisition of wealth through the precious metals of the Americas was the ambition to profit from the riches of the Far East. The first efforts from the mid-sixteenth century led to the establishment of the East India Company in 1600, to trade with India, the Malay archipelago, and ultimately China, paving the way for greater territorial influence and settlements in the eighteenth and nineteenth centuries.

By the beginning of the sixteenth century, France was also poised to be a major player in the projection of Europe overseas. Although less well advanced in such matters than a lead country like Spain, its large land base, an extensive coastline, large number of effective ports, and a population that was willing to settle overseas meant that France has 'the majority of the necessary ingredients for imperial success' (Scammell 1981: 438). French privateering in the Spanish Caribbean in the mid-sixteenth century, intense activity by fishermen from the French Atlantic ports in the Grand Banks fisheries of north-eastern North America, and the activities of fur traders from the early sixteenth century formed part of the background to French colonization on the North American mainland. Settlements were established in the Lesser Antilles in the Caribbean in the early sixteenth century, but the main colonial impulse came further north. In 1608 Samuel de Champlain founded a post at Quebec on the St Lawrence River, and the French settlement of the St Lawrence valley began, with other settlements founded at Trois Rivières in 1634 and Montreal in 1642. Agricultural settlements, based on the seigneurial and long-lot system, expanded along the valley between the major town settlements. The agricultural settlement of Acadia began in the 1630s on the marshes around the Bay of Fundy (Harris, 1990). Trade forts were established on the shores of the Great Lakes and along the Ohio–Mississippi valleys in the 1680s and 1690s. French explorers reached the Gulf of Mexico in 1682, and French settlements were established on and to the east of the Mississippi delta in the late seventeenth century. Elsewhere in the world, French imperial and commercial ambitions and achievements were limited, notwithstanding the presence of French vessels in the Indian Ocean and the Far East in the early sixteenth century. The zenith of French imperial achievement was yet to come, accelerating through the eighteenth century to a peak in the nineteenth.

The reasons for the great burst of European imperial and colonial energy from the late fifteenth century onwards are inevitably complex and open to continuing debate. Jones reminds us that the ventures of discovery and exploitation from the late fifteenth century were rather a continuation of what he describes as 'an old endeavour to pierce the void apparently surrounding Europe' going back to the tenth century (Jones, 1987: 70), and Meinig's point about the continuation of the energy of the Spanish reconquest of Spain from Islam transferred overseas is also pertinent. European imperialism of this period also involved expansion across land, notably to the east and north-east, with the Urals being crossed by the Russians in 1480, and the journeys of imperial discoveries and dominance continued to the Pacific.

## Mapping and Recording the Other

The mapping of the new areas of European geopolitical and commercial interests was an important and integral part of the expeditions and voyages of discovery, resulting in significant changes and additions to the range and depth of geographical knowledge. In the seventeenth century, the rapidly developing science and art of cartography in the Netherlands, for example, was closely tied to geopolitical and commercial ambitions. Thus the Dutch East India Company, founded in 1602, initially supplied maps for its ships that were based on earlier Portuguese and Spanish maps. In time, though, it began to use maps produced by such famous cartographers as Ortelius and Mercator, and in 1617 appointed Gerritz, an experienced geographer, to incorporate new geographical information into its maps, for example of the western coast of Australia, and to develop a more independent cartographic source at a time of conflict with England. A major centre for geographical and cartographic information was begun in 1632 in Batavia, a central point for Dutch involvement in Asia, particularly in its search for sources of gold and silver and for other valuable trading commodities such as spices. The Governor-General, Hedrick Brouwer, made recommendations for the advancement of navigation and exploration at the request of his successor, Antonio van Dieman, who in turn appointed four cartographers as fleetmasters. The information gleaned from the Dutch voyages of exploration in South Asia and the Pacific were rapidly incorporated into Blaeu's New World Map of 1645 (Zandvliet, 1988: 80).

The impact of these activities on Dutch culture can be seen from the fact that 'The houses of Amsterdam merchants and regents were decorated with maps of the world and of Asia, proudly symbolizing worldwide commerce and knowledge. Yet at the same time Dutch Calvinist culture was steeped in the belief that earthly wealth was relative and maps served as a reminder of this inevitable fact'. (Zandvliet, 1988: 82).

From the mid-eighteenth century, conspicuously with the scientific exploration of the Pacific by French and British expeditions, and later in the century by the invasion of Egypt by France, which was accompanied by detailed scientific and archaeological surveys, a phase of imperialism commenced which was to last to the early years of the twentieth century—an era in which scientific (including geographical) investigation and imperialism marched closely together. Although the principal motive for these expeditions was scientific exploration, supported financially by both scientific societies and national governments, and aided by the new forms and instruments of navigation and measurement (such as chronometers that would enable very accurate determination of longitude), taxonomies of plants and animals, and their visual representation, the knowledge which they gained was in various ways used to national advantage. A major outcome of these voyages was the production of improved maps and charts of the Pacific Ocean, its land masses and islands, and a greater knowledge of the natural resources of land and sea which led in turn to their exploitation. The role of these European scientific expeditions in the histories of geographical thought, particularly those of Cook to the Pacific, have been evaluated and debated by Smith (1985), Carter (1987), Stoddart (1986), Livingstone (1992), Gregory (1994), Mackay (1985), and others. These and other scholars have reviewed such problems and concepts as the significance for a modernizing geography of the development of empirical modes of scientific investigation (which none the less overlapped with residues of late Renaissance belief systems and myths), the developing forms of natural history and anthropology, and the experiences of these European impacts by the non-European inhabitants of these 'other' worlds.

The major expeditions in the Pacific were the voyages of the British sailor James Cook, in 1768–71, 1772–5 and 1766–9. Cook (1728–79) had been master of a Royal Navy squadron in the late 1750s which had been involved in the mapping of the St Lawrence River and the coasts of Nova Scotia and Newfoundland. Subsequently from 1763 to 1768 he continued the survey of the coast of Newfoundland, and during this process he observed an eclipse of the sun in 1766, which attracted him to the attention of the Royal Society, planning an expedition to Tahiti to observe the transit of Venus across the face of the sun, and he was appointed commander of the expedition (Kemp, 1988: 199). Cook's first expedition included not only the making of astronomical observations, but also a search for the great southern continent Terra Australis, the circumnavigation of the North and South Islands of New Zealand, discovered by Tasman in 1642, and the mapping of their coastlines. In April 1770 he reached south-east Australia, and sailed northward to Botany Bay, and ultimately sailed up the east and then the north-east coasts of Australia before sailing for New Guinea and the East Indies (Kemp, 1988: 200). The voyage in the vessel *Endeavour* was also notable for the scientific work of the naturalist Joseph Banks, who collected and catalogued many thousands of plant and animal specimens, notably with the aid of the botanist Danial Solander. Following in the earlier tradition of explorers of the sixteenth and seventeenth centuries such as Drake, the Pacific explorers of the eighteenth century took paid artists with them to make pictorial records of plants, animals, and people. Thus very detailed pictorial records were produced on all three of Cook's voyages, on the voyage of HMS *Investigator* in the circumnavigation of Australia by Matthew Flinders (1801–5), and by amateur artists, including naval personnel, involved in the First Fleet and the early settlement at Port Jackson (Sydney). The volume of output was considerable: for example, Sydney Parkinson, one of the natural history artists employed by Banks on the *Endeavour*, produced nearly 1,400 drawings, and Ferdinand Lukas Bauer, employed to work with the naturalist Robert Brown on Flinders' 1801–5 voyage, produced 2,064 sketches (Whitehead, 1988).

With the end of Spanish seapower in the Pacific by the mid-eighteenth century, another nation which became involved in scientific exploration, recording, and mapping at this time was France, notably through the expedition of Louis de Bougainville, who crossed the Pacific, entering that ocean in 1768 as part of a circumnavigation of the world that was completed in 1769 and included the founding (and withdrawal from) a settlement on the Falkland Islands, the annexation of Tahiti, and the expeditions of Marion Dufresne, who landed at Van Dieman's Land in 1772, and of La Pérouse, who crossed the Pacific in 1787. The outcome of the La Pérouse expedition was by all accounts disappointingly modest, but the *Atlas du Voyage de La Pérouse*, published in Paris in 1797, took full advantage to 'convey a delusory impression of heroic achievements' (Terry, 1988: 206.)

The tradition of pure and applied scientific exploration of overseas territories continued into the nineteenth century. The French scientific missions of this period have been well documented and analysed, including those to Egypt (1798–1801), Algeria (1839–42), Greece (1829–31), and Mexico (1864–7). Godlewska (1994) has outlined the growing intensity of the relationship between geographical knowledge and the ambitions of the French state, including imperialistic ambitions, in the late eighteenth century, especially under Napoleon, a process which was partly

a consequence of developments internal to science and to shifts in disciplinary territory. These encouraged geographers to expand their focus from the production of cartography on the one hand and regional description on the other to field- and text-based research into the material basis of the state. This coincided with and was related to changes in the nature of warfare and an evolving sense of the role and capacity of the state to monitor and control ever more local and particular aspects of social, economic, and political life. (Godlewska, 1994: 39)

The direct link between geographical knowledge and imperialism came through a commitment to the perceived superiority of French culture and language to those of the regions controlled and through the practical surveys and maps undertaken in connection with the conquering forces' needs to have detailed information of the new lands of empire for the purposes of control. A telling statement from Napoleon indicates how science fitted national and imperial purpose: 'True victories, the only ones which leave no regret, are those made over ignorance. The most honourable occupation and the most useful to nations is to contribute to the extension of human ideas. The true power of the French republic must henceforth consist in not allowing there to be new ideas which we do not control' (Terry, 1988: 153). The statistics and logistics of production are in one sense impressive: Godlewska cites a report by Berthier in 1802 that in a single war in Europe the engineer-geographers of Napoleon's campaign had produced 7,278 printed maps, 51 atlases, and 600 descriptive memoirs (Godlewska, 1994: 41). The enormous *Description d'Égypte*, produced between 1809 and 1828 in twenty-three volumes, is a classic example of the use of geography and other sciences in the service of imperialism, and one which was continued further afield, and extensively cited by Said (1978) in his major work in orientalism.

Heffernan (1994a), in his analysis of the archives of the French Ministère de l'Instruction Publique, has shown the strong influence of a powerful centralized state on the development and production of scientific knowledge, including colonial geographical knowledge, in nineteenth-century France. Fact-finding missions to research the commercial intentions of colonial rivals were established in the Restoration Monarchy from 1815 to 1830, and after the revolution of July 1830 there was even stronger governmental influence. This was effected by various agencies, including the Comité des Travaux Historiques (founded 1834) and the Service des

Missions, a scientific research council founded in 1842, though in 1845 the two institutions were merged, with the Service des Missions becoming 'the major source of funds for French fieldwork and research overseas' (Heffernan, 1994a: 24). Applications for funding reached a peak between 1890 and 1900, and the main countries for which funds were granted in the period between 1870 and 1914 were Italy, Germany, Russia, the United States of America, Algeria, Tunisia, Morocco, Egypt, Turkey, Greece, the Holy Land, Lebanon, and Syria. Significant other missions were funded for research in Latin America, West and Central Africa, India, South-East Asia, and China.

The major disciplines involved were archaeology, history, medicine, exploration, natural sciences, education, anthropology, geology, and languages; archaeology in particular, through the study of decayed but formerly great civilizations, pointed to the future potential for such regions if exposed to the civilizing influence of France (Heffernan, 1994a: 36).

Spain also, while less of a colonial power than before, but in the interests of its South American colonies and its possessions in the Philippines, sent a maritime expedition to the Pacific in the late eighteenth century: the Malaspina expedition of 1789–94, which was initiated by two navy captains Alejandro Malaspina and José Bustamente. They planned to follow the routes of Cook and La Pérouse, make scientific studies, compile material and plant collections, study the Spanish colonies, and report on the Russian interest in the north-east Pacific and the British settlement at Botany Bay, and were provided with information from the La Pérouse expedition and with instruments by the Royal Society through the agency of its hydrographer. The expedition left in December 1791, and reached Manila in March 1792. Malaspina travelled eastwards to Sydney, where the specialists undertook scientific observations, collected plants, and made sketches and maps. The ships returned to Cadiz in September 1794, having achieved some important scientific aims, including the measurement of the force of the earth's gravitational field by use of an oscillating pendulum, the collection and classification of 16,000 plants, and the ultimate creation of more than 800 works of art. The Italian-born Malaspina suffered from his critical comments on the nature of the Spanish colonies, and was arrested and imprisoned until released in 1803 as a result of the Napoleonic invasion of Iberia (Waldersee, 1988).

## Environments and European Impacts

The environments encountered by European explorers, officials, and settlers on their first experiences of the unknown territories that they had set out to investigate, conquer, and settle were for the most part unlike anything they had ever encountered before. Little wonder that in their initial experiences of these new environments and their inhabitants they were prone to produce exaggerating and misleading accounts of weird and wonderful people, beasts, and birds. Early accounts by the Venetian traveller Marco Polo in the late thirteenth century of people of the Nicobar and Andaman islands of South-East Asia had spoken of people with dogs' heads and of people with tails, and similar myths continued into the era of exploration in the fifteenth, sixteenth, and seventeenth centuries, for example of ape-like people, probably as a result of confusion with certain types of ape, including the orang-outan (Savage, 1984). The perceptions of the environments and cultures which Europeans sought to modify, inhabit and use to their advantage were complex, and both spontaneous and more carefully planned colonial developments often failed disastrously, both from the perspective of European colonial ambition and the futures of the environments and peoples affected. We are still a long way from being able to make certain quantitative estimates of these effects, not least, obviously, because of the enormous difficulties that are involved in estimating the size and the precise nature of pre-European populations and environments. The picture, however, is becoming gradually clearer as a result of intense research into both the destructive and the constructive aspects of European effects on peoples and environments overseas.

The lands that they conquered and colonized were rapidly infused with increasing numbers of their fellow countrymen and women, and with the animals and diseases that they brought with them. It is clear that demographic disaster of a very high order followed the progressive colonization of the New World of North, Central, and

South America by Europeans from the late fifteenth century, through the effects both of war and of epidemic disease. In Pawson's phrase about the main effect of British imperialism, 'the key weapon of British settler colonization was shared with other Europeans. It was disease. The transmission of deadly infections to the New World has recently been described as a component of "ecological imperialism", an attractive phrase which none the less masks an essentially biological process with a spurious mantle of intentionality' (Pawson, 1990: 532).

In the New World the major epidemic killer diseases were smallpox, measles, typhus, plague, influenza, yellow fever, and diphtheria, reducing the native population of the Americas between 1492 and 1650 from c.50 million to c.5 million, though the effects on population varied widely, including the ability of indigenous populations to recover from these major epidemics, with a higher rate of survival in the Middle American and Andean Highlands than in the Caribbean and tropical lowlands (Newson, 1993). In addition there were the obvious direct effects of warfare, together with the effects, especially before the mid-sixteenth century, of exploitation of the indigenous labour force by a process of enslavement, notably for gold and silver mining (Newson, 1993: 262–4) and the effects of venereal diseases on long-term fertility.

Much research has been conducted into the detailed impact of Spanish conquest and settlement on the Caribbean and Central and South America. Estimates of the mortality figures for Hispaniola in the period 1493 to 1518 vary from 60,000 to eight million, but the main fact is that of the virtual extinction of the indigenous population by pneumonia and smallpox, diseases which also affected the Spaniards (Lovell, 1992b: 428). The population of central Mexico was devastated by smallpox in the period 1518–1605, that of Guatemala by either smallpox or measles (contemporary accounts speak of 'plague') in 1519–20, with smallpox also a major factor in mortality in the central Andes in 1524 (Lovell, 1992b).

One of the major moral and demographic effects of the imperial, especially commercial, ambitions of European states from the sixteenth to the nineteenth century was the inhumane and ruthless forced movements of peoples on a massive scale.

Wolf (1982) estimated that between 1451 and 1600 275,000 slaves were sent to America and Europe from Africa, increasing to 1,341,000 during the seventeenth century, and peaking in the eighteenth century, with more than six million slaves being sent from Africa between 1701 and 1810 (Wolf, 1982: 195–6). Notwithstanding the abolition of the slave trade by Britain in 1807, the trade continued from Africa on a major scale until well into the mid-nineteenth century. In addition, conditions of slavery were imposed upon indigenous peoples in many of the regions colonized by the European powers from the sixteenth century.

The effects of slavery on the populations of the countries from which and to which they were taken are, in broad terms, well known, and include the decimation of regions of West Africa in particular, the transport of slaves in conditions of unspeakable horror across the Atlantic, the inhumane conditions in which they were forced to work, the admixture of races, and the complex processes of attempted abolition of slavery with its political and social consequences, with which we still live. The profits of Western imperial capitalism were substantially made, especially in the sugar industry, through this massive exploitation of a slave labour force: one of the worst examples of human inhumanity to fellow humans in the history of the world.

The detailed demographic conditions of slave societies varied regionally. Marshall (1996a: 283) has suggested that up to 15 per cent of slaves sent from Africa to the British Caribbean in the eighteenth century might die on the voyage, and up to 20 per cent in the first three years of plantation work, with slave populations having a low reproductive rate. From the early nineteenth century Indian indentured labourers were substituted in the West Indies, and 'free' labourers migrated from India to Malaya in the nineteenth and twentieth centuries, together with a massive migration of Chinese labourers to Malaya in particular.

The epidemic and morality consequences of European contact with indigenous populations was replicated through almost all later phases of European imperialism, partly on account of many of these populations possessing no immunity from Old World diseases, and partly because of the consequential effect, related to greater population movement and mixing, of the intensification of existing diseases. The indigenous inhabitants of New England were much reduced by smallpox in the mid-seventeenth century, and the same disease had deadly effect on the Aboriginal population of Australia in 1789, as did measles in Fiji in 1875 (Pawson, 1990: 533). In the British empire a contrast has been suggested between the territories, such as Australia, New Zealand, and the Pacific islands, where European diseases came in with devastating effect, and Africa and India, where existing diseases intensified, partly through urbanization and concentration of populations and partly through irrigation schemes, especially in the Indian subcontinent. Thus

one of the unintended consequences of some irrigation schemes was to create a favourable environment for the anopheles mosquito, the spreader of malaria. India was ravaged by cholera (twenty-three million people may have died in epidemics between 1865 and 1949), by the plague (an estimated twelve million victims died in the great pandemic that started in 1896), and by malaria, influenza and tuberculosis.

There were parallel increases from the mid-nineteenth century in Africa of sleeping sickness, malaria, and rinderpest in animals (Fieldhouse and Marshall, 1996: 142). As will be shown later, a great deal of attention was given, especially in the nineteenth and twentieth centuries, to the problem of European acclimatization, especially to tropical climates, and to the treatment and eradication of the major tropical diseases. The main concern was with the reduction of the mortality rates among Europeans, rather than the indigenous populations, but the danger from contagious diseases inevitably involved much scientific endeavour to reduce such diseases overall, through vaccination and other modes of prevention.

While Western understanding of tropical diseases progressed through the work of the schools of tropical medicine, founded in most of the imperial European states in the late nineteenth century, insufficient attention was paid to the cornucopia of indigenous medical knowledge and to the encouragement of medical training. For a very long time, the advancement of standards of health in imperial territories was based on the assumed superiority of Western medicine and a denial of the existence of traditional and effective local medical practices. Local agents were used in vaccination programmes and as assistants to Western doctors, as exemplified in the establishment of the Dokter Djawa School of training in Indonesia in 1851 by the Dutch, though the scope of training was limited to Western methods and instruction was in Dutch (Verma, 1995). Verma indicates similarities and differences between colonial attitudes to medical training between the Dutch East Indies and British India: the similarity being that Indian medical education was essentially Western, the difference being that Indian demand for greater training was accepted by the British, and medical colleges were founded in Calcutta and Madras in 1835, Bombay in 1845, and Lahore in 1860. By contrast, the provision of university-level medical education in Indonesia was delayed until 1927 (Verma, 1995).

Conflict between different traditions was also evidenced in attitudes to the provision of higher standards in public health, especially in congested urban areas. Thus, as Yeoh (1992) has shown, the colonial intention at the end of the nineteenth century in Singapore was to use British expertise, effective by this time in the dense industrial areas of Britain, to reduce the high mortality rates (32.9–51.1 per thousand between 1893 and 1910), driven by tuberculosis, beriberi, and malarial fevers, cholera, enteric fever, diarrhoea, and smallpox. Colonial authorities emphasized the role of Asiatic, especially Chinese, domestic practices as major factors contributing to the high mortality rates, and initiated sewage and reconstruction schemes as solutions. While British authorities acknowledged and tolerated the presence of traditional Chinese health and medical provision, such systems were used as a form of passive resistance to municipal controls, leading to greater acknowledgement and tolerance of these different forms of medicine and health provision.

A major aspect of the European imperial encounter was that of a concern with the effects of climate on settlement, including acclimatization—the fitness of white Europeans for settlement and labour in the tropical and equatorial regions—together with European assumptions made of the abilities of indigenous populations to progress towards self-determination (not entirely an environmental question, but connected, almost inevitably, with European concepts of race). Acclimatization was not merely a question of the fitness of humans to settle in humid and arid zones: it was also bound up with the propensity of certain animals and plants to survive and prosper in such places, and both of these aspects were reflected in the large numbers of acclimatization societies that developed in Europe and America in the late nineteenth and early twentieth centuries, and the importance given to such issues in major scientific societies, including geographical societies (Livingstone, 1987; Livingstone, 1992: 232–41; Bell, 1995a).

European colonization also affected the natural environments of these countries, and set in motion major changes in the environments that were first encountered at 'discovery' or early periods of settlement. While the initial effects are sometimes difficult to estimate, on the basis of the uncertainty of the degree of modification of natural environments by pre-European inhabitants, the longer-term effects were undoubtedly of major proportions. As far as the Americas in the late medieval and early modern period are concerned, Denevan (1992) has contested the 'pristine' view of pre-European landscapes and environments, that is, the view that such environments were barely altered from their natural state by a thinly scattered population, a view that he ascribes to the invention of nineteenth-century romanticist and primitivist American writers. He asserts, in contrast, that 'By 1492 Indian activity throughout the Americas had modified forest extent and composition, created and expanded grasslands, and rearranged micro-relief via countless artificial earthworks. Agricultural fields were common, as were houses and towns and roads and trails', adding the interesting argument that because of the massive reduction in the size of the native population, 'The landscape of 1750 was more "pristine" (less humanized) than that of 1492' (Denevan, 1992: 370–1). Watts (1995: 272–3) is more cautious about the pre-European environmental impacts of indigenous peoples on the island Caribbean, but makes a similar point to Denevan about the abandonment of cultivation sites, as a result of population decimation and their covering by forest, in this instance in Barbados, St Kitts, Montserrat, Nevis and Antigua, Martinique, and Guadeloupe. The major ecological shock in the West Indies came in the form of the extensive development of sugar-cane plantations with the labour provided by slaves, the process starting in Barbados and St Kitts in the 1640s, thereafter spreading to the Greater Antilles and the remaining small islands. The process involved extensive clearing of forest and, as a result, an extensive loss of existing plant species. Because sugar was very much a crop traded internationally, it was subject to fluctuations in demand and price and also to the effects of decline in soil quality (Watts, 1995: 275–6). The plantation system depended heavily on slave labour. Following the abolition of the slave trade in 1807 and the British abolition of slavery in the empire, indentured labourers from India were introduced in Trinidad, British Guiana, and Mauritius, and at a later period in Natal, Fiji, and Malaya (Christopher, 1988: 178–9). New crops were introduced to the plantations developed by the British in India and Malaya in the nineteenth and early twentieth centuries: tea, planted in Assam from the 1830s, extended to the foothills of the Himalayas, the Nilgri hills of southern India, and to Ceylon, while the rubber-tree, disseminated by botanical gardens from the Amazon via Kew to South-East Asia was grown on major rubber plantations developed in Malaya, Ceylon, and North Borneo within the British Empire (Christopher, 1988: 180–2).

Forest clearance, for settlement, timber for fuel and various types of house and agricultural construction, and as a prelude to agriculture of various kinds, was a major feature of European colonization and environmental management overseas. In North America, for example, the European settlement, preceded by forest-clearing by indigenous inhabitants, accelerated the process from the early seventeenth century, initially along the eastern seaboard, but accelerating in the eighteenth and nineteenth centuries as the settlement frontier moved westwards, so that Williams (1990) has estimated that 113.7 million acres of forest had been cleared before 1850. In Australia, woodland and forest clearance accelerated with the first European settlement, and before 1850 environmental modification by European settlers was primarily through the development of subsistence agriculture and the use of raw materials for the early settlements like Sydney. After 1850, environmental change was strongly conditioned by the gold and mineral rushes, intensification of cereal and animal food production for home and overseas markets, and by land settlement policies, all within a changing context of understanding of what might be possible in the harsh environments of Australia (Heathcote, 1975; Powell, 1988).

In India the principal imperial influence on the environment came through the development of commercial plantations, irrigation schemes, and the use of forest resources. Pre-existing irrigation schemes were adapted and extended by British engineers in the nineteenth and early twentieth centuries, notably in the Punjab and Madras, in order to expand the economy and food production and to provide a greater measure of famine relief (Christopher, 1988: 172–5). Similarly, pre-imperial systems of forest use and, indeed, conservation, were adapted by the British in India. Thus Grove (1995) has shown that the East India Company imitated the forest conservation and promotion policies of conquered Indian states, and many such forests were earning large revenue sums for the British by the 1870s. Concern at the deforestation of India led to the establishment of a Forestry Committee to examine the state of teak forests in Malabar, with reference to the availability of timber for the British navy. In 1846 a new forest conservancy was established in the Bombay Presidency (Grove, 1995: 436), and ultimately an Indian Forest Service, which was copied in other British colonies (Christopher, 1988: 188).

## Administration and Imperial Policy, Including Classical Precedents

The modes and systems of administration in the early European colonies and in the fully-fledged and developed imperial territories were greatly varied. The classical model of the Roman Empire was frequently invoked, mainly as a justification for the control of territories overseas, though in the case of British enthusiasts for the invocation of this model contrasts were also made, including the tyrannical and exploitive nature of the Roman Empire and its territorial contiguity compared with that of the British Empire (Butlin, 1995: 173; Betts, 1971: 153–4). Similar use was made of this Roman model by the French, with particular reference to their North African territories, the Germans (also with reference to the Holy Roman Empire), and the Italians.

The systems by means of which the European states administered their imperial territories were extremely complex, and mirrored the complexity of the phenomenon of imperialism itself. Lowe (1994: 74) has suggested that search for simple motives for European imperialism in the nineteenth century is fraught by the obvious fact that

no European empire seems to have served any definable purpose. The British empire alone was a curious hotch-potch of crown colonies, protectorates and self-governing dominions, scattered across the whole globe. By about 1900 the French empire included vast (and unproductive) tracts of West Africa, some more valuable lands in North Africa, and Indo-China. The Germans, who ruled over a disparate collection of territories in Africa, as well as a few Pacific islands, were unsure whether their colonies should serve as areas of white settlement or as regions for economic exploitation.

Administrative systems thus reflected this geographical diversity and uncertainty of purpose, the histories of colonization and imperial development, and the changing ideologies of imperialism of individual countries.

The British system ranged from direct to indirect governance. By the late nineteenth century it has been estimated that Britain ruled about a quarter of the world's population, and did so by means of what Stockwell (1996: 149) has called 'a gangling bureaucracy'. This focused on monarch and Parliament in London, and administration through the Colonial, India, and Dominions Offices. Evolution of government for some colonies was via the representative system (with Royal Governor) through to responsible government and ultimately Dominion status, while for others their status was that of crown colonies (Ceylon and Trinidad, for example) or protectorates, the latter dating from the late nineteenth century. India was initially governed through the East India Company, with British sovereignty over India recognized from 1813, though some princely or native states were linked to the British Crown by treaties (Marshall, 1996b: 152–3).

## Ethnological and Cultural Perspectives

A very important ingredient in European imperialism throughout the ages was the very strong sense of the superiority of European belief systems and cultures to those of the peoples and areas conquered and colonized. This superiority found particular expression through religion and the attempted proselytization of both local and major regional belief systems by various forms of Christianity, allied in most cases to military and economic conquest and exploitation. Additionally, Western medicine and education in countries colonized and controlled by European states was provided by their missions and missionary societies, and used as part of a complex process of proselytization and imperial control.

In Latin America in the sixteenth, seventeenth, and eighteenth centuries, Franciscan, Jesuit, and Dominican Catholic missionaries operated within the territories of the Spanish and Portuguese empires. The Spanish missionary orders were almost exclusively staffed by Spaniards, who in the early years of conquest and mission attempted to combine conversion to Christianity with the preservation of the commendable traits of these 'noble savages' from the ravages of Spanish military and settlers, notwithstanding the brutality of the conquistadores, the multiplicity of indigenous languages and belief systems, and the difficulties of the environment. Early ideals and successes were moderating by the last quarter of the sixteenth century, when receipt of the sacraments was low among indigenous Christians in Mexico City, for example, and quarrelling among the religious orders had become commonplace. The gap in European colonist behaviour between Christian precept and practice was everywhere apparent to the indigenous population, and a move begun by the Jesuits in particular to more material and business interests, including the ownership and management of estates, a process of material acquisition and prosperity that was considerably exceeded by the secular church (Scammell, 1981: 356–7). Newson (1993: 272–4) has suggested that, in addition to the effects of European diseases unwittingly introduced by the missionaries in the New World, they also attempted to sedentarize many semi-nomadic groups. In consequence, notwithstanding the innovations of new crops and livestock, they actually effected a reduction in food production and in diet, and insisted on monogamous marriage practice. Spanish mission activity also spread to the Philippines and to Cambodia, China, and Japan.

The role of the Portuguese missionary orders and secular church was world-wide, and by the mid-sixteenth century, greatly influenced by the activities of the Jesuits (founded in 1534). They had arrived in India in 1542, were effective in Brazil where they founded São Paulo in 1554, and reached China in 1582 and Japan from the late sixteenth century. In India, Jesuits and Franciscans were active in the region of Goa, where by the early seventeenth century a number of churches had been established and a small percentage of the population converted (Scammell, 1981: 285). Overall the impact of Portuguese religious orders was very much less than that of the Spanish.

The eighteenth century was a quiet period as far as Catholic missionary work was concerned, but there was a major revival in the nineteenth century. In the reformed churches the Society for Promoting Christian Knowledge was founded in 1698–9, to disperse bibles and religious tracts at home (Britain) and abroad, and the Society for the Propagation of the Gospel in foreign parts (SPG) in 1701, though European missionary work at this time, other than through the Roman Catholic Church, was also undertaken by the Moravians, originating in Bohemia in the fifteenth century.

The eighteenth and the early nineteenth century saw the growth of many European missionary societies, established to follow in the steps of the explorers, armies, and administrators and to minister to the newly incorporated indigenous populations of the growing European overseas empires. The background was not merely that of the Bible following the flag and merchants: the Christian religion in particular had acquired new energy. This was evidenced by High Anglican piety and John Wesley's Methodist movement—but two parts of the Evangelical Revival in Britain—and by revivalism in Germany, the Reveil of Switzerland in the 1830s which also spread to France, and the parallel developments in Norway. 'The immediately obvious result of this release of new Christian energy was the rapid geographical expansion of the work' (Neill, 1975: 252).

The foundations in Britain of the Methodist Missionary Society in 1786, the London Missionary Society in 1795, the Church Missionary Society in 1799, the British and Foreign Bible Society in 1804, and the development of the missionary arms of the Wesleyan Methodists in 1818 and the Church of Scotland in 1825, had parallels in other European countries. The first mission in Germany—the Berlin Society—was founded in 1824, the Basel Mission in Switzerland in 1815, and similar societies in Denmark in 1821, France in 1822, Sweden in 1835, and Norway in 1842 (Neill, 1975: 252). Much progress was achieved by the translation of the Bible into many languages: about 70 by the end of the eighteenth century and about 520 (Bible, New Testament, and smaller fractions of the Bible) by the end of the nineteenth. The process of development of missionary objectives was through preaching, teaching in schools, and through medical missions, agricultural and horticultural projects, printing and publishing ventures, industrial schools and handicrafts. Initially the missionaries were men, but by the mid-nineteenth century women were also missionaries, and through time the early patriarchal system changed and widened towards an indigenous ministry (Neill, 1975: 237). The work of missionaries seems to have been most effective in regions of large, poor populations, such as in Hindu societies beset by famine, land problems, and debt in late nineteenth-century India, and the process of mission included the significant role 'as effective solvents of non-European societies; they acted as channels for the inflow of western knowledge and values, calling into question traditional authority and institutions, challenging customary ideas, for example about marriage or the status of women, and influencing patterns of local political influence or power' (Porter, 1994a: 136).

Missionary work was also undertaken by the priests and chaplains attached to the major trading companies such as the East India Company. Missionary activity in the Calcutta region of India, for example, was enabled by the renewal of the charters of the East India Company in 1813 and 1833, and the foundation of the diocese of Calcutta in 1814, partly to control the growing number of missionaries and also to offset the possible effect of too much freedom of activity in antagonizing the Muslims and Hindus (Porter, 1994a: 132). The pattern of British missionary activity in India in the early nineteenth century differed from that in the West Indian Colonies and in the Cape Colony in South Africa, where there was much conflict between planter and missionary (Porter, 1994a: 132).

European soldiers, administrators, and settlers did not, of course, find the lands of empire devoid of belief systems. In North Africa they found Islamic civilizations, and Islam had also been transported to parts of West Africa and the coast of East Africa. The resistance of such groups to the impact of Christianity was very much tied up with their resistance to outside colonization and imperial intrusion, and ultimately with the aggressive promotion of independence movements from the late nineteenth to the mid-twentieth century.

The relationship between missionary activity and the advance and decline of empire still needs further investigation. As Porter has indicated,

These debates [on links between missionary activity and the capitalist intensions of empire] have been highly illuminating, but missionaries' integration with the process of empire-building, it sometimes seems, was significant for scarcely more than its contribution to the extension of capitalist markets. In such ways religion is widely presented as the flimsiest of ideological stucco on the imperial edifice'. (Porter, 1991: 6–7)

Key features for re-examination cited by Porter include the complexity of the relations between government and missions (including the reservations by missionary societies about imperial policy which reflect a gulf between the two in the years leading up to the First World War) and, interestingly, the beginnings of more sympathetic relations with non-European belief systems (Porter, 1991). Prasch (1989) has reviewed the debate in Victorian England about the appropriateness of Christianity or Islam as civilizing influences for Africa, but has none the less reminded us that the Islamicist sympathies among small numbers of Victorian clergy and lay people were still based on a racist perspective: 'Islam was seen as an acceptable religion for Africa because both Islam and African were seen as lower forms, incapable of attaining the apex of Western civilization' (Prasch, 1989: 72–3).

## World-Systems and the Commercial Geographies of Empire

One of the most frequently discussed reasons for European imperial expansion, especially in the nineteenth century, is that of economic gain. There is no doubt that, throughout the evolving process of overseas exploration, trade, settlement, and conquest, economic benefits were anticipated and realized. Vance (1970) has outlined and developed a stage model of trade development from a European base: the first involved an exploratory stage in which new lands were found and visited, and their economic potential assessed, and some materials such as precious metals and valuable plants sent back to the homelands; the second involved the establishment of 'factories' of European merchants in small port settlements in the colonies; the third phase involved larger-scale settlement by Europeans in North America, parts of Latin America, South Africa, Australia and New Zealand, or smaller-scale settlements allied to strong colonial administration and varied degrees of military presence in, for example, India and Africa. The consequences were gradual intensification of trade between homeland and colonial territories, a process incorporating the development of mercantilist and capitalist systems.

Mercantilism, according to Dodgshon (1987: 295) was a form of regulated trade, incorporating the use of privileged trading companies such as the Dutch East India Company, the English East India Company, and the Hudson's Bay Company, and the use of protectionist measures, including tariffs on imports, in order to foster economic growth. An intensification of the search for profit from the import of natural resources and products such as sugar and tea from low-cost, slave- or coerced-labour plantations from colonial territories was part of the linked process of intensification of a capitalist system of economic and social relations. The change from a mercantilist/early capitalist to a fully-fledged capitalist system of resource exploitation took place in Britain in the early to mid-nineteenth century, with the abandonment of tariffs and preferential duties, and the move to create a world market for imported food and raw materials and for manufactured exports. Additionally, overseas investment of surplus European capital in infrastructural and other developments was seen as an attractive possibility for financial gain.

The rhetoric of the economic gains to be had from imperial engagement and investment was strong and widespread, and the establishment in Europe in the late nineteenth century of many societies of 'commercial geography', usually linked to the commercial ambitions of large port and trading cities, is an interesting indicator of this. Schneider (1990) has shown in a study of this

kind of 'municipal imperialism' in France in the period 1870–80 that the potential use of geographical information about supplies of raw materials and markets in colonial territories led to an intensification of commercial interest within geographical societies in France and the additional development of societies and a commission of commercial geography. There were parallel developments in most other European imperial countries, notably Germany, though in Britain the few attempts to establish formal societies of commercial geography, national not regional, got no farther than initial conception, much of the work being left to the provincial societies (MacKenzie, 1995a).

Did, however, the rhetoric of the economic benefits of empire match the reality? Much depends on the period under review and the basic premisses of the theoretical stances adopted. Generally speaking the answer is that it did not. As far as the period from 1860 to 1914 is concerned, Porter (1994a: 40–1), outlining the essential features of 'metropolitan' explanations of European imperialism, has argued that 'with few exceptions colonial trade remained throughout this period an insignificant proportion of metropolitan commerce', and that 'Europe's long-term overseas investment also went overseas to areas outside the formal colonial empires. French capital went to Russia, Italy's to the Balkans and Middle East, British to the United States and Latin America, German to all of these areas.' The reasons given by Porter for the gap between advocacy of imperial investment and the reality thereof include: the diffuse nature of banking and finance in Europe; the lack of businessmen in the ranks of the advocates of imperial investment, especially the colonial societies; the coincidence of agitation by chambers of commerce with downturns in trading conditions, which partly account for the short-lived nature of such movements; and the varying competitions for trade and investment which were motivated by political ambitions (Porter, 1994a: 42–4).

Additional features of the analysis of the links between European economies and those which formed part of their empires are such questions as the significance of 'peripheral' factors such as the nature of the societies colonized, local conditions, and interaction with Europeans in the imperial territories themselves. Another feature of the historiography of imperial explanations is that of the Wallerstein 'world-system' thesis, which links the historic growth of a capitalist system of production and trade in a heartland of north-west Europe from the late fifteenth century to a configuration in which that core draws large areas of the less developed world into a global economic hierarchy, reaching a fully developed form in the late nineteenth and early twentieth centuries (Wallerstein, 1974, 1980, 1989). This approach, with its detailed and complex illustrations and documentation, has been subject to much criticism, including objections to its Eurocentricity and consequent underestimation of the development of other non-European major trading cores, together with its emphasis on the economic (Porter, 1994: 59–60), and also objections to the chronology, pace, and scales of evolution of major economic and cultural systems (Dodgshon, 1987; 1993).

## Geographical Societies and Imperial Connections: The Experiences of the Nineteenth and Twentieth Centuries

The nature of the relationship between geographical knowledge, the geographical societies of Europe, and the nationalist and colonial ambitions of their respective countries at the time of most rapid imperial development in the nineteenth and early twentieth centuries is a complex, continually contested, yet extremely interesting question. This was a period of rapid change in the nature of geographical knowledge, through exploration, survey, and mapping, and also of change in imperial ideologies and opportunities, and a time when geographers both supported and to a lesser degree opposed the concepts and ambitions of imperialism.

Bridges has contended, for example, that the Royal Geographical Society and its explorers were progenitors of imperialism and participants in the process through their interest in what he terms 'infrastructures', that is primarily the activities of the members and explorers who provided the organization for new colonial territories, including the infrastructures of roads, railways, telegraphs, and administration, and this statement is generally true of the other geographical societies of Europe (Bridges, 1973). Similar comments have also been made about imperial contexts of the activities of the Hakluyt Society, founded in 1846 for the

publication of scholarly editions of accounts of historic voyages and travels, and with strong links to the Royal Geographical Society (Bridges and Hair, 1996: 231). The potential application of the geographical knowledge acquired by the many explorers derived from the maps that they made, their knowledge of routes, access, and transport potential (by water and land, including railways), assessments of economic potential, and appraisals of the character of the indigenous inhabitants of potential colonial territories.

The links between national and local geographical societies and the imperial endeavours and achievements of the major European powers were complex and varied, and recent work has begun to provide new insights into this important question. Two major collections on the relations between geography and empire (Godlewska and Smith, 1994; Bell, Butlin, and Heffernan, 1995) include data on specific European geographical societies in this context.

The dates of foundation of the major geographical societies of the European imperial powers were: Paris Geographical Society, 1821; Berlin Geographical Society, 1828; Royal Geographical Society (London), 1830; Italian Geographical Society (Florence), 1867; Royal Dutch Geographical Society (Amsterdam), 1873; Lisbon Geographical Society, 1875; Royal Geographical Society of Madrid, 1876; Belgian Geographical Society, 1876; Royal Scottish Geographical Society, 1884. All of these societies were involved in the support of scientific research, including exploration, map-making, and resource survey, together with the organization of meetings, international conferences, and publications, including their own journals. The exploration tradition was very deep-rooted in the Royal Geographical Society of London, which was preceded by the African Association (founded 1788), and evidenced by expeditions such as those of Speke and Burton to discover the source of the Nile, starting in 1857. Expeditions from the European powers and sponsored by the major geographical societies had mixed motives: although the acquisition of new and important geographical knowledge was a major goal, so also was the review of both economic and strategic prospects for the home country. Thus the Sumatra expedition of the Royal Dutch Geographical Society (1877–9) had an economic motive—the discovery of a route for the coal from the Ombilin field—but a political motive was the control of part of Sumatra through which the coal would be transported (Van der Velde, 1995: 86).

In France and Italy geographical knowledge was closely linked to national image and ambition, including imperial aspiration. France's promotion of geography after the Franco-Prussian War of 1870–1 as a means of redress of this great setback is well known, and reflected in part by the increase in members of the Paris Geographical Society. Heffernan (1995: 222–3) has shown that both geography and history were engaged in the consequent programme of national educational reform, which included the creation of chairs of geography in the universities, including a chair of colonial geography at the Sorbonne. Members of the Paris Geographical Society were also active in the association of clubs and societies which formed the Parti Colonial, one of the major promoters of French colonial ambition (Heffernan, 1995: 225).

The Italian Geographical Society from the point of unification of Italy in 1870 supported an expedition to Eritrea, and in the mid-1870s to East Africa, Morocco, and Tunisia. As Atkinson (1995: 268–9) has pointed out, 'Although the banner of science and notions of trading and commercial links featured prominently in the public profile of such expeditions, the journeys were supported by a society enmeshed in the overseas ambitions of Italy's political class', trends continued into the fascist period of the 1930s.

The German experience is similar to that of Italy but different from that of Britain and France: national unification was not achieved until 1871; there was no long tradition of overseas colonization and control; the extent of colonial conquest in the nineteenth and early twentieth centuries was very modest; and the contribution to the German economy was little more than marginal. That is not to say, of course, that colonialism as an ideal and as a prospect was not important: throughout the late nineteenth and early twentieth centuries arguments for and against overseas investment and territorial control occupied politicians, commercial traders, bankers, and scientists, and led to the formation of special-interest societies, some specifically for the advancement of colonial trade and settlement, others having such items high on their agendas, and many of which paralleled in purpose and achievement similar organizations in France and Britain. Sandner and Rössler (1995) have shown that there was support from prominent German geographers in 1882 for the foundation of the German Colonial Society, though no evidence that geographers were supportive of liberal anti-colonial sentiment. The links between geography and imperialism were sustained into the period of fascism in the 1930s and 1940s (Murphy, 1994).

## The Empire Strikes Back: Anti-Imperialism and Independence in the Nineteenth and Twentieth Centuries

The demise of the European colonial empires, which had further developed in the nineteenth and early twentieth centuries from pre-existing bases to their zenith at the beginning of the Second World War, was remarkably rapid: they had largely disappeared by the mid-1960s through complex processes of opposition and disengagement. There were two main waves of decolonization in the twenty-five years following the end of the Second World War: the first involved the Islamic Middle East and the Far East (before 1950); and the second mainly involved states in Africa and the Caribbean (Dickenson et al., 1996: 47–8).

The European states with colonial territories in 1939 were Britain, France, the Netherlands, Italy, Belgium, Spain, and Portugal. Some of these states, notably Britain, France, and the Netherlands, had already accepted the need for progress of dependent territories towards varied measures of self-government, but the precise details of national ideologies in this respect varied, with only Britain being committed to the notion of total independence of colonial territories, albeit from 1931 within the framework of a Commonwealth. Elsewhere continuing links with metropolitan territories and the maintenance of incorporation of indigenous cultures within a broader 'civilizing mission' were part of post-colonial philosophies. There was also great resistance to the notions of national independence of colonial states by resident European colonial settlers, which clashed with the energetic and determined advance of the cause of self-rule by religious and secular factions.

The process of European disengagement was complex, and often bloody. The Second World War itself had profound effects on this process, including the conquest of colonial territories in the Pacific and South-East Asia by the Japanese and the establishment of nationalist-based governments which were the force for independence after the Japanese surrender in 1945. Conflict and resistance to British rule in India during this period led to the transfer of power to India and Pakistan in 1947, and from 1957, when Ghana was given independence, there followed a rapid process of British decolonization in Africa. The Fourth French Republic from 1946 rearranged French imperial structures into metropolitan and overseas departments, but had to disengage from Vietnam after the French defeat at Dien Bien Phu in 1954. Laos, which had been under French rule since 1896, became independent in 1953, and there was a rapid withdrawal of France from West Africa between 1958 and 1960, after the collapse of the Fourth Republic in 1958. In North Africa, France gave up its power in Tunisia and Morocco in 1956, and also in Algeria in 1962 at the end of eight years of bitter strife.

Decolonization by Belgium, the Netherlands, and Portugal took place in the same period. Indonesia's declared independence of 1945 was recognized by the Dutch in 1959; the Congo became independent from Belgium in 1960, but was engaged in a long and bloody crisis based on regional and tribal loyalties from 1960 to 1965; Portugal was the last of the European powers to be forced to withdraw from colonial activity, mainly by the overthrow of the Portuguese domestic dictatorship in 1974 and by the strength of guerrilla movements in her African territories, which had achieved independence by the end of 1975. German and Italian colonial territories were lost as a result of the Second World War.

In the Middle East, a combination of powerful nationalist movements, a progression to independence from mandate status, and varied degrees of reduction of European economic and military commitment led to independence for Syria and the Lebanon in 1944 and 1945, while Israel's struggle for statehood was partially successful in 1947, but with consequences that continue to the present day.

The reasons for the speed of European decolonization and its attendant processes are virtually impossible to generalize, for they differ even between the territories of individual European states. Common features clearly included the effects of the Second World War, including the role of troops from colonial territories, intimations of independence, for example by Britain, from the early 1930s, changes in global geopolitics through the rise of the Soviet Union and the United States as superpowers, the weakening of the racist and cultural ideologies, and the increasing military cost of holding on to empire, combined with the powerful guerrilla and independence movements within the colonies themselves.

Equally difficult of analysis is the role of geographical knowledge in this process. If the work of professional and amateur geographers had contributed in complex and disputed ways to the process of colonization and imperialism over time, in what ways did it contribute to its demise? At this point in time the answer would seem to be 'very little', for it may be argued that in the post-war period, at least in Europe, geography, at least by the 1960s and 1970s,

in a conscious attempt to ride the discipline of its unsavoury imperialist connotations and resulting lack of rigour and respectability . . . entered an abstract world of mathematical modelling and quantitative spatial science. In the aftermath of widespread social protest across the western world around 1968, many of its practitioners also recaptured the western fascination with a recently-defined 'Third World', through a more politically engaged orthodoxy in which the geographer's science was employed to champion the rights of the world's poor and dispossessed. (Bell, Butlin, and Heffernan, 1995: 7)

While geographers such as Kropotkin and, later, Dresch actively promoted an anti-imperialist stance (Bell, Butlin, and Heffernan, 1995: 6–7), they seem in their time to be isolated figures surrounded by a geographical community either actively or benignly supportive of the imperial cause.

More recently geographers have begun to engage with the debates and post-colonial critiques initiated by such scholars as Edward Said, notably in his seminal work *Orientalism* (1978). A wide range of debates around a fascinating and important array of topics has ensued, including the Said theories themselves (MacKenzie, 1995b); gender and empire (Bell, 1995b; McEwan, 1995 and 1996); the analysis of paintings, photographs, and accounts of travel; and the impact of imperial perspectives and propaganda on teaching and texts

(Ryan, 1995; Driver and Maddrell, 1996; Ploszajska, 1996). None the less, an analysis of the apparent absence of anticolonial critiques among the ranks of professional geographers in Europe in the mid-twentieth century is an interesting and important question still awaiting an answer.

REFERENCES

ATKINSON, D., 'Geopolitics, Cartography and Geographical Knowledge: Envisioning Africa from Fascist Italy', in M. Bell, R. A. Butlin, and M. J. Heffernan (eds.), *Geography and Imperialism 1820–1940* (Manchester, 1995), 265–97.

BELL, M., 'Edinburgh and Empire: Geographical Science and Citizenship for a "New" Age, ca 1900', *Scottish Geographical Magazine*, 111(3) (1995a), 139–49.

—— '"Citizenship now charity": Violet Markham on Nature, Society and the State in Britain and South Africa', in M. Bell, R. A. Butlin, and M. J. Heffernan (eds.), *Geography and Imperialism 1820–1940* (Manchester, 1995b).

BETTS, R. F., 'The Allusion to Rome in British Imperialist Thought of the Late Nineteenth Century', *Victorian Studies*, 15 (1971), 153–61.

BROC, N., *La Géographie de la Renaissance*, (Paris, 1986).

BRIDGES, R. C., 'Europeans and East Africans in the Age of Exploration', *Geographical Journal*, 139 (1973), 220–31.

—— and HAIR, P. E. H., 'Epilogue: The Hakluyt Society and World History', in R. C. Bridges and P. E. H. Hair (eds.), *Compassing the Vaste Globe of the Earth: Studies in the History of the Hakluyt Society 1846–1996* (London, 1996), 225–39.

BUTLIN, R. A., 'Historical Geographies of the British Empire, *c.* 1887–1925', in M. Bell, R. A. Butlin, and M. J. Heffernan (eds.), *Geography and Imperialism 1820–1940* (Manchester, 1995), 151–88.

BUTZER, K. W., 'The Americas before and after 1492: An Introduction to Current Geographical Research', in K. W. Butzer (ed.), *The Americas before and after 1492: Current Geographical Research. Annals of the Assocation of American Geographers*, 82(3) (1992), 345–68.

CARTER, P., *The Road to Botany Bay: An Essay in Spatial History*, (London, 1987).

CHRISTOPHER, A. J., *The British Empire at its Zenith*, (London, 1988).

DEAGAN, K. and CRUXENT, J. M., 'From Contact to *Criollos*: The Archaeology of Spanish Colonization in Hispaniola', in W. Bray (ed.), *The Meeting of Two Worlds: Europe and the Americas 1492–1650* (Oxford, 1993), 67–104.

DENEVAN, W. M., 'The Pristine Myth: The Landscape of the Americas in 1492', in K. W. Butzer (ed.), *The Americas before and after 1492: Current Geographical Research. Annals of the Association of American Geographers*, 82(3) (1992), 369–385.

DICKENSON, J., GOULD, W., CLARKE, C., MATHER, S., PROTHERO, R. M., SIDDLE, D., SMITH, C., and THOMAS-HOPE, E., *A Geography of the Third World*, 2nd edn. (London and New York, 1996).

DODGSHON, R. A., 'The Modern World System: A Spatial Perspective', *Peasant Studies*, 6(1) (1977), 8–19.

—— *The European Past: Social Evolution and Spatial Order*, (London, 1987).

—— 'The Early Modern World-System: A Critique of its Dynamics', in H.-J. Nitz (ed.), *The Early Modern World-System in Geographical Perspective*, (Stuttgart, 1993), 26–41.

DRIVER, F., and MADDRELL, A. M. C., 'Geographical Education and Citizenship: Introduction', *Journal of Historical Geography*, 22(4) (1996), 371–2.

EARLE, C., 'From Cabot to Cartier: The Early Exploration of Eastern North America 1497–1543, in K. W. Butzer (ed.), *The Americas before and after 1492: Current Geographical Research—Annals of the Association of American Geographers*, 82(3) (1992), 500–21.

FERNÁNDEZ-ARMESTO, F., *Columbus*, (Oxford, 1992).

FIELDHOUSE, D., 'The Diaspora of the Africans and the Asians', in P. J. Marshall (ed.), *The Cambridge Illustrated History of the British Empire*, (Cambridge, 1996), 280–95.

—— and MARSHALL, P. J., 'Health and Disease', in P. J. Marshall (ed.), *The Cambridge Illustrated History of the British Empire*, (Cambridge 1996), 142–3.

GODLEWSKA, A., 'Napoleon's Geographers (1797–1815): Imperialists and Soldiers of Modernity', in A. Godlewska and N. Smith (eds.), *Geography and Empire*, (Oxford, 1994) 31–54.

—— and SMITH, N. (eds.), *Geography and Empire*, (Oxford, 1994).

GREGORY, D., *Geographical Imaginations*, (Oxford, 1994).

GROVE, R. H., *Green Imperialism: Colonial Expansion, Tropical Island Edens and the Origins of Environmentalism*, (Cambridge, 1995).

HARLEY, J. B., *Maps and the Columbian Encounter*, (Milwaukee, 1990).

—— 'Rereading the Maps of the Columbian Encounter', in K. W. Butzer (ed.), *The Americas before and after 1492: Current Geographical Research. Annals of the Association of American Geographers*, 82(3) (1992), 522–35.

HARRIS, C., 'French Landscapes in North America', in M. Conzen (ed.), *The Making of the American Landscape*, (Boston, 1990), 63–79.

HEATHCOTE, R. L., *Australia*, (London, 1975).

HEFFERNAN, M. J., 'A State Scholarship: The Political Geography of French International Science during the Nineteenth Century', *Transactions of the Institute of British Geographers*, NS 19 (1994a), 21–45.

—— 'The Science of Empire: The French Geographical Movement and the Forms of French Imperialism, 1870–1920', in A. Godlewska and N. Smith (eds.), *Geography and Empire*, (Oxford, 1994) 92–114.

—— 'The Spoils of War: Société de Géographie de Paris and the French empire, 1914–1919', in M. Bell, R. A. Butlin, and M. J. Heffernan (eds.), *Geography and Imperialism 1820–1940* (Manchester, 1995), 221–64.

JONES, E. L., *The European Miracle: Environments, Economies and Geopolitics in the History of Europe and Asia*, 2nd edn. (Cambridge, 1987).

KEMP, P. (ed.), *The Oxford Companion to Ships and the Sea*, (Oxford, 1988).

LIVINGSTONE, D., 'Human Acclimatization: Perspectives on a Contested Field of Enquiry into Science', *History of Science*, 25 (1987), 358–94.

—— *The geographical tradition*, (Oxford, 1992).

LOVELL, W. G., *Conquest and Survival in Colonial Guatemala*, rev. edn. (Montreal and Kingston, 1992a).

—— '"Heavy Shadows and Black Night": Disease and Depopulation in Colonial Spanish America', in K. W. Butzer (ed.), *The Americas before and after 1492: Current Geographical Research. Annals of the Association of American Geographers*, 82(3) (1992b), 426–43.

LOWE, J., *The Great Powers, Imperialism and the German Problem, 1865–1925*, (London, 1994).

MCEWAN, C., '"The Mother of All Peoples": Geographical Knowledge and the

Empowering of Mary Slessor', in M. Bell, R. A. Butlin, and M. J. Heffernan (eds.), *Geography and Imperialism 1820–1940*, (Manchester, 1995), 125–50.

—— 'Gender, Culture and Imperialism', *Journal of Historical Geography*, 22(4) (1996), 489–94.

MACKAY, D., *In the Wake of Cook: Exploration, Science, and Empire 1780–1801*, (London, 1985).

MacKENZIE, J., 'The Provincial Geographical Societies in Britain, 1884–1914', in M. Bell, R. A. Butlin, and M. J. Heffernan (eds.), *Geography and Imperialism 1820–1940*, (Manchester, 1995a).

—— *Orientalism, History, Theory and the Arts*, (Manchester, 1995b).

MADDRELL, A. M. C., 'Empire, Emigration and School Geography: Changing Discourses of Imperial Citizenship, 1880–1925', *Journal of Historical Geography*, 22(4) (1996), 373–87.

MARSHALL, P. J., 'Health and Disease', in P. J. Marshall (ed.), *Cambridge Illustrated History of the British Empire*, (Cambridge, 1996a), 142–3.

—— 'British Colonial Government', in P. J. Marshall (ed.), *Cambridge Illustrated History of the British Empire*, (Cambridge, 1996b), 152–3.

MEINIG, D. W., *The Shaping of America: A Geographical Perspective on 500 years of history. i. Atlantic America, 1492–1800*, (New Haven and London, 1986).

MURPHY, D. T., 'Space, Race, and Geopolitical Necessity: Geographical Rhetoric in German Colonial Revanchism, 1919–1933', in A. Godlewska and N. Smith (eds.), *Geography and Empire*, (Oxford, 1994), 173–87.

NEILL, S., *A History of Christian Missions*, (Harmondsworth, 1975).

NEWSON, L. A., 'The Demographic Collapse of Native Peoples of the Americas, 1492–1650', in B. Bray (ed.), *The Meeting of Two Worlds: Europe and the Americas 1492–1650* (Oxford, 1993), 247–89.

PAWSON, E., 'British Expansion Overseas c. 1730–1914', in R. A. Dodgshon and R. A. Butlin (eds.), *An Historical Geography of England and Wales*, 2nd edn., (London, 1990), 521–44.

PHILLIPS, C. R., 'The Growth and Composition of Trade in the Iberian Empires, 1450–1750', in A. D. Tracy (ed.), *The Rise of Merchant Empires*, (Cambridge, 1990), 34–101.

PLOSZAJSKA, T., 'Constructing the Subject: Geographical Models in English Schools, 1870–1944', *Journal of Historical Geography*, 22(4) (1996), 388–98.

PORTER, A. N., 'Religion and Empire: British Expansion in the Long Seventeenth Century', Inaugural Lecture in the Department of History, King's College, University of London, 20 Nov. 1991.

—— (ed.), *European Imperialism, 1860–1914* (London, 1994a).

—— (ed.), *Atlas of British Expansion Overseas*, (London, 1994b).

POWELL, J. M., *An Historical Geography of Modern Australia*, (Cambridge, 1988).

PRASCH, T., 'Which God for Africa: The Islamic–Christian Missionary Debate in Late-Victorian England', *Victorian Studies*, 33 (1989), 51–73.

PREM, H. J., 'Spanish Colonization and Indian Property in Central Mexico, 1521–1620', in K. W. Butzer (ed.), *The Americas before and after 1492: Current Geographical Research. Annals of the Assocation of American Geographers*, 82(3) (1992), 444–60.

RYAN, J. R., 'Imperial Landscapes: Photography, Geography and British Overseas Exploration, 1858–1872', in M. Bell, R. A. Butlin, and M. J. Heffernan, (eds.), *Geography and Imperialism 1820–1940* (Manchester, 1995), 53–79.

SAID, E. W., *Orientalism: Western Conceptions of the Orient*, (London, 1978).

SANDNER, G., and RÖSSLER, M., 'Geography and Empire in Germany, 1871–1945', in A. Godlewska and N. Smith (eds.), *Geography and Empire*, (Oxford, 1994), 115–30.

SAVAGE, V. R., *Western Impressions of Nature and Landscape in Southeast Asia*, (Singapore, 1984).

SCAMMELL, G. V., *The World Encompassed: The First European Maritime Empires c. 800–1650* (London, 1981).

SCHNEIDER, W. H., 'Geographical Reform and Municipal Imperialism in France, 1870–80', in J. M. Mackenzie (ed.), *Imperialism and the Natural World*, (Manchester, 1990), 90–117.

SMITH, B., *European Vision and the South Pacific*, (New Haven, 1985).

STOCKWELL, A. J., 'Power, Authority and Freedom', in P. J. Marshall (ed.), *Cambridge Illustrated History of the British Empire*, (Cambridge, 1996), 147–85.

STODDART, D., *On Geography and its History*, (Oxford, 1986).

TERRY, M., 'Terre Napoleon', in W. Eisler and B. Smith (eds.), *Terra Australis: The Furthest Shore*, (Sydney, 1988), 153–6; catalogue section, 205–15.

VANCE, J. E. Jr., *The Merchant's World: The Geography of Wholesaling*, (Englewood Cliffs, NJ, 1970).

VELDE, P. VAN DER, 'The Royal Dutch Geographical Society and the Dutch East Indies, 1873–1914: From Colonial Lobby to Colonial Hobby', in M. Bell, R. A. Butlin, and M. J. Heffernan (eds.), *Geography and Imperialism 1820–1940* (Manchester, 1995), 80–92.

VERMA, R., 'Western Medicine, Indigenous Doctors and Colonial Medical Education', *Itinerario*, 19(3) (1995), 130–41.

WALDERSEE, J., 'The Malspina Expedition 1789–1794', in W. Eisler and B. Smith (eds.), *Terra Australis: The Furthest Shore*, (Sydney, 1988), 149–52.

WALLERSTEIN, I., *The Modern World-System*, ii. *Capitalist Agriculture and the Origins of the European World-Economy in the Sixteenth Century*, (London, 1974).

—— *The Modern World-System*, ii. *Mercantilism and the Consolidation of the European World-Economy, 1600–1750*, (London, 1980).

—— *The Modern World-System*, iii. *The Second Era of Great Expansion of the Capitalist World-Economy, 1730–1840s* (London, 1989).

WATTS, D., 'Ecological Responses to Ecosystem Shock in the Island Caribbean: The Aftermath of Columbus, 1492–1992', in R. A. Butlin and C. N. Roberts, (eds.), *Ecological Relations in Historical Times*, (Oxford, 1995).

WEE, H. VAN DER, 'Structural Changes in European Long-Distance Trade, and Particularly in the Re-export Trade from South to North, 1350–1750', in J. D. Tracy (ed.), *The Rise of Merchant Empires: Long-Distance Trade in the Early Modern World 1350–1750* (Cambridge, 1981), 14–33.

WHEATLEY, P., *The Golden Khersonese*, (Kuala Lumpur, 1961).

WHITEHEAD, P., 'Natural History Drawing on British Eighteenth Century Expeditions to the Pacific', in W. Eisler and B. Smith (eds.), *Terra Australis: The Furthest Shore*, (Sydney, 1988), 135–40.

WILLIAMS, M., 'The clearing of the forests', in M. P. Conzen (ed.), *The Making of the American Landscape*, (Boston, 1990), 146–68.

WOLF, E. R., *Europe and the People without History*, (Berkeley and Los Angeles, 1982).

YEOH, B., 'Municipal Sanitary Ideology and the Control of the Urban Environment in Colonial Singapore', in A. R. H. Baker and G. Biger (eds.), *Ideology and Landscape in Historical Perspective*, (Cambridge, 1992), 148–72.

ZANDVLIET, K., 'Golden Opportunities in Geopolitics: Cartography and the Dutch East India Company during the Lifetime of Abel Tasman', in W. Eisler, and B. Smith (eds.), *Terra Australis: The Furthest Shore*, (Sydney, 1988), 67–84.

# Chapter 7

# The Changing Political Map: Geography, Geopolitics, and the Idea of Europe since 1500*

## M. Heffernan

## Introduction

The political geography of Europe has been in a constant state of flux over the past 500 years. The continent's borders and frontiers have changed so frequently that a substantial (and probably rather tedious) volume would be required to provide even a crude outline of the major events. Conceptually, such a project would be extremely challenging, for the idea of 'Europe' and the broader notion of 'political space' are not universal, transcendent concepts; rather, they are complex ideas whose meanings have changed enormously over the past half-millennium. The assumption that the political map of Europe in 1500 can be regarded simply as a different version of the comparable map today is thus theoretically and historically absurd.

This conceptual problem provides the starting point for this chapter. I want to consider the changing *idea* of political space in Europe over the past 500 years, in terms both of the continent's external limits and of its internal political divisions. What follows, then, is an attempt at a *historical geography*

of the idea of Europe since the Renaissance. I emphasize this phrase because I believe the idea of Europe needs to be understood *historically* and *geographically*.

## The Idea of Europe: Towards a Historical Geography

The future direction of the European project is a matter of overwhelming importance. Unfortunately, discussion has been dominated by the bureaucratic and financial questions involved in the process of European integration. As a result, debate tends to generate more heat than light. It is important, therefore, to step back from the technical detail and ask larger questions, notably: 'What has Europe meant in the past and what might it mean in the future?'

This enquiry is scarcely unheralded, of course, for several histories of the European project have grappled with this very question (Albrecht-Carrié, 1965; Barraclough, 1963; Brugmans, 1970; Chabod,

* I would like to thank Gerry Kearns, David Slater, Graham Smith, and Peter Taylor for useful advice and comments. The maps were drawn by Peter Robinson.

1965; Duroselle, 1966, 1990; Foerster, 1967; Fuhrmann, 1981; Gollwitzer, 1951b; Hay, 1968; Morin, 1991; Voyenne, 1964). Excellent though they are, these accounts are limited in two respects.

First, most are written from an uncritically 'pro-European' perspective, as if the concept of a united Europe is both intrinsically desirable and historically inevitable; the only alternative to the manifest dangers of traditional nationalism. This is an understandable, indeed a laudable, interpretation. After all, the terrible events which disfigured European history in the early twentieth century were clearly motivated by racism and intolerance, the vicious offspring of old-style nationalism. Yet the belief that the awful things which have taken place within Europe represent the negation of the European ideal, aberrations from a pre-ordained march of European civilization towards greater peace and harmony, is mere wishful thinking. The darker episodes in Europe's past can equally be interpreted as integral to Europe's collective experience and identity; as rooted in the very idea of Europe itself. While European union may well become a desirable alternative to traditional nationalism, it is dangerous to assume this, as both the idea of Europe and the idea of the traditional nation-state are predicated on a common assumption that political identity must have a territorial dimension. Placing undue faith in the idea of European union leaves unresolved the deeper moral, political and ideological questions about the nature of territorial politics. If European union is to provide a serious alternative to the existing, moribund geopolitical order, then the idea of Europe needs to be subjected to critical and dispassionate analysis. As Gerard Delanty has noted (1995: p. ix), Europe should no longer be seen as a 'self-positing spiritual entity that unfolds in history and never needs to be explained'.

Secondly, most histories of the European idea underestimate the role of geography as an element shaping human consciousness. Geography is generally viewed as the immutable and independent physical background; an immobile and relatively unimportant stage upon which the rich historical drama takes place (although see Baechler *et al.*, 1988; Dodgshon, 1987; Fisher, 1966; Halecki, 1950; Jones, 1987; Louis, 1954; McNeil, 1963, 1974; Seton-Watson, 1985). The 'trivializing' of geography is partly motivated by an understandable desire to avoid accusations of geographical determinism. But it is also a manifestation of a deeper Western intellectual prejudice which not only emphasizes the historical at the expense of the geographical but also assumes that the former is naturally the realm of ideas whereas the latter is the preserve of the un-interpreted physical fact or empirical detail and thus ontologically insignificant (Agnew, 1989; Soja, 1989). In many surveys of the European idea, virtually everything of interest is automatically regarded as a 'historical' issue (including matters as intrinsically geographical as debates about the spatial configuration of states, for example). The prosaic, the dull, and (above all) the physical are deemed the proper concerns of geography. All the geographer can offer, it would seem, is factual information about river basins and mountain ranges the political significance of which must be left to other, more sophisticated commentators.

Geography is worthy of much more serious consideration than this. It is necessary to challenge the popular, limited understanding of the term 'geography'; to shift the emphasis away from the external, material world and towards the inner realm of the human mind. There is no reason why geography should not be seen as an intellectual arena of ideas and beliefs. This is not to deny the enduring power of the material world, nor is it to undervalue the importance of the interaction between external nature and internal consciousness; it is, rather, to underline the importance of intellectual conceptions of geographical space in the formation of human consciousness (Sack, 1980).

How Europeans have understood, and defined, their geopolitical order is therefore an important geographical question, albeit one that has rarely been asked. We need to re-examine Europe's political geography bearing in mind the arguments developed by Henri Lefebvre (1991) that space is not an independent given but a mutable and ever-changing product of economic, social, cultural, and political processes. Geography should be seen as a central element in the formation of European consciousness and identity.

Paradoxically, this argument has thus far only been developed with respect to European understanding of 'other', non-European regions (Said, 1978). While numerous studies now exist on the idea of the Orient or Asia or Africa in the European geographical imagination, no comparable studies exist for Europe itself. How Europeans have creatively imagined themselves *geographically* is a surprisingly ignored theme. Why this should be so is an interesting question. Part of the answer surely lies in the Eurocentric tendency to see Europe as an 'old' continent, the kernel of an ongoing civilization rooted in the ancient past which can only be understood historically. Non-Europeans, on the other hand, are still interpreted as 'people without history', existing in space (or nature) rather than time and explicable in geographical rather than historical terms (Wolf, 1982).

This chapter is no more than a hesitant step towards a broader geographical reassessment of the idea of Europe. Little can be achieved in a short essay and I shall discuss only a few of the many themes which have structured the debate about the political geography of Europe from the Renaissance to the present. As this is an exercise in historical geography, I have chosen not to dwell on the institutional details of European economic and political union after 1945. This is partly because these developments have been expertly surveyed elsewhere but also because my primary concern is with the earlier visions of Europe which influenced, to a greater or lesser extent, the post-1945 process. Extracting a selection of simple themes from the swirling flow of ideas over 500 years is a hazardous business. What follows is but a personal perspective and alternative accounts could (and should) be developed. If one were to compare the idea of Europe since 1500 to a sonorous, complex, and discordant piece of music, then the following represents merely a sample of the more insistent motifs.

## Christian Geopolitics and the Idea of 'Europe' on the Eve of the Great Discoveries

How was medieval and early Renaissance Europe understood geopolitically? How might a political dimension be added to Clifford Darby's memorable portrait of the continent in the Age of Columbus (Darby, 1961; Fig. 7.1)? Most accounts agree that in the Classical world of Greece and Rome, the Mediterranean (rather than the three surrounding continents of Asia, Africa, and Europe) was seen as the central and most advanced of the world's regions, a pre-eminence explained by reference

to environmental and climatic theories derived from Aristotle and Hippocrates (Dilke *et al.*, 1987; Dion, 1977; Glacken, 1967; Romm, 1992; Staszak, 1995). The notion of Europe, or of any land continent, had little significance in classical geography (J. Fischer, 1957). Significantly, the word 'Europe' does not appear in the Bible (Boer, 1993: 19).

A concern with global continental divisions emerged only in the post-Roman, Christian era, particularly from the seventh century AD when any residual sense of Mediterranean unity was shattered by the first great wave of Muslim expansion. The famous T-and-O mappae mundi, developed in the early seventh century by Isidore, Bishop of Seville, depict a circular world divided into three unequal segments—Asia (representing half the earth) and Africa and Europe (representing one-quarter each). These three zones were supposedly inhabited by the descendants of Shem, Ham, and Japheth respectively, the three sons of Noah (Harvey *et al.*, 1987; Woodward, 1985).

From this period, Islam and Christianity were on a collision course. Following the rise of Charlemagne, the emergence of the Holy Roman Empire and the launching of the first Crusades, an implicit association began to emerge between Christianity and the area which becomes Europe; between *Christianimus* (the faith) and *Christiantas* (the region) (Hay, 1980). By the twelfth century, maps devoted solely to the depiction of 'Europe', such as Lambert's *Liber Floridus*, were beginning to appear (Boer, 1993: 29–32; Brincken, 1973; Derolez, 1968).

Although the word 'Europe' was rarely used (P. Burke, 1980; Gollwitzer, 1951*a*; Schmidt, 1966; Talmor, 1980), by the early fourteenth century 'Christendom' had become a meaningful geopolitical concept (Bartlett, 1993; Hay, 1968; Meyer, 1989). Its relative unity was based on what Southgate (1993: 131) calls the 'European superculture' derived from a common 'official' language of Latin and a general adherence to Christianity, both of which produced artefacts, symbols, and practices (coins, statues, icons, legal charters, and educational systems) which were widely understood and accepted. These were the building blocks of what Denys Hay calls the 'continental ideology' (Hay, 1980: 1; Rubin, 1992).

To be sure, the idea of a common, fourteenth-century European identity would have been limited to a small, educated élite who were in other respects

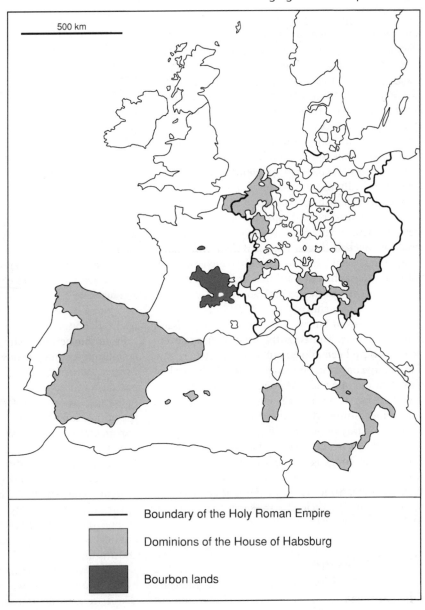

500 km

Boundary of the Holy Roman Empire

Dominions of the House of Habsburg

Bourbon lands

Fig. 7.1. Western Europe, c.1520.

divided by mutual suspicion and antagonism (Balzaretti, 1992; Leyser, 1992; Reuter, 1992). The fact that one of the first proposals for a united 'Europe' (devised by the French jurist Pierre Dubois as a means to defend Christianity against the Ottoman threat) dates from precisely this period is itself eloquent testimony that a sense of common 'Europeanness' was still relatively weak (Heater, 1992: 1–14). The cohesive power of Christianity in this period was, moreover, strictly limited because the medieval Church had been split between the Western Latin and Eastern Byzantine branches from the mid-eleventh century. It was only after the resurgence of Muslim power, culminating with the fall of Constantinople to Mohammed II in 1453 and the ensuing collapse of the Byzantine Church, that a more culturally and religiously homogeneous sense of a Western, Christian Europe emerged. By the late fifteenth century, the idea of *Respublica christiana* was frequently evoked in papal proclamations with specific reference to the ominous external threat posed by an Islamic 'other' into whose covetous hands fell the last vestige of Christian Greece (Morea) in 1460. The 'enemy' was now at the gates and the Christian realm was on the defensive. In 1521 Belgrade succumbed and by the end of that decade Vienna itself was under siege.

A more unified Christendom, though it was never without its internal divisions and disputes, developed in the midst of a terrible and seemingly irresistible external threat. These changes can be discerned in the world maps of the early Renaissance in which 'Europe' was usually depicted as a small, vulnerable, and exclusively Christian region (Edgerton, 1987). One of the best examples is Andreas Walsperger's *Weltkarte* of 1448, rediscovered a century ago in the Vatican library (Kretschmer, 1891).

Hesitant though it was, the emergence of Latin Christendom engendered the first stirrings of a new concept of territory and political space; of a Christian geopolitics in which the spatial extent of Christendom assumed new significance. This was revealed, *inter alia*, by the angry fourteenth- and early fifteenth-century debate about the correct location of the Holy See, a dispute sparked off by Pope Clement V's decision to transfer his residence from Rome to Avignon. This was subsequently justified on the grounds that Avignon was closer to the geographical centre of the Christian world, an argument which clearly implies a spatial vision of Christendom.

However, the growth of Latin Christendom in the early centuries of this millennium (in the face of fierce Muslim resistance) demonstrates that the medieval idea of political space was still radically different from the modern concept. This theme has been explored by Robert Bartlett in his work on Christian expansion from the mid-tenth to the mid-thirteenth century, an era in which Christendom virtually doubled in size. Bartlett concludes that the territorial growth of 'Catholic Europe' did not lead to the kind of regional distinctions between core and periphery which have been such a feature of later forms of imperial conquest: 'The net result of this [medieval] colonialism', he claims,

was not the creation of 'colonies', in the sense of dependencies, but the spread, by a kind of cellular multiplication, of the cultural and social forms found in the Latin Christian core. The new lands were closely integrated with the old. Travellers in the Middle Ages going from Magdeburg to Berlin and on to Wroclow, or from Bruges to Toledo and on to Seville, would not be aware of crossing any decisive social or cultural frontier.

Driven by consortia of Latin priests, Frankish knights, merchants, and travellers, '[h]igh medieval colonization was . . . a process of replication, not differentiation' (Bartlett, 1993: 306). The 'borders' between the Christian and Muslim worlds were also, it seems, transitory and far from clear-cut (Bartlett and Mackay, 1989). A similar argument about the non-territorial nature of early Christianity has also been developed by Joan Taylor (1993).

There are a number of reasons for this weakly developed sense of Christian territoriality. Economically, the idea of private property was less than fully developed under the feudal system. As a result, land and territory were not sharply demarcated (Tribe, 1978). Politically, the medieval order was based on multiple loyalties and complex allegiances operating in an overlapping and essentially a-spatial fashion (M. Mann, 1986). The idea of compartmentalized political space was, therefore, somewhat alien to the medieval Christian worldview and the familiar modern notion of a hierarchical spatial order, of cores and peripheries, was less evident. The essential characteristic of medieval Europe's external limits and internal organization was a relatively undifferentiated economic and political space, lacking clear, identifiable borders and frontiers.

The spatial fluidity of medieval geopolitics was by no means unique to the Christian world. It was also discernible in Islamic culture, where even the idea of 'area', familiar enough in Arabo-Islamic mathematics, had not developed as a geographical concept (Brauer, 1995). Arguably, the reluctance to conceptualize space as a commodity to be mapped and delineated persisted in the non-European world well into the modern era when, as we shall see in the next section, a very different European understanding of political space began to emerge (Harley and Woodward, 1992, 1994).

## Secular Geopolitics and the Idea of Europe in the Early Modern Period

The demise of Christian geopolitics was determined by the momentous and interwoven economic and intellectual changes associated with the rise of European capitalism and the expansion of European influence into the 'New World'. Europe's increasing engagement with, and knowledge of, the distant regions of the globe allowed educated Europeans to develop an entirely new worldview and a new assessment of their relative position within the global scheme. This is best illustrated by post-Renaissance world maps, particularly those based on the Mercator projection, in which the land surface of the world was depicted as if naturally grouped around a western European core. This was no longer a medieval, Christian image of the world viewed from Jerusalem or Rome. Here was the globe as seen from the brash new maritime cities of northern Europe and the Atlantic—from Lisbon, London, or Hamburg. The old idea of Christendom had given way to the new, secular notion of Europe.

The secularization of the European self-image was also influenced by the breakdown of the fragile unity of the Christian Church following the Reformation and the emergence of thriving Christian communities (Catholic and Protestant) in other continents. The idea that Europe was the sole arena of a united Christian faith was no longer tenable. At the same time, the modern system of European nation-states began to emerge which changed the meaning of political space (J. Anderson, 1986; Elliot, 1992; M. Mann, 1984; Pounds and Ball, 1964; Shennan, 1974; Tilly, 1975; Wallerstein, 1974, 1980). It is the cartographic image which again betrays this shift. Early modern maps were profoundly political documents, carefully constructed self-portraits of the new European nation-states designed to legitimate their territorial authority (Akerman, 1995; Buisseret, 1992; Harley, 1988; Kain and Baigent, 1992). As I have already suggested, political frontiers and borders were previously understood only in the vaguest terms as zones-of-transition rather than sharp lines of demarcation. Rarely were such divisions expressed cartographically on international agreements before the eighteenth century (Hertslet, 1875–91; Konvitz,

1987: 33). Gradually, however, national frontiers were surveyed, mapped, and then policed on the ground. The great Cassini survey of France, inaugurated in the late seventeenth century by Jean-Baptiste Colbert and endlessly updated through the eighteenth, was simply the most ambitious example of a broader development: nation-building through mapping (Pelletier, 1990; see also Dion, 1947; Guenée, 1986; Nordman, 1986; Nordman and Revel, 1989; Pounds, 1951, 1954; Sahlins, 1990; N. Smith, 1969; Weber, 1986).

This 'fixing' of national spaces on the European map took place in the midst of widespread violence and warfare, particularly during the Reformation, the Counter-Reformation and the Thirty Years War. These traumatic events engendered a new international geopolitics best exemplified by the idea of the balance of power. One of the earliest prophets of this concept was Niccolò Machiavelli, an official in the Florentine chancellery, whose *Il Principe* (1513) first laid down the principles which were to govern modern statecraft and international relations (Pocock, 1975; Skinner, 1981). Although primarily concerned with the labyrinthine political intrigues of Italian city-states, Machiavelli's ideas were clearly relevant to the political system of Europe as a whole (Dionisotti, 1971). His vision of international relations was based on the metaphor of a set of scales. The ideal situation was a balance of power between the rival dynasties—on the one hand, the French Valois and on the other, the Habsburgs who dominated Spain, most of Italy, the Low Countries, Austria, the German lands, and Burgundy. Relatively independent nation-states, notably England, were of critical importance for they held the balance of power. This finely-tuned system of checks and balances was the best, perhaps the only way to achieve peace and stability in a Europe where clashes between dynastic houses and tensions between religious faiths threatened to wreak unprecedented havoc.

The balance of power was initially developed as an idea which might keep the Habsburgs in check. Following the collapse of the Spanish Armada in 1588 and the general diminution of Habsburg influence during the seventeenth century, the same notion was then used to challenge the expansive ambitions of France, especially under Louis XIV (G. Parker and Smith, 1978). Increasingly the idea of Europe came to be defined in a somewhat utopian fashion as the political arena in which the balance of power between rival nation-states might conceivably operate and in which war and destruction might ultimately become impossible. Thus was born the modern belief that the idea of Europe is somehow naturally associated with the cause of peace and harmony.

The defining moment in the development of the new European geopolitics was the Treaty of Westphalia in 1648, which brought an end to the Thirty Years War and established the idea of international law along lines laid down by the Dutch jurist Hugo Grotius (Bull, Kingsbury, and Roberts, 1990; Cutler, 1991; P. J. Taylor, 1996: 95–8). The Treaty of Westphalia was, according to Peter Taylor (1996: 21), 'the foundation treaty of the inter-state system'; the rock upon which the secular idea of Europe as an arena of separate nation-states held in a balance of power was constructed. Threats to the geopolitical balance were depicted as threats to the very idea of Europe itself. When Louis XIV's armies attacked the Dutch United Provinces in 1672, pamphlets streamed from Europe's printing presses denouncing France's disruption of the balance of power, the essence of what it meant to be European. French counter-propaganda, on the other hand, spoke of the Sun King as the defender of the older idea of *Respublica christiana*. English Whigs and the Dutch royal house under William of Orange—united constitutionally following the 'Glorious Revolution' of 1688—saw themselves as struggling to preserve the religious freedom and commercial liberty of Europe against the intolerance of an older, Catholic geopolitical order. It was 'in the course of the seventeenth and early eighteenth centuries', claims Denys Hay (1968: 116), '[that] Christendom slowly entered the limbo of archaic words and Europe emerged as the unchallenged symbol of the largest human loyalty'. Schmidt (1966: 178) goes further and offers what he calls 'conclusive evidence that the term Europe established itself as [an] expression of supreme loyalty in the fight against Louis XIV. It was associated with the concept of a balanced system of sovereign states, religious tolerance, and expanding commerce.' As a political expression, Europe emerged as a secular, if not a Protestant, idea of political liberty which could be preserved by a just balance between rival nation-states. Although increasingly deployed through the sixteenth and seventeenth centuries,

it found real currency and influence only after *c.*1700 (P. Burke, 1980; Hazard, 1990). Significantly, the Treaty of Utrecht, signed in 1713 at the end of the War of the Spanish Succession, is the last European accord to make reference to *Respublica christiana* (Hay, 1968: 118–19; Outram, 1995: 65).

The most prophetic vision of Europe within the early modern secular geopolitical imagination was formulated by the Duc de Sully (Maximilien de Béthune), Henri IV's shrewd finance minister (Barbiche, 1978; Buisseret, 1968). After the assassination of Henri in 1610, Sully retired from public life and, supported by cardinal Richelieu, developed a 'Grand Design' for European unity and peace which appeared in several forms during the middle decades of the seventeenth century, usually attributed to his former royal patron (Buisseret, 1984; Heater, 1992, 15–38; Najam, 1956; Ogg, 1921; Soueleyman, 1941). Whether Sully was genuinely concerned to promote European reconciliation or motivated instead by a desire to diminish the power of the Habsburgs and reflect well on the ambitions of the Bourbons is a matter of serious debate. His legacy has been lasting, however, not least because of his pragmatic acceptance of geopolitical realities and his advocacy of peace on both moral and political grounds. Sully was no utopian dreamer—he even described himself as 'cold, cautious and unenterprising' (Ogg, 1921: 25). If states were no longer influenced by religion, he reasoned, they might yet be persuaded by political economy. The ultimate objective was 'to divide Europe equally among a certain number of powers and in such a manner that none of them might have cause either to envy or fear from the possessions or power of others' (Ogg, 1921: 41). Once Habsburg power was limited to its Iberian heartland (together with its non-European colonial possessions), Sully envisioned fifteen of the existing nation-states combining to form a new European order. Six of these were hereditary monarchies (France, Spain, England, Denmark, Sweden, and Lombardy), five were elective monarchies (the Holy Roman Empire, the Papacy, Poland, Hungary, and Bohemia), and four were republics (the Venetian, the Italian Ducal, the Swiss, and the Belgian). Each state would provide proportional quantities of men and resources to establish a common European army, obviating thereby the need for rival national armies. A European general council, as well as several regional

**Fig. 7.2.** Sully's Grand Design.

sub-councils, were also envisaged, the former to be located 'in the centre of Europe' on the borders of Frankish and Germanic lands. Metz, Luxembourg, Nancy, Cologne, Frankfurt, Heidelberg, Speyer, Strasbourg, Besançon, Basle, and Trier were all suggested as possible seats of the general council (Fig. 7.2).

It would be wrong to interpret Sully's project in entirely secular terms. As a Protestant, he was well aware of the divisive potential of religious beliefs. Partly for this reason, he insisted on the need to preserve the existing status quo between Catholic and Protestant Europe. He also specifically rejected the possibility that non-Christian nationalities could be classed as part of Europe. In Sully's world-view, the Ottoman Empire remained a principal 'other' against which Europeans could seek common cause (Neumann and Welsh, 1991). In

one crucial respect, however, Sully's analysis was based on a form of reasoning which owed nothing to the traditions of Christian geopolitics. Russia, Sully insisted, had no part in the European order. The exclusion of Russia was less to do with its religious distinctiveness (it was, after all, a Christian place) than with Sully's firm belief that the Russian people were culturally inferior. 'I say nothing of Muscovy and Russia', he wrote: the people of

[t]hese vast countries . . . being in part still idolators, and in part schismatics, such as Greeks and Armenians, have introduced so many superstitious practices in their worship, that there scarce remains any conformity with us among them; besides, they belong to Asia at least as much as to Europe. We may indeed almost consider them as a barbarous country, and place them in the same class as Turkey, though for these five hundred years,

we have ranked them among the christian powers. (Ogg, 1921: 31–2)

As we shall see in the next section, Sully's exclusion of Russia on cultural (rather than specifically religious) grounds had considerable significance for Enlightenment conceptions of Europe's political space.

## Europe and its Others: The Enlightenment and the Rise of Cultural Geopolitics

During the eighteenth century the national frontiers of the European map became more clearly inscribed in the geographical imagination of the continent's peoples. At the same time, the intellectual upheavals of the Enlightenment significantly altered the debate about Europe's extent and internal geopolitical order (Livey, 1981). Of critical importance was the emergence of a cultural (or what Agnew and Corbridge (1995, 52–6) call a 'civilizational') geopolitics, hinted at in Sully's blueprint for European union. This was to supplement, and to an extent replace, existing ideas about the nature and distinctiveness of Europe. This is not to suggest that the idea of Europe as the privileged haven of liberty, tolerance, reason, science, and industry was less important after 1700. However, it was only in the eighteenth century that secular theories of the human condition and of Europe's place within the global order were fully developed. Early examples were the climatic and environmental theories put forward by Charles Louis de Secondat, the Baron de Montesquieu, to explain Europe's apparent moral and political superiority over the despotic Orient, most famously in his *De l'esprit des lois* of 1748 (Glacken, 1967: 551–654; Springborg, 1992).

After 1750, a new lexicon of words and concepts, hitherto unknown or differently used, began to appear in moral and political debate and in the newly-constructed dictionaries and encyclopaedias. Three of the most important expressions were the interrelated terms 'culture', 'civilization', and 'progress', a secular trinity of Enlightenment thought which featured with increasing regularity in discussions of Europe's nature and distinctiveness. 'Culture' and 'civilization', despite subtle differences in French, German, and English, were used from the

1760s to indicate the peculiarly European characteristics of dynamism, energy, and development mixed with logic, reason, and self-restraint. These qualities came to be seen as the principal causes of European superiority over the more 'primitive' peoples of the Americas, Africa, and Asia (Bauman, 1985; Elias, 1994; Febvre, 1974; Kroeber and Kluckhohn, 1952; Vogt, 1996; R. Williams, 1976). By 1766, the French term *civilisation* (first coined a decade earlier by the physiocratic thinker the Marquis de Mirabeau) had been twinned with its familiar adjective *européenne* in a treatise on French colonies in North America by the Abbé de Baudeau in order to contrast Europe at the highest stage on the ladder of human progress with the Americas at the most primitive and savage. The same word and concept was soon being used in English, notably by Adam Smith, whose economic theories, summarized in *The Wealth of Nations* (1776), can be interpreted as part of this broader Enlightenment concern with civilizational differences (Dockès, 1969). The eighteenth-century notion of 'civilization' was thus closely connected to the debates about the nature and potential of human progress which took place around the same time, particularly the materialist theories of human development conceived as the endless march of civilization onward and upward through a series of stages from barbarism to enlightenment (Nisbet, 1980).

This cultural rhetoric soon dominated definitions of Europe's geographical limits, its internal geopolitical identity, and the very nature of government and statecraft. This was connected to a more general Enlightenment shift in the language of politics. It is now widely accepted that modern notions of collective, national political identity emerged in Europe during the eighteenth century when the complex array of national icons, symbols, and cultural practices first began to influence the values and outlooks of the great mass of European peoples (Colley, 1992). Largely as a result, eighteenth-century nation-states began to define themselves by reference to the number, wealth, and contentedness of their subjects or citizens. New and more subtle political and constitutional relationships developed which amounted, according to Michel Foucault (1991), to a fundamental change in the character of government and state power. Brute force and violence were no longer the principal forms of governmental authority (although they remained the ultimate sanction); power henceforth resided in a more complex dialectic between those who governed and those who were governed, between rulers and ruled (M. Mann, 1993; Clarke, 1994). This relationship was based on mutual consent and a collective consensus or common will, what Jean-Jacques Rousseau termed *la volonté générale* (P. Burke, 1992). Modern nation-states were, therefore, far more concerned than their predecessors with the people that lived within their borders. This should not be seen as a benign concern on the part of the ruling élite with the welfare and happiness of ordinary folk. Rather, the social and political contract which developed between the state and the citizen (even before the emergence of participatory democracy) was part of a broader reconfiguration of European political culture and a realization that people were both the form and the instrument of government, the ultimate source of national power. This idea directly informed the Enlightenment utopias, such as David Hume's vision of the 'perfect commonwealth' (Hume, 1752). After the eighteenth century, then, nation-states were more readily identified by reference to their inhabitants rather than their rulers.

Debate about the geopolitical character and extent of Europe reflected these changes. The earlier preoccupation with the relationships between kings, princes, and governments gradually shifted towards a cultural, even a quasi-anthropological, concern with the different peoples of Europe, the most important indicators of the continent's internal order and external limits. The concept of the balance of power, and its conceptual link with Europeanness, remained firmly in place but Enlightenment geopolitical debate was now articulated in a cultural register. At the same time, the relationships between European states became ever more complex. Britain, increasingly conscious of its extra-European imperial status, still maintained the balance between Bourbon and Habsburg, but this relatively simple arrangement was disturbed by the emerging eastern powers of Prussia under Frederick the Great and Russia under first Peter and then Catherine the Great. Gradually a more complex version of the European balance of power emerged based on a 'pentarchy' of five major powers, Britain, France, Austria, Prussia, and Russia, an arrangement hesitantly inaugurated by the Treaties of Utrecht (1713).

This new geopolitical order raised a fundamental question which has bedevilled the debate about the nature and extent of Europe ever since: where were Europe's eastern limits? As we have seen, Sully felt able to draw a clear line between Russia and the rest of Europe in the mid-1600s based on cultural or civilizational criteria. Yet by the early eighteenth century this division seemed somewhat problematic as Peter the Great's Russia was playing a more important role in European affairs. Peter accepted the inherent superiority of Western culture and civilization and sought to change the face of Russia from a backward Slavic, Asiatic power to a modern, enlightened, European nation, whilst simultaneously expanding its territorial extent through almost constant warfare (Sumner, 1973). As Mark Bassin (1991) has shown, this required energetic geopolitical propaganda to assert Russia's status as a European power which, like Britain and France, possessed a non-European empire. The only difference was that Russia's extra-European empire was contiguous with the European core. The critical question, therefore, was where best to locate the dividing line between European Russia and its Asian empire, a border which Peter believed would mark the eastern limits not only of Russia but of Europe as a whole (Slezkine, 1994). Older ideas that Europe and Asia were divided along fluvial lines, notably by the River Don, were quickly rejected in favour of the north–south spine of the Urals. This clear and immobile barrier became European Russia's official line of cultural and political demarcation and was mapped and re-mapped at a variety of scales throughout the eighteenth century (France, 1985). By the time Catherine the Great ascended the throne she was able confidently to assert (in French, of course) that 'La Russie est une puissance européenne' (Bassin, 1991: 9; see also M. S. Anderson, 1954; Cahnman, 1952; Halecki, 1952; Hartley, 1992; Kristof, 1968; McNeil, 1964; Palmer, 1970; W. H. Parker, 1960; Pesonen, 1991; Shaw, 1996; Tazbir, 1977; Thaden, 1984; Wandycz, 1992; Wanklyn, 1941; Webb, 1952).

Peter's ambitions to forge a new, European identity for Russia were greeted with only limited enthusiasm in the west. Some were clearly convinced by his reforms. The first Enlightenment scheme for European peace and unity, devised during the negotiations at Utrecht by the well-meaning but pedantic Abbé de Saint-Pierre—'the great bore of eighteenth-century France' (Heater, 1992: 66; see also Drouet, 1912; Perkins, 1959)—included Russia within the European orbit, and even sought some accommodation with the Ottoman Empire (Saint-Pierre, 1712, 1714). These ideas were taken up by Jean-Jacques Rousseau nearly half a century later in his prophetic analysis of Europe's potential for peace and unity (Rousseau, 1761; see also Nuttal, 1927; Roosevelt, 1990: 185–220). For Rousseau, European unity was ultimately desirable but would only be achieved at great cost and through violent revolution. In the mean time, a degree of superficial unity was already in place but this was not something to be welcomed:

'Whatever people say, there are not today any more Frenchmen, Germans, Spaniards, even English', Rousseau observed in the early 1760s: 'There are only Europeans. They all have the same tastes, the same feelings, the same customs because none has received a national shape by any exclusive institutions . . . They are at home wherever there is money to steal and women to seduce.' (Partially quoted in Joll, 1980: 11–12; Heater, 1992: 79)

Notwithstanding Rousseau's efforts, Saint-Pierre's expansive (and long-winded) vision of a cosmopolitan Europe stretching from the Atlantic to the Urals was widely disparaged. Frederick the Great wrote sarcastically to Voltaire that 'The Abbé de Saint-Pierre . . . has sent me a fine work to re-establish peace in Europe. The thing is very practicable: all it lacks to be successful is the consent of all Europe and few other such small details' (quoted in Heater, 1992: 84–5). In a powerful recent account, Larry Wolff (1994) has argued that Enlightenment debates about civilization and barbarism had the effect of firmly excluding Russia (together with Poland, Hungary, and the Ottoman Empire) from the realm of European culture. It was during the eighteenth century, claims Wolff, that the familiar east–west division of Europe first emerged. This distinction became a geographical shorthand for a general moral and cultural dichotomy between 'backward' and 'civilized' regions and replaced an older north–south vision of the continent, characteristic of the Christian and secular geopolitics of the medieval, Renaissance, and early modern eras. Despite the predominantly Christian faith of its inhabitants, eastern Europe was 'invented' during the eighteenth century as western Europe's new constituting 'other'; as a barbarous realm beyond the limits of the map of civilization (Chiriot, 1989). As the Muslim Ottoman threat declined to political irrelevance, so a new external 'threat' was imagined in the Christian East against which 'enlightened' Europe could define itself (an alternative view of the 'east–west' and 'north–south' divisions is offered by Okey, 1992: 110–11).

## The Geopolitics of the Masses: The Idea of Europe after 1789

During the nineteenth and early twentieth centuries, the idea of Europe became infinitely more complex and disputed. The French Revolution was a defining moment in this respect. The modern political landscape owes much of its shape and character to the events which followed 1789. The revolutionary political culture which developed in France after 1789 was based in large part on new ways of understanding time and space. Jules Michelet, the most sympathetic nineteenth-century historian of the Revolution, was at his most rhapsodic on this theme: before 1789, he claimed, 'the world was a prison'; after it 'geography itself was annihilated. There were no longer any mountains, rivers, or barriers between men . . . Time and space, those natural conditions to which life is subject, were no more' (quoted in White, 1973: 151). Hyperbole perhaps, but the revolutionaries did introduce a new decimalized calendar dating from the establishment of the Republic rather than the birth of Christ and structured around 'republican' months derived from the natural seasons in place of the old Gregorian sequence (Baczko, 1984).

Space too was imagined afresh after 1789. As Mona Ozouf (1988: 126) has observed

[t]here is no end to the list of spatial metaphors associated with the Republic or with Revolutionary France: from the beginning of the Revolution a native connivance linked rediscovered liberty with reconquered space. The beating down of gates, the crossing of castle moats, walking at one's ease in places where one was once forbidden to enter: the appropriation of a certain space, which had to be opened and broken into, was the first delight of the Revolution.

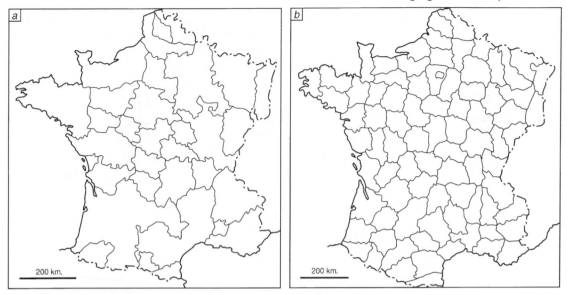

**Fig. 7.3.** Revolutionized space: (*a*) The administrative geography of *ancien régime* France; (*b*) The administrative geography of Republican France.

The reorganization of the French administrative system was but one manifestation of the new spatial order. Beginning in the 1790s, the territorial units of the *ancien régime* were swept aside. Into their place came a new, rational political space based on a larger number of small *départements*, each administered by a state official, the *préfet*, nominated by the central government in Paris (Fig. 7.3). The borders of the new *départements*, though modified to take into account natural features, were initially based on a simple Cartesian grid laid over a map of the national territory (Konvitz, 1987: 44–5). The objective was to enhance the power of the revolutionary core at the expense of an ideologically suspect periphery where regional *Parlements* dominated by local aristocrats and Catholic clerics posed an obvious threat to central authority (a threat which was to manifest itself in bloody counter-revolutionary wars across western and southern France during the mid-1790s). For the revolutionaries, Paris was to control the national space as the heart controls the workings of the body. Paris alone was to supply the revolutionary energy which would sustain the entire nation. The geography of the revolutionary state had to reflect that simple fact (Konvitz, 1987: 32–62, 1990; Margadant, 1992; Nordman and Vic-Ozouf Marignier, 1989*a*, *b*; Vic-Ozouf Marignier, 1989; Vovelle, 1993).

The centralization of French political power was scarcely unprecedented, of course, but the new concepts of political and administrative space which developed after 1789 intensified this process. Administratively and geopolitically, the First Empire extended the revolutionary spatial order to other parts of Europe. By 1812, the zenith of Napoleonic power, the entire map of Europe had been transformed and new 'French' *départements* stretched across the north European plain beyond the mouth of the Elbe and southwards as far as Rome (Fig. 7.4; see Broers, 1996: 174–5). Paraphrasing Alexis de Tocqueville, Sutherland (1985: 438) has argued that '[t]he First Empire was no mere episode in French history . . . beneath the spectacular details of Napoleon's biography the period witnessed the completion of the millennial process of centralization'. These words were written with respect to France alone but they might equally refer to Europe in general. Napoleon's domination of the continent was based on a bureaucratic and governmental revolution designed to make this sprawling empire function as an integrated political whole.

As Balzac once observed '[t]o organize is a word of the Empire' (Woolf, 1991: 83). The Napoleonic Empire raised state organization to new levels of sophistication. Although the First Empire is firmly associated in the modern historical imagination

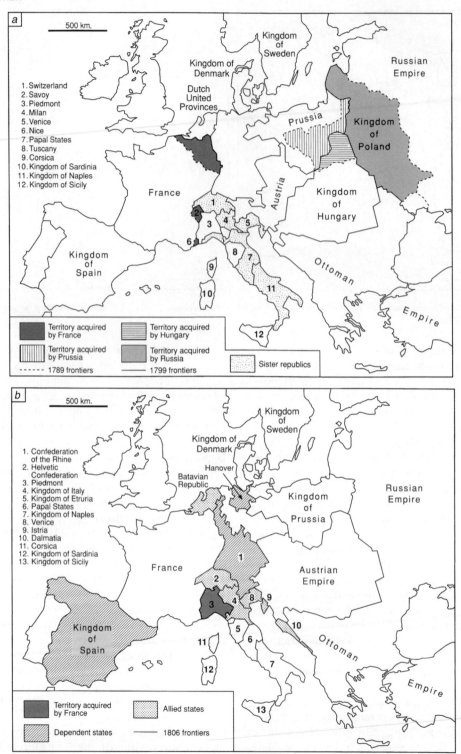

a

500 km.

1. Switzerland
2. Savoy
3. Piedmont
4. Milan
5. Venice
6. Nice
7. Papal States
8. Tuscany
9. Corsica
10. Kingdom of Sardinia
11. Kingdom of Naples
12. Kingdom of Sicily

Kingdom of Sweden

Kingdom of Denmark

Dutch United Provinces

Russian Empire

Prussia

Kingdom of Poland

France

Austria

Kingdom of Hungary

Kingdom of Spain

Ottoman

Empire

| Territory acquired by France | Territory acquired by Hungary |
| Territory acquired by Prussia | Territory acquired by Russia |
| ----- 1789 frontiers | ——— 1799 frontiers | Sister republics |

b

500 km.

1. Confederation of the Rhine
2. Helvetic Confederation
3. Piedmont
4. Kingdom of Italy
5. Kingdom of Etruria
6. Papal States
7. Kingdom of Naples
8. Venice
9. Istria
10. Dalmatia
11. Corsica
12. Kingdom of Sardinia
13. Kingdom of Sicily

Kingdom of Sweden

Kingdom of Denmark

Hanover

Batavian Republic

Kingdom of Prussia

Russian Empire

France

Austrian Empire

Kingdom of Spain

Ottoman

Empire

| Territory acquired by France | Allied states |
| Dependent states | ——— 1806 frontiers |

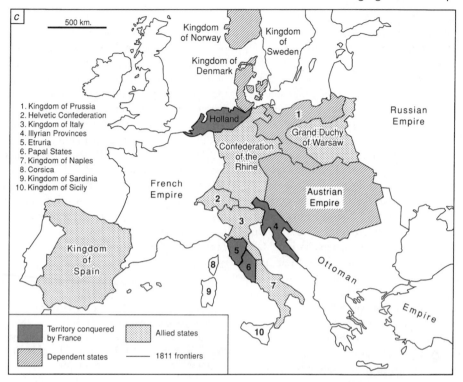

**Fig. 7.4.** Revolutionary and Napoleonic Europe: (*a*) Europe, 1789–1799; (*b*) Europe, 1800–1806; (*c*) Europe, 1807–1812.

with romantic military campaigns, sweeping cavalry charges, and dashing young officers, its fleeting success probably owed more to horn-rimmed bureaucrats at their high chairs, armed with nothing more dangerous than ledgers, quill pens and ink pots. The Empire's most enduring legacy may have been the transformation of state-craft into a 'social science' based on an information revolution of statistics, censuses, and endless government inquiries (Le Bras, 1986; Perrot and Woolf, 1984; Woolf, 1989*a*).

The revolutionary and Napoleonic period strengthened the idea of Europe and witnessed the first concerted attempt to unite the continent culturally and politically (Broers, 1996; Woolf, 1989*b*, 1991, 1992). But it has left an ironic and paradoxical legacy. A united Europe, so often invoked before and since 1789 as a means of assuring peace and harmony between different peoples, came momentarily into being only as a result of warfare and revolution (just as Rousseau had predicted a generation earlier). The idea of Europe, in this period, arguably owed more to war than to peace (Levy, 1983). Moreover, the Napoleonic attempt to create a united Europe was based on a revolutionary and imperial mind-set which elevated a national political culture to a European, ideed a universal, level. The idea of Europe as a balance of power between rival states had no place in the Napoleonic order; this was a Europe forged in the image of a single, revolutionary and imperialist state whose civilization would henceforth speak for all, whose culture possessed inherent and unquestionable benefits for all men and women. French revolutionary civilization could—indeed should—be imposed throughout Europe, and ultimately the world, for the good of all humankind. Thus was the idea of Europe conflated with the idea of France and thus were acts of imperial conquest and domination turned into acts of liberation.

In the midst of these epoch-making events, new schemes for universal peace appeared, including Jeremy Bentham's *Plan for an Universal and Perpetual Peace* (1789), the fourth and final essay in his *Principles of International Law* (1780–9), and

Immanuel Kant's *Zum ewigen Frieden* (1795) (Heater, 1992: 93–4; Hinsley, 1967; Raumer, 1953). These were, however, universal rather than specifically European projects, and it was not until the first collapse of the Napoleonic regime in 1814 that a fresh vision of European unity was proposed, the first and most important of the nineteenth century. Its architect was the visionary technological and utopian philosopher Claude Henri de Saint-Simon. His project for a 'reorganized' and united Europe, based initially on an Anglo-French union, was radically different from preceding proposals. Geopolitically, its most important feature was its emphasis on the power of modern technology, transport and communications. Foreshadowing Marx, Saint-Simon felt that the old European nation states were destined to wither away. Into their place would come a new European political space which would be determined not by the balance of power between nation-states (he was scathing about the Treaty of Westphalia and the form of international politics it inaugurated) but by the power of science and technology to create an integrated economic union. Governments would ultimately be no more than administrative agencies; the real power would lie with an élite class of European *industriels* and scientists (Saint-Simon and Thierry, 1814; see Ionescu, 1976: 83–98; K. Taylor, 1975: 129–36). Saint-Simon's managerial view of technology as a means of creating new political spaces, whether national or international, had enormous impact on subsequent writings on the European theme (Eichthal, 1840; Pecqueur, 1842) and clearly influenced economists such as Friedrich List (List, 1904; see also Henderson, 1957, 1962, 1983).

The European geopolitical order constructed at the Congress of Vienna was, however, quite different from that envisaged by Saint-Simon and, of course, fundamentally at odds with Napoleonic imperialism (Fig. 7.5). The traditional idea of Europe as an arena of nation-states held in a balance of power reasserted itself with a vengeance. But the forces unchained by 1789 could not be suppressed. Nationalism had become a genuinely mass ideology and had been immeasurably strengthened in all parts of the continent by the Napoleonic experience. Henceforth, popular nationalism was the dominant force shaping the political geography of Europe.

To begin with, there was no contradiction between nationalism and 'Europeanism', as the idea of the balance of power seemed to reconcile these two concepts. Indeed, it was the universalist pretensions of French republicanism and imperialism, against which all European nationalists had united, which had posed the greatest threat to the idea of Europe. As Pim der Boer (1993: 68) has noted: 'in the revolutionary mentality, there was hardly any place for Europe between citizenship of the world and one's own nation'. Edmund Burke, the father of modern British conservatism and a passionate opponent of the Revolution, was thus able to mobilize a distinctly European argument in his attack on French republicanism. Five years after his initial broadside against the Revolution (Burke, 1983), Burke set forth his vision of Europe which France threatened to destroy:

At bottom [religion, laws and manners throughout Europe] are all the same. The writers on public law have often called this aggregate of nations a commonwealth. They had reason. It is virtually one great state having the same basis in general law . . . [N]o citizens of Europe could be altogether in exile in any part of it . . . When a man travelled or resided for health, pleasure, business or necessity from his own country, he never felt himself quite abroad. (Quoted in Boer, 1993: 67)

The European dimension of early nineteenth-century nationalism was clearly evident during the great revolutions of 1848, the 'springtime of nations' which swept away the last vestiges of the European *ancien régime* in the name of liberalism, democracy and popular nationalism. The fact that these extraordinary upheavals took place in so many regions at virtually the same time demonstrates that this youthful nationalism was still part of a broader European phenomenon. Yet nationalism contained within itself the seeds of its own transformation into a more competitive and disputatious force which was ultimately to undermine all sense of a common European consciousness and bring about the virtual destruction of the entire continent. The inexorable rise of Bismarck's Prussia after 1848, culminating in victory over France in 1870 and the establishment of a united German Empire, transformed the European balance of power and the nature of European nationalism. A revolutionary and liberating force in 1848, nationalism generated ever more divisiveness and rancour after 1870 not only within Europe but around the world, as the European powers sought competitive advantage by conquering more and more colonial

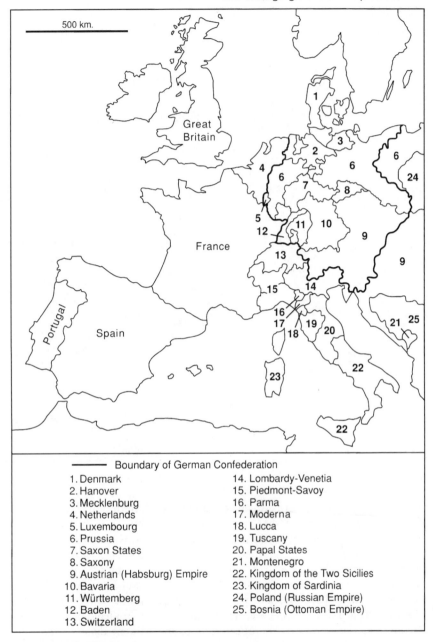

500 km.

Great
Britain

France

Portugal

Spain

Boundary of German Confederation

1. Denmark
2. Hanover
3. Mecklenburg
4. Netherlands
5. Luxembourg
6. Prussia
7. Saxon States
8. Saxony
9. Austrian (Habsburg) Empire
10. Bavaria
11. Württemberg
12. Baden
13. Switzerland

14. Lombardy-Venetia
15. Piedmont-Savoy
16. Parma
17. Moderna
18. Lucca
19. Tuscany
20. Papal States
21. Montenegro
22. Kingdom of the Two Sicilies
23. Kingdom of Sardinia
24. Poland (Russian Empire)
25. Bosnia (Ottoman Empire)

**Fig. 7.5.** Western Europe, 1815.

territory. European nationalism and European imperialism thus reached their twin peaks in the same period (Hobsbawm, 1987, 1992a; A. D. Smith, 1991). The political map of the continent on the eve of the First World War reflects the geopolitics of *fin-de-siècle* nationalism and imperialism (Fig. 7.6).

## In Europe's Name: The Geopolitics of Destruction 1914–1945

It is tempting to claim that the twin forces of nationalism and imperialism transmogrified during the 1920s and 1930s into fascism, catapulted Europe

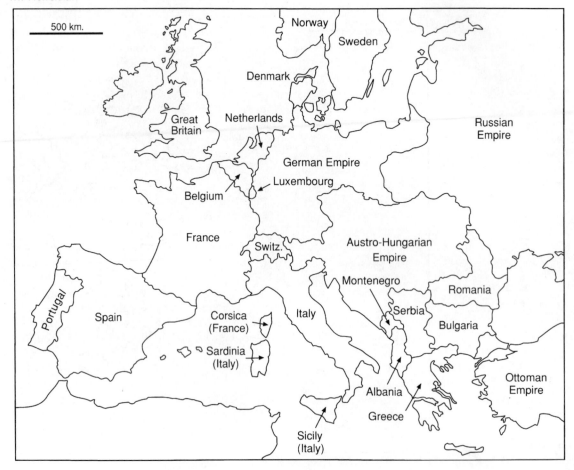

**Fig. 7.6.** Europe, 1914.

towards virtual destruction between 1914 and 1945. As I have suggested already, those with an optimistic view of Europe sometimes insist that this tragic period represents the very antithesis of the European idea, the high-point of confrontational nationalism. Jean-Baptiste Duroselle (1966: 261) has insisted that '[d]e 1914 à 1945, il n'y a plus d'Europe'. This reflects, I believe, an unduly elevated and partial view of the European idea. Sadly, brutal and inhuman acts were perpetrated in Europe's name during the early twentieth century. We must face the unpalatable fact that the idea of Europe cannot unproblematically be associated with peace, tolerance, and understanding (the ideals Professor Duroselle so eloquently and passionately championed). The more sombre thesis of Ernst Nolte (1987) comes closer to the truth. Between 1917 and 1945, Europe was gripped by a continuous civil war which, although it was fought by rival nation-states, was actually a struggle between rival ideologies and proto-ideologies—fascism, communism, and liberal democracy. At the end of this titanic contest for the European soul, fascism was destroyed but only through a temporary and unsustainable alliance between the other two ideologies. The cost of this victory was a continent divided into a liberal-democratic, capitalist west dominated by the USA and a communist east dominated by the USSR. The east–west division of the continent, a cultural divide dating back at least as far as the eighteenth century, was now a harsh and impenetrable geopolitical reality, an iron curtain stretching, in Churchill's most memorable phrase, 'from Stettin in the Baltic to Trieste in the Adriatic'.

The extent to which these upheavals were influenced by wider changes in geopolitical reasoning is a moot point. Certainly, the all-important ideological division of the continent after 1945, including the division of Germany itself, owed far more to the relative positions of Soviet, American, and British tanks at the time of the German capitulation than it did to geopolitics or negotiation. Yet the language and theories of geopolitics were further transformed during the first half of the twentieth century. Indeed, it was only in this period that the term 'geopolitics' enters the lexicon of academic and political debate, the first use of the word being attributed to the Swedish geographer Rudolf Kjellén in 1899 (Holdar, 1992; see also O'Tuathail, 1996). The tense and violent years between 1890 and 1945 were, therefore, an era of vigorous (and often partisan) debate about Europe's internal political geography and its position within the world.

The *fin-de-siècle* flowering of geopolitical debate is perhaps best revealed by the writings of Halford Mackinder in Britain and Friedrich Ratzel in Germany (Bassin, 1987a; Blouet, 1987; Kearns, 1993; O'Tuathail, 1992; W. H. Parker, 1982; W. Smith, 1980). Many of the concepts developed at this time, such as Mackinder's much-quoted 'heartland thesis' (Mackinder, 1904, 1919) and Ratzel's ideas of the organic state and *Lebensraum* (Ratzel, 1896, 1897, 1901), were early attempts to understand how the future geopolitical order of Europe and the wider world was destined to develop. Two regions were the focus of particularly intense discussion in western Europe before 1914: German central Europe and Russia.

The idea that there existed within Europe three 'natural' regions ranged from west to east was a relatively new concept (Szucs, 1988). As I have suggested, before the middle of the nineteenth century Europe was normally divided by a simple but geographically variable binary line, normally east–west. The Germanic notion of an intervening zone— *Mitteleuropa* or *Zwischeneuropa*—developed gradually after 1850 (Brechtefeld, 1996; Schultz, 1989). To begin with, the concept was understood in economic rather than political terms. Robin Okey (1992: 114) has shown that of the seventy-nine German atlases available in 1856 only four depicted the political geography of an area identified as *Mitteleuropa*, whereas twenty-four included thematic economic maps organized on this basis. However, a more overtly political understanding of *Mitteleuropa* as a greater Germany began to develop after

unification in 1871 (Partsch, 1904). The debate about *Mitteleuropa* was immensely complex, as the concept was evoked for a bewildering variety of reasons by both right and left with little consensus about its geographical limits (Fig. 7.7). It was also understood in different ways outside Germany. Some saw it as a solution to Germany's legitimate territorial ambitions; others (and this was a more common view) insisted that the very idea of central Europe was a dangerous stepping-stone towards German domination of the entire continent. The most widely read analysis was provided by a Lutheran pastor, Friedrich Naumann, during the First World War (Naumann, 1916; see also Penck, 1915; Hassinger, 1917; F. Fischer, 1967). Naumann foresaw a German central Europe sealed off from the enemy peoples to east and west by two great 'Chinese walls'. This, he argued, was the only solution to the chronic incompatibility between German *Kultur* and Anglo-French *civilisation*, a mutual exclusivity which had been exercising the best philosophical minds in Paris and Berlin (Bergson, 1915; Scheler, 1915). Naumann's argument struck an unexpected chord with a war-weary German public and was surprisingly well received in Allied countries (Chisholm, 1917; Lyde, 1916–17).

The problematic status of Russia was the other focus of western European geopolitics before the First World War (Neumann, 1993; Hauner, 1990). For long periods of the nineteenth century, Russia and the west were at loggerheads, most obviously during the Crimean War and in the era of the 'Great Game' of Anglo-Russian rivalry in Asia. As a result, Russia was generally depicted as an alien, Asiatic power with sinister ambitions towards Europe. Following the establishment early in the twentieth century of an anti-German Triple Alliance of Russia, France, and Britain, Russia was welcomed back into the European fold. This enthusiasm was not reciprocated in all quarters, however, for a late nineteenth-century upsurge of romantic Slavic nationalism rejected the materialism and commercialism of the west. Nikolay Alexsandrovic Berdyaev, for example, wrote of Russia as occupying a unique spiritual position as a world unto itself, neither east nor west (Berdyaev, 1924). In western Europe, however, the main problem was the apparent inability of the Tsarist empire to reform itself and to introduce a proper democratic system. Without such reforms, many believed that the

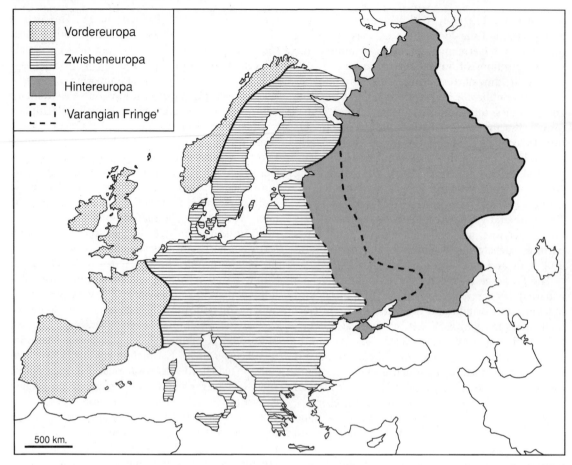

**Fig. 7.7.** A German vision of a tripartite Europe, 1915. (*Sources*: A. Penck, 'Politische-geographische Lehren des Krieges', *Meereskunde*, 9–10 (1915), 12–21; K. A. Sinnhuber, 'Central Europe—Mitteleuropa—Europe Centrale', *Transactions and Papers of the Institute of British Geographers*, 20 (1954), 15–39.)

regime was destined to collapse. This would have far-reaching, indeed global, implications. According to Mackinder, Russia's Eurasian position gave it unique strategic advantages with respect to the Asian land-mass, the pivot region of 'the world island'. Whoever controlled eastern Europe and Russia would ultimately control the entire world.

During the First World War, a third theme began to influence western geopolitical debate, particularly in the Allied countries. Much effort was expended pondering large theoretical questions about the role of international frontiers and boundaries and about the most appropriate geopolitical configuration for Europe and the wider world (Brigham, 1919; Holdich, 1916, 1918; D. W. Johnson, 1917a; Lyde, 1915). Discussion continued through

the inter-war years, beginning with the major surveys of the new world order that had been created in Paris in 1919 (Bowman, 1921; Brunhès and Vallaux, 1921; Newbigin, 1920; Lapradelle, 1928; Ancel, 1938; see also Kristof, 1959; Prescot, 1965: 1–32; Schöttler, 1995). This provoked wider debate about the very basis and organization of the liberal nation-state in the post-war era. The range of geopolitical proposals was truly bewildering, some of the most insistent being the advocacy of language as the means of building culturally unified states within a broader European union (Dominian, 1917; see also Chadwick, 1945; Cornish, 1936), calls for smaller (and hence more democratic and 'naturally' peaceful) European states (Bryce, 1922; Fawcett, 1922; Simon, 1939), and projects for the

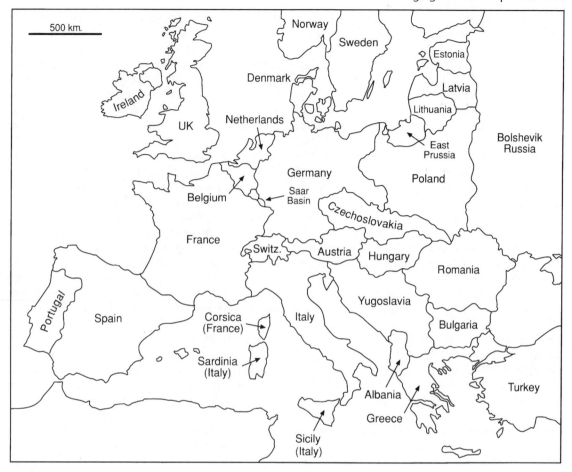

Fig. 7.8. Europe, 1923.

decentralization and devolution of political power (Fawcett, 1919; Peake, 1919).

The impact of these ideas on political leaders was negligible. The unprecedented horrors of the war coupled with the manifest failure to devise genuinely imaginative solutions to the continent's political problems generated a widespread sense of unease and despondency, the dominant mood of European geopolitical discussion during the 1920s and 1930s (Eksteins, 1989). The Bolshevik Revolution in Russia in 1917 demonstrated the fragility of the old autocracies and raised the spectre of a Europe organized according to the revolutionary ideals of Marx and Lenin. Russia was once again plunged into economic and political isolation, rejected by (and in turn rejecting) a hostile west which was far from certain about its own future.

The sense of Europe's decline was reinforced by the enormous cost of the war. The new Europe—the product of an uneasy compromise between American idealism and European *realpolitik*—was weaker and more complex than it had been in 1914 (Fig. 7.8). An extra 20,000 kilometres of international frontier had been added to the political map. These new frontiers would henceforth be traversed only with the aid of passports and a plethora of official paperwork. Twice the number of currencies were in circulation (twenty-seven compared to fourteen). Overall, the carnage and destruction of the 'Great War' cost the equivalent of 6.5 times the global national debt from 1700 to 1914 and set back real levels of European industrial output by at least a decade (Kitchen, 1988: 28; Knox and Agnew, 1994: 176–7).

Demographically, Europe's vitality had been hugely reduced, perhaps for ever. Nine million men had been killed in the fighting between 1914 and 1918 and perhaps fifty million civilians had died from starvation and disease, including those, weakened by the deprivations of war, who succumbed to the 1918 influenza epidemic (Gilbert, 1989: 541). In some regions, up to 15 per cent of the labour force had been wiped out; 63 per cent of all German men aged between 20 and 30 in 1914 had been killed by the end of 1918 (Kitchen, 1988: 22). Nor could one take encouragement from those who survived the terrible ordeal. The broken and brutalized survivors of war, at least twenty million souls, were all too evident in the shuffling armies of beggars and drifters which eked out a meagre existence on the mean streets of Berlin, Paris, and London (Bourke, 1996; Diehl, 1987).

Fear of Europe's decline provoked numerous calls for greater unity as the only means to stave off further degeneration (Demangeon, 1920; Spengler, 1918–22; see also Stirk, 1989, 1996, 18–50). Richard Coudenhove-Kalergi's Pan-European Union, established in the mid-1920s, was the most successful inter-war organization devoted to European unity. It developed a technological and economic agenda to solve Europe's chronic political problems (Wiedemer, 1993). Although recognizably part of a broader inter-war enthusiasm for pan-regionalism (which surfaced in both fascist and communist geopolitics), the Pan-European Union was the clearest manifestation of the liberal-democratic idea of a united Europe (Whittlesey, 1940; see also O'Loughlin and Wusten, 1990). It represented, according to Peter Bugge (1993: 101) 'an astonishing mixture of large-scale Utopianism, potent political analysis and clear-sighted pragmatism'. The urbane and multi-lingual Coudenhove-Kalergi (a native of cosmopolitan Bohemia, raised in Vienna with an Austrian diplomat father and a Japanese mother) was passionately opposed to communism and later became equally hostile to fascism. He saw no place for Russia in his scheme for a united Europe so long as it remained under the Bolshevik yoke. More surprisingly, he was also ambivalent about Britain, which he insisted was an imperial, non-European power though culturally part of the European clan. Like the great French geographer Albert Demangeon (1923), Coudenhove-Kalergi felt that as long as Britain clung to

its empire, it would never be able to play its full role in Europe, though he accepted the need for white Europeans to discharge their 'colonial duties' amongst the 'lesser' peoples of the non-European world (Fig. 7.9). Ultimately, Britain could act as a bridge between Europe proper and the outside world, particularly the USA (Coudenhove-Kalergi, 1923; see also T. Mann, 1935; Noel-Baker et al., 1934; Ortega y Gasset, 1930).

The Pan-European Union received fleeting though genuine support from some political leaders. Inspired by these ideals, Aristide Briand, the veteran French Foreign Minister, issued his famous memorandum of 17 May 1930 inviting the other European powers to begin formal talks with a view towards European economic and political unification. This remarkable initiative was received with a polite lack of enthusiasm in London and foundered completely when Briand fell from power and when Gustav Streseman, the German Foreign Minister and the other enthusiast for the scheme, died from exhaustion and overwork (Pegg, 1983: 140–8; see also Herriot, 1930; Jouvenel, 1930).

The rise of European fascism, first in Italy and then in Germany, destroyed the liberal-democratic vision of Europe which Coudenhove-Kalergi represented. Into its place came a more disturbing geopolitical imagination. As Jeffrey Herf (1984) and others have demonstrated, fascism drew on complex and somewhat contradictory intellectual inspirations: a mystical, atavistic and deeply romantic preoccupation with the *Volk* and its place within nature; and a crushingly modernist, technocratic concern with speed, energy and power (see also Mosse, 1980). These twin sources jostled uncomfortably within fascist thinking but both profoundly influenced the geopolitical theories which developed during the 1930s under the dominating influence of Karl Haushofer in Germany and Ernesto Massi in Italy.

Much has been written on fascist enthusiasm for geopolitical theorizing, on the complicity of professors and intellectuals in the broader fascist project, and on the moral bankruptcy of the work carried out in Germany and Italy under the label 'geopolitics' (Atkinson, 1996; Burleigh, 1988; Fahlbusch, Rössler, and Siegrist, 1989; Herb, 1989; Heske, 1986, 1987; Korinman, 1990, 1991; Kost, 1988, 1989; Kristof, 1960; Murphy, 1996; O'Sullivan, 1986; G. Parker, 1985; Rössler, 1989). It is tempting to see

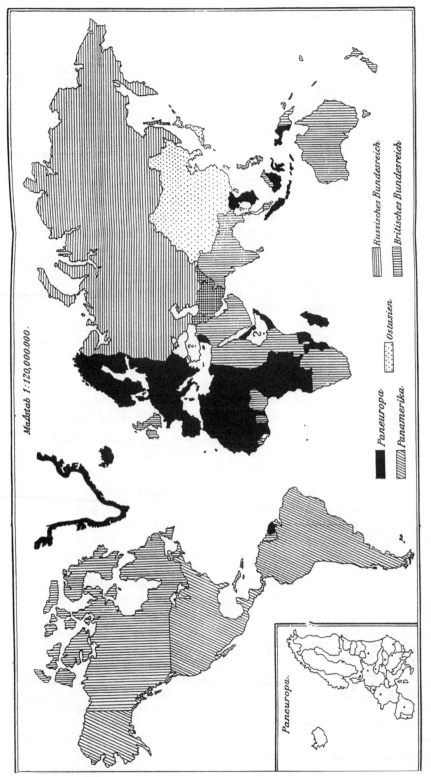

Fig. 7.9. Richard Coudenhove-Kalergi's Paneuropa, 1923. (*Source*: R. N. Coudenhove-Kalergi, *Pan-Europa* (Vienna, 1923).)

*Maßstab 1 : 120,000.000.*

*Paneuropa*
*Panamerika.*
*Ostasien*
*Russisches Bundesreich*
*Britisches Bundesreich*

*Paneuropa.*

inter-war geopolitics as a peculiar and generally unsavoury intellectual development, the product of the extreme political climate of this period. Certainly one can identify a characteristic aesthetic which geopolitical atlases and journals (*Zeitschrift für Geopolitik* in Germany and *Geopolitica* in Italy) all shared: a simple, populist style of writing mixed with bold black-and-white maps of striking symbolic representations (Fig. 7.10). But it would be wrong to interpret inter-war geopolitics as a specifically fascist development or to draw too clear a distinction between the *fin-de-siècle* geopolitics of Kjellén, Ratzel, and Mackinder and the 'new' geopolitics which emerged during the 1920s. The latter thrived in several European countries, including liberal France where a powerful school developed partly (but only partly) in opposition to the German writers (G. Parker, 1987; see, for example, Ancel, 1936; Goblet, 1936). Even in the fascist states, the relationship between the prophets of geopolitics and the political leadership was not always cosy (Bassin, 1987*b*). Moreover, the big themes of *fin-de-siècle* geopolitics, particularly central Europe and Russia, remained focal issues in inter-war geopolitical debate. These twin concerns had always overlapped but by the 1930s they fused into a brutally simple question which dominated German geopolitical writings in the 1930s and 1940s: who was to dominate east-central Europe?

The existing concept of *Mitteleuropa*, mixed with the Ratzelian idea of *Lebensraum*, filtered with various modifications into the rhetoric of Nazi geopolitics as expressed both in the beer gardens and in the university lecture halls. Once Nazi domination of central Europe was assured, the east seemed the 'natural' sphere into which Germany should expand. Long before 1918, romantic German nationalists (like their Russian equivalents) were describing the eastern borderlands of Europe as a realm of spiritual purity, unsullied by the evil materialism of Britain, France, and the USA (Schubert, 1938). To win control of this area, Germany was destined to become embroiled in a life-or-death conflict for living space with Soviet Russia. Not surprisingly, the idea of *Mitteleuropa* was widely attacked, particularly in France. Joseph Aulneau, for example, was adamant that central Europe was an evil fiction which 'n'a vécu que dans l'imagination des conquérants ou des écrivains' (Aulneau, 1926: 8; see, for various perspectives, Ancel,

1936–45; Lemonon, 1931; L'Héritier, 1928; Martonne, 1930; Masaryk, 1918; Ormsby, 1935; Sinnhuber, 1954).

The extent to which abstract geopolitical notions like *Mitteleuropa* and *Lebensraum* directly influenced the murderous course which Hitler set for Germany after 1933, is, however, a complex and difficult question (Haushofer, 1930, 1937; see also Dickenson, 1943; Droz, 1960; H. C. Meyer, 1946, 1955; W. Smith, 1986). Suffice it to say that at the high-point of Nazi power in June 1941 a German-dominated *Mitteleuropa* had become a reality and the Wehrmacht was poised to unleash its full force in Operation Barbarossa, the all-out assault on Soviet Russia (Fig. 7.11). We are, once again, confronted by the cruel paradox that a form of European unity had been achieved by force rather than persuasion. Although the Nazi rhetoric was overwhelmingly nationalist and German-focused, Hitler spoke on occasion of a new European order and certainly saw himself as Europe's standard-bearer against the evils of communism: 'This is not a war we are fighting just on behalf of the German people', he once bellowed, 'but a struggle for the whole of Europe and the whole of civilized humanity' (quoted in Bugge, 1993: 110). Bound together by modern railways and roads, Hitler's Europe was far more integrated economically and geopolitically than had been possible under Napoleon and at far greater cost to the continent's people (M. L. Smith and Stirk, 1990).

This leads us to the most disturbing aspect of the fascist geopolitical imagination. The Nazi vision of Europe was, at its very core, a racial concept. Racial theories had been central to the European imperial world-view during the nineteenth century, particularly after Darwin. To some extent, the scientific racism which flourished all over Europe in the late nineteenth century can be seen as an extension of the civilizational debates of the eighteenth century, the dark side of the Enlightenment urge to classify, categorize, order, and control. Although initially focused on non-European peoples, by 1900 spurious racial arguments were widely used to explain the characteristics of different European peoples, including the Prussians whose alleged anti-democratic, militarist culture was frequently understood as an inherent racial trait (D. W. Johnson, 1917*b*; Lyde, 1913, 1931; see also Pick, 1993: 88–96). After 1914, racial criteria were also discussed as one

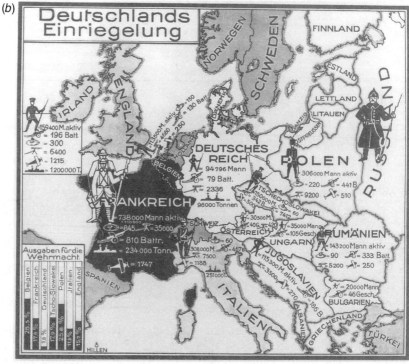

Fig. 7.10. German geopolitical maps, 1929: (*a*) 'Germany's mutilation'; (*b*) 'Germany's imprisonment'. (*Source*: F. Bran and A. Hallen-Ziegfeld, *Geopolitischer Geschichtsatlas*, ii. *Die Neuzeit* (Dresden, 1929), 37 and 40.)

**Fig. 7.11.** Nazi Europe, June 1941.

way to devise a more stable European geopolitical order (H. J. Johnson, 1919; Newbigin, 1917). But it was under the Nazis that the racial interpretation of Europe reached its horrific apogee with the attempt to obliterate from the face of the continent those groups, Jews and others, who were deemed to have no place within the European order that Hitler sought to construct. This strategy was based on its own warped spatiality, the significance of which we are only now beginning to understand (Clarke *et al.*, 1996; Cole and Smith, 1995; Freeman, 1995; Gilbert, 1982).

It is generally argued that the momentum towards greater European integration after 1945 was propelled by a desire to avoid a repetition of 1939–45, an era which claimed at least forty-six million lives (Gilbert, 1989: 1). This is perfectly true, for the architects of western European unity—Jean Monnet, Robert Schuman, Konrad Adenauer, and Alcide Gasperi—were clearly motivated by this honourable objective. But one must also accept that post-1945 Europe was a continent divided into two ideologically distinct sectors, each dominated by a rival superpower whose strategic considerations and geopolitical visions were global rather than specifically European, particularly after the large-scale development of nuclear weapons. The idea of Europe after 1945 was geopolitically constrained; frozen between the two global superpowers of the USA and the USSR. Notwithstanding the wholesale relocation of Poland, it is significant that the geopolitical changes which took place after 1945 were meagre in comparison with the post-1918 transformation (Fig. 7.12).

**Fig. 7.12.** Europe, *c.*1948.

Geopolitics now operated on a much larger, global scale through military alliances such as NATO and the Warsaw Pact. The national territorial geopolitics of Europe was suddenly of far less significance.

The immediate post-war debate about western European unity was also constrained intellectually and politically. In contrast with the flagrant disregard for economic considerations which Keynes (1920) lamented in the post-1918 European order, the agenda for western European integration after 1945 was almost entirely dominated by practical economic matters overseen by precisely the kind of managerialist bureaucracies which Saint-Simon had predicted in 1814—the Organization for European Economic Co-operation (1948), the European Coal and Steel Community (1951), the European Economic Community (1957), and the European Atomic Agency Community (1957) (Archer, 1990; Gillingham, 1991; Lipgens, 1982; Milward, 1984; Stirk, 1996; Urwin, 1989, 1991; Waever, 1993; A. Williams, 1991). Although designed to advance an ideological objective—the creation of an Atlantic Europe bound together and collectively linked to the USA—western European integration after 1945 was determined by a gradualist and pragmatic agenda which, as Alan Milward (1992) has argued, strengthened rather than diminished the power, sovereignty, and independence of the various European nation-states involved. It is partly for this reason that virtually no serious intellectual history of the idea of Europe after 1945 has been written (Milward *et al.*, 1993: 185).

# Conclusion

Since 1989 the European debate has soared and plunged on a roller-coaster in which general euphoria has been followed by widespread despair. The end of communist rule in eastern Europe and the reunification of Germany breathed new life into the European project. After *perestroika* and *glasnost*, the Soviet leadership sought a new place for Russia within a 'Common European House'. The old idea of *Mitteleuropa*, prophetically reconsidered in the mid-1980s by Milan Kundera (1984), has also become the subject of intense interest (Chilton and Ilyin, 1993; Malcolm, 1989; Stirk, 1994; Waever, 1993: 186–7). For some, a revived and reunited *Mitteleuropa* offered a sublime resolution of European history which might heal the 'scar of Yalta'; for others, it would represent the triumph of Germany's vision for Europe (Bender, 1987; Betz, 1990; Croan, 1989; Evans, 1992; Garton Ash, 1990, 1993; Gellner, 1990; Hobsbawm, 1991; Judt, 1990; Rupnik, 1990; Schöpflin and Wood, 1989).

This initial burst of energy and enthusiasm quickly dissipated, however, as the stresses and strains of the new order began to manifest themselves across the continent (Crouch and Marquand, 1992; Kumar, 1992). The re-emergence of genocidal ethnic conflict in the former Yugoslavia and the threat of something even worse in the former Soviet Union has intensified the sense of despondency. A *fin-de-siècle* fear of decline, so characteristic of the 1890s, has resurfaced in the 1990s in the doom-laden predictions of European environmental exhaustion and demographic decrepitude. At the time of writing, there are many in the former Soviet empire who look back with affection and nostalgia for the certainties of the Cold War. The European project is now gripped by a conceptual confusion which reveals itself even in the language of political debate. Take, for example, the vexed question of 'federalism', the dreaded 'f-word' of European politics. This is used by some to invoke a centralizing, universalizing, and homogenizing superstate; by others to suggest the traditional doctrine of balance; and by still others to imply decentralized, local or regional autonomy (G. Smith, 1995).

Fear of the Soviet Union was the single most important factor promoting economic and political co-operation in western Europe after 1945. While recent tensions within the European Union have clearly been connected to the difficulties associated with economic convergence, the terms of the Maastricht Treaty, the dangers of a single European currency, and the threats to national sovereignty, it is equally clear (though less regularly debated) that the growing doubts about the European project stem from a fundamental question: what purpose does Europe serve in the absence of a perceived Soviet threat (Dalby, 1988)? To put it another way: can Europe exist without an obvious constituting 'other'?

This is an enormous question with which to conclude a short chapter. It could be argued, however, that just as Russia and eastern Europe rose to replace the fading threat of the Ottoman Empire as Europe's defining 'other' after the eighteenth century, we now face the prospect of an expanded Europe, in which west and east are increasingly integrated, that will once again define itself in almost medieval terms against a non-Christian, Muslim world to the south. This would be a disturbing development, particularly if the racism and xenophobia which such a distinction implies were to gather momentum within Europe itself to be unleashed against the continent's minorities.

We cannot assume that a united Europe will automatically produce a more cosmopolitan, inclusive, and tolerant political order than the traditional collection of nation-states. Both geopolitical orders rest on a common assumption that politics and political identities must inevitably be territorial. This has been manifestly true in the case of the conventional nation-states, but, as Gerard Delanty (1995) suggests, it is no less true in respect of the European project. Lurking behind the rhetoric of unity and peace, the idea of Europe has always been about the politics of exclusion and division; the drawing of lines between those within and those without (Balibar, 1991; Husbands, 1988). Delanty concludes that it is only by rejecting this exclusionary geopolitics, particularly with respect to immigrant communities, that the idea of Europe will retain any moral or political purpose in the coming century (Miall, 1993, 1994). This means challenging and undermining our assumptions about territorial identities (Rokkan and Urwin, 1982).

Recent debates about the possibilities of de-territorialized internationalism and cosmopolitanism, ideas heralded by Hedley Bull's anarchical vision of a 'new medievalism' (Bull, 1977), point the way towards new forms of political identity both with respect to Europe and in response to a broader process of globalization (Camerilli and Falk, 1992; Cerutti, 1992; Coakley, 1994; Habermas, 1992; Halliday, 1988; Hamm, 1992; Hobsbawm, 1992*b*; Hodgson, 1993; Marquand, 1994; Nelson *et al.*, 1992; Ohmae, 1990, 1995; O'Brien, 1992; Pocock, 1991; Pogge, 1992; Ruggie, 1993; A. D. Smith, 1992;

Taylor, 1994, 1995). For Stjepan Mestrovic (1991, 1993, 1994) this would mean a postmodern Europe of hybridity and multiple identities while Christopher Harvie (1994), in a more prosaic vein, offers us a glimpse of how a 'regional Europe' might begin to operate in ways which could transcend traditional nation-states. Whether these ideas represent the final demise of spatialized political identity or simply another form of European geopolitics must be left for another essay.

## References

AGNEW, J., 'The Devaluation of Place in Social Science', in J. Agnew and J. S. Duncan (eds.), *The Power of Place: Bringing together the Geographical and Sociological Imaginations*, (London, 1989), 9–29.

—— and CORBRIDGE, S., *Mastering Space: Hegemony, Territory and International Political Economy*, (London, 1995).

AKERMAN, J. R., 'The Structuring of Political Terrain in Early Printed Atlases', *Imago Mundi*, 47 (1995) 138–55.

ALBRECHT-CARRIÉ, R., *The Unity of Europe: An Historical Survey*, (London, 1965).

ANCEL, J., *Géopolitique*, (Paris, 1936).

—— *Manuel géographique de la politique européenne*, i (1936). *L'Europe centrale*; ii, Pt. 1 (1940). *L'Europe germanique et ses bornes*; ii, Pt. 2 (1945). *L'Allemagne* (Paris, 1936–45).

—— *Géographie des frontières*, (Paris, 1938).

ANDERSON, J. (ed.), *The Rise of the Modern State*, (Brighton, 1986).

ANDERSON, M. S., 'English Views of Russia in the Seventeenth Century', *Slavonic and East European Review*, 39 (1954), 143–53.

ARCHER, C., *Organizing Western Europe*, (London, 1990).

ATKINSON, D., 'Geopolitics and the Geographical Imagination in Fascist Italy' (unpublished Ph.D. thesis, University of Loughborough, 1996).

AULNEAU, J., *Histoire de l'Europe centrale*, (Paris, 1926).

BACZKO, B., 'Le Calendrier républicain: décréter l'éternité', in P. Nora (ed.), *Les Lieux de mémoire*, i. *La République* (Paris, 1984), 37–83.

BAECHLER, J., HALL, J. A., and MANN, M. (eds.), *Europe and the Rise of Capitalism*, (Oxford, 1988).

BALIBAR, E., '*Es gibt keinen Staat in Europa*: Racism and Politics in Europe Today', *New Left Review*, 186 (1991), 5–19.

BALZARETTI, R., 'The Creation of Europe', *History Workshop Journal*, 33 (1992), 181–96.

BARBICHE, B., *Sully*, (Paris, 1978).

BARRACLOUGH, G., *European Unity in Thought and Practice*, (Oxford, 1963).

BARTLETT, R., *The Making of Europe: Conquest, Colonization, and Cultural Change 950–1350* (London, 1993).

—— and MACKAY, A., *Medieval Frontier Societies*, (Oxford, 1989).

BASSIN, M., 'Imperialism and the Nation State in Friedrich Ratzel's Political Geography', *Progress in Human Geography*, 11 (1987*a*), 473–95.

BASSIN M., 'Race contra Space: The Conflict between German Geopolitik and National Socialism', *Political Geography Quarterly* 6 (1987*b*), 115–34.

—— 'Russia between Europe and Asia: The Ideological Construction of Geographical Space', *Slavonic Review*, 50 (1991), 1–17.

BAUMAN, Z., 'On the Origins of Civilization: A Historical Note', *Theory, Culture and Society*, 2 (1985) 7–14.

BENDER, P., 'Mitteleuropa: Mode, Modell oder Motiv', *Die neue Gesellschaft*, 4 (1987), 297–304.

BERDYAEV, N. A., *The New Middle Ages*, (London, 1924).

BERGSON, H., *La Signification de la guerre*, (Paris, 1915).

BETZ, H. E., 'Mitteleuropa and post-modern European identity', *New German Critique*, 50 (1990), 173–92.

BLOUET, B. W., *Sir Halford Mackinder: A Biography*, (College Station, Tex., 1987).

BOER, P. den, 'Europe to 1914: The Making of an Idea', in K. Wilson and J. van den Dussen (eds.), *The History of the Idea of Europe*, (London, 1993), 13–82.

BOURKE, J., *Dismembering the Male: Men's Bodies, Britain and the Great War*, (London, 1996).

BOWMAN, I., *The New World: Problems of Political Geography*, (New York, 1921).

BRAUER, R. W., 'Boundaries and Frontiers in Medieval Muslim Geography', *Transactions of the American Philosophical Society*, 85 (6) (1995), 1–73.

BRECHTEFELD, J., *Mitteleuropa in German Politics, 1848 to the Present*, (London, 1996).

BRIGHAM, A. P., 'Principles in the Determination of Boundaries', *Geographical Review*, 7(4) (1919), 201–19.

BRINCKEN, A. D., 'Europa in der Kartographie des Mittelalters', *Archiv für Kulturgeschichte*, 55 (1973), 289–304.

BROERS, M., *Europe under Napoleon 1799–1815*, (London, 1996).

BRUGMANS, H., *L'Idée européenne 1920–1970*, (Bruges, 1970).

BRUNHÈS, J., and VALLAUX, C., *La Géographie de l'histoire: géographie de la paix et de la guerre sur terre et sur mer*, (Paris, 1921).

BRYCE, J., *Modern Democracies*, (London, 1922).

BUGGE, P., 'The Nation Supreme: The Idea of Europe 1914–1945', in K. Wilson and J. van den Dussen (eds.), *The History of the Idea of Europe*, (London 1993), 83–149.

BUISSERET, D., *Sully and the Growth of Centralized Government in France 1598–1610*, (London, 1968).

—— *Henry IV, King of France*, (London, 1984).

—— (ed.), *Monarchs, Ministers and Maps: The Emergence of Cartography as a Tool of Government in Early Modern Europe*, (Chicago, 1992).

BULL, H., *The Anarchical Society: A Study of Order in World Politics*, (London, 1977).

—— KINGSBURY, B., and ROBERTS, A. (eds.), *Hugo Grotius and International Relations*, (Oxford, 1990).

BURKE, E., *Reflections on the Revolution in France and on the Proceedings in Certain Societies in London relative to that Event*, (Harmondsworth, 1983 [1790]).

BURKE, P., 'Did Europe Exist before 1700?', *History of European Ideas*, 1 (1980), 21–9.

—— 'We, the People: Popular Culture and Popular Identity in Modern

Europe', in S. Lash and J. Friedman (eds.), *Modernity and Identity*, (Oxford, 1992), 293–308.

BURLEIGH, M., *Germany Turns Eastwards: A Study of Ostforschung in the Third Reich*, (Cambridge, 1988).

CAHNMAN, W., 'Frontiers between East and West', *Geographical Review*, 49 (1952), 605–24.

CAMILLERI, J., and FALK, J., *The End of Sovereignty? The Politics of a Shrinking and Fragmenting World*, (Aldershot, 1992).

CERUTTI, F., 'Can there be a Supranational Identity?', *Philosophy and Social Criticism*, 18(2) (1992), 147–62.

CHABOD, F., *Storia dell'idea d'Europa*, (Bari, 1965).

CHADWICK, H. M., *The Nationalities of Europe and the Growth of National Ideologies*, (Cambridge, 1945).

CHILTON, P., and ILYIN, M., 'Metaphor in Political Discourse: The Case of "Common European House"', *Discourse and Society*, 4 (1993), 7–31.

CHIRIOT, D. (ed.), *The Origins of Backwardness in Eastern Europe: Economics and Politics from the Middle Ages until the Early Twentieth Century*, (Berkeley and Los Angeles, 1989).

CHISHOLM, G. G., 'Central Europe: A Review', *Scottish Geographical Magazine*, 33 (1917), 83–8.

CLARKE, D., DOEL, M., and McDONOUGH, F. X., 'Holocaust Topologies: Singularity, Politics, Space', *Political Geography*, 15 (1996), 457–89.

CLARKE, J., *The Language of Liberty 1660–1832: Political Discourse and Social Dynamics in the Anglo-American World*, (Cambridge, 1994).

COAKLEY, J., 'Approaches to the Resolution of Ethnic Conflicts: The Strategy of Non-Territorial Autonomy', *International Political Science Review*, 15 (1994), 297–314.

COLE, T., and SMITH, G., 'Ghettoization and the Holocaust: Budapest 1944', *Journal of Historical Geography*, 21(3) (1995), 300–16.

COLLEY, L., *Britons: Forging the Nation 1707–1837*, (New Haven, 1992).

CORNISH, V., *Borderlands of Language in Europe and their Relation to the Historic Frontier of Christendom*, (London, 1936).

COUDENHOVE-KALERGI, R. N., *Pan-Europa*, (Vienna, 1923).

CROAN, M., 'Lands In-Between: The Politics of Cultural Identity in Contemporary Eastern Europe', *Eastern European Politics and Society*, 3 (1989), 176–97.

CROUCH, C., and MARQUAND, D. (eds.), *Towards Greater Europe? A Continent without an Iron Curtain*, (Oxford, 1992).

CUTLER, A. C., 'The "Grotian Tradition" in International Relations', *Review of International Studies*, 17 (1991), 41–65.

DALBY, S., 'Geopolitical Discourse: The Soviet Union as Other', *Alternatives*, 13 (1988), 415–42.

DARBY, H. C., 'The Face of Europe on the Eve of the Great Discoveries', in G. R. Potter (ed.), *The New Cambridge Modern History*, i. *The Renaissance 1493–1520*, (Cambridge, 1961), 20–49.

DELANTY, G., *Inventing Europe: Idea, Identity, Reality*, (London, 1995).

DEMANGEON, A., *Le Déclin de l'Europe*, (Paris, 1920).

—— *L'Empire britannique: étude de géographie coloniale*, (Paris, 1923).

DEROLEZ, A. (ed.), *Liber floridus*, (Ghent, 1968).

DICKENSON, R. E., *The German Lebensraum*, (London, 1943).

DIEHL, J. M., 'Victors or Victims? Disabled Veterans in the Third Reich', *Journal of Modern History*, 59 (1987), 705–36.

DILKE, O. A. W., MILLARD, A. R., and AUJAR, G., 'Cartography in Ancient Europe and the Mediterranean', in J. B. Harley and D. Woodward (eds.), *The History of Cartography*, i. *Cartography in Prehistoric, Ancient and Medieval Europe and the Mediterranean*, (Chicago, 1987), 103–280.

DION, R., *Les Frontières de la France*, (Paris, 1947).

—— *Aspects politiques de la géographie antique*, (Paris, 1977).

DIONISOTTI, C., *Europe in Sixteenth-Century Italian Literature*, (Oxford, 1971).

DOCKÈS, P., *L'Espace dans la pensée économique du XVIe au XVIIIe siècle*, (Paris, 1969).

DODGSHON, R. A., *The European Past: Social Evolution and Spatial Order*, (London, 1987).

DOMINIAN, L., *Frontiers of Language and Nationality in Europe*, (New York, 1917).

DROUET, J., *L'Abbé de Saint-Pierre: l'homme et l'œuvre*, (Paris, 1912).

DROZ, J., *L'Europe centrale: evolution historique de l'idée de 'Mitteleuropa'*, (Paris, 1960).

DUROSELLE, J.-B., *L'Idée d'Europe dans l'histoire*, (Paris, 1966).

—— *Europe: A History of its Peoples*, (London, 1990).

EDGERTON, S. Y., 'From Mental Matrix to Mappamundi to Christian Empire: The Heritage of Ptolemaic Cartography in the Renaissance', in D. Woodward (ed.), *Art and Cartography: Six Historical Essays*, (Chicago, 1987), 10–50.

EICHTHAL, G. d', *De l'unité européenne*, (Paris, 1840).

EKSTEINS, M., *Rites of Spring: The Great War and the Birth of the Modern Age*, (London, 1989).

ELIAS, N., *The Civilizing Process: The History of Manners and State Formation and Civilization*, (Oxford, 1994).

ELLIOT, J. H., 'Europe of Composite Monarchies', *Past and Present*, 137 (1992), 48–71.

EVANS, R. J. W., 'Frontiers and National Identities in Central Europe', *International History Review*, 14(3) (1992), 480–502.

FAHLBUSCH, M., RÖSSLER, M., and SIEGRIST, D., 'Conservatism, Ideology and Geography in Germany 1920–1950', *Political Geography Quarterly*, 8 (1989), 353–67.

FAWCETT, C. B., *Provinces of England: A Study of Some Geographical Aspects of Devolution*, (London, 1919).

—— 'Some Geographical Factors in the Growth of the State', *Scottish Geographical Magazine*, 38 (1922), 221–32.

FEBVRE, L., 'Civilization: Evolution of a Word and a Group of Ideas', in P. Burke (ed.), *A New Kind of History from the Writings of Febvre*, (New York, 1974), 219–57.

FISCHER, F., *Germany's Aims in the First World War*, (London, 1967).

FISCHER, J., *Oriens-Occidens-Europa: Begriff und Gedanke 'Europa' in der Späten Antike und im Frühen Mittelalter*, (Wiesbaden, 1957).

FISHER, C. A., 'The Changing Dimensions of Europe', *Journal of Contemporary History*, 1(3) (1966), 3–20.

FOERSTER, R. M., *Europa: Geschichte einer politischen Idee*, (Munich, 1967).

FOUCAULT, M., 'Governmentability', in G. Burchell, C. Gordon, and P.

Miller (eds.), *The Foucault Effect: Studies in Governmentality*, (Chicago, 1991), 87–104.

FRANCE, P., 'Western Civilization and its Mountain Frontiers (1750–1850)', *History of European Ideas*, 6(3) (1985), 297–310.

FREEMAN, M., *Atlas of Nazi Germany: A Political, Economic and Social Anatomy of the Third Reich*, 2nd edn. (London, 1995).

FUHRMANN, M., *Europa—zur Geschichte einer kulturellen und politischen Idee*, (Constance, 1981).

GARTON ASH, T., 'Mitteleuropa', *Daedalus*, 119(1) (1990), 1–22.

—— *In Europe's Name: Germany and the Divided Continent*, (London, 1993).

GELLNER, E., 'Ethnicity and Faith in Eastern Europe', *Daedalus*, 119(1) (1990), 279–94.

GILBERT, M., *Atlas of the Holocaust*, (London, 1982).

—— *Second World War*, (London, 1989).

—— *First World War*, (London, 1994).

GILLINGHAM, J., *Coal, Steel, and the Rebirth of Europe, 1945–1955: The Germans and French from Ruhr Conflict to Economic Community*, (Cambridge, 1991).

GLACKEN, C., *Traces on the Rhodian Shore: Nature and Culture in Western Thought from Ancient Times to the End of the Eighteenth Century*, (Berkeley and Los Angeles, 1967).

GOBLET, Y., *The Twilight of Treaties*, (London, 1936).

GOLLWITZER, H., 'Zur Wortesgeschichte und Sinndeutung von Europa', *Saeculum*, 2 (1951*a*), 161–5.

—— *Europabild und Europagedanke*, (Munich, 1951*b*).

GUENÉE, B., 'Des limites féodales aux frontières politiques', in P. Nora (ed.), *Les Lieux de Mémoire*, ii. 2. *La Nation*, (Paris, 1986), 11–33.

HABERMAS, J., 'Citizenship and National Identity: Some Reflections on the Future of Europe', *Praxis International*, 12(1) (1992), 1–19.

HALECKI, O., *The Limits and Divisions of European History*, (London, 1950).

—— *Borderlands of Western Civilization: A History of East Central Europe*, (New York, 1952).

HALLIDAY, F., 'Three Concepts of Internationalism', *International Affairs*, 64 (1988), 187–98.

HAMM, B., 'Europe: A Challenge to the Social Sciences', *International Social Science Journal*, 44 (1992), 3–22.

HARLEY, J. B., 'Secrecy and Silences: The Hidden Agenda of Cartography in Early Modern Europe', *Imago Mundi*, 40 (1988), 111–30.

—— and WOODWARD, D. (eds.), *The History of Cartography*, ii. 1. *Cartography in the Traditional Islamic and South Asian Societies*, (Chicago, 1992).

—— —— (eds.), *The History of Cartography*, ii. 2. *Cartography in the Traditional East and Southeast Asian Societies*, (Chicago, 1994).

HARTLEY, J. M., 'Is Russia Part of Europe? Russian Perceptions of Europe in the Reign of Alexander I', *Cahiers du Monde Russe et Soviétique*, 33 (1992), 369–86.

HARVEY, P. D. A., WOODWARD, D., CAMPBELL, T., and HARLEY, J. B., 'Cartography in Medieval Europe and the Mediterranean', in J. B. Harley and D. Woodward (eds.), *The History of Cartography*, i. *Cartography in Prehistoric, Ancient, and Medieval Europe and the Mediterranean*, (Chicago, 1987), 283–509.

HARVIE, C., *The Rise of Regional Europe*, (London, 1994).

HASSINGER, H., 'Das geographische Wesen Mitteleuropas, nebst einigen grund-sätzlichen Bemerkungen über die geographischen Naturgebiete Europas und ihre Begrenzung, *Mitteilungen der Geographischen Gesellschaft Wien*, 60 (1917), 437–93.

HAUNER, M., *What is Asia to Us? Russia's Asian Heartland Yesterday and Today*, (London, 1990).

HAUSHOFER, K., 'Mitteleuropa und der Anschluss', in F. Kleinwaechter and H. Paller (eds.), *Der Anschluss*, (Vienna, 1930), 147–59.

—— 'Mitteleuropa und die Welt', *Zeitschrift für Geopolitik*, 14 (1937), 1–4.

HAY, D., *Europe: The Emergence of an Idea*, 2nd edn. (Edinburgh, 1968).

—— 'Europe Revisited: 1979', *History of European Ideas*, 1 (1980), 1–6.

HAZARD, P., *The European Mind: The Critical Years 1680–1715*, (New York, 1990).

HEATER, D., *The Idea of European Unity*, (Leicester, 1992).

HENDERSON, W. O., 'A Nineteenth-Century Approach to a West European Common Market', *Kyklos*, 10 (1957), 448–59.

—— *The Origins of the Common Market*, (London, 1962).

—— *Friedrich List: Economist and Visionary 1789–1846*, (London, 1983).

HERB, G. H., 'Persuasive Cartography in Geopolitik and National Socialism', *Political Geography Quarterly*, 8 (1989), 289–303.

HERF, J., *Reactionary Modernism: Technology, Culture and Politics in Weimar and the Third Reich*, (Cambridge, 1984).

HERRIOT, E., *The United States of Europe*, (London, 1930).

HERTSLET, E., *The Map of Europe by Treaty; Showing the Various Political and Territorial Changes which have Taken Place since the General Peace of 1814* (4 vols.; London, 1875–91).

HESKE, H., 'German Geographic Research in the Nazi Period', *Political Geography Quarterly*, 5 (1986), 267–82.

—— 'Karl Haushofer: His Role in German Geopolitics and Nazi Politics', *Political Geography Quarterly*, 6 (1987), 135–44.

HINSLEY, F. H., *Power and the Pursuit of Peace*, (Cambridge, 1967).

HOBSBAWM, E. J., *The Age of Empire, 1875–1914*, (London, 1987).

—— 'The Return of Mitteleuropa', *Guardian*, 11 Oct (1991).

—— *Nations and Nationalism since 1780*, (Cambridge, 1992a).

—— 'Ethnicity and Nationalism in Europe Today', *Anthropology Today* 8(1) (1992b), 3–8

HODGSON, G., 'Grand Illusion: The Failure of European Consciousness', *World Policy Journal* 10(2) (1993), 13–8.

HOLDAR, S., 'The Ideal State and the Power of Geography: the Life and Work of Rudolf Kjellén', *Political Geography*, 11 (1992), 307–23.

HOLDICH, T. H., *Political Boundaries and Boundary Making*, (London, 1916).

—— *Boundaries in Europe and the Near East*, (London, 1918).

HUME, D., 'The Idea of a Perfect Commonwealth', reprinted in D. Claeys (ed.), *Utopias of the British Enlightenment*, (Cambridge, 1994) (1752), 55–69.

HUSBANDS, C., 'The Dynamics of Racial Exclusion and Expulsion: Racist Politics in Western Europe', *European Journal of Political Research*, 16 (1988), 701–20.

IONESCU, G. (ed.), *The Political Thought of Saint Simon*, (Oxford, 1976).

JOHNSON, D. W., 'The Role of Political Boundaries', *Geographical Review*, 4 (1917a), 208–13.

—— *The Perils of Prussianism*, (New York 1917b).

JOHNSON, H. J., 'The Anthropologists Approach: Race, Language and Nationality in Europe', *Sociological Review*, 11 (1919), 37–46.

JOLL, J., 'Europe: An Historian's View', *History of European Ideas*, 1 (1980), 7–19.

JONES, E. L., *The European Miracle: Environments, Economies, and Geopolitics in the History of Europe and Asia*, 2nd edn. (Cambridge, 1987).

JOUVENEL, B. de, *Vers les États Unis d'Europe*, (Paris, 1930).

JUDT, T., 'The Rediscovery of Central Europe', *Daedalus*, 119(1) (1990), 23–54.

KAIN, R. J. P., and BAIGENT, E. (eds.), *The Cadastral Map in the Service of the State: A History of Property Mapping*, (Chicago, 1992).

KEARNS, G., 'Fin de siècle geopolitics: Mackinder, Hobson and Theories of Global Closure', in P. J. Taylor (ed.), *Political Geography of the Twentieth Century: A Global Analysis*, (London, 1993), 9–30.

KEYNES, J. M., *The Economic Consequences of the Peace*, (New York, 1920).

KITCHEN, M., *Europe between the Wars: A Political History*, (London, 1988).

KNOX, P., and AGNEW, J., *The Geography of the World Economy*, (London, 1994).

KONVITZ, J. W., *Cartography in France, 1660–1848: Science, Engineering, and Statecraft*, (Chicago, 1987).

—— 'The State, Paris and Cartography in Eighteenth- and Nineteenth-Century France', *Journal of Historical Geography*, 16(1) (1990), 3–16.

KORINMAN, M., *Quand l'Allemagne pensait le monde: grandeur et décadence d'une géopolitique*, (Paris, 1990).

—— *Continents perdus: les précurseurs de la géopolitique allemande*, (Paris, 1991).

KOST, K., *Die Einflüsse der Geopolitik auf Forschung und Theorie der Politischen Geographie von ihren Anfängen bis 1945. Ein Beitrag zur Wissenschaftsgeschichte der Politischen Geographie und ihre Terminologie unter Berücksichtigung von Militär- und Kolonialgeographie*, (Bonn, 1988).

—— 'The Conception of Politics in Political Geography and Geopolitics in Germany until 1945', *Political Geography Quarterly*, 8 (1989), 369–85.

KRETSCHMER, K., 'Eine neue mittelalterliche Weltkarte der vatikanischen Bibliothek', *Zeitschrift für Erdkunde*, 26 (1891), 371–406.

KRISTOF, L. D., 'The Nature of Frontiers and Boundaries', *Annals of the Association of American Geographers*, 49 (1959), 269–82.

—— 'The Origins and Evolution of Geopolitics', *Journal of Conflict Resolution*, 4(1) (1960), 15–51.

—— 'The Russian Image of Europe', in C. A. Fisher (ed.), *Essays in Political Geography*, (London, 1968), 345–87.

KROEBER, A. L., and KLUCKHOHN, C., *Culture: A Critical Review of Concepts and Definitions*, (New York, 1952).

KUMAR, K. 'The 1989 Revolutions and the Idea of Europe, *Political Studies*, 40 (1992), 439–61.

KUNDERA, M., 'The Tragedy of Central Europe', *New York Review of Books*, 26 Apr. (1984), 33–8.

LAPRADELLE, P. de, *La Frontière: étude de droit international*, (Paris, 1928).

LE BRAS, H., 'La Statistique Générale de la France', in P. Nora (ed.), *Les Lieux de Mémoire*, ii. 2. *La Nation* (Paris, 1986), 317–53.

LEFEBVRE, H., *The Production of Space*, (Oxford, 1991).

LEMONON, E., *La Nouvelle Europe centrale et son bilan économique*, (Paris, 1931).

LEYSER, K., 'Concepts of Europe in the early and high Middle Ages', *Past and Present*, 137 (1992), 25–47.

LEVY, J. S., *War in the Modern Great Power System, 1495–1975* (Lexington, Ky., 1983).

L'HÉRITIER, M., 'Régions historiques: Europe centrale, Orient méditerranéen et question d'Orient', *Revue de Synthèse Historique*, 45 (1928), 43–67.

LIPGENS, W., *A History of European Integration*, i. *1945–1947* (Oxford, 1982).

LIST, F., *National Systems of Political Economy*, (London, 1904).

LIVEY, J., 'The Europe of the Enlightenment', *History of European Ideas*, 1 (1981), 91–102.

LOUIS, H., 'Über den geographischen Europabegriff', *Mitteilungen der geographischen Gesellschaft in München*, 39 (1954), 73–93.

LYDE, L. W., *The Continent of Europe*, (London, 1913).

—— *Some Frontiers of Tomorrow: An Aspiration for Europe*, (London, 1915).

—— 'Europe v. Middle Europe', *Sociological Review*, 9 (1916–17), 88–93.

—— *Peninsular Europe: Some Geographical Peregrinations, Ancient and Modern*, (London, 1931).

MACKINDER, H. J., 'The Geographical Pivot of History', *Geographical Journal*, 23 (1904), 421–42.

—— *Democratic Ideals and Reality: A Study in the Politics of Reconstruction*, (London, 1919).

McNEIL, W. H., *The Rise of the West*, (Chicago, 1963).

—— *Europe's Steppe Frontier, 1500–1800* (Chicago, 1964).

—— *The Shape of European History*, (Oxford, 1974).

MALCOLM, N., 'The "Common European Home" and Soviet European Policy', *International Affairs*, 65(4) (1989), 659–76.

MANN, M., 'The Autonomous Power of the State: Its Origins, Mechanisms and Results', *Archives of European Sociology*, 25 (1984), 185–213.

—— *The Sources of Social Power*, i. *A History of Power from the Beginning to AD 1760* (Cambridge, 1986).

—— *The Sources of Social Power*, ii. *The Rise of Classes and Nation-States, 1740–1914* (Cambridge, 1993).

MANN, T., *Achtung Europa*, (Munich, 1935).

MARGADANT, T. W., *Urban Rivalries in the French Revolution*, (Princeton, 1992).

MARQUAND, D., 'Re-inventing Federalism: Europe and the Left', *New Left Review*, 203 (1994), 17–26.

MARTONNE, E. de, *L'Europe centrale*, (2 vols.; Paris, 1930).

MASARYK, T. G., *The New Europe*, (London, 1918).

MESTROVIC, S., *The Coming Fin de Siècle: An Application of Durkheim's Sociology to Modernity and Postmodernism*, (London, 1991).

—— *The Barbarian Temperament: Towards a Postmodern Critical Theory*, (London, 1993).

—— *The Balkanization of the West: The Confluence of Postmodernism and Postcommunism*, (London, 1994).

MEYER, H. C., 'Mitteleuropa in German Political Geography', *Annals of the Association of American Geographers*, 36 (1946), 178–94.

—— *Mitteleuropa in German Thought and Practice 1815–1945* (The Hague, 1955).

MEYER, J. W., 'Conceptions of Christendom: Notes on the Distinctiveness of the West', in M. L. Kohn (ed.), *Cross National Research in Sociology*, (London, 1989).

MIALL, H., *Shaping the New Europe*, (London, 1993).

—— (ed.), *Minority Rights in Europe*, (London, 1994).

MILWARD, A S., *The Reconstruction of Western Europe, 1945–1951* (London, 1984).

—— *The European Rescue of the Nation-State*, (London, 1992).

—— LYNCH, F. M. B., ROMERO, F., RUGGERO, R., and SØRENSEN, V., *The Frontier of National Sovereignty: History and Theory 1945–1992* (London, 1993).

MORIN, E., *Concepts of Europe*, (New York, 1991).

MOSSE, G. L., *Masses and Man: Nationalist and Fascist Perceptions of Reality*, (New York, 1980).

MURPHY, D., *The Heroic Earth: The Flowering of Geopolitical Thought in Weimar Germany, 1924–1933* (Kent, Ohio, 1996).

NAJEM, E. W., 'Richelieu's Blueprint for Unity and Peace', *Studies in Philology*, 53 (1956), 25–34.

NAUMANN, F., *Central Europe* (London, 1916 [1915]).

NELSON, B., ROBERTS, D., and VEIT, W. (eds.), *The Idea of Europe: Problems of National and Transnational Identity*, (Oxford, 1992).

NEUMANN, I. B., 'Russia as Central Europe's Constituting Other', *East European Politics and Societies*, 7 (1993), 348–69.

—— and WELSH, J. M., 'The Other in European Self-Definition: An Addendum to the Literature on International Society', *Review of International Studies*, 7 (1991), 327–48.

NEWBIGIN, M. I., 'Race and Nationality', *Geographical Journal*, 50(5) (1917), 313–29.

—— *Aftermath: The Geographical Study of the Peace Terms*, (Edinburgh, 1920).

NISBET, R., *History of the Idea of Progress*, (London, 1980).

NOEL-BAKER, P., *et al.*, *Challenge to Death*, (London, 1934).

NOLTE, E., *Der europäische Bürgerkrieg 1917–1945: Nationalsozialismus und Bolschewismus*, (Berlin, 1987).

NORDMAN, D., 'Des limites d'État aux frontières nationales', in P. Nora, *Les Lieux de Mémoire*, ii. 2. *La Nation* (Paris, 1986), 35–61.

—— and REVEL, J., 'La Formation de l'espace français', in A. Burgière and J. Revel (eds.), *Histoire de la France*, i. (Paris, 1989).

—— and VIC-OZOUF MARIGNIER, M., *Atlas de la Révolution française*, iv. *Le Territoire: réalités et représentations*, (Paris, 1989a).

—— *Atlas de la Révolution française*, v. *Le Territoire et les limites administratives*, (Paris, 1989b).

NUTTAL, E. M., *A Project for Perpetual Peace: Rousseau's Essay*, (London, 1927).

O'BRIEN, R., *Global Financial Integration: The End of Geography*, (London, 1992).

OGG, D. (ed.), *Sully's Grand Design of Henry IV: From the Memoirs of Maximilien de Béthune, Duc de Sully, (1559–1641)* (London, 1921).

OHMAE, K., *The Borderless World: Power and Strategy in the Inter-linked Economy*, (London, 1990).

—— *The End of the Nation-State: The Rise of Regional Economies*, (London, 1995).

OKEY, R., 'Central Europe/Eastern Europe: Behind the Definitions', *Past and Present*, 137 (1900), 102–33.

O'LOUGHLIN, J., and WUSTEN, H. van der, 'The Political Geography of Pan-regions', *Geographical Review*, 80(1) (1990), 1–20.

ORMSBY, H., 'The Definition of Mitteleuropa and its Relation to the Concept of *Deutschland* in the Writings of Modern German Geographers', *Scottish Geographical Magazine*, 51 (1935), 337–47.

ORTEGA Y GASSET, J., *The Revolt of the Masses*, (London, 1930).

O'SULLIVAN, P., *Geopolitics*, (New York, 1986).

O'TUATHAIL, G., 'Putting Mackinder in his Place: Material Transformations and Myth', *Political Geography*, 11 (1992), 100–18.

—— *Critical Geopolitics*, (London, 1996).

OUTRAM, D., *The Enlightenment*, (Cambridge, 1995).

OZOUF, M., *Festivals and the French Revolution*, (Cambridge, Mass., 1988).

PALMER, A., *The Lands Between: A History of East-Central Europe since the Congress of Vienna*, (London, 1970).

PARKER, G., *Western Geopolitical Thought in the Twentieth Century*, (London, 1985).

—— 'French Geopolitical Thought in the Inter-War Years and the Emergence of the European Idea', *Political Geography Quarterly*, 6 (1987), 145–50.

—— and SMITH, L. (eds.), *The General Crisis of the Seventeenth Century*, (London, 1978).

PARKER, W. H., 'Europe: How Far?', *Geographical Journal*, 126 (1960), 278–97.

—— *Mackinder: Geography as an Aid to Statecraft*, (Oxford, 1982).

PARTSCH, J., *Mitteleuropa: Die Länder und Völker von den Westalpen und dem Balkan bis an den Kanal und das kurische Haff*, (Gotha, 1904).

PEAKE, H. J., 'Devolution: A Regional Movement. A: Provinces of England. B: European Aspects', *Sociological Review*, 11 (1919), 97–113.

PECQUEUR, C., *De la paix, de son principe et de sa réalisation*, (Paris, 1842).

PEGG, C. H., *Evolution of the European Idea, 1914–1932* (Chapel Hill, NC, 1983).

PELLETIER, M., *La Carte de Cassini: l'extraordinaire aventure de la carte de France*, (Paris, 1990).

PENCK, A., 'Politisch-geographische Lehren des Krieges', *Meereskunde*, 9–10 (1915), 12–21.

PERKINS, M. L., *The Moral and Political Philosophy of the Abbé de Saint-Pierre*, (Geneva and Paris, 1959).

PERROT, J.-C., and WOOLF, S., *State and Statistics in France 1789–1815* (London, 1984).

PESONEN, P., 'The Image of Europe in Russian Literature and Culture', *History of European Ideas*, 13(4) (1991), 399–409.

PICK, D., *War Machine: The Rationalization of Slaughter in the Modern Age*, (New Haven, 1993).

POCOCK, J. G. A., *The Machiavelli Moment: Florentine Political Thought and the Atlantic Republican Tradition*, (Princeton, 1975).

—— 'Deconstructing Europe', *London Review of Books*, 13(24) (1900), 6–10.

POGGE, T. W., 'Cosmopolitanism and Sovereignty', *Ethics*, 103 (1992), 48–75.

POUNDS, N. J. G., 'The Origins of the Idea of Natural Frontiers in France', *Annals of the Association of American Geographers*, 41 (1951), 146–57.

—— 'France and "les limites naturelles" from the Seventeenth to the Twentieth Centuries', *Annals of the Association of American Geographers*, 44 (1954), 51–62.

—— and BALL, S. S., 'Core Areas and the Development of the European State System', *Annals of the Association of American Geographers*, 54 (1964), 43–50.

PRESCOT, J. R. V., *The Geography of Frontiers and Boundaries*, (London, 1965).

RATZEL, F., 'Die Gesetze des räumlichen Wachstums der Staten', *Petermanns Geographische Zeitschrift*, 42 (1896), 97–107.

—— *Politische Geographie*, (Munich and Leipzig, 1897).

—— 'Der Lebensraum. Eine biogeographische Studie', in K. Bücher and K. Fricker (eds.), *Festgaben für Albert Schäffle zur siebenzigsten Wiederkehr seines Geburtstags am 24. Februar 1901* (Tübingen, 1901), 101–90.

RAUMER, K. von, *Ewiger Friede. Friedensrufe und Friedenspläne seit der Renaissance*, (Munich, 1953).

REUTER, T., 'Medieval Ideas on Europe and their Modern Historians', *History Workshop Journal*, 33 (1992), 176–80.

ROKKAN, S., and URWIN, D. W. (eds.), *The Politics of Territorial Identity*, (London, 1982).

ROMM, J. S., *The Edges of the Earth in Ancient Thought*, (Princeton, 1992).

ROOSEVELT, G. G., *Reading Rousseau in the Nuclear Age*, (Philadelphia, 1992).

RÖSSLER, M., 'Applied Geography and Area Research in Nazi Society: Central Place Theory and Planning, 1933 to 1945', *Environment and Planning D: Society and Space*, 7 (1989), 419–31.

ROUSSEAU, J.-J., *Extrait du projet de paix perpétuelle de Monsieur l'Abbé de Saint-Pierre*, (Amsterdam, 1761).

RUBIN, M., 'The Culture of Europe in the Later Middle Ages', *History Workshop Journal*, 33 (1992), 162–75.

RUGGIE, J. G., 'Territoriality and Beyond: Problematizing Modernity in International Relations', *International Organization*, 47 (1993), 139–74.

RUPNIK, J., 'Central Europe or Mitteleuropa?', *Daedalus*, 119(1) (1900), 249–78.

SACK, R. D., *Conceptions of Space in Social Thought: A Geographic Perspective*, (London, 1980).

SAHLINS, P., 'Natural Frontiers Revisited: France's Boundaries since the Seventeenth Century', *American Historical Review*, 95 (1990), 1423–51.

SAID, E. W., *Orientalism*, (London, 1978).

SAINT-PIERRE, C.-I.-C. de, *Mémoires pour rendre la paix perpétuelle en Europe*, (Cologne, 1712).

—— *A Project for Setting an Everlasting Peace in Europe. First Proposed by Henry IV of France, and Approved by Queen Elizabeth, and Most of the Princes of Europe, and Now Discussed at Large, and Made Practicable by the Abbé de St. Pierre, of the French Academy*, (London, 1714).

SAINT-SIMON, C.-H. de, and THIERRY, A., *De la réorganisation de la Société européenne, ou de la nécessité et des moyens de rassembler les peuples de l'Europe en un seul corps politique, en conservant à chacun son indépendance nationale*, (Paris, 1814).

SCHELER, M., *Der Genius des Krieges und der deutschen Krieg*, (Berlin, 1915).

SCHMIDT, H. D., 'The Establishment of "Europe" as a Political Expression', *Historical Journal*, 9(2) (1966), 172–8.

SCHÖPFLIN, G., and WOOD, N. (eds.), *In Search of Central Europe*, (Oxford, 1989).

SCHÖTTLER, P., 'The Rhine as an Object of Historical Controversy in the

Inter-War Years: Towards a History of Frontier Mentalities', *History Work-shop Journal*, 39 (1995), 1–21.

SCHUBERT, W., *Europa und die Seele des Ostens*, (Munich, 1938).

SCHULTZ, H. D., 'Fantasies of *Mitte*: *Mittellage* and *Mitteleuropa* in German Geographical Discussion in the 19th and 20th Centuries', *Political Geography Quarterly*, 8 (1989), 315–40.

SETON-WATSON, H., 'What is Europe, Where is Europe? From Mystique to Politique', *Encounter*, 60(2) (1985), 9–17.

SHAW, D. J. B., 'Geographical Practice and Its Significance in Peter the Great's Russia', *Journal of Historical Geography*, 22(2) (1996), 160–76.

SHENNAN, J. H., *The Origins of the Modern European State 1450–1725* (London, 1974).

SIMON, E. D., *The Smaller Democracies of Europe*, (London, 1939).

SINNHUBER, K. A., 'Central Europe—Mitteleuropa—Europe Centrale', *Transactions and Papers of the Institute of British Geographers*, 20 (1954), 15–39.

SKINNER, Q., *Machiavelli*, (Oxford, 1981).

SLEZKINE, Y., *Arctic Mirrors: Russia and the Small Peoples of the North*, (Ithaca, NY, 1994).

SMITH, A. D., *National Identity*, (Harmondsworth, 1991).

—— 'National Identity and the Idea of European Unity', *International Affairs*, 68 (1992), 55–76.

SMITH, G. (ed.), *Federalism: The Multiethnic Challenge*, (London, 1995).

SMITH, M. L., and STIRK, P. M. R. (eds.), *Making the New Europe: European Unity and the Second World War*, (London, 1990).

SMITH, N., 'The Idea of the French Hexagon', *French Historical Studies*, 3 (1969), 139–55.

SMITH, W., 'Friedrich Ratzel and the Origins of *Lebensraum*', *German Studies Review*, 3 (1980), 51–68.

—— *The Ideological Origins of Nazi Imperialism*, (Oxford, 1986).

SOJA, E., *Postmodern Geographies: The Reassertion of Space in Critical Social Theory*, (London, 1989).

SOUELEYMAN, E., *The Vision of World Peace in Seventeenth-Century France*, (New York, 1941).

SOUTHGATE, B. C., '"Scattered over Europe": Transcending National Frontiers in the Seventeenth Century', *History of European Ideas*, 16 (1993), 131–7.

SPENGLER, O., *The Decline of the West*, (2 vols.; London, 1918–22).

SPRINGBORG, P., *Western Republicanism and the Oriental Prince*, (Cambridge, 1992).

STASZAK, J.-F., *La Géographie avant la géographie: le climat chez Aristote et Hippocrate*, (Paris, 1995).

STIRK, P. M. R. (ed.), *European Unity in Context: the Inter-War Period*, (London, 1989).

—— (ed.), *Mitteleuropa: History and Prospects*, (Edinburgh, 1994).

—— *A History of European Integration since 1914*, (London, 1996).

SUMNER, B. H., *Peter the Great and the Emergence of Russia*, (New York, 1973).

SUTHERLAND, D. M. G., *France 1789–1815: Revolution and Counter-Revolution*, (London, 1985).

SZUCS, J., 'Three Historical Regions of Europe', in J. Keane (ed.), *Civil Society and the State: New European Perspectives*, (London, 1988), 291–332.

TALMOR, E., 'Reflections on the Rise and Development of the Idea of Europe', *History of European Ideas*, 1 (1900), 63–6.

TAYLOR, J. E., *Christians and Holy Places*, (Oxford, 1993).

TAYLOR, K. (ed.), *Henri Saint-Simon (1760–1825): Selected Writings on Science, Industry and Social Organization*, (London, 1975).

TAYLOR, P. J., 'The State as Container: Territoriality in the Modern World System', *Progress in Human Geography*, 18 (1994), 151–62.

—— 'Beyond Containers: Internationality, Interstateness, Interterritoriality', *Progress in Human Geography*, 19 (1995), 1–15.

—— *The Way the Modern World Works: World Hegemony to World Impasse*, (Chichester, Sussex, 1996).

TAZBIR, J., 'Poland and the Concept of Europe in the Sixteenth–Eighteenth Centuries', *European Studies Review*, 7 (1977), 29–45.

THADEN, E. C., *Russia's Western Borderlands, 1710–1870* (Princeton, 1984).

TILLY, C., *The Formation of Nation States in Western Europe*, (Princeton, 1975).

TRIBE, K., *Land, Labour and Economic Discourse*, (London, 1978).

URWIN, D. W., *Western Europe since 1945* (London, 1989).

—— *The Community of Europe: A History of European Integration*, (London, 1991).

VIC-OZOUF MARIGNIER, M., *La Formation des départements et la représentation du territoire français à la fin du XVIIIe siècle*, (Paris, 1989).

VOGT, E. A., '*Civilisation* and *Kultur*: Keywords in the History of French and German Citizenship', *Ecumene*, 3(2) (1996), 125–45.

VOVELLE, M., *Le Découverte de la politique: géopolitique de la Révolution française*, (Paris, 1993).

VOYENNE, B., *Histoire de l'idée européenne*, (Paris, 1964).

WAEVER, O., 'Three Competing Europes: German, French, Russian', *International Affairs*, 66(3) (1900), 477–94.

—— 'Europe since 1945: Crisis to Renewal', in K. Wilson and J. van der Dussen (eds.), *The History of the Idea of Europe*, (London, 1993), 151–214.

WALLERSTEIN, I., *The Modern World System*, i. *Capitalist Agriculture and the Origins of the European World-Economy in the Sixteenth Century*, (London, 1974).

—— *The Modern World System*, ii. *Mercantilism and the Consolidation of the European World-Economy 1600–1750* (London, 1980).

WANDYCZ, P., *The Price of Freedom: A History of East Central Europe from the Middle Ages to the Present*, (London, 1992).

WANKLYN, H. G., *The Eastern Marchlands of Europe*, (London, 1941).

WEBB, W. P., *The Great Frontier*, (Boston, 1952).

WEBER, E., 'L'Hexagon', in P. Nora (ed.), *Les Lieux de Mémoire*, ii. 2. *La Nation*, (Paris, 1986), 96–116.

WHITE, H., *Metahistory: The Historical Imagination in Nineteenth-Century Europe*, (Baltimore, 1973).

WHITTLESEY, D., 'A Utopia for Europe', *New Republic*, 12 (Feb.) (1940), 35–41.

WIEDEMER, P., 'The Idea behind Coudenhove-Kalergi's Pan-European Union', *History of European Ideas*, 16(4) (1993), 827–33.

WILLIAMS, A., *The European Community: The Contradictions of Integration*, (Oxford, 1991).

WILLIAMS, R., *Keywords: A Vocabulary of Culture and Society*, (London, 1976).

WOLF, E. R., *Europe and the People without History*, (Berkeley and Los Angeles, 1982).

WOLFF, L., *Inventing Eastern Europe: The Map of Civilization on the Mind of the Enlightenment*, (Stanford, Calif., 1994).

WOODWARD, D., 'Reality, Symbolism, Time and Space in Medieval World Maps', *Annals of the Association of American Geographers*, 75 (1900), 510–21.

WOOLF, S., 'Statistics and the Modern State', *Comparative Studies in Society and History*, 31(3) (1989a), 588–604.

—— 'French Civilization and Ethnicity in the Napoleonic Empire', *Past and Present*, 124 (1900), 96–106.

—— *Napoleon's Integration of Europe*, (London, 1991).

—— 'The Construction of a European World-View in the Revolutionary-Napoleonic Years', *Past and Present*, 137 (1900), 72–101.

# Chapter 8

# Changes in Population and Society, 1500 to the Present*

### P. E. Ogden

Corpses posed terrible problems, as churchyards over-flowed and the air reeked of death. The authorities ordered mass graves, lined with quicklime. It was impossible to bury all immediately, and bodies piled up in the streets. Pepys came across a body in an open yard, with nobody to bury it, 'the plague making us as cruel as dogs one to another'. (Porter, 1994: 83, on the plague in London in 1665)

'Every man desires to live long,' Swift observed, 'but no man would be old.' If current trends continue, many of us may live well past 80. How many of us will live past 100 and how well we will live at advanced ages are two much more uncertain questions. (Kannisto *et al.*, 1994: 806)

## The Demographic Revolution and the Revolution in Demography

European demographic change in the recent past has been revolutionary. Population has increased rapidly. Average life expectancy at birth has increased from perhaps 25–30 years to 75–80; the impact of infant mortality has been utterly transformed. Crude birth and death rates have declined

from between 30 and 40 per thousand to around 10, and family size has fallen dramatically. Changes on this scale have sparked both an extensive literature and a lively series of debates about exactly what happened and how the demographic revolution was linked to industrialization and urbanization. The demographic rates conceal changes also in the affective nature of society: the value attached to children, the nature of the family, the meaning of old age. In addition, human mobility, internal and international, has taken on a variety of characteristics, reflecting and to some extent determining the nature of demographic change and certainly bringing about a major transformation of the influence of European culture at the world scale.

By no means least, of course, was the absolute increase in population which, as Tables 8.1 and 8.2 show, has been considerable, though variable, throughout the continent. The total population of the continent increased from around 67 million in 1500 to 295 million by 1900 and 498 million by 1990. The European share of world population reached

* I am most grateful to Dr Ray Hall and Dr John Walton for their comments on an earlier draft of this chapter.

**Table 8.1.** European population growth 1500–1990 (excluding Russia and republics of the former Soviet Union)

| Year | Total population (millions) | % of world total |
|------|------|------|
| 1500 | 67 | 14.5 |
| 1600 | 89 | 15.4 |
| 1700 | 95 | 14.0 |
| 1750 | 111 | 14.4 |
| 1800 | 146 | 15.3 |
| 1850 | 209 | 16.8 |
| 1900 | 295 | 18.1 |
| 1950 | 395 | 15.6 |
| 1990 | 498 | 9.4 |

*Source*: Livi-Bacci, 1992: 31; after Biraben, 1979.

a peak of around 18 per cent in 1900 and thereafter has fallen back sharply in the face of rapid growth in the less developed world, especially after 1950.

Geography was also of the first importance, and one of the products of recent research has been to identify some of the geographical contrasts within Europe at a variety of scales. The European demographic regime was distinctive at the world scale.

Nevertheless, there has been much variation between countries and groups of countries historically, though over the last century in particular there has been a tendency towards a certain demographic homogeneity. Much experience of changing fertility and mortality in the long term has been shared, but geographical contrasts within countries—for example, between urban and rural places or amongst cultural regions—have been marked. We are still, though, far from understanding the full complexity of past demographic geographies, and the geography as well as the chronology of change is a recurrent theme in this chapter.

Our knowledge has, indeed, increased greatly over the last two or three decades with the rapid increase in the techniques and theory of historical demography. The study of population does of course have deep roots (Dupâquier and Dupâquier, 1985). There has been both a growth in empirical work, and innovation in methodology, particularly concerned with the exploitation of sources (Ogden, 1987) but in addition an attempt to link demographic processes to the mainstream changes in economic and social organization. Wrigley (1981), for example, has made a plea for the 'logical status of population history' to be considered carefully, that is, whether population history is relegated to the role of a dependent variable or whether population trends themselves

**Table 8.2.** Estimated population growth in some major European countries 1600–1993

| Country | Population | | | | Indexed growth | | | | |
|---------|------|------|------|------|------|------|------|------|------|
| | 1600 | 1750 | 1850 | 1993 | 1750[a] | 1850[b] | 1850[c] | 1993[d] | 1993[e] |
| England | 4.1 | 5.7 | 16.5 | 49.2[f] | 1.39 | 289 | 402 | 298 | 1200 |
| Holland/NL | 1.5 | 1.9 | 3.1 | 15.2 | 127 | 163 | 207 | 490 | 1013 |
| Germany[g] | 12.0 | 15.0 | 27.0 | 81.1 | 125 | 180 | 225 | 300 | 676 |
| France | 19.0 | 25.0 | 35.8 | 57.7 | 132 | 143 | 188 | 161 | 304 |
| Ireland | 2.2[h] | 3.2 | 8.2[i] | 3.6 | 145 | 256[j] | 164 | 44 | 164 |
| Italy | 12.0 | 15.7 | 24.8 | 57.8 | 131 | 158 | 207 | 233 | 482 |
| Spain | 6.8 | 8.4 | 14.5 | 39.1 | 124 | 173 | 213 | 270 | 575 |

*Notes*: [a] 1600 = 100; [b] 1750 = 100; [c] 1600 = 100; [d] 1850 = 100; [e] 1600 = 100; [f] England and Wales 1991; [g] 1993 figure includes the former GDR; [h] 1687; [i] 1841; [j] 1841.

*Sources*: Livi-Bacci, 1992: 62, 69; *Population et Sociétés*, 282 (1993).

may be considered to influence wider social and economic change. Despite all the recent advances in the field, our knowledge of so much of the history of European demography remains obscure, and each new wave of empirical work has tended to involve some quite significant revision of received wisdom. Major strides have been made in both our understanding of the variety of demographic change over the last five centuries and also the nature of the interrelationships between demographic change and other variables. Progress has been made of late, for example, in reassessing the relative roles of mortality and fertility (and thereby nuptiality) in determining overall rates of population increase; in the causes of mortality decline; in the history of the family; or in the nature of mobility in the past.

Nevertheless, in some senses, historical demography is in its infancy and there remains much scope for further research. Assembling the material for this chapter makes two limitations very evident. First, the available primary sources have been both a help and a hindrance. We are much aided by, for example, the ecclesiastical or civil registers of vital events in many countries, sometimes from as early as the sixteenth century, or by the population censuses from the early nineteenth century (for a full discussion of sources see Willigan and Lynch, 1982) but there are many difficulties in the extent, reliability, and technical exploitation of the sources which are only gradually being overcome. Secondly, the geographical coverage of work in the field is highly skewed towards north-west Europe and especially England, France, Scandinavia, and Germany. The variety of work that has been achieved certainly indicates considerable geographical diversity in both the patterns and causes of demographic change, and here we shall draw on a wide variety of material. But we must always treat with care existing generalizations, often based on the experiences of a limited range of countries or indeed of small areas within them, before assuming that they necessarily apply throughout the continent. An interesting conundrum is that many studies treat individual countries and many others treat individual communities or parishes so that it is at the regional, sub-national level that the greatest gaps, and probably some of the most interesting variations, lie.

# Demographic Systems and Pre-industrial Patterns

## The Malthusian Model

At the core of the attempt to understand what has influenced the European pattern of life and death are still the basic questions posed by Malthus in the late eighteenth century. It is the long-term adjustment between fertility, mortality, and rising or falling living standards which holds the key to population change rather than simply, for example, the sporadic impact of subsistence and mortality crises, though the latter had of course dramatic consequences at certain times and places. The general elements of the Malthusian system are shown in Fig. 8.1. This shows ways in which population decline or increase is a function of the interplay between food prices, real income, nuptiality, and fertility (the preventive check) on the one hand and mortality (the positive check) on the other. The exact workings of this system are still a matter of discussion, though a number of recent studies have added considerably to our knowledge. Seccombe (1983: 33–4) considers that 'until roughly 1750 . . . the great peasant populations of western Europe appear to have lived and laboured within a homeostatic regime of demographic checks and balances, so that population growth, in the long run, was very slight'. The term 'homeostatic' which we may apply to this system was not a 'placid equilibrium, but a continuous process of equilibration—of disruption, crisis and rectification—where mortality crises periodically decimated the rural masses and, in their wake, fertility rates increased in response to the eased availability of arable land and earlier age at marriage'. As Livi-Bacci (1992: 100) has pointed out, traditional societies were inefficient in the sense that 'in order to maintain a low level of growth, a great deal of fuel (births) was needed and a huge amount of energy was wasted (deaths)'.

Work by Wrigley and Schofield (1981) for England is the most exhaustive review to date of the mechanics underlying population change. They show that, for the period up to the Industrial Revolution, much of Malthus's thinking was on the right lines. The normal check to population growth was not exerted through wars, epidemics, and famine, but rather through the control of

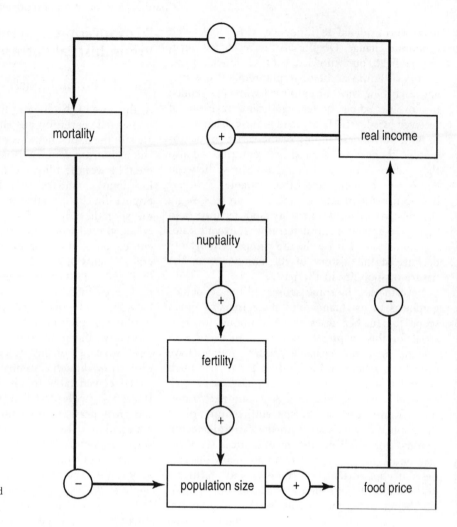

**Fig. 8.1.** A model of the Malthusian demographic system. (*Source*: Wrigley and Schofield, 1981: 458.)

fertility primarily via the delay and deterrence of marriage. Fertility changes had perhaps twice the effect on population growth than mortality had in England (Coleman and Salt, 1992: 33). Marriage rates and birth rates were strongly related to changes in real wages, death rates much less so. Most English mortality crises were exogenous, that is, related to external events rather than to declining real wages. As Wilson and Woods (1991) point out, it is only for England that we can establish very long-term demographic indices of reasonable accuracy. The pattern of fertility and life expectancy over four centuries is shown in Fig. 8.2. While marital fertility remained fairly stable until the decline in the third quarter of the nineteenth century, overall fertility fluctuated consider ably, rising for example towards the end of the eighteenth century, in line with a rise in the proportion of women married. How far these findings for England may be applied to other parts of Europe is far from being entirely resolved. Though there is enough evidence to suggest that England may have been exceptional rather than typical, the light that these data throw on the demographic processes means that at least we know what sorts of questions to ask in other areas. Other studies in Europe covering long periods confirm the existence of cycles of population change, with population decrease as well as increase recorded for several countries, though with variable balances being struck between Malthus's positive

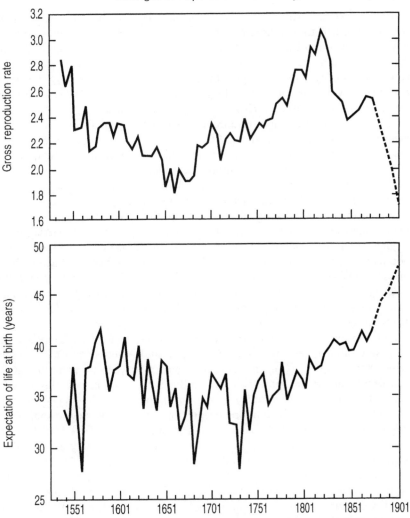

**Fig. 8.2.** Long-term evolution of fertility and mortality rates for England 1541–1901. (*Source*: Wrigley and Schofield, 1981: 231.)

and preventive checks (Galloway, 1988) and with generally more weight being given to the importance of mortality—especially in southern and eastern Europe—than in the English case. In Languedoc in southern France, for example, from the fourteenth century population grew more rapidly than resources and, in the absence of technological improvements, positive checks intervened in the shape of mortality, thus differing from the English case (Livi-Bacci, 1992: 85).

## Marriage and Household

There is, thus, much of interest in trying to discover the extent to which Europe was characterized by different patterns of marriage and household formation, the origins of those patterns, and the relationship between them and the broader evolution of industrialization and social change. Hajnal (1965) suggested that to the west of a line from the then Leningrad to Trieste marriage was late (26–7 for men and 23–4 for women) and a high proportion remained celibate throughout life (10–15 per cent of men, and rather more women, were still unmarried by age 59). This pattern, which included a predominantly nuclear family structure, seems to have been general throughout western Europe by the sixteenth century. Thus, in England by the end of the seventeenth century, mean age at first marriage in England was 28 for men and 26 for women.

More than 25 per cent never married at all. The west European pattern thus kept fertility relatively low and made it more adaptive to economic trends. Ambitious claims have been made for its significance: by slowing population growth it may have had the very important effect of increasing per capita distribution of resources and keeping the balance of population and resources reasonably favourable. It made average income higher than it would otherwise have been and facilitated the emergence of the Industrial Revolution. As Chambers (1972: 57) observed, 'in this way was laid the groundwork of a society which learned not only the economic but also the demographic techniques that made continuous economic growth possible'. In England and in much of north-western Europe population growth was low, creating a 'low-pressure' demographic system.

In eastern Europe by contrast, families were larger, marriages more numerous and took place at earlier ages, and households tended towards the complex rather than the nuclear. Population densities were often higher, mortality higher too, and crises more frequent (Coleman and Salt, 1992: 3). The genesis of the European pattern is difficult to date with accuracy, but Smith (1979, 1981, 1983), for example, has delved much more deeply into its origins and has suggested that it is much older than first assumed. Certainly, for a variety of reasons the division between east and west has been maintained over long periods (Rallu and Blum, 1991; Ni Bhrolchain, 1991). For England and north-west Europe, the crucial finding of research was to explode the myth that the large, extended pre-industrial household had given way to the nuclear family at the hands of industrialization (Hareven, 1991).

[T]he typical domestic group has been small and simple in structure at least since medieval times . . . In the European peasant family . . . day-to-day family life was centred around the conjugal pair [and] at no time in the past one thousand years . . . has kinship been the dominant basis for social organisation, certainly in Britain and probably anywhere in Western Europe. (Anderson, 1979: 50–1)

Nevertheless, detailed geography matters too. Within western Europe the small conjugal family was very common but far from universal: 'married brothers and their families often shared the same house in parts of France and Italy . . . while

stem-family systems were quite widely found in Austria, France and Germany. In parts of eastern Europe, huge and complex households predominated with extensions both laterally and vertically' (Wrigley, 1977: 78; and see Wall, Robin, and Laslett, 1983; Alderson and Sanderson, 1991). Mediterranean regions also had distinctive and diverse household forms, though the differences from the west European pattern should not be exaggerated (Kertzer and Brettell, 1987).

## Mortality Crises

Although much recent emphasis has been given to studies of marriage and fertility, work on mortality crises and long-term trends in mortality in pre-industrial Europe has continued. Schofield and Reher (1991: 1; and Perrenoud, 1991: 18–19) show that what we know is but an outline and the variation in and decline of mortality is a highly complex process, though the arguments are too often presented in simplistic terms. In the early modern period mortality might fluctuate violently from year to year, though these fluctuations gradually diminished (Livi-Bacci, 1992: 47). We have exceptionally long estimates of mortality for England and these suggest that between 1580 and 1640 expectation of life at birth worsened from nearly 40 to just above 30 years (Schofield and Reher, 1991: 3; and see Fig. 8.2). Dearth, epidemics, and war all contributed to varying mortality and Thomas Hobbes's view in 1651 of 'the life of man, solitary, poor, nasty, brutish and short' has ample justification in the demographic facts. A major cause were crises of subsistence, related to harvest failure or climatic changes. These were generally less in England than in France, for example, where agriculture was more intensive and specialized (Biraben and Blum, 1988). The sixteenth, seventeenth, and early eighteenth centuries were characterized by subsistence crises with adverse demographic consequences several times each century. 'The great crises of 1693–4 and 1709–10 doubled the number of deaths in France relative to normal years and left a lasting mark on both the demographic structure and historical memory of the populations affected' (Livi-Bacci, 1992: 81).

Want was also accompanied by disease, though the latter also appeared without necessary relation to the economic climate. There was little protection against, or treatment for, a whole range of epidemic and endemic diseases which ravaged European populations: plague, smallpox, tuberculosis, malaria, typhus, typhoid, scarlet fever, dysentery, influenza, and so forth. Bubonic plague holds a particular place of ignominy: it was one of the major causes of death from the fourteenth to the seventeenth centuries. 'The great "Atlantic Plague" of 1596–1603, which gnawed at the coasts of western Europe, possibly cost one million lives, two-thirds of them in Spain alone. In France between 1600 and 1670 plague carried off between 2.2 and 3.3 millions' (Kamen, 1984: 34). Plague hit towns particularly badly: Mantua lost 70 per cent of its population in 1630, Naples and Genoa nearly half of theirs in 1656. Barcelona lost nearly a third in 1589 and about 45 per cent in 1651. The plague killed in London until it finally disappeared in the later seventeenth century: major epidemics in 1563, when perhaps a fifth of the population perished, in 1593, in 1603, 1625, 1636 and of course 1665 when at least 80,000 of the 400,000 inhabitants died (Slack, 1968: 62–3; Porter, 1994: 84; Finlay, 1981). As Porter points out in his compelling description of the disease, plague deaths in London in 1665 were equivalent to the total populations of the 'next five towns in the kingdom—Norwich, Bristol, Newcastle, York, and Exeter. It was a disaster of nuclear proportions.' This latter epidemic also hit northern France, the Low Countries, and the Rhine valley, after which its incidence in western Europe declined sharply except for occasional re-emergence in, for example, eighteenth-century Provence (Livi-Bacci, 1992: 48). The plague arrived in Marseilles on 20 June 1720 and of the population of 90,000, 39,334 (43.3 per cent) died; by early August it was ravaging Aix, where 7,534 (31.4 per cent) of the 24,000 inhabitants perished (Biraben, 1972: 237). Outbreaks occurred in eastern Europe until the mid-nineteenth century (Flinn, 1981: 56). The disappearance of plague is not adequately explained, but the importance of quarantine (as the movement of people and goods was very significant in its spread) and of other public health measures has been much discussed (Slack, 1985; Flinn, 1981). One of its particularly interesting features is that it was largely external to the socio-demographic system, acting independently of the social structures, levels of economic development, and population density (Livi-Bacci, 1992: 48), though its very impact did bring about adaptations in the system.

Other diseases were readily able to fill the gap left by the plague. Smallpox gradually became a major killer in England from the early seventeenth century—by the mid-eighteenth century in provincial towns in England it was alone responsible for 10–19 per cent of deaths, a level matched in many other European countries, for example 14 per cent in rural parishes in Finland (Coleman and Salt, 1992: 23). It struck elsewhere and earlier: in Geneva roughly 107 children out of every thousand died of smallpox at the end of the sixteenth century (Perrenoud, 1991: 27). Its control and virtual eradication through effective immunization in the late eighteenth and early nineteenth centuries in much of Europe removed a major threat to human life. Tuberculosis became one of the most important causes of adult death, not least in the growing cities (Mercer, 1990: 156). Typhus was lice-borne and had long been associated in Europe with famine and 'with people living or searching for food in close proximity to rats which could be carriers of infected lice' (Mercer, 1990: 156). Cholera was a rather late arrival on the scene, spreading to all European countries during the 1820s—killing 100,000 in France in the 1832 outbreak, for example—and leading to the imposition of quarantine regulations and cordons sanitaires in a similar way to those imposed against plague (Bourdelais, 1991: 119). A dramatic outbreak occurred in Hamburg as late as the 1890s (Evans, 1987). It is also worth noting the way in which Europe not only received but also exported disease as trade and migration grew (McNeill, 1976).

## Demographic Transition

From around the mid-eighteenth century, many countries began to experience a new phase in their demographic history, with rapid population growth and fundamental changes in both fertility and mortality. From the waste and disorder of *ancien régime* Europe eventually emerged a more controlled and orderly system. The process known as 'demographic transition'—the decline of both fertility and mortality rates—had affected most of the continent by the early part of the twentieth century, and it is to these—by no means wholly understood or agreed—processes that we owe the current convergence which characterizes the demographic regimes of the late twentieth century. The progress of the transition was, though, geographically uneven.

For Europe as a whole, the annual rate of growth rose from around 0.15 per cent between 1600 and 1750 to 0.63 per cent between 1750 and 1850 (Livi-Bacci, 1992: 68). This involved most countries (Table 8.2) but was greater in England than elsewhere. Britain witnessed one of the most dramatic shifts in population growth in Europe: a population of around 5.3 million in 1695, possibly no more than the medieval maximum (Coleman and Salt, 1992: 35) had become 9.2 million by the census of 1801. By 1901 the population of England and Wales stood at 32.5 million. Almost every European country experienced unprecedented growth in the nineteenth century: Norway from 1 million in the 1820s to 2.3 million by the 1900s; Sweden from 2.3 million in 1800 to 5.1 million by 1900; Italy from 17 million to 33.6 million (Reinhard, Armengaud, and Dupâquier, 1968: 684–7). France was a notable exception, where an early decline in fertility from the later eighteenth century led to much slower population growth than in neighbouring countries (Ogden and Huss, 1982).

In Europe in general, our knowledge of the mechanisms which brought about the dramatic reduction in both mortality and fertility from the later nineteenth century (and earlier in some cases) have been much increased over the last twenty-five years.

## The Decline of Mortality

Classic descriptions of the mortality transition (Schofield and Reher, 1991: 1) identify three periods in the decline of mortality in Europe. First, during the eighteenth century, the incidence of the crisis mortality discussed above diminished drastically in most of Europe. There were considerable gains in life expectancy, and population increase characterized many countries. This was followed by a period when, with the reduction of epidemic disease, endemic infectious diseases, and others such as cholera, became relatively more common, and improvements in life expectancy slowed. The squalor of industrial and urban living conditions in the cities of nineteenth-century Europe hardly helped (Hardy, 1993). The second period of decline—indeed the beginning of sustained decline in most

countries—did not begin until the later decades of the nineteenth century. Reductions in child and infant mortality were particularly marked, though all age-groups benefited from decline of diseases like diphtheria, whooping cough, scarlet fever, cholera, tuberculosis and, especially in the early twentieth century, diarrhoea, pneumonia, and influenza. The third period of mortality decline began after 1945 and 'spread throughout the world . . . inextricably, though not exclusively, linked to the discovery and use of sulpha drugs and antibiotics' (Schofield and Reher, 1991: 1). As these authors go on to demonstrate, we know less about mortality than about fertility, though much recent work has added many inflections to the periodicity suggested above and has also sown the seeds for many vigorous discussions of the causes of change.

The decline of crisis mortality—from diseases such as the plague (which had largely disappeared by the early eighteenth century), smallpox, and typhus (in substantial decline by the early nineteenth) is clearly established: Fig. 8.3 shows examples of the way in which mortality 'peaks' were gradually ironed out. It is not, however, yet clear how far this early decline affected the whole of the continent (Schofield and Reher, 1991: 3) and certainly crisis mortality did continue in some areas well into the nineteenth century or later. The last great subsistence crisis under the joint effects of famine and typhus hit all of Europe in 1816–17 (Livi-Bacci, 1992: 68) and Finland, for example, remained vulnerable to mortality crises until the end of the nineteenth century, a peak mortality rate of 78 per thousand being recorded in the famine of 1868 (Vallin, 1991: 42; Pitkanen, 1993); while the combined effects of famine and terror caused perhaps over four million deaths in the Ukraine as late as the 1920s and 1930s. The catastrophic Irish famine of the 1840s and associated epidemics caused probably in excess of one million extra deaths (O'Grada, 1994: 179). In addition, the influenza pandemic of 1918 was 'the last major mortality crisis in Europe . . . almost an anachronism' (Vallin, 1991: 43). Nor should we forget the effects of war: in the First World War an estimated 9.45 million people died in military action (Winter, 1985: 75), to which we must add the injured who subsequently died, civilian deaths, and births lost as a result of the disruption caused to family life by the war. The Second World War may have caused

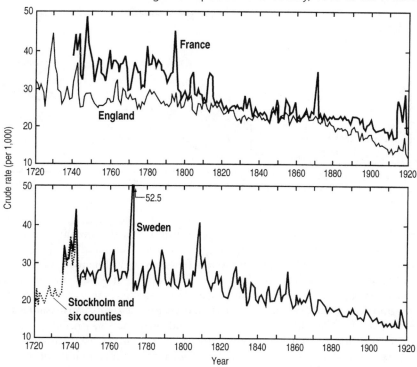

**Fig. 8.3.** Decline of crisis mortality for France, England, and Sweden 1735–1920. (*Source*: Vallin, 1991: 40.)

up to 55 million deaths world-wide, and within Europe the most affected countries were Poland, the USSR, Yugoslavia, and Germany (Rallu and Blum, 1991: 18).

Underlying levels of non-crisis mortality seem to have fluctuated considerably. In England, for example, there were large swings, and though there were certainly improvements 'truly "secular" improvements only took place after 1870' (Schofield and Reher, 1991: 4). In the later eighteenth century, there was some increase in life expectancy in France (from 25 in 1740–9 to 40 in 1840–9), in Sweden (from 37 in 1750–9 to 45 in 1840–9), and in Denmark (35 in 1780–9 to 44 in 1840–9) (Livi-Bacci, 1992: 70), but data for southern and eastern Europe are few and far between, though life expectancies of around 30 in the mid-nineteenth century suggest that improvements prior to this date were slow. Schofield and Reher (1991: 4) also point out the extent of sub-national regional variation: in Spain in 1860 life expectancy varied from 25 to over 40 and in Sweden infant mortality in 1861 was 290 per thousand in Stockholm and as low as 100 in many rural areas. These variations may well have been present for long periods. In England, geographical

differences persisted: Hardy (1993) has documented the ravages of disease in the Victorian city, while Woods (1984) has shown that as late as 1861 expectation of life at birth for males in Liverpool was 26, compared to 57 in Okehampton in Devon, a difference as great as the national difference between 1840 and 1960. Differences between countries in levels of infant mortality were probably greatest around 1850 (Vallin, 1991: 52), though in most countries advance was slow and little progress was made before 1900. In England, for example, rates stayed at around 150 until 1900 (Saito, 1996: 546; Woods, 1985, 1994). Some examples of the long-term evolution of expectation of life at birth are given in Fig. 8.4.

The secular decline in mortality from the later nineteenth century eventually affected the whole of Europe and brought with it a decline, though by no means complete extinction, of regional differences which persist up to the present. Major strides were made in quite short periods and the decline of death went hand in hand with the decline of fertility. Child mortality, then infant mortality, declined quickly and contributed most to the gains in life expectancy between 1870 and 1930. Control over

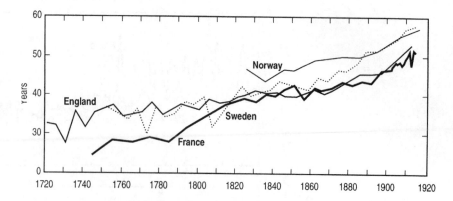

**Fig. 8.4.** Life expectancy at birth in France, England, Norway, and Sweden 1720–1920. (*Source*: Vallin, 1991: 48.)

diseases such as respiratory tuberculosis, pneumonia, bronchitis, and influenza made large contributions to increased life expectancy, though as Vallin (1991) and Caselli (1991) argue there were differences in the way diseases contributed to the mortality transition from country to country. Considerable geographical differences in life expectancy on the eve of the First World War were still in evidence (Table 8.3): the difference between Denmark at the top (57.7 years) and Hungary at the bottom (37.5 years) is nearly as large as the total gain achieved by the early starters over the previous two centuries. In addition, Russia, which is not included in Table 8.3, had an infant mortality rate of 250 per thousand in 1910 and a life expectancy of around 30 years (Vallin, 1991: 49).

Debates about the reasons for the decline of mortality focus, crudely summarized, on four issues (Perrenoud, 1991: 19): nutrition, public health, medicine, and biological factors which may have modified the relationship between humans and the organisms which cause disease, though other factors such as living standards and working conditions, urbanization, education,

**Plate 8.1.** 'A Child's Funeral' 1879, by Albert Edelfelt (1854–1905). The Museum of Finnish Art Ateneum, Helsinki. Photo: The Central Art Archives.

Table 8.3. Life expectancy at birth ($e_0$) on the eve of the First World War and in 1995

| Country | Exact dates | $e_0$ | $e_0$ (1995) | |
|---|---|---|---|---|
| | | All | Male | Female |
| Denmark | 1911–15 | 58 | 73 | 78 |
| Norway | 1911–21 | 57 | 74 | 80 |
| Sweden | 1911–20 | 57 | 76 | 81 |
| Netherlands | 1910–20 | 56 | 74 | 80 |
| Ireland | 1910–12 | 54 | 73 | 78 |
| England and Wales | 1910–11 | 53 | 74 | 79 |
| Switzerland | 1910–11 | 52 | 75 | 81 |
| France | 1908–13 | 50 | 74 | 82 |
| Germany | 1910–11 | 49 | 73 | 79 |
| Italy | 1910–12 | 47 | 74 | 80 |
| Finland | 1911–20 | 46 | 72 | 79 |
| Spain | 1910 | 42 | 73 | 81 |
| Bulgaria | 1899–1902 | 40 | 68 | 74 |
| Austria | 1900–01 | 40 | 73 | 80 |
| Hungary | 1900–01 | 37 | 65 | 74 |

Note: Selected European countries for which life tables are available.
Source: Vallin, 1991: 47; Population et Sociétés, 304 (1995).

infant-feeding practices and hygiene, politicians, planners and reformers and even climate have been variously called in evidence (Schofield and Reher, 1991: 10). Much recent work has followed from the attempt by McKeown (1976) to attribute the major part of the mortality transition to nutrition, which increased resistance to infection, on the grounds that both medical advance and public health measures came after rather than before the decline of significant numbers of diseases. On the other hand, authors such as Preston (1976) argue for the importance of the role of the state in 'organizing public defence against disease, providing basic health facilities and educating the population in accord with scientific advances in health care' (Schofield and Reher, 1991: 8). The various essays in the latter volume suggest to the editors that the argument needs to be more subtle: 'a disease-by-disease analysis tends to suggest that declines in mortality were often achieved by both public health measures and nutritional improvement' (Schofield and Reher, 1991: 13).

In addition, the role of medicine may be more significant than it has recently been customary to admit:

before the discovery of antibiotics and sulpha drugs during the middle part of the present century, physicians had almost no effective weapons with which to combat disease and infection directly; and by then much of the battle against infectious disease had been won . . . Yet the situation is not as simple as it might seem. Jenner's smallpox vaccine, the use of the diphtheria antitoxin, the establishment of tuberculosis sanatoria to isolate patients, and the discoveries of Pasteur, all led either directly or indirectly to significant declines in mortality . . . Medical men were also important in so far as they were behind most public-health and sanitation policies, and were leaders in the movement for reform and education in these areas. (Schofield and Reher, 1991: 14–15)

P. E. Ogden

In addition, the changing nature of disease may also have been important, for example Wrigley and Schofield (1981; and see Perrenoud, 1991: 22), addressing long-term change in England; and other authors working in Scandinavia (Fridlizius, 1984: see Perrenoud, 1991: 22) assert that exogenous factors influenced by natural or biological changes rather than economic or social ones were responsible for the origins of mortality decline, establishing a 'new equilibrium between pathogenic agents and their human hosts' (Perrenoud, 1991: 22). Finally, we should not underestimate the importance of climate in affecting both long- and short-term changes and geographical variation: for example, infant and child mortality patterns in Europe show considerable correlation with summer climates where diarrhoea and other intestinal diseases were more important than elsewhere (Schofield and Reher, 1991: 16; Galloway, 1986).

## The Decline of Fertility

As Gillis, Tilly, and Levine (1992: 1) suggest,

Some revolutions begin with the roar of cannons and are commemorated with parades and fireworks. Others take place without fanfare and run so silently that they are ignored even by historians charged with recording great social changes. The rapid and comprehensive fertility decline that began in Europe in the 1870s had an effect every bit as dramatic as two other great transformations, urbanization and industrialization.

We know rather more about this decline of fertility from the late nineteenth century and its regional incidence than we do about mortality. In particular, much has been achieved by the Princeton European Fertility Project, which has surveyed fertility trends in Europe from the later nineteenth century (Coale and Watkins, 1986; Watkins, 1991) and has allowed much firmer conclusions about both the timing and geographical incidence of the fertility transition. Fig. 8.5 gives a diagrammatic summary based on the experience of seventeen countries of the changing relationship between the total fertility rate (number of children per woman) and the expectation of life at birth ($e_0$), whose evolution has been discussed above, and the resulting rate of population growth. The ellipses thus 'contain' the countries at each period as indicated. Two important points emerge. First, the pre-transitional regime of relatively high fertility (4.5 to 6.5 children per woman) has given way to the regime of the

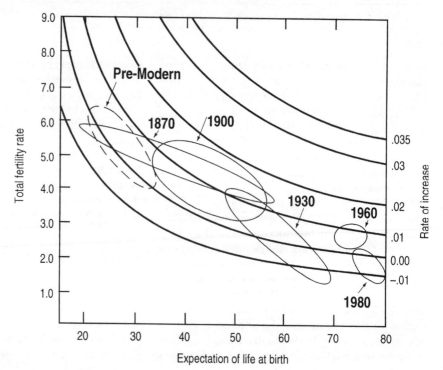

**Fig. 8.5.** The 'strategic space of growth' for seventeen European countries in the nineteenth and twentieth centuries. (*Source*: Coale and Watkins, 1986: 27.)

192

two-child low fertility family and high life expectancy by the 1980s. Secondly, the strong geographical variations evident from the breadth of the ellipses in 1870 or 1900 gradually give way to a much less varied pattern amongst the countries (and therefore smaller ellipses in the diagram) by the mid-twentieth century.

The essential difference between fertility change during this transition and in previous periods was that it was no longer only the Malthusian preventive check of changes in marriage age and the propensity to marry that were most important but rather regulation of fertility within marriage. 'Beginning in the 1870s . . . women were starting to stop before the end of their fertile years, signalling not only a fundamental strategic change in patterns of family limitation but new attitudes toward fertility itself' (Gillis, Tilly, and Levine, 1992: 2). Thus, the nature of and access to, means of family limitation—either via sexual abstinence or by the use of appliance methods of birth control—have been recognized as important facilitating aspects of the secular decline in marital fertility. In addition, other key factors in the decline were the reduction in the number of large families, in the number of illegitimate births, and in the frequency of marriage, an increase in the average age at marriage, and the growth of emigration (Festy, 1979: 180; Wilson and Woods, 1991).

The work of the Princeton project (Coale and Watkins, 1986; Watkins, 1990, 1991) points to a number of quite novel conclusions, some of which depend on the fact that rather unusually analysis was conducted at the regional scale within each country. First, the timing of the decline in fertility—the point at which it registered a 10 per cent drop relative to a previous stable level—was remarkably constrained (see Fig. 8.6). As Knodel and Van de Walle (1979: 235–6) point out,

the striking factor that the countries of Europe had in common when fertility declined was time itself . . . with the exception of the forerunner, France, and a few stragglers, such as Ireland and Albania, the dates of the decline were remarkably concentrated. The momentous revolution of family limitation began in two-thirds of the province-sized administrative areas of Europe during a thirty-year period, from 1880 to 1910.

(See also Fig. 8.7.) The extent of early decline in France, which has contributed to its reputation as

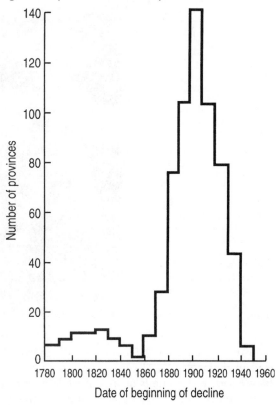

**Fig. 8.6.** Distribution by decade of number of provinces of Europe experiencing 10 per cent decline in marital fertility. (*Source*: Coale and Watkins, 1986: 38.)

something of a demographic enigma, was remarkable, but by 1900 it was no longer alone: low and apparently controlled marital fertility was already a feature of much of England and several other provinces in southern Belgium, Germany, and Italy and by 1930 the geographical spread was very wide. There was something of a core and periphery, with Ireland, Scotland, and parts of Wales to the north-west and parts of Spain, Portugal, and Italy to the south, along with parts of Scandinavia and much of central and eastern Europe, declining later than the core. In many cases, therefore, population increase was sustained well into the twentieth century. A typical example of the way in which the transition took place is shown for Austria in Fig. 8.8.

Regional variation within the countries was thus of great significance and the explanations which have emerged from the Princeton studies are of particular interest in that they do not necessarily support the general notion of urbanization and

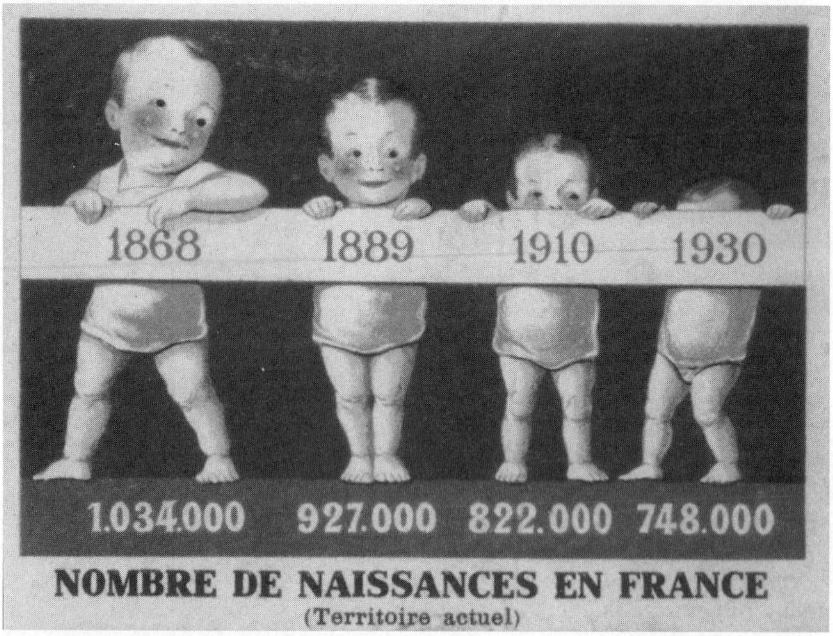

**Plate 8.2.** 'Births in France 1868–1930'—cover of a pamphlet issued by the Alliance Nationale pour l'Accroissement de la Population Française: F. Boverat, *La Crise des naissances: ses conséquences tragiques et ses remèdes* (Paris, 1932).

industrialization being the root causes which would explain the geographical pattern and trends. Livi-Bacci (1992: 123) reminds us that 'if we look at the entire process, we see that no population has maintained high levels of fertility for long in the face of increasing well-being and declining mortality. The demographic transition has clearly been an integral part of the transformation of European society.' Yet, for Knodel and Van de Walle (1979) or Seccombe (1983: 26), the striking feature is that fertility decline took place under very diverse socio-demographic conditions and there was no straightforward correlation between development and demographic change. They emphasize the importance of cultural factors—such as common dialect and common customs—rather than socio-economic variations in determining the pace of change with respect to, for example, the geographical diffusion of attitudes to fertility and contraceptive practice. 'There is greater similarity in fertility trends among provinces within the same region but with different socio-economic characteristics than is true among provinces with similar socio-economic characteristics but located in different regions' (Knodel and Van de Walle, 1979: 236). Gradually, though, low fertility came to characterize most provinces and demographic behaviour shifted from the sub-national, provincial scale to

the point where national boundaries became more important. Thus, the

reduction in within-country demographic diversity was paralleled by a trio of macro-level changes: the integration of national markets, the expansion of state functions and nation-building. It might seem that these processes would have little to do with such private behaviours as marriage and childbearing. But [they] increasingly drew local communities into national networks. (Watkins, 1990: 262)

Though the concept of demographic transition which broadly covers the decline of mortality and fertility described above has been subject to much critical attention of late (see Gillis, Tilly, and Levine, 1992; Schofield, 1985; Seccombe, 1983; Kirk, 1996), it nevertheless provides a convenient framework within which to see the population growth and then relative stability which characterizes the demographic history of much of Europe over the last two centuries. Certainly for much of north-west Europe by the inter-war period, it was the fear of population decline rather than growth which attracted comment (Kirk, 1946; Teitelbaum and Winter, 1985; Quine, 1996). It is also worth noting the effect of changes in fertility and mortality on patterns of kinship and family. Zhao (1996: 269–70) has shown the way in which fertility

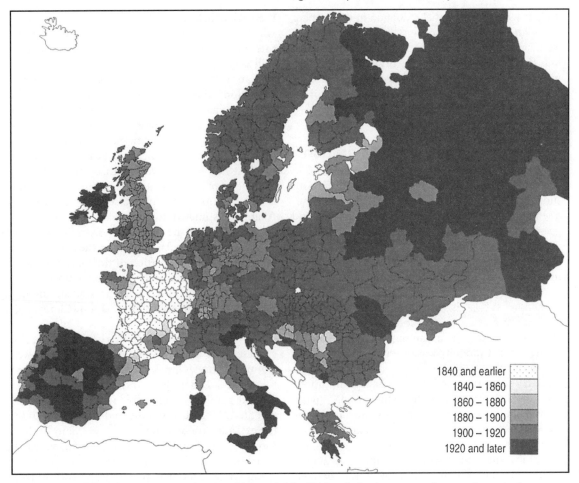

**Fig. 8.7.** Estimated dates for European provinces at which marital fertility has declined 10 per cent from maximum (and never returned to that level). (*Source*: Coale and Watkins, 1986: map 2.1 at end of vol. ii. 1.)

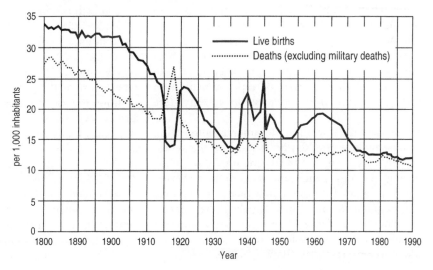

**Fig. 8.8.** Crude birth and death rates, Austria 1880–1990. (*Source*: Findl, 1991: 226.)

decline 'led to a substantial reduction in the number of children, siblings, cousins, uncles and aunts, nephews and nieces available to an individual', whilst improvements in mortality 'brought about an increasing vertical extension in a person's kin connections. The kinship network that emerged from these changes had probably never existed before.'

## Patterns of International Migration

The increase in geographical mobility over both short and long distances has been a major consequence of industrialization and urbanization and goes hand in hand with the demographic changes discussed above. Yet recent research has emphasized that we should not be deluded by the dramatic migrations of the nineteenth and twentieth centuries into believing that mobility is a new phenomenon. Whilst it is true that some historical communities were inward-looking and relatively immobile, there is a mounting body of evidence to suggest that mobility was the rule rather than the exception for many and that a simple correlation between migration and some concept of 'modernization' is inappropriate (Moch, 1992, 1996). Movements were of great diversity, including seasonal and short-term as well as permanent migrations, often over great distances (Lucassen, 1987). Certainly, the movement of people has had a major influence on the social geography of continents as well as of individual communities and was an integral part of the process of European colonization both directly through emigration from Europe and indirectly through the European role in organizing the slave trade. Here we discuss international movements which were set against the background, especially in the later eighteenth and nineteenth centuries, of rapid urbanization and movement from countryside to city (see Chapter 14).

International migration movements, both within Europe and overseas, did not begin with the demographic transition, but they were greatly amplified during the nineteenth and twentieth centuries. Fig. 8.9 shows the extent to which international migrations took on an increasingly global character, eloquent evidence of the rise of the 'modern world-system' (Wallerstein 1974; Moch, 1996). Much recent research (Bailyn, 1986a, 1986b) on the period before 1815 has emphasized the way in which migration grew out of traditional patterns of internal migration and of change in the domestic as well as the international economy. The contributors to Canny (1994) have shown the variety of early migrations linked to the expansion of colonial empires. Spanish migration to the Americas from the sixteenth century onwards involved the movement of about 750,000 emigrants during the three centuries of colonial rule, with particular peaks in the first half of the seventeenth century. The Portuguese were also early leaders in migration both to Africa and Asia and to a lesser extent across the Atlantic. Estimated total migration between 1500 and 1760 was of the order of 1–1.5 million, drawn from a smaller population than in the case of Spain, involving a high proportion of the male population particularly. It was also less skilled and encountered much more difficult circumstances, including high mortality, in the places of destination (Canny, 1994: 270–1).

The Dutch had many similarities with the Portuguese, not least because much of their seventeenth-century empire had been seized from them. The Dutch East India Company shipped around 1 million people abroad between the early seventeenth century and the mid-nineteenth to fuel the extended trading empire in Asia. Many of these, though, were Scandinavians and Germans, as the Dutch lessened the domestic effect of migration by using foreign labour both in Europe and overseas. Meanwhile, the British were also on the move: the English to the Chesapeake and Caribbean in the seventeenth century and growing numbers of Scots and Irish during the eighteenth. Bailyn (1986a) has used an emigration register for the years 1773–6 to show the complexity of these migration flows to North America in great detail. Migration was everywhere part of the growing British Empire, and in addition the use of African slaves was central to the British as well as to the Portuguese and Dutch overseas expansion. About 8 million African slaves arrived in the Americas before 1820—the great majority being brought to the Caribbean and South America—compared to 2.3 million Europeans (Moch, 1996: 7). The French too used slaves for the development of the Caribbean colonies, but in general the French themselves took much less of a role in international migration than their European neighbours (Ogden and White, 1989).

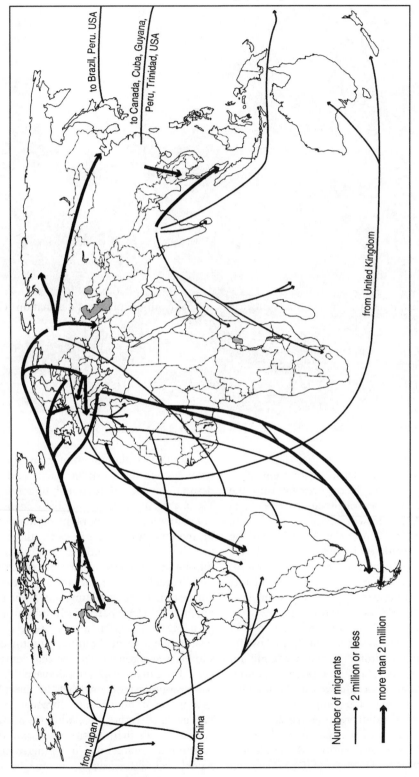

to Brazil, Peru. USA

to Canada, Cuba, Guyana, Peru, Trinidad, USA

from United Kingdom

from Japan

from China

Number of migrants

——→ 2 million or less

━━▶ more than 2 million

Fig. 8.9. Global voluntary migrations 1815–1914. (*Source:* Segal, 1993: 17.)

**Plate 8.3.** Immigrants admiring the Statute of Liberty as they enter New York harbour, 1887.

The nineteenth century was to see these early migrations continue and intensify, with many other origins and destinations being brought into the picture and with an increasingly complex relationship with population growth and industrialization. Above all, migration across the Atlantic and the growth of the 'Atlantic economy' was to dominate (Vecoli and Sinke, 1991; Hatton and Williamson, 1994a, 1994b; Baines, 1991). Between 1850 and 1913 more than 40 million people emigrated from Europe to the New World: 18 million from Great Britain and Ireland, 11 million from Italy, 6.5 million from Spain and Portugal, 5.2 million from Austria-Hungary, 4.9 million from Germany, 2.9 million from Poland and Russia, and 2.1 million from Sweden and Norway (Livi-Bacci, 1992: 124). 'Although a significant and growing share ultimately returned home, this mass migration represented an unparalleled population transfer that had profound effects on the global distribution of population, income and wealth' (Hatton and Williamson,

1994a: 533). At the same time many more left their rural homes for work in towns and cities. The destinations of international migrants shown in Fig. 8.9 again reveal the global dimensions of the phenomenon. The geography of areas of departure in 1900 is shown in Fig. 8.10: the British Isles, Scandinavia, and southern and parts of central Europe are all areas of particularly strong emigration. Not all, however, began to participate in mass emigration at the same time: it was the countries of northern Europe, particularly Britain and Ireland, Scandinavia and Germany, that were earlier exporters of people in mid-century, and the countries of southern and central Europe that joined the flow later. The causes of emigration were, like its chronology, complex and involved the interplay of conditions at home and abroad. Hatton and Williamson's recent (1994a) re-evaluation of the evidence leads them to the conclusion that natural increase at home, which we have documented above, and differences in income between origin

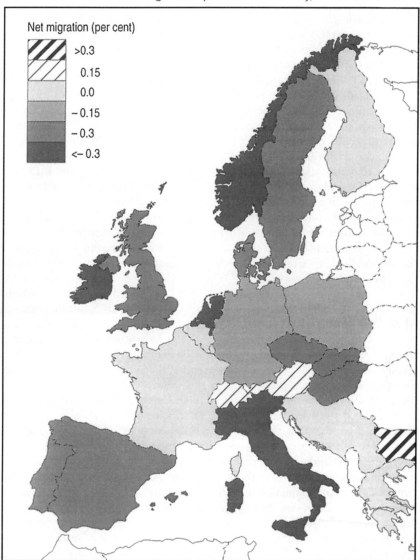

Net migration (per cent)

| | |
|---|---|
| ⊘ | >0.3 |
| ⊘ | 0.15 |
| | 0.0 |
| | −0.15 |
| | −0.3 |
| | <−0.3 |

**Fig. 8.10.** Net migration around 1900 (per thousand). (*Source*: Rallu and Blum, 1993: 9.)

and destination were important, as was the industrial revolution. They also support the view that the influence of friends and relatives among previous emigrants abroad was important, 'creating persistence and path dependence in emigration flows' (p. 557). This reinforces earlier work by, for example, Erickson, who argues that 'prospective emigrants quickly learned to distrust the emigrant's guides and collections of letters from satisfied customers produced by the land speculators, and came to believe only such information as was conveyed in private, personal letters from close friends or members of the family'

(as summarized in Fender, 1992: 41). There was also clearly a cycle to emigration reflecting the changing economic conditions at home and abroad. Return was a more frequent aspect of the mass migrations of the nineteenth century than is commonly assumed. So too was regional variation in the intensity of departure, as Baines (1985) has shown in detail for Britain, and a complexity of migration paths within often complex labour markets (Thistlethwaite, 1991: 26). The growing role and importance of women migrants has also been recognized in recent scholarship (Gabaccia, 1996).

A particular case was Ireland: particular in the volume of emigration, in its demographic impact on Ireland itself, in its lack of return flows, in its cultural impact on places to which migrants went, and above all in its brutal causality—the Great Famine of the 1840s and its aftermath. The population of Ireland declined from 8.2 million in 1841 to 4.5 million by 1900, the only European country to leave the nineteenth century with a smaller population than at the start and with at least 4 million Irish emigrants between 1850 and 1914, enormous by international standards (O'Grada, 1994: 178, 214, 224; Miller, 1985). All European emigrations have had a major impact on the global cultural map but few are as impressive as the Irish: some 5 million people live in the island today but 'the world-wide community of Irish, by descent and birth, totals more than seventy million' (Houston and Smyth, 1993: 338).

## Into the Twentieth Century

Many of the distinctive traits of European demography seen in a global pespective at the end of the twentieth century have their roots in the trends discussed above. The century itself, though, has seen major changes. In general two characteristics dominate: first, the growing national and regional convergence towards greater homogeneity; and secondly, the fact that the Europe of 2000 shares many common traits with the rest of the developed world but stands in sharp distinction from the developing countries. The European demographic regime is distinctive in a number of ways, the principal of which is that low population growth rates characterize the late twentieth century and reflect a 'mature' demographic regime of relatively low rates of migration, low fertility and mortality, and new attitudes towards marriage, family, contraception, and divorce.

First, if we follow through the international migration trends discussed above, we see that the twentieth century was marked by a decline in emigration overseas for the majority of countries especially after 1914, an increasing recourse to immigration especially after 1945, and to a general regulation of population flows unheard of in the nineteenth century. The European countries gradually split between those who continued to export labour, but increasingly to neighbouring states—for example Italy, Spain, Portugal, Yugoslavia, and Greece; and those who turned to immigration as a way of bolstering economic production either from European neighbours or from further afield—for example France from North Africa, Germany from Turkey, or Britain from the Commonwealth, especially the Caribbean and South Asia, thus establishing a new phase of relations between Europe and the Third World (Cohen, 1987). More recently still, with the redrawing of the European map during the 1990s, there has been considerable immigration from central Europe towards the west, especially Germany, and new migrations from the developing world into southern Europe (King, 1993).

Secondly, the decline of mortality, whose origins were noted above, made its most spectacular progress during the twentieth century. Death rates declined for all infectious diseases, with particularly notable declines in respiratory tuberculosis and other respiratory diseases. Most deaths are now attributable to non-communicable causes such as circulatory disease, bronchitis, or cancer. Infant mortality declined from 150–200 per thousand in 1900 to rates that are below 10 in much of western and southern Europe by the 1990s, with much of that progress made since 1950 (Fig. 8.11a). Life expectancies have also risen to a European average of over 70 for men and 77 for women, an experience shared by most countries (Table 8.3). Mortality contrasts remain nevertheless, with the countries of the ex-Soviet Union and central Europe doing significantly less well (Coleman, 1996) and with significant differences by region and by social group.

A significant and sustained upturn in marriage and fertility in many countries during the years after 1945 ensured a much higher rate of population increase than had been anticipated. Nevertheless, the trend towards low fertility re-emerged in most of Europe from the 1960s (Fig. 8.11b) and, despite continuing geographical differences, the century has seen the completion of the demographic transition. Countries such as Italy, Spain, or Greece saw later declines but now have fertility at historically very low levels, with some of the lowest rates in the contemporary world. Average household size for much of Europe is below three persons; there has been an increase in living alone, in lone parenthood, a consequence in part of increased divorce, and a decline of marriage and increase of

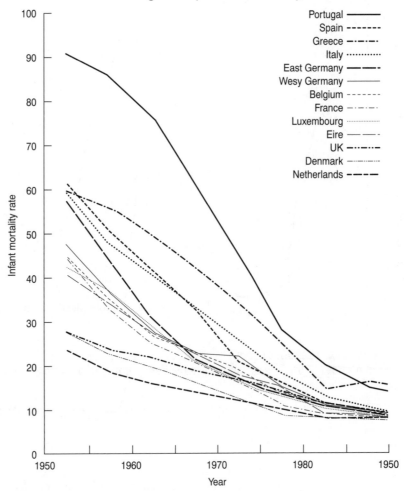

Portugal ———
Spain - - - - - -
Greece —·—·—·
Italy ············
East Germany ——— -
Wesy Germany ———
Belgium - - - - -
France —··—··—
Luxembourg ············
Eire ——— ·—
UK —··—··—
Denmark —·········—
Netherlands — — — —

Infant mortality rate

Year

**Fig. 8.11a.** Trends in infant mortality rates for selected European countries 1950–1990. (*Source*: Noin, 1993: 40.)

cohabitation. In addition, there has been a substantial ageing of the population. Indeed, for Van da Kaa (1987), these trends suggest that the experience of the last quarter of the century may be described as a 'second demographic transition' one of whose principal features is population stability and indeed potential decline. Projections suggest that western Europe's population will be in absolute decline by 2005 and southern Europe's by 2010; the populations of northern and eastern

Europe will continue to grow slowly. Europe's share of world population is expected to continue to decline—from 15.6 per cent in 1950 and 9.4 per cent in 1990 to 6.1 per cent by 2025 (Hall and White, 1995: 11). Though there remains much scope for change in European populations and much continuing diversity, the present marks both a distinct period of the history of European population and an unprecedented degree of control over our demographic fortunes.

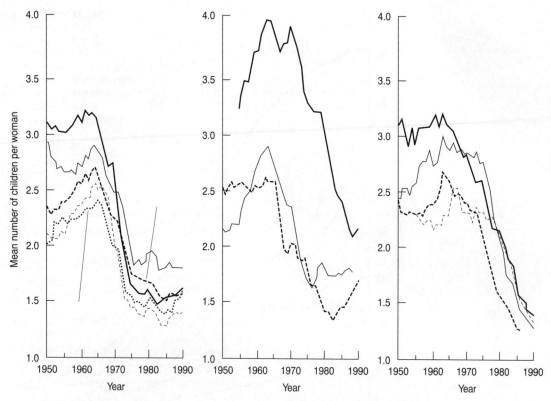

**Fig. 8.11b.** Trends in fertility in selected European countries 1950–1990. (*Source*: Sporton, 1993: 53.)

## REFERENCES

ALDERSON, A. S., and SANDERSON, S. K., 'Historic European Household Structures and the Capitalist World Economy', *Journal of Family History*, 16(4) (1991), 419–32.

ANDERSON, M. (ed.), 'The Relevance of Family History', in C. Harris (ed.), *The Sociology of the Family: New Directions for Britain*, Sociological Review Monograph, 28, (Keele, 1979), 49–73.

BAILYN, B., *Voyagers to the West: Emigration from Britain to America on the Eve of the Revolution*, (London, 1986a).

—— *The Peopling of British North America: An Introduction*, (London, 1986b).

BAINES, D., *Migration in a Mature Economy*, (Cambridge, 1985).

—— *Emigration from Europe 1815–1930*, (Basingstoke, 1991).

BIRABEN, J.-N., 'Certain Demographic Characteristics of the Plague Epidemic in France, 1720–1722', in D. V. Glass and R. Revelle (eds.), *Population and Social Change*, (London, 1972), 233–41.

—— 'Essai sur l'évolution du nombre des hommes', *Population*, 34 (1979), 13.

—— and BLUM, A., 'Géographie et intensité des crises', in J. Dupâquier (ed.), *Histoire de la Population Française*. ii. *De la Renaissance à 1789*, (Paris, 1988), 192–219.

BOURDELAIS, P., 'Cholera: A Victory for Medicine?', in R. Schofield, D. Reher, and A. Bideau (eds.), *The Decline of Mortality in Europe*, (Oxford, 1991), 118–130.

CANNY, N. (ed.), *Europeans on the Move: Studies on European Migration 1500–1800*, (Oxford, 1994).

CASELLI, G., 'Health Transition and Cause-Specific Mortality', in R. Schofield, D. Reher, and A. Bideau (eds.), *The Decline of Mortality in Europe*, (Oxford, 1991), 68–96.

CHAMBERS, J. D., *Population, Economy and Society in Pre-industrial England*, (Oxford, 1972).

COALE, A. J., and WATKINS, S. C. (eds.), *The Decline of Fertility in Europe*, (Princeton, 1986).

COHEN, R., *The New Helots: Migrants in the International Division of Labour*, (Aldershot, 1987).

COLEMAN, D. (ed.), *Europe's Population in the 1990s*, (Oxford, 1996).

—— and SALT, J., *The British Population: Patterns, Trends and Processes*, (Oxford, 1992).

DUPÂQUIER, J., and DUPÂQUIER, M., *Histoire de la démographie. La statistique de la population des origines à 1914*, (Paris, 1985).

EVANS, R. J., *Death in Hamburg: Society and Politics in the Cholera Years 1830–1910*, (Oxford, 1987).

FENDER, S., *Sea Changes: British Emigration and American Literature*, (Cambridge, 1992).

FESTY, P., *La Fécondité dans les Pays Occidentaux de 1870 à 1970*, INED Cahier 85, (Paris, 1979).

FINDL, P., 'Austria', in J.-L. Rallu and A. Blum, *European Population*, i. *Country Analysis*, (London, 1991), 225–36.

FINLAY, R., *Population and Metropolis: The Demography of London 1580–1650*, (Cambridge, 1981).

FLINN, M. W., *The European Demographic System 1500–1820*, (Brighton, 1981).

FRIDLIZIUS, G., 'The Mortality Decline in the First Phases of the Demographic Transition: Swedish Experiences', in T. Bengtsson, G. Fridlizius, and R. H. Ohlsson, *Pre-industrial Population Change*, (Stockholm, 1984), 74–114.

GABACCIA, D., 'Women of the Mass Migrations: From Minority to Majority, 1820–1930', in D. Hoerder and L. P. Moch (eds.), *European Migrations: Global and Local Perspectives*, (Boston, Mass., 1996), 90–111.

GALLOWAY, P. R., 'Long-Term Fluctuations in Climate and Population in the Pre-industrial era', *Population and Development Review*, 12(1) (1986), 1–24.

—— 'Basic Patterns in Annual Variations in Fertility, Nuptiality, Mortality and Prices in Pre-industrial Europe', *Population Studies*, 42(2) (1988), 275–303.

GILLIS, J. R., TILLY, L. A., and LEVINE, D. (eds.), *The European Experience of Declining Fertility: A Quiet Revolution*, (Cambridge, Mass., and Oxford, 1992).

HAJNAL, J., 'European Marriage Patterns in Perspective', in D. V. Glass and D. E. C. Eversley (eds.), *Population in History*, (London, 1965), 101–43.

HALL, R., and WHITE, P. E. (eds.), *Europe's Population: Towards the Next Century*, (London, 1995).

HARDY, A., *The Epidemic Streets: Infectious Disease and the Rise of Preventive Medicine 1856–1900*, (Oxford, 1993).

HAREVEN, T. K., 'The History of the Family and the Complexity of Social Change', *American Historical Review*, 96 (1991), 95–124.

HATTON, T. J. and WILLIAMSON, J. G., 'What Drove the Mass Migrations from Europe?', *Population and Development Review*, 20(3) (1994*a*), 533–60.

—— —— (eds.), *Migration and the International Labour Market 1850–1939*, (New York, 1994*b*).

HOUSTON, C. J., and SMYTH, W. J., 'The Irish Diaspora: Emigration to the New World 1720–1920', in B. J. Graham and L. J. Proudfoot (eds.), *An Historical Geography of Ireland*, (London, 1993), 338–63.

KAMEN, H., *European Society 1500–1700*, (London, 1984).

KANNISTO, V. *et al.*, 'Reductions in Mortality at Advanced Ages: Several Decades of Evidence from 27 countries', *Population and Development Review*, 20(4) (1994), 753–810.

KERTZER, D. I., and BRETTELL, C., 'Advances in Italian and Iberian Family History', *Journal of Family History*, 12(1–3) (1987), 87–120.

KING, R. (ed.), *The New Geography of European Migrations*, (London, 1993).

KIRK, D., *Europe's Population in the Inter-war Years*, (Princeton, 1946).

—— 'The Demographic Transition', *Population Studies*, 50(3) (1996), 361–87.

KNODEL, J., and VAN DE WALLE, E., 'Lessons from the Past: Policy Implications of Historical Fertility Studies', *Population and Development Review*, 5 (1979), 217–45.

LIVI-BACCI, M., *A Concise History of World Population*, (Oxford, 1992).

LUCASSEN, J., *Migrant Labour in Europe 1600–1900*, (London, 1987).

McKEOWN, T., *The Modern Rise of Population*, (London, 1976).

McNEILL, W. M., *Plagues and Peoples*, (London, 1976).

MERCER, A., *Disease, Mortality and Population in Transition*, (Leicester, 1990).

MILLER, K. A., *Emigrants and Exiles: Ireland and the Irish Exodus to North America*, (Oxford, 1985).

MOCH, L. P., *Moving Europeans. Migration in Western Europe since 1650* (Bloomington and Indianapolis, 1992).

—— 'Introduction', in D. Hoerder and L. P. Moch (eds.), *European Migrations: Global and Local Perspectives*, (Boston, Mass., 1996), 3–18.

NI BHROLCHAIN, M., 'East–West Marriage Contrasts, Old and New', in J.-L. Rallu and A. Blum (eds.), *European Population*, ii. *Demographic Dynamics*, (London, 1991) 461–79.

NOIN, D., 'Spatial Inequalities in Mortality', in D. Noin and R. Woods (eds.), *The Changing Population of Europe*, (Oxford, 1993), 38–48.

O'GRADA, C., *Ireland: A New Economic History 1780–1939*, (Oxford, 1994).

OGDEN, P. E., 'Historical Demography', in M. Pacione (ed.), *Historical Geography: Progress and Prospect*, (Beckenham, Kent, 1987), 217–49.

—— and HUSS, M.-M., 'Demography and Pronatalism in France in the Nineteenth and Twentieth Centuries', *Journal of Historical Geography*, 8 (1982), 283–98.

—— and WHITE, P. E. (eds.), *Migrants in Modern France: Population Mobility in the Later Nineteenth and Twentieth Centuries*, (London, 1989).

PERRENOUD, A., 'The Attenuation of Mortality Crises and the Decline of Mortality', in R. Schofield, D. Reher, and A. Bideau (eds.), *The Decline of Mortality in Europe*, (Oxford, 1991), 18–37.

PITKANEN, K. J., *Deprivation and Disease: Mortality Decline During the Great Finnish Famine of the 1860s*, (Helsinki, 1993).

PORTER, R., *London: A Social History*, (London, 1994).

PRESTON, S. H., *Mortality Patterns in National Populations with Special Reference to Recorded Causes of Death*, (New York, 1976).

QUINE, M. S., *Population Politics in Twentieth Century Europe*, (London, 1996).

RALLU, J.-L. and BLUM, A., 'European Population', in A. Blum and J.-L. Rallu (eds.), *European Population*, ii. *Demographic Dynamics*, (London, 1991), 3–48.

REINHARD, M., ARMENGAUD, A., and DUPÂQUIER, J., *Histoire générale de la population mondiale*, (Paris, 1968).

SAITO, O., 'Historical Demography: Achievements and Prospects', *Population Studies*, 50(3) (1996), 537–53.

SCHOFIELD, R., 'Through a Glass Darkly: The Population History of England as an Experiment in History', *Journal of Interdisciplinary History*, 15(4) (1985), 571–94.

—— and REHER, D., 'The Decline of Mortality in Europe', in R. Schofield, D. Reher, and A. Bideau (eds.), *The Decline of Mortality in Europe*, (Oxford, 1991), 1–17.

SECCOMBE, W., 'Marxism and Demography', *New Left Review*, 137 (1983), 22–47.

SEGAL, A., *An Atlas of International Migration*, (London, 1993).

SLACK, P., *The Impact of Plague in Tudor and Stuart England*, (London, 1985).

—— 'Metropolitan Government in Crisis: The Response to Plague', in A. L. Beier and R. Finlay (eds.), *London 1500–1700* (London, 1986), 60–81.

SMITH, R. M., 'Some Reflections on the Evidence for the Origins of the "European Marriage Pattern" in England', in C. Harris (ed.), *The Sociology of the Family*, (Keele, 1979), 74–112.

—— 'Three Centuries of Fertility, Economy and Household Transformation in England', *Population and Development Review*, 7 (1981), 595–622.

—— 'Hypothèses sur la Nuptialité en Angleterre aux XIIIe–XIVe siècles', *Annales, Economies, Sociétés, Civilisations*, 38 (1983), 107–36.

SPORTON, D., 'Fertility: The Lowest Level in the World', in D. Noin and R. Woods (eds.), *The Changing Population of Europe*, (Oxford, 1993), 49–61.

TEITELBAUM, M. S., and WINTER, J. M., *The Fear of Population Decline*, (London, 1985).

THISTLETHWAITE, F., 'Migration from Europe Overseas in the Nineteenth and Twentieth Centuries', in R. J. Vecoli and S. M. Sinke (eds.), *A Century of European Migrations, 1830–1930* (Urbana and Chicago), 17–49 (reprinted from XIe Congrès International des Sciences Historiques, Rapports (Uppsala, 1960), 5 (1991), 32–60.

VALLIN, J., 'Mortality in Europe from 1720–1914: Long-Term Trends and Changes in Patterns by Age and Sex', in R. Schofield, D. Reher, and A. Bideau (eds.), *The Decline of Mortality in Europe*, (Oxford, 1991), 38–67.

VAN DA KAA, D. J., 'Europe's Second Demographic Transition', *Population Bulletin*, 42(1) (1987), 3–57.

VECOLI, R. J., and SINKE, S. M. (eds.), *A Century of European Migrations, 1830–1930*, (Urbana and Chicago, 1991).

WALL, R., ROBIN, J., and LASLETT, P. (eds), *Family Forms in Historic Europe*, (Cambridge, 1983).

WALLERSTEIN, I., *The Modern World-System: Capitalist Agriculture and the Origins of the European World-Economy in the Sixteenth Century*, (New York, 1974).

WATKINS, S. C., 'From Local to National Communities: The Transformation of Demographic Regimes in Western Europe, 1870–1960', *Population and Development Review*, 16 (1990), 241–72.

—— *From Provinces into Nations: Demographic Integration in Western Europe 1870–1960*, (Princeton, 1991).

WILLIGAN, J. D., and LYNCH, K. A., *Sources and Methods of Historical Demography*, (New York, 1982).

WILSON, C., and WOODS, R. I., 'Fertility in England: A Long-Term Perspective', *Population Studies*, 45 (1991), 399–415.

WINTER, J. M., *The Great War and the British People*, (Basingstoke, 1985).

WOODS, R. I., 'Mortality Patterns in the Nineteenth Century', in R. Woods and J. Woodward (eds.), *Urban Disease and Mortality in Nineteenth Century England*, (London, 1984), 37–44.

—— 'The Effects of Population Redistribution on the Level of Mortality in Nineteenth Century England and Wales', *Journal of Economic History*, 45 (1985), 645–51.

—— 'La Mortalité infantile en Grande Bretagne: un bilan des connaissances historiques', *Annales de Démographie Historique*, (1994), 119–33.

WRIGLEY, E. A., 'Reflections on the History of the Family', *Daedalus*, 106 (1977), 71–85.

—— 'Population History in the 1980s', *Journal of Interdisciplinary History*, 12(2) (1981), 207–26.

—— and SCHOFIELD, R., *The Population History of England 1541–1871: A Reconstruction*, (London, 1981).

ZHAO, Z., 'The Demographic Transition in Victorian England and Changes in English Kinship Networks', *Continuity and Change*, 11(2) (1996), 243–72.

# Chapter **9**

# The Changing Cultural Geography of Europe since 1500

W. R. Mead

## New Dimensions and New Perspectives

The outward and visible signs of European civilization are in its cultural features—its arts, learning, architecture, institutions. The inner and spiritual signs are in their values and meanings. Major changes in the character of these features throughout most of Europe date from about 1500—the time of the High Renaissance. New concepts of the natural world, new technologies, and new means of expression modified the intellectual climate. Changes took place against the background of a Europe which consisted of an old-established mosaic of relatively small political units—city-states, principalities, and countries (or provinces) which were in general identified by the names of their inhabitants and by the languages which they spoke rather than by the precise territories which they occupied. It was a Europe with few formal political boundaries and in which the concept of the nation had yet to develop.

After 1500 significant shifts in the centre of gravity of trading occurred, Mediterranean Europe yielding supremacy to the North-West Atlantic seaboard. The authority of northern Europe's Hanseatic League declined simultaneously. Antwerp and Amsterdam adopted the mantle of Venice and Genoa. The cultural map of Europe, inseparable from the resource base of its constituent parts, responded to the changing relative strength of the individual economies. In part, these shifts had their reflection in regional variations in the impact of the High Renaissance. For example, historians have recognized a Northern Renaissance, inseparable from the 'Europe of antiquity', but also having its own distinctive features.

Thus, while the High Renaissance was regarded as a golden age in Spain and Portugal, it would be an exaggeration to apply the same metaphor to the England of the day. At the same time, bearing in mind the increasingly complex web of commercial contacts between the Mediterranean and north-west Europe, France occupied a strong intermediate position, and its artistic life reflected this accordingly.

All in all, by about 1600 a change in the psyche of western Europe is recognized by historians. Geographically, the image of Europe had acquired a shape and dimensions approximating to those by which it is known today. It was also a Europe seen increasingly in a world perspective. In the field of cartography, scientist and artist strove together to pull into shape the grids of latitude and longitude. The two-dimensional maps which were drawn with increasing precision in mathematical terms were complemented by the texts of artists explaining the construction of three-dimensional vistas.

Mathematicians who were seeking solutions to the problem of converting a spherical into a flat surface, were in process of bequeathing their names to future geographical texts—Gerhardus Mercator, a father figure. The map as an ornament—it was even reproduced on the walls of Vermeer's interiors—soon became the map in the service of the state, with the instrument-maker the servant of the surveyor and cartographer.

Refinements in the measurement of time accompanied those in the measurement of space. The great mechanical contrivances for measuring the round of the year as well as the hours of the day, to be found in places of worship such as Lund cathedral, were gradually complemented by an infinity of domestic timepieces produced by high-class instrument makers, skilled watchmakers, and local artisans. Space and time eventually came together in the workshop of John Harrison (1693–1776), who in his declining years finally received a reward for the decades of labour he had devoted to the invention of the marine chronometer. With it, longitude could at last be measured accurately and maps of the continent and the world finally pulled into shape.

All such processes of change, inseparable from the interaction between the new intellectual aristocracy and technical talent, resulted in a more portentous concept of culture—high culture, as it came to be called. Lesser cultural traditions retreated to the margins of economies and societies. High culture became an autonomous European feature. For long, it remained all of a piece, with eastern Europe far from marginalized, as the Jagellonian dynasty attests. It was urban-centred, court-centred, and cloister-centred, with men of letters and of science, artists and musicians, authors and architects, seemingly moved by a common spirit.

And they generated their own expressions of this spirit in the form of waves of fashion which have crossed the continent at varying intervals, but with increasing speed. A range of adjectives has emerged to define these waves, running from the baroque and rococo, through the classical and the romantic to the ever more finely differentiated contemporary. All have left their distinctive legacies upon the cultural map of Europe. Accompanying them, there have been alien ripples of fashion—Indian, Chinese, Japanese, and American—which have acquired their own European character as they have been absorbed into the system. Once absorbed, they have become an integral part of a pan-European culture, illustrative of the capacity for assimilation identified by Paul Valéry as one of Europe's distinguishing characteristics.

In all fields of geographical activity, the problems of constructing models to explain and illustrate the origins and development of these waves of fashion lack effective solutions. In theory, it should be possible to map the changing intensity of occurrence of particular cultural elements or trends, but they can often only be explained in terms of complex processes of interaction between individuals and places. Cultural developments are so often the result of casual linkages, though once established they are inclined to be organic in growth.

A centre–periphery model makes much sense, bearing in mind the promotional role played by city-states, capital cities and, eventually, by nation-states. Mantua might claim that a third dimension in art, architecture, and music was acquired within its precincts by the simultaneous activity of Mantagna, Alberti, and Monteverdi. Network analysis offers possibilities. There was a distinct network of artists and patrons in the Florence of the Medicis. For the cartographers, Ortelius's *Book of Friends* provides a record of his European contacts. Somewhat later, astronomers had their contact systems as the links between Kepler, Tycho Brahe, and Galileo attest. In eighteenth-century Venice, the relations between Ricci, Bellotto, Canaletto, Guardi, and Tiepolo were intertwined, with Vivaldi and Scarlatti in succession belonging to their circle. Sometimes, there have been cross-cultural meeting places right down to individual cafés and their clienteles. Vienna, in the latter days of the Habsburgs, was a nexus city for the arts. It had its varous cultural circles—indeed, its concentric circles. Europe-wide, there is a geography of cities and city districts where artistic groups have tended to converge and where 'schools' of activity have waxed and waned.

Norman Hampson (1968: 129), discussing the processes of diffusion in Europe during the period of the Enlightenment, has observed that 'while [its] frontiers spread outwards in space and downwards through the social hierarchy, the centre itself was evolving all the time. We are left', he concluded, 'with a crude mathematical formula, more appropriate to tracing the spread of an influenza epidemic than to the infinitely subtle and varied discrimination of a complex of partially inconsistent discoveries, assumptions and attributes, through societies which could only assimilate them in local terms.'

Europe defies treatment as a *Gesamt kunstwerk*, yet, even accepting a secular west and a theocratic east, there is a cultural unity in which the diverse forms that have evolved since the Renaissance are held. In what follows, cultural considerations are first regarded as inseparable from a common European discipline and system of instruction—discipline in the figurative arts and instruction in the schoolroom. Italy has been the fountainhead of the former: Erasmus stands as the symbol of the latter. The constituents of culture are intellectual, though they have their material expressions. Examples from literature, art, and music will exemplify the former: architecture, in the broadest sense, provides illustrations of the latter. A gustatory element in European cultural geography may also be identified. The concept of nationhood is inseparable from the complex web of all of these. In addition, cultural influences yield their own particular landscapes—landscapes of pleasure and landscapes of pain. And, if death and transfiguration are not exactly re-enacted in the cultural sense, decay and regeneration are.

## The Legacy of Erasmus

Culture implies instruction. From the Renaissance onwards, with the spirit of enquiry abroad, the educational map of Europe was transformed. The transformation was closely linked to the rapidly widening distribution of the printing press. Some 1,100 presses were probably shared by some 250 towns in 1500. Mainz was a primary centre of diffusion and Gutenberg was the man whose movable type radically improved the early designs. The machine was set up in a Europe which was without significant political boundaries, in which scholars enjoyed freedom of movement. Erasmus, praising in his books the pursuit of learning above that of pleasure, moved between a score of universities. The links between scholarship and book production was illustrated equally by the experience of Wyclif and his translation of the Pentateuch. He left England for Cologne, Worms, and beyond, and his printers were in Antwerp.

At the same time, schools began to multiply. Italian cities already had academies: France established them in its provincial cities. In England, they were created on the initiative of both the Crown and the nobility. All needed books—not the extravagant books produced in Venice, but the products of the printing presses of the north. Frankfurt established its book fair at the end of the seventeenth century.

A map of the distribution of universities at the beginning of the eighteenth century reflects their increase in numbers since the Renaissance (see Fig. 9.1). Their growth, stimulated by Lutheran and Calvinist efforts, was paralleled by the establishment of learned societies, which were to become a hallmark of European culture. The torch-bearers were the Royal Society of London (1662) and the Académie Royale des Sciences (1666).

The latter was central to a cluster of academies which provided European models—academies for the Dance (1661), Belles Lettres (1663), Music (1669), and Architecture (1671). Prominent among the more specialized institutions to be established were the botanical gardens. Pisa, Florence, Padua, and Bologna preceded London and the Jardin des Plantes of Paris. There were already five German art schools when the Royal Academy of London was founded in 1768. It was in London that the applied arts subsequently called forth the internationally influential British Society for the Encouragement of Arts and Manufactures. In the Scandinavian countries, the Patriotic Society of Sweden (1774), the Society for the Betterment of Norway (1809), and the Economic Society of Finland (1797) provided the Lutheran clergy, many of whom occupied immense parishes on the frontiers of European settlement, with a network of contacts whereby the material and spiritual culture of Europe was transmitted to their scattered congregations.

Pedagogical initiatives proceeded side by side with the diffusion of the educational ideas of men such as Rousseau. Among the pioneers was Pestalozzi of Switzerland, whose concern with studies of the familiar home area attracted the attention of the (then) schoolmaster Carl Ritter. Such developments were inseparable from the widening interest

**Fig. 9.1.** The distribution of European universities at the beginning of the eighteenth century. (Based on Sir Ernest Barker, *The European Inheritance* (Oxford, 1956), with additions and amendments.) This map is interestingly complemented by a map of the spread of academies in Europe (David Goodman and Colin A. Russell (eds.), *The Rise of Scientific Europe* (Open University Press, 1991), pp. 244–5).

in the natural world. It was a world made fashionable by Buffon's *Histoire naturelle* (1749–89) systematized by Linnaeus, and popularized by a growing band of field naturalists. Societies for botanists, geologists, and antiquarians were followed by those who chose to call themselves geographers. Geographical societies were established across Europe. Paris took the lead in 1822, Berlin followed in 1825, London in 1830, and St Petersburg in 1843. The French social philosopher Saint-Simon believed that these rapidly multiplying societies represented a source of spiritual power as well as of intellectual attraction. An English word had to be

found to cover those who were engaged in their pursuits. 'Scientist' entered the dictionary in 1840. And while French remained the language of diplomacy and English was fast becoming the language of commerce on the linguistic map of Europe, German became an increasingly indispensable language for scientists.

In the eighteenth century, education was limited to the few and was presumed to take in most fields of knowledge. The shift to education of the many in the latter part of the nineteenth century was accompanied by changes in the curriculum. The United Kingdom introduced compulsory education in 1872, France followed in 1882, and the rest of Europe ran in parallel. The map of literacy in Europe, given new meaning through this legislation, provided the mass of population with the possibility of acquiring at least some inkling of past cultural achievements, of developing an appreciation of the arts, and eventually of understanding something of the various encoding means which they represent. Everything anticipated a new wave of fashion in pedagogical and cultural pursuits—the experimental.

## Art

Art, literature, and music are of the quintessence of what is popularly conceived as culture. Art in all its manifestations and music are different from literature in that understanding of them transcends linguistic and national boundaries. The face of Europe bears the imprint of a greater range of styles and fashions in the arts than that of any other continent. Furthermore, the styles and fashions have succeeded each other at an increasing tempo. One aspect of their evolution is illustrated in Fig. 9.2. Their principal source is the Mediterranean world of the Italian Renaissance. Flowing out from it are the tributaries which have their distinctive forms and colours—colours which have a place in cultural expression artistically and symbolically.

The science of colouring was already yielding its own literature in the sixteenth century—science with art, indeed, for it was considered that colour created its own sensations. A vocabulary of colour was slowly constructed to define the contents of the artist's palette. A search for the representation of light accompanied the concern with colour. Again an understanding of light drew artist and scientist together. The refraction of light was a topic for discussion by philosophers such as Descartes as well as for scientists such as Newton. Artists no less than Bacon's pedagogues were 'merchants of light'.

The experience of Venice was absorbed into the works of Nuremberg's Dürer: El Greco brought the experience of Venice and Rome to Spain: Rubens carried the baroque style to his homeland. The ingenious employment of perspective, the sweeping rhythms, the billowing clouds, the striking contrasts of light and shade of Michelangelo and Titian

were widely emulated. Geography itself was given symbolic expression in Bernini's sculptural celebration of the four continents in the Piazza Nevona and in Tiepolo's Würzburg murals.

In the High Renaissance, there was a shift in subject matter from biblical and classical allegories in favour of portraiture. Monarchy and the court spectacular were summed up in Van Dyke's portraiture: the intellectuals of the age in Holbein's painting of Erasmus and, later, in Franz Hals's portrait of Descartes. The citizenry were assembled in Rembrandt's *Night Watch* (1642), which not only underlined the shift from the religious to the secular but also marked the rise of talents in north-west Europe to rival those of the Mediterranean.

With the name of Italy a byword among the European cognoscenti, it was natural that it should be the climax for those who took the Grand Tour. Canaletto, Guardi, and their contemporaries guaranteed that Venice would remain a continuing source of attraction beyond the High Renaissance. Nor did Italy fail to benefit from the rise of interest in landscape painting. Through Salvator Rosa and his French counterpart Claude, travellers acquired a new appreciation of the country scene through the eye of the painter. 'One sees nature no more,' was the remark of Goethe on his Italian travels, 'nothing but pictures.' Antiquarians also joined the tourists, setting a fashion for the reappraisal of classical ruins and artefacts.

New forms of illustration enabled the traveller to take home a fuller record of his journey. As the eighteenth century advanced, the printing presses produced a steady stream of richly illustrated books which were decorated by leading artists of the day. The popularization of art awaited the invention of lithography in Munich in 1798. Print-publishing plants were soon established in Dresden, Berlin, and Vienna. The word 'illustration' was given currency. The trade of publisher was born. By the time that their products were gaining a market, public art galleries were being founded. In the wake of the Louvre (1793) national galleries were opened in Madrid, Berlin, Munich, and London, with a second generation following in such cities as Dresden, Brussels, Vienna, and Amsterdam. The galleries also stimulated a demand for the work of sculptors to embellish public and private places.

As an art market emerged, collectors and dealers began to classify their wares geographically as well

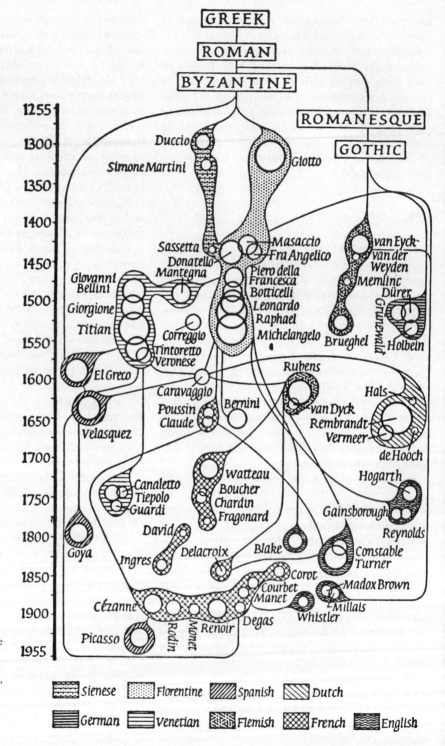

**Fig. 9.2.** Eric Newton's representation of the chief schools of European painting and the threads of influence between schools and artists. He had in mind 'a map of a river system', though it could 'give no indication of the force of the current'. (Reproduced by permission of Penguin Books Ltd. from Eric Newton, *European Painting and Sculpture*, 4th edn. (Harmondsworth, 1966), p. 240.)

GREEK
ROMAN
BYZANTINE
ROMANESQUE
GOTHIC

1255
1300 — Duccio
Simone Martini — Giotto
1350
1400
Sassetta — Masaccio / Fra Angelico — van Eyck / van der Weyden
1450 — Donatello / Mantegna — Piero della Francesca — Memlinc / Dürer
Giovanni Bellini — Botticelli — Grünewald
1500 — Giorgione — Leonardo / Raphael
Titian — Correggio — Michelangelo — Brueghel / Holbein
1550 — Tintoretto / Veronese
El Greco — Caravaggio — Rubens — Hals
1600 — Poussin / Claude — Bernini — van Dyck / Rembrandt / Vermeer
Velasquez — de Hooch
1650
1700 — Watteau / Boucher / Chardin / Fragonard — Hogarth
Canaletto / Tiepolo / Guardi — Gainsborough
1750 — David — Reynolds
Goya — Ingres — Delacroix — Blake — Constable / Turner
1800
1850 — Corot — Madox Brown
Courbet / Manet — Millais
Cézanne — Degas — Whistler
1900 — Picasso — Rodin / Monet / Renoir
1955

Sienese   Florentine   Spanish   Dutch
German   Venetian   Flemish   French   English

212

as stylistically. A virtual hierarchy of place-names evolved in association with artists and their work. National 'schools' of art were identified by critics and admirers of what were named 'old masters' of the Italian, Spanish, Flemish, and French schools. Romanticism, continent-wide in its artistic expression, had a high point in the German school (embracing the neighbouring Swiss). Lower in the hierarchy of place-names and later in the history of painting, artists came to be associated with particular locations. Formal 'schools' in such cities as Dresden and Düsseldorf were complemented by informal summer gatherings of artists in resorts as diverse as Dieppe, Skagen, and St Ives. Artists' quarters in cities suggested a new urban classification of little artistic oases such as the Latin Quarter in Paris and London's Chelsea. The rise of landscape painting, especially in watercolour, reflected a genre which also had its regional specialization, but which was a west European rather than an east European phenomenon. Association with renowned artists eventually endowed particular areas with a virtual cultural sanctity—the Suffolk of Constable: the Provence of Cézanne and Van Gogh.

The French school anticipated a situation in which there was a greater concern with colour than with (what Degas called) 'lines and styles'. Impressionism implied an escape from visual realities to the forces behind them. The shifting and changing colour harmonies of Renoir, the experiments of the *pointillistes* were new forms of 'intellectual art' which anticipated a growing concern with perception. Some critics consider latter-day France to have been a source of developments in art to be the equivalent of Florence during the Renaissance. And France had Rodin to strengthen such a claim.

The plastic arts have had their own independent geographies, their focus changing in place and time with invention and innovation. The arrival of Chinese porcelain to Western Europe spurred the search for a European equivalent. It was produced in Saxony in 1709–10. To the old-established majolica of Florence and faience of Delft were therefore added the fine wares of Meissen and Sèvres, Dresden and Limoges. From Worcester and Derby in the west to St Petersburg in the east with Copenhagen and Rörstrand in between, the products of ceramic factories were endowed with royal insignia.

Designs reflected the exotic influence of the Orient (with the willow pattern and Indian Tree advancing across the continent) as well as the discoveries of the age (with Wedgwood drawing inspiration from the decoration of the Portland vase). Luxury products remained, but the manufacture of everyday ware responded to the new mass market. The production of cutlery and silverware, not forgetting Sheffield plate, was a parallel development. All such products represented a cultural advance at the domestic level for the rapidly expanding middle classes.

Changes in glass manufacture ran concurrently, the shifting centres of production not independent of the supply of raw materials as well as of individual enterprise. Specialization began at the core of Europe—Venice, Bohemia, Silesia. The twentieth century witnessed the entry into the market of high-class competitors on the periphery of the continent, from Waterford in Ireland to Örrefors in Sweden.

## Literature

Literature also has a geography of its own. At the national level, literary atlases have been compiled and the geographical content of literary works has been explored. While the visual arts overcome linguistic barriers, most of Europe's literature can only make a wider impact through translation. Nevertheless, changes in literary form and style, having identifiable sources of origin and routes of diffusion, are generally unimpeded by language. Examples from the drama, the novel, and folk literature must suffice.

After 1500, drama was the first literary form to cross language barriers. It is coincidental that Shakespeare should have been born in an era which already had a strong theatrical tradition, but it is less easy to explain why his plays, in manifold translations, should have had the widest continental impact of any European dramatist. His Spanish contemporaries—Calderón and Lope de Vega—never received the same widespread staging, and this despite the status of Spain in Europe and the impact of Cervantes. The plays of Shakespeare also competed successfully with those of the French dramatists—Racine, Molière, Corneille—despite the dominance of the French language.

Move forward and the spotlight shifts inconsequentially to Scandinavia, with Ibsen and Strindberg having a revolutionary impact upon *fin-de-siècle* drama. It would be interesting to trace the theatrical contact patterns that led to the first performances of their plays throughout western Europe. Berlin was a critical centre. The speed with which Ibsen and Strindberg gained a European audience contrasted with that with which Chekov's comedies of manners were accepted. In the twentieth century Brecht and Pirandello were among those who shifted the focus of drama southwards. Simultaneously, the film emerged to challenge the theatre. Western Europe took the lead in consciously treating the film as an art form. The success of the film studio became a new criterion for measuring national prestige: the cinema became a universal feature of the urban scene.

The novel-reading public differs from the theatre audience. 'The novels with which England is inundated', as Rousseau expressed it, were not without significance in his own France. They reached a momentary climax in Weimar and then took root and flourished throughout the continent. The flowering paralleled the rise in literacy, the cheapening of books, and, later, the establishment of public libraries—in short, the democratization of literature. It is understandable that the socially-motivated novels of Charles Dickens should have penetrated eastern as well as western Europe.

Hugo, Balzac, and Dumas in their different ways repeated the success of Dickens before the realism of Zola commanded the attention of the continent. Not surprisingly, once adequately translated, the nineteenth-century Russian novelists, with their sense of space and their giant key characters—'craving for the superlative', in the words of Pasternak—swept all before them.

Beside such major authors the numerous and widely scattered authors concerned with the archetypal rural life of Europe, important though they may have been in their respective homelands, have made only a limited impact beyond. Such has been the case with the work of Pereda in Spain, of Giovanni Verga in his 'heat-hammered island' of Sicily, of Alexia Kivi's *Seven Brothers* in Finland, of Czeslaw Nitossa's *Dolina issy* in rural Lithuania and, later, of the series of novels set in the Icelandic countryside by Halldór Laxness.

Parallel with the rise of the novel was the growth of interest in folklore and folk legend, a genre which had an appeal in the nursery as well as the library. It had roots in central Europe and was largely inspired by publication of J. G. Herder's *Volkslieder* (1778–9) and the *Deutsche Sagen* (1814–18) of the brothers Grimm. Vud Karadzic collected Serbian materials (1814 his first publication).

Epic fragments were gleaned from the backwoods of Finland and Karelia and published as *Kalevala* (*Kalevipoeg* its Estonian counterpart). Rhineland fragments yielded *Niebelungen*. Asbjørnsen and Moe trekked from farmstead to farmstead collecting Norwegian material. Different in character, but making their own impact, were the translations of the Norse sagas mostly from manuscripts held in Copenhagen.

The identification of places with literary figures intensified. It is H. C. Andersen's Denmark, Joyce's Dublin. From the 'shining glade' of Tolstoy's Yasnaya Polyana to the bleak heights of Haworth parsonage the shrines of the literati draw their pilgrims. It is the same with the homes of most European geniuses. In all their thousands they have become a part of a remunerative twentieth-century cultural industry.

# Harmonica Mundi

In the western hemisphere from the Renaissance onwards the sounds of music increased and diversified. The refinement, growth, and diffusion of a musical culture was technical as well as artistic. Even the simplest sounds began to change. For the labourer in the field, the sonic landscape was born of church bells. Their harmonious rings and mathematically intricate peals attained a maximum complexity with the construction of mechanically driven carillons. To the east, the simpler, often dissonant clangours of orthodox bells spread from Greece to Russia. For the youthful Stravinsky they were the loudest source of diurnal sound.

Italy was the cradle of European musical culture, with Monteverdi gathering instruments together in the orchestra, spanning the religious and the secular, converting the *drama di musica* into the opera. The evolution of music is inseparable from court entertainment. The court spectaculars, representing the harmonies of the universe and confirming the divinity of monarchy, became distinctive features during the sixteenth and seventeenth centuries. The tradition was established in Italy and spread through France to England. The masques of the Stuart court were linked directly through men such as Inigo Jones to those of the Medicis. At the same time, the origins of musical development confirmed that Italian language and notational systems would command the page of the musical score.

The emergence of music as high culture—the province of court, city, and ceremonial—encouraged the refinement of musical instruments. In the process, traditional instruments retreated to the geographical margins of settlement and the production of musical instruments began to develop its own geography. Italy became the home of the strings, France became renowned for its woodwinds. Iberia specialized in the mandolin and guitar. The zither of Russia, born of a variety of ancestors in the eastern Mediterranean, was modified to become the kantele of Finland.

North of the Alps, where Lutheran communities developed their own musical ritual, the chorale emerged as a distinctive form and the organ an essential feature. A geographically dispersed organ-building industry grew in response. Brass, silver, and percussion, initially associated with military and ceremonial functions, gradually acquired new functions. Beyond their employment in the civic and national orchestras that multiplied throughout the continent, they displayed a new pattern of distribution, geographically and socially. In the brass bands of mills and collieries they began to represent a working-class culture. At the same time, the reappraisal of the bagpipe gave a national instrument to Scotland while the harp—indeed choirs of harps masculine in composition—became a symbol of Wales.

Systems of notation and the frame of the clefs, first given printed form in Paris, c.1530, confirmed the link between music and mathematics. They gave rise to increasingly precise compositional rules. The lengths of strings and pipes, the disposition of pedals and stops, all recalled Kepler's *Harmonica Mundi* (1619), and the sixteenth-century concept of music as 'a divine science'. The metronome, invented in the early nineteenth century in the Netherlands, established a temporal discipline. The refinement of the contrapuntal form reflected a geographical shift of musical invention northwards,

with the name of the Bach family focusing it upon the German principalities.

Both opera and ballet sprang from baroque Italy, being introduced to Paris and early Hanoverian England before establishing themselves in the courts of the Austrian Empire and the German principalities. Italian musical influences reached St Petersburg as early as 1725, when the first ballet school was founded. Opera came later with Glinka its pioneer, deriving his inspiration from his Italian experiences, and the work of Bellini and Donizetti. *Mitteleuropa* claimed Mozart, Hayden, and Beethoven: England, Handel. The northernmost outlier of musical life was established by Gustavus III at Drottningholm. By the latter part of the eighteenth century, composers were sharing a common European heritage not least through their travels round the continent.

With the rise of Romanticism, the style, content, and function of composition changed. Music embraced poetry in the *Lied*, with Schubert offering a model. Momentarily, the arts formed a collectivity, in particular during the brief German romantic period of *Sturm und Drang*. National romanticism found its roots in the movement. Partly as a result, composers became increasingly associated with the name of their homelands. Geography and music hung together. Thus it became Verdi's Italy, Dvořák's Bohemia, Greig's Norway, Tchaikovsky's Russia, De Falla's Spain, Elgar's England, Sibeilius's Finland, and (later) Bartok's Hungary. Chopin became the apotheosis of Poland; Wagner of the united Germany—Wagner musically elevating the Nordic gods; Offenbach caricaturing those of the Mediterranean.

Simultaneously, a Europe-wide multiplication of places of musical entertainment occurred. Orchestras began to bear the names of their home cities. Pioneered by Bayreuth (1876), specialized musical festivals were to multiply a thousandfold within the span of a century. Schools of music complemented those of artists and architects. At the turn of the century, for all the Paris of *la belle époque*, Vienna wore the musical crown imperial, with Austria's Wörthersee a stimulus to Brahms, Mahler, and Berg. And in the east, 'abstract city' though it might be to Dostoevsky, St Petersburg generated a special radiance. Its ballet was a virtual court spectacular; its atmosphere was only capable of being captured in words by the symbolist Andrei Byely. The diaspora of Russian artistic talent—Stravinsky, Prokofiev, Rachmaninov, Diaghilev—as a consequence of the revolution, had parallels in the arts other than music. *Ex orient lux.*

At the same time, folk-song, folk-dance, and folk instruments were being rediscovered in the byways and backwoods of the continent to which they had retreated. Rediscovery did more than support ethnographic distinctiveness, for collectors and composers used their findings from the 'little tradition' to advance the 'great tradition' itself. And it was in *Mitteleuropa*, breaking with both traditions, that the musical avant-garde sought a new freedom of expression in atonality. Berg and Schoenberg—Strauss also in *Elektra*—set in motion the idea of expressionism.

## The European as Epicurean

There is another feature of the culture of Europe which, by its nature, is antecedent to most of the arts. Symbolic testimony is provided in the still-life compositions of the fruits of the earth, the fish of the sea, the fowls of the air, let alone the tankards and goblets painted by Dutch masters. It is the culture of the table.

As long as geographies have been written, the vine and the olive have been presented as essentially cultural symbols. Disease has caused changes in the distribution and character of the vine, and the olive groves have suffered neglect and decay. Yet both remain engraved on the mental map of Europeans. Beyond the limits of the vine, the distillation of spirits from grain and the potato has prevailed. The end product gradually became the whiskies of Scotland and Ireland, the gins of the Netherlands, the akvavits of Scandinavia, and the vodkas of Poland and Russia. In general, a beer culture complemented that of *aqua fortis*, the humble hop making its own contribution to the agricultural landscape. Rhineland beer festivals were to become more formidable than any festivals linked to the vine.

Since the Renaissance, Europe's farm stock has been endowed with cultural qualities in its own right, through elaborate hybridization and the establishment of agricultural shows. Equally, what Braudel called 'Carnivorous Europe' has its humble expression. The Christmas pig long remained a feature of northern Europe. The autumnal goose feasts of Sweden and Denmark and the goose fairs of England continue as relics of an earlier tradition.

'Epicures of Fish', as the seventeenth-century traveller Thomas Nugent called them, have tended to be more closely associated with the Atlantic than with the Mediterranean coasts. Salting, drying, pickling, and smoking enabled the fish of high latitudes to be carried to Catholic Europe. Canning and refrigeration extended the market in the later nineteenth century, before which Norway had established a significant ice trade. Europe's fishmongers have commonly coupled the harvest of game with their sales. For all that the Dutch contributed to the gastronomic geography of Europe in the seventeenth century, it was the French that took command of it. A whole European culture is represented in the rites that surround the serving of a meal. Brillat-Savarin, with his *Almanach des Gourmands* (1817) made sure that the language of the menu remained French, which Alexandre Dumas confirmed for the literary world with his *Grand Dictionnaire de Cuisine*. *Haute* was soon to be added to *cuisine* no less than to *couture*. In the twentieth century, Michelin stars were conceived to challenge the performance of restaurateurs, and the very phrase 'vintage year' was taken over as a metaphor from the wine harvest.

The Dutch were also to lose another feature of their gustatory reputation. For all their worship of sugar, and their historic reputation as 'dental cripples', they have ceded pride of place to the Belgians and Swiss in confectionery. In all this, the north seems a world away from the south as summed up in the honey and sesame of Greece.

And while the Dutch craved sweetness, the Londoners worshipped coffee. It was strange that London's seventeenth-century coffee-houses should decline while the French took over the concept of the café and established a café culture which was to be repeated continent-wide. In continental eyes, tea became the British institution because of India, though the word *salon* had to be brought across the Channel to lend an air of refinement to places where it was served.

All of these gustatory refinements reflect the escape from the medieval threat of famine, though Church calendars continued to mark the times of feasting and fasting. Shrove Tuesday provided a street scene for Breughel's Battle of *Carnival and*

*Lent.* The ecclesiastical division of Europe denied Protestants the *mardi gras.* For the Catholic south, the pagan midsummer bonfires of the north, which the church discreetly linked to the Eve of St John, had no meaning.

Lent does not appear to have affected the tobacco culture that spread across Europe from the seventeenth century onwards. The pipe was enshrined by both Dutch and French artists of distinction: the snuffbox enjoyed a passing fashion before the intrusive cigarette. The scent of cigar smoke, a part of Europe's olfactory geography, was to reflect national differences in commodity taxation. Tobacco culture has its own European museum in Bergerac.

Perhaps the ultimate in epicureanism has been the Europe of the essences. The perfume industry is a feature of Europe inseparable from the rise of a flower culture. The attar of Balkan roses and the oils of Provençal lavender—even the curiously named eau-de-Cologne—were of the essence of European culture until chemists invented more exotic perfumes for France to claim its own.

## The Dwelling Place

The High Renaissance was characterized by an increased intellectual concern with the earth as a dwelling place. A new understanding of nature was matched by new approaches to the visual arts and by new architectural achievements. In the architectural field, builders had to accommodate themselves to the available constructional materials. Few were more fortunate than those of the Italian cities, with access to a rich diversity of rocks. In Europe at large, where the softer limestones and sandstones prevailed, builders were able to indulge freely in the succession of fashionable architectural forms. Though the brick and tile of the Netherlands made for appealing domestic architecture, the formidable brick façades of the great ecclesiastical buildings of the North German plain reflected the constraints of the claylands. Granites posed an even greater challenge. 'Foot pavements of granite' similar to those in England (save that the English were mostly of sandstone) was the ambition of Alexander I for St Petersburg, but it was half a century after his death before Rusia began to import equipment for cutting and splitting its granites. Stone,

slate, shingle, and thatch for roofing all reflected local and regional materials and the designs of craftsmen. Copper sheathing, weathering colourfully, became popular throughout Europe for the roofs of prestigious buildings. Lead was its older, more malleable, but less attractive competitor.

It was the baroque style that came to dominate the new buildings that sprang out of the High Renaissance. Its bold pillars, cupolas, and sculptural detail had their origins in the papal lands. The cities of Spain and Portugal, the Netherlands, and Poland responded to the fashion no less than Vienna, Prague, and Munich. In London, the Great Fire opened the way for a baroque reconstruction. In both the capitals and provincial towns of Scandinavia (for this was the age of greatness in the North) its effects were felt (see Fig. 9.3).

The transformation of the baroque into the rococo was marked by a more elaborate decoration. Its fullest expression was witnessed in central Europe's great monastic churches and in the residences and pleasure domes of its aristocracy. The West's Versailles and Fontainebleau were mirrored in the East's Peterhof and Strelne; Oxfordshire's Blenheim had its counterpart in St Petersburg's Tauride. And while to the architect it was probably a *Schloss* that he had designed, fashion decreed that it should bear the names Sanssouci, Bellevue, Montbijou, or Monrepos, with the features planned around it becoming boulevards, promenades, and belvederes.

From the seventeenth century onwards, the planning of new towns, the creation of new districts, and the transformation of old became cultural undertakings in their own right. Independently of practical considerations such as the widening of streets to reduce fire risk, a more formal treatment of urban space resulted from the diffusion of ideas of perspective. The siting of ecclesiastical and public buildings, monuments, and other ornamental features paid increasing attention to the visual connection with relief, waterways, and road systems. Such was the case with the development of Berlin under Schinkel, of St Petersburg under Italian guidance, and of Edinburgh under the Adamses.

The tides of architectural fashion flowed faster in the nineteenth century. The heavy Biedermeyer and bourgeois architecture of Second Empire France were challenged by the neo-Gothic. A whiff of art nouveau stirred in Belgium, took off in Vienna as *Jugendstil,* and eventually attained a peak of

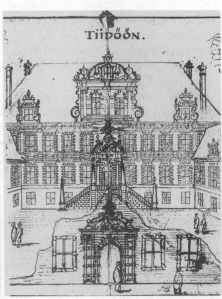

**Fig. 9.3.** Baroque architectural features were well established in seventeenth-century Sweden. Examples are provided by the façades of two manor houses—(*a*) Vibyholm, 1626, and (*b*) Tidö, 1688. (*Source*: Erik Dahlberg, *Suecia antiqua et hodierna*, Stockholm, 1714.)

restrained decorative whimsy in early twentieth-century Helsinki. The next step was to replace the harmony of form with the dissonance resulting from experiment. The twelve tones in music (dissonance to many) that sprang out of Austria had their architectural counterpart in the Cubist movement, the origins of which are associated with Czechoslovakia.

A new aesthetic trend entered the urban scene with the development of colour washes and paints to accompany or succeed other preservatives. Tar, especially for weather-proofing, was employed for Norway's dark stave churches and the 'Stairway to Heaven' churches of Karelia. Where copper was processed, it provided a preservative. *Rödfärg* coloured wooden buildings in Sweden and Finland, with red replaced by ochre for the residences of 'persons of quality'. Stucco invited cream and colour washes. Pastel shades were commonly employed on the neo-classical buildings of Russia, Poland, Finland, Estonia, and Latvia, though only Russia itself was sufficiently adventurous to employ blue, pistachio, and raspberry. Nowhere was the palette employed more freely than in the external decoration of Russia's historic Orthodox churches.

The regionally distinctive buildings of the European countryside have acquired a reputation as quintessentially cultural features. Reference to the chalets of Switzerland, Austria, and Bavaria, must suffice. Their timbering, elaborate fretwork, and

verandahs were copied in many summer villas and hunting lodges throughout Europe.

The conscious search for building styles which harmonize with the countryside is not new, but it has taken time for it to be accepted. In Sweden and Finland, some architects have gone so far as to ascribe an inherent moral quality to the granite which is basic to their geography. It suggests an austerity appropriate to the Lutheran north. Sweden's Hugo Lindberg spoke for the avant-garde of his day when he declared that it was 'tasteless to transplant . . . the luscious flower of Italian architecture . . . to our bleak north'.

For generations after the High Renaissance, the profile of the European city (often sketched for baroque atlases) continued to be dominated by cultural features—the spires and domes of churches, the towers of castles, and the commanding rooftops of palaces. With the rise of industry, factory chimneys competed with the dreaming spires; in turn, the chimneys have been subordinated to the intrusive skyscraper. In many respects, the silhouette of a city, etched against the skyline, sums up its character and its cultural geography.

## Landscapes of Pleasure

The cultural map of Europe is inseparable from the concern with pleasure and the spread of leisure across the social spectrum. 'Pleasure's landscapes' embrace all of the estates, parks, and gardens that the wealthy have laid out and the public spaces subsequently established by civic authorities. The search for pleasure in European landscapes, which began with the aristocratic clientele on the Grand Tours, has had its ultimate expression in the package tour, available to most, with the Mediterranean the continuing destination.

For the more sedentary inhabitants of urban areas, pleasure gardens such as London's Vauxhall and Ranelagh provided the earliest attractions (Copenhagen's Tivoli gardens took over from them at the time of their demise). Aches and pains, here and there assuaged by holy wells such as those of Wales, were offered new cures from the eighteenth century onwards through the establishment of spas. The high noon of individual watering places varied in time. Bath had soon to experience competition from a dozen other hydropathic resorts. Spas proliferated in central Europe wherever mineral springs issued from limestone rocks. Baden, Carlsbad, Marienbad, with their royal patronage, had their rivals in France, though none was as successful in commercializing its waters as Vichy. Even the Scandinavian countries discovered remedial waters, those of southern Norway providing Henrik Ibsen with a theme for his *Enemy of the People*. With the coming of the railway, coastal resorts slowly developed to complement the inland spas. France and Italy began to exploit their Rivieran littorals. France added the Atlantic to the Mediterranean experience, with such fashionable centres as Biarritz challenging Deauville. Black Sea resorts were cultivated by Russians whose passports denied them the pleasure of journeying farther afield. No country in Europe developed more seaside resorts than England, with the pier a distinguishing feature spanning the tidal zone—a zone unfamiliar to Mediterranean or Baltic shores.

For the more energetic, there has always been the Europe of the hunters, changing its character with the passage of time. The elk hunters of the Finnish forests and the bear hunters of Norway and Sweden were still drawn from the farming fraternity until the twentieth century. The aristocratic hunting lodge (a shooting box in British terminology) became an integral part of the European scene, with the wild boar and deer as quarries. By the eighteenth century hunting in England had become fox-hunting with the likes of Mr Jorrocks being eased out in Victorian times through the territorial definition of the packs and the social refinement of the participants.

Simultaneously, throughout most of Europe, from the grouse moors of Scotland to the marshlands of Turgenev's *Sketches*, the calendar for the shooting of game birds was slowly formalized. And if S. T. Aksakov's *Russian Gentleman* could indulge his sporting instincts at 'the flash of a fish', his English counterpart could only cast flies over the trout stream if he owned it.

The historic equine culture has made its own contribution to pleasure's landscape. The jousting ground might disappear, but the horse retained a place in the Mediterranean bullring. Austria enjoyed the dignity of Vienna's old-established Lipizaner showplace; Italy, the licence of Siena's historic *palio*. The racecourse contrived to marry high culture with low throughout Europe. From Ireland's Curragh to Longchamps and beyond, the calendar of annual race meetings has not ceased to multiply. The high noon of horse-drawn carriages, in such public parks as the Bois de Boulogne and Hyde Park, disappeared before the combustion engine, though the association of the horse with all forms of national and local pageantry remains a part of popular culture. Over the last century there has developed a ludic geography in which the variety of activities has multiplied, in which their distributions first sharpened nationally and then blurred internationally, and in which all have established their place in the seasonal round of the sporting calendar.

Indeed, Europe has become very much a playground in which more energy is spent in sporting pursuits than its Renaissance ancestors had to spend in keeping body and soul together. As sport has become the epitome of popular culture, every physical feature of the landscape has been turned to account in its interest. In his *Aesthetic Education of Man*, Schiller wrote that man is only fully a human being when he plays. But, given professionalism, when does play become work?

The same might be asked of the enthusiasts who, from the High Renaissance onwards, began their collections of *objets d'art* and cabinets of curiosities. Ambition moved collectors as diverse as Ferranti Imperato in sixteenth-century Milan, Ole Würm in

Copenhagen, Buffon with his *cabinet du roi* in France, Sir Hans Sloane in Chelsea, and Sir William Hamilton in Naples. It was but a step forward to the establishment of museums of high culture—pioneered by Rome in 1732 and London in 1753. Later, they were to be complemented by museums of popular culture. In 1894, at Skansen in Sweden, Arthur Hazelius created the first folk museum, and endowed the vernacular with a new cultural value.

## Landscapes of Pain

The cultural map of Europe is also an expression of the military machine—constructively and destructively. The schools of surveyors, with their ABC books of fortification plans, represent the constructive side. Luxembourg city illustrates the layer upon layer of fortification, the rabbit warrens of underground passages burrowed through limestone contrasting with the residual watchtowers and Vauban's solid bastions. The tunnels of the 'Maginot Line' are for tourists today: likewise, Norway's tunnelled salient at Heggli in Stjødalen. In the west, Waterloo—Byron's 'deadly Waterloo' and deadlier Verdun—are balanced in the east by Borodino—Tolstoy's deadly Borodino and deadlier Stalingrad. Place-names are etched in the memory of Europe through a litany of battlefields—the Agincourt of Henry V, the Lutzen of Gustavus Adolphus, the Poltava of Charles XII, the Sedan of Émile Zola, the Sarajevo of the Archduke Ferdinand. Alma, Balaclava, Inkerman, and Sebastopol are still remembered because of their alphabetical sequence. Bastides in the west and Kremlins in the east may belong to history, but they are firm reminders in the present-day landscape of past siege and struggle. Westwards, poppies have taken over the residual trenches of Flanders. Eastwards, encroached upon by woodlands, are the hallowed defences blasted in the granite along the Fenno-Russian border.

'Pain's landscape', to employ the phrase of R. S. Thomas, is written in all of these features. It is also expressed in a European landscape alive with monuments and memorials to wars and warriors, each country having its own peculiar iconography. Poland has its cross at Giby. Hungary, Serbia, Montenegro, and Greece have monuments to commemorate liberation from Ottoman rule. In western

Europe, while the frequency of memorials reflects quantitatively the sacrifices of warring countries, the quality is not inseparable from the wealth of the combatants. It is rare for a village in France, Britain, or Belgium not to have a war memorial.

From the immense Menin Gate, through the serried ranks of tombstones in the 'warrior province of Picardy' to the great mausoleum above Belfort—all are a source of amazement to long-term neutrals such as the Swedes or the Swiss. Eternal flames burn; 'last posts' are sounded. In Russia, where the death toll on the eastern front during the Second World War was greater than that on the western front a generation earlier, local traditions may differ. At the national level they are the same. 'Sites of memory, sites of mourning': Tony Winter's words could be writ large over the immense memorial to the victims of what was Stalingrad.

International war cemeteries, places of pilgrimage, are widely distributed in Europe. In north Norway, the cemeteries in such settlements as Harstad and Tromsø, displaying headstones of different designs in discreetly separated national plots, are maintained with different degrees of care. Older cemeteries in their own right have become places of cultural pilgrimage—from Bunfields to Highgate, from Père Lachaise to St Petersburg's Tikhvinskoye. Baedeker was even pointing out a unique Jewish cemetery in Prague over a century ago. 'Europe is a cemetery', wrote the French historian Georges Sorel.

Military action has reduced the stock of Europe's cultural monuments. Pain's landscape could be written in maps of destruction, perhaps taking their model from those of landscape devastation compiled for the *département* of Marne after the First World War. Sometimes, individual cultural symbols have been selected for destruction. The deliberate dynamiting of the great donjon of Coucy by Ludendorf in 1917 is echoed in the destruction of the royal castle of Warsaw and of Peterhof in the Second World War. On a larger scale were the deliberate aerial raids on Dresden and Bath, Coventry and Charlottenburg. Europe remains full of ironies. While the churches of Orthodox Russia and of Armenia have been reprieved, those of Transylvania have suffered. While the Moorish relics of Spain are visited by multitudes of tourists, European heritage sites in Yugoslavia are reduced to rubble. The cultural geography of the continent is incomplete without an excursion into Europe's heart of darkness.

## The Nationalization of Culture

The principal changes in the cultural map of Europe since the Renaissance owe much to the rise of nationhood and the steady increase in the number of independent states (see Chapter 7). Nationality is inseparable from language and culture. Culture clarifies nationhood, and helps to legitimize it. The days before the emergence of separate states with their clearly defined boundaries were in some respects a time of geopolitical infancy.

The identification of culture is inseparable from education—education in a particular language and usually in a particular area. When education and the book were in Latin, literature was European and there was a universalist approach (save in the Latinophobe territories of Russia). The classical tradition may have persisted for some two centuries after the Renaissance, but the vernacular languages were gnawing at the roots. The appearance of the Bible in the vernacular (the language of William Tyndale's ploughboy) represented a major change of tradition, the accidentals taking over from the universal.

Side by side with Latin, though in a more narrowly defined area, Italian at first dominated, but by the end of the sixteenth century it had yielded to French. French, commanding intellectual and court circles, became the language of the arts. Most British literature—from Shakespeare and Milton, through Pope and Swift to Locke and Hume—circulated in mainland Europe in French. So, too, did the Spanish classics. The rules of French literary style prevailed, though the Royal Society made its own pronouncements on the correct use of scientific English. Austria, Hungary, Poland, and Russia all adopted the French model. When Frederick III founded his Academy of Sciences in 1700, it bore a French title and from 1745 its proceedings were printed in French. Gibbon published his history of Switzerland in French. Alexander von Humboldt wrote mostly in French, though the penmanship might be in the German script.

In the nineteenth century France continued to maintain a linguistic lead, with publishing houses capitalizing on the tradition and dominating the west European market. The Portuguese author Boade Quieras declared that he owed everything to French culture and to 'the packing cases of books from France' which invaded the University of Coimbra.

By this time, European scholars had already responded through the publications of J. G. von Herder and his followers to the ideas of *Volk*, *Volkstam*, and *Volkgeist* (perhaps best translated as collective memory). The word 'folklore' entered the English dictionary in 1846. Folk music, art, craft, and dance were all fostered by the Romantics. From Slovenia and Romania through Poland to the Nordic countries artists recorded the 'folk' costumes of the countryfolk.

It was, however, the concern with languages that was critical. Standardization of alphabets and scripts was accompanied by the codification of the major languages in grammars and by the preparation of dictionaries. The concern for minority languages followed. From Greece and Romania, Serbia and Bohemia, to Finland and Norway, the language of the people was resuscitated. Language struggles became a European phenomenon, with one after another of the lesser tongues being legally recognized. Hungarian was one of the earliest to secure a victory when, in 1842, it was constitutionally accepted as the language of the Diet.

Once they were established, the myth-making associated with particular language groups became a part of the received culture that was to pass from one generation to the next. It began in the schoolroom, where there were already in the eighteenth century books about the homeland for children. Simultaneously, the map of the home country began to circulate. Myths were also made for children by authors such as Finland's Zachris Topelius with his *Book about our Land*. It was said of his Danish contemporary H. C. Anderson that he could make a myth out of a milestone.

The theories of cultural determinists lent support to the national cause. Thus, Hippolyte Taine looked to geographical environment as a powerful force favouring nationality. Landscape types were believed to generate human qualities. The mountains of such countries as Switzerland and Norway were considered to breed both a sturdy independence and a moral authority. National schools of history-writing made their own contribution—from Carlyle and Macaulay in Britain and Michelet in France, through Geijer in Sweden to Karamzin in Russia. Joseph Nadler linked geography and history in his *History of Germanic Races and Landscapes* (1912). Archival collections, libraries, and theatres were fostered to further the national image.

Nor must the role of the Church—schismatic north–south, east–west—be disregarded. The break with Rome strengthened the relation between Church and State in northern Europe. In England, the Crown became 'the Defender of the Faith'. It was Sweden as a Protestant kingdom which became 'the Lion of the North'. The seven provinces of Holland were regarded as being supported by the 'patriotic scriptures'. 'Faith, Tsar, and State' became a holy trinity in Russia. The archetypal clouds of Greek saints, transported to Byzantine Russia, became powerful icons to challenge the authority of the Western Churches as well as the expansion of Islam on the eastern borders.

All of these changes engendered the internal and external personalization of nation-states. In the nineteenth century there emerged in popular parlance Mother Russia, Mother Sweden and 'little mother Denmark'. It was 'Mother Germany' before, together with Austria and Holland, Germany became a Fatherland. Spain was 'the little fatherland'. France, yet to become Marianne, was *la belle France*. Britain was accorded a dual personality—Britannia and John Bull. Comic maps of Europe played upon these personifications as well as upon their heads of state. In contrast to the strange mixture of sexes attributed to countries, languages remained mother tongues in the company of the mother church and mother earth.

While the cultural map of Europe was enriched by the fuller realization of its ethnographic diversity, national awakening was divisive as well as cohesive. The fraternity of the European intelligentsia was weakened: the cosmopolitan aristocratic culture was reduced. 'Cosmopolitanism is all twaddle', declared Turgenev's Judin; 'without nationality there is no art nor truth in life.' The Versailles settlement represented the high water mark of national self-determination. New states were confirmed. New place-names entered the atlas: some, such as those of Alsace-Lorraine, were restored to earlier forms. New boundaries were drawn (to create new problems because it was impossible to accommodate all claims). New flags, new currencies were created, new stamps printed.

The nationalization of culture was inevitable. The problem arose later with its transformation into popular nationalism. Popular nationalism was a creed for majorities, in turn encouraging cultural uniformity and, in authoritarian states, cultural programming.

## Tod und Verklärung

Few European historical atlases have paid attention to cultural features other than religion and language. The changing geopolitical scene dominates them: the changing cultural scene has been absolutely subordinated. In general, this is because information from the past is rarely adequate to enable the construction of anything more than simple distribution maps of critical elements in the cultural advance of the continent. The *Atlas of the Renaissance* (1994) is unique in setting the scene for a continent in the process of leaving behind a late medieval cultural uniformity with strong religious roots and entering an age of cultural differentiation with a humanist bias. Contemporary sources of information offer the possibility of more elaborate compilations, such as the *Atlas of Sweden* (1993) with its detailed nation-wide distributional maps of every feature of cultural activity. Among the custodians of the cultural heritage of Europe whose life and activities are summed up in such maps, four possible groups may be identified. First, there are those whose 'culture' is of continuing concern to social anthropologists. They consist of the diminishing number of people who remain at the grass roots of society, from whom folk culture has been derived and among whom the residue of it resides. In the vernacular they bear such names as *paysan, Bauer, bonde, talonpoja, rustico*. Information obtained from such groups constitutes the material from which elaborate maps such as those in the *Finnish Ethnographic Atlas* (1994) are constructed. Secondly, there is the absolutely dominant urban citizenry, whose financial contributions determine to a large extent to which cultural activities investment is directed. Thirdly, there are the active contributors to the cultural pool, the artists and scholars of all kinds who may or may not be dependent upon private, civic, or commercial patronage. Fourthly, there are the administrators who deal with the management of cultural institutions. They are not only the custodians of national monuments, but are also responsible for mobilizing national achievements in the arts and sciences so that the distinctiveness of their national culture may be kept before the international community.

The material heritage varies greatly in individual countries. The inheritance of the Nordic countries, for example, is totally recorded and hedged about with protective legislation. Contrastingly, the multitude of residual features, often too numerous even to document, that typify much of the continent, can only be selectively maintained. Furthermore, it is often ensembles rather than individual features to which attention needs to be directed. If a European city suffers a natural catastrophe a cry goes up from the entire continent—'Save Florence . . . Venice . . . Dubrovnik'. The cry is the louder if the setting is in the Mediterranean, because the reverence for the south prevails, with Greece and Italy the continuing metaphors for classical and Renaissance cultures respectively. And now that the idea of a 'European city of the year' has emerged, civic administrators muster their cultural legacies and attributes as competitive potential.

*Tod und Verklärung*, the title of Richard Strauss's composition, is relevant to any appreciation of the changing cultural geography of Europe. Death represents the recurrent wars and upheavals that have interrupted cultural progress and that have been destructive of much of Europe's material heritage. It stands for the Dark Ages that preceded the Renaissance of classical antiquity: for the Thirty Years War as forerunner to the Age of Enlightenment. It stands for the passing fashions in the arts as wave after artistic wave crosses the continent. Europeans have displayed a resilience in the face of disaster, almost as if destruction itself was a challenge to creativity. And creativity may imply more than regeneration—even transfiguration.

REFERENCES

*The works given here are trimmed to twenty out of the dozens of publications that have been consulted.*

ALDSKOGIUS, H., and ARNBERG, ULLA (eds.), 'Cultural Life, Recreation and Tourism', *National Atlas of Sweden*, (Stockholm, 1993).

BARKER, Sir Ernest, *et al.*, *The European Inheritance*, (Oxford, 1956).

BLACK, C., *et al.*, *Atlas of the Renaissance*, (London, 1994).

BRANCH, M., and HAWKSWORTH, C. (eds.), *The Rise of Tradition*, (London, 1994).

BURCKHART, J., *The Civilization of the Renaissance in Italy*, (London, 1944).

BURKE, P., *Venice and Amsterdam: A Study in Seventeenth-Century Cultural Elites*, (London, 1974).

CLOUT, H., *After the Ruins*, (Exeter, 1996).

COSGROVE, D., and DANIELS, S., *The Iconography of Landscape*, (Cambridge, 1988).

DA KOSTA KAUFMAN, T., *Court, Cloister and City: The Art and Culture of Central Europe 1450–1800*, (London, 1995).

GOODMAN, D., and RUSSELL, C. A., *The Rise of Scientific Europe 1500–1800*, (Open University, 1991).

HALE, J., *The Civilization of Europe in the Renaissance*, (London, 1994).

HAMPSON, N., *The Enlightenment*, (London, 1968).

HOBSBAWM, E. J., *The Age of Revolution 1789–1848*, (London, 1980).

LOWENTHAL, D., *The Past is a Foreign Country*, (Cambridge, 1985).

NEWTON, E., *European Painting and Sculpture*, (London, 1941).

NORBERT-SCHULTZ, C., *Meaning in Western Architecture*, (London, 1975).

SCHAMA, S., *The Embarrassment of Riches*, (London, 1991).

SCHORSKE, C., *Fin de Siècle Vienna*, (New York, 1980).

SMITH, A. D., *The Ethnographic Origins of Nations*, (Oxford, 1986).
SMITH, Preserved, *A History of Modern Culture*, (2 vols.; London, 1934).
WILLIAMS, R., *Culture and Society 1780–1950*, (London, 1958).
WINTER, A., *Sites of Memory and Sites of Mourning*, (Oxford, 1995).

# Chapter 10

# Rural Europe since 1500: Areas of Innovation and Change*

## H. Clout

## Rural Europe c.1500

As in earlier times, Europe remained an emphatically rural continent in the age of the great discoveries. Human life was threatened by pestilence, war, and famine, hence the key imperative was to produce enough food for its population to survive. This meant ensuring a supply of grain for daily bread, since meat was a luxury for all except the urban élite (Braudel, 1981: 104). Agricultural systems varied according to physical circumstances, social organization, local traditions, and access to urban markets and trade routes. Despite important regional differences, cereal yields were low, with seed : yield averages for both wheat and rye standing at 1:4, and four-fifths of the workforce was involved, to a greater or lesser degree, in food production (Pounds, 1990: 225). Much of the landscape would have appeared relatively untamed to our eyes, with stretches of waste or forest surrounding cultivated areas (Darby, 1957: 20). More land was worked when population pressures increased, but the margin of cultivation retreated, and settlements declined or were abandoned when population numbers declined. As a result, the cultivated area had expanded between 1100 and 1300, then retreated until the sixteenth century when moors, marshes, heaths, and woods were reclaimed, only to contract in the seventeenth century, and be replaced by expansion from 1750 to 1850 (Grigg, 1980: 32). Towards the end of the nineteenth century the amount of cultivated land declined again, largely because cereals from the New World could be imported more easily than ever before. Such pulsations in land use were reinforced by price fluctuations, with reclamation taking place especially when cereal prices were high, as during the sixteenth century and from 1750 to 1850 (Slicher van Bath, 1963: 199).

Land clearance was the traditional way of increasing food supply and continued to be practised after 1500 in many parts of Europe at times of population pressure. If labour were to be removed from food production, devoted to crafts or other non-agricultural functions, and, probably, be concentrated in large settlements, there were two options for ensuring adequate food supply. Local farming had to be made more productive, or

* I wish to express my sincere thanks to Hugh Prince for his advice during the preparation of this chapter.

transport systems had to be improved to ensure supply from more distant areas. The present chapter will explore both options, but especially the challenge of 'intensification', which might involve increasing the frequency of cropping, growing higher-yielding crops, adopting new techniques, or concentrating on specific crops rather than growing a wide range in an attempt at self-sufficiency (Overton and Campbell, 1991: 17).

Urbanization was both a stimulus for raising agricultural productivity and a response to it (Hohenberg and Lees, 1985: 72). Across successive centuries urban and industrial growth depended on an intricate and uneven transformation of agricultural activity, both in time and space, a transformation which originated in medieval Flanders (and to a lesser extent in Lombardy), whence it was diffused to neighbouring regions (Jones and Woolf, 1969: 6; Duby, 1972: 194). This chapter is concerned primarily with such areas and, to a lesser extent, with sections of eastern Europe which came to serve western markets. The slower pace of agricultural change in peripheral parts of the continent is analysed in Chapter 11.

Agricultural intensification in Europe has been shown to be a remarkably complex phenomenon, extending over a broader span of time than had been thought by earlier scholars (Wrigley, 1987a: 65–71; Overton, 1990: 185–9; Price, 1994: 72–109). For example, Overton and Campbell (1991: 48) remark that the 'so-called "new husbandry" of the eighteenth and nineteenth centuries took over two hundred years to assemble and a further hundred years to become sufficiently widely adopted to make a decisive impact on overall production'. This transformation not only involved new crops and techniques, but also the transition from feudalism to capitalism, whereby servile farming and self-sufficiency were overtaken by individualism and production for the market. The spatial relations of agricultural activity were duly refashioned with respect to landholdings and inter-regional connections, and the physical 'facts' of climate, hydrology, slopes, and soils were reappraised in the light of changing social organization, economic demand, and biological and technological innovation (Bertrand, 1975: 68–74). The 'natural' environment offered remarkable scope for transformation through drainage, irrigation, fertilization, and other processes. The 'discipline of distance' must be interpreted in terms of the mode of transport employed, with water routes and overland routes offering different opportunities in past centuries, and the appearance of the railway and the steel-hulled steamship in the nineteenth century enabling spatial relations to be reshaped (Price, 1983).

At the start of the sixteenth century most of Europe was composed of small, relatively self-sufficient settlements; however, surpluses were required to settle taxes, tithes, and other commitments whether in kind or in cash (Laslett, 1965). In north-west Europe much farmland was cultivated in open-field strips, with two cereal crops being followed by a fallow period within the unimproved triennial rotation (Smith, 1978: 191–241). Enclosed land was rare except in the Low Countries, England, and western France. Craft activities and services employed some members of farming households, especially during slack periods in the agricultural year. Any food surpluses were traded at local market towns. Only a few cities exceeded 20,000 inhabitants and these commanded 'islands' of intensive food production where land values and the value of peasants' time were raised, so technical improvements spread relatively easily (Parker and Jones, 1975: 20). Over a dozen of these nodes could be recognized c.1500, including parts of Catalonia and the Rhine valley, as well as Flanders, the Netherlands, and southern England (Slicher van Bath, 1960: 147). Europe's few really large cities called upon supply networks which extended far into the surrounding countryside. Notable features in the pattern of urban settlement included dense clusters of cloth-making towns in Flanders and on the north Italian plain, and the city of Paris (already with c.200,000 inhabitants in 1400), which headed the urban hierarchy of western Europe. By contrast, London had only 55,000 inhabitants in 1520.

Flanders formed the most striking nodal region of early modern Europe, contrasting with a suite of peripheries with which it had few trading relations (Hechter and Brustein, 1980: 1061–94). Flanders commanded its own local system of agricultural specialization but also drew in grain, wine, dairy products, and cattle from elsewhere. Towns in the northern Netherlands and south-eastern England formed secondary supply bases for this region, which was served by water routes in the form of the English Channel, the Bay of Biscay, the Baltic Sea and the Rhine and Meuse rivers (De Vries, 1974: 2). To the east of the Elbe a broad stretch of territory supplied cereals through the Baltic to the Flemish core, while oxen were driven from Poland, Hungary, and Jutland for fattening near the cities of Flanders, the Netherlands, and northern Germany (Wallerstein, 1974: 68; Braudel, 1981: 193). French wine was shipped through the Bay of Biscay and the Channel to Flanders, while German wines were sent along the Rhine. Drawing on these supplies and benefiting from the proximity of urban demand centres, the Flemish countryside was able to emerge as a model of agricultural experimentation and intensification whose innovations were to be adapted and adopted in other parts of north-west Europe. By contrast, intensive agriculture in Lombardy depended on irrigation and was not suitable for emulation in temperate Europe (Zamagni, 1993: 7; Houston, 1964: 436).

Adoption of variants of the Flemish model was a slow and complicated spatial process that has been likened by Le Roy Ladurie (1994: 19) to the hands of a watch operating in reverse, involving first the northern Netherlands, then England, and then France. Agricultural ideas were then transmitted from England into Germany, Denmark, Sweden, and even Russia, with visits by north Europeans to the great improving estates of lowland England and their subsequent reports being instrumental (Abel, 1980: p. x). Of course, the diffusion of information did not operate with such spatial or chronological precision; however, the analogy of turning the clock back offers a convenient means of reviewing the processes that contributed to four centuries of agricultural change (Bairoch, 1973: 460).

## Areas of Innovation and Change

### Flanders

Flanders had experienced a sharp rise in population and food supply from the eleventh to the thirteenth centuries. Its cultivated surface increased owing to coastal reclamation, and early disintegration of the feudal economy enabled Flemish farmers to experiment. But local supplies fell short of demand, and cereals had to be imported from northern France (Blomme, 1993: 28). Between the late thirteenth and

the late fifteenth centuries three methods were devised to improve productivity within the established triennial rotation (Slicher van Bath, 1960: 133). First, land was fallowed less frequently, with beans or cereals being grown on the former fallow third. Secondly, 'convertible husbandry' was practised, with arable turned over to pasture for a number of years, enabling more livestock to be raised and more manure generated. Thirdly, fodder crops were grown on the original fallow third, or turnips were planted once cereals had been harvested. Such innovations were more numerous prior to 1400 than in later years (Verhulst, 1990: 119). The subsequent process of adoption was complicated, with some parts of Flanders retaining bare fallow until the sixteenth century but others having long accepted complex systems which included fodders, pulse crops, turnips, flax, and dye plants (Blomme, 1993: 29). Cultivation of clover had the particular virtue of maintaining nitrogen in the soil (Chorley, 1981: 71–93).

The 'Flemish model' was composed of a range of related variants which enabled fodders and industrial crops to be grown as well as cereals, thereby facilitating the early rise of craft industries, high population densities, and formidable fragmentation of landholding (Mendels, 1975: 179). Many more livestock were raised, which yielded greater quantities of manure to fertilize the soil. Night soil and other urban refuse were collected scrupulously from Flemish towns and carried in carts or in canal barges for liberal application in the fields. The countryside around Ghent was fertilized not only with local manures but also with supplies from Bruges and Courtrai (Goossens, 1993: 291). By virtue of the value attributed to manure, foreign visitors were struck by the cleanliness of Flemish streets and speculated on how Flemish agricultural techniques might be incorporated in their home regions.

### Northern Netherlands

The main aspects of agricultural improvements diffused rapidly from Flanders to Holland and other parts of the northern Netherlands. In Braudel's (1984: 178) words: 'since land was scarce agriculture . . . had to stake everything on productivity'. In arable districts emphasis was placed on innovative crops, complex rotations, and heavy manuring. The detail of these responses was not only a reflection of the so-called physical environment but also related to the northward flight of refugees after 1570 from Flanders (the Spanish Netherlands), who contributed substantially to urban growth further north (Grigg, 1980: 157). The Netherlands experienced a distinctive 'agricultural cycle' whereby farming flourished between 1590 and 1670, unlike the relative stagnation in other parts of Europe during the seventeenth century (De Maddalena, 1972: 312). This was indeed a golden age in which Dutch trade flourished and many merchants were willing to invest heavily in promoting land improvement and agricultural innovation (Van der Wee, 1993: 61). Local urban growth and agricultural specialization depended on an ability to import vast quantities of grain from the Baltic region, and in particular from Poland, where noble estates and their enserfed peasantry produced rye and wheat for the Dutch market (Peet, 1972: 3). Everything 'depended upon an ability to command the surplus of literally hundreds of thousands of serfs and millions of hectares of land' (De Vries, 1974: 170). Imports from Poland continued to rise during the seventeenth century but after 1700 increasing amounts of grain were obtained from Flanders, England, and France.

With the installation of sea walls and sluices and using horse-driven waterwheels, large areas of Dutch coastal marsh had been reclaimed in medieval times and in the sixteenth century. After 1600 urban capital was invested in new technology in the form of windmills with a greater lift than before. Attention was directed away from the coastlands to shallow inland lakes formed by peat-digging (Kain and Baigent, 1992: 19). After five years' work and the use of forty-nine windmills, Jan Leeghwater (1575–1650) and his collaborators drained the Beemster in 1612 and set out 200 new farmsteads on reclaimed land (Wagret, 1968: 84). This impressive result convinced investors that such schemes could indeed succeed. In the first half of the seventeenth century most of the lakes north of Amsterdam were drained and by 1665 throughout the Netherlands 31,600 hectares of former lakes had been reclaimed (Grigg, 1980: 150). The advice of 'Jan Wind', as he was nicknamed, was sought for drainage projects in France, England, the German lands, Denmark, and around the Baltic. Humphrey

Bradley advised on numerous schemes in France, and in 1630 Cornelius Vermuyden launched his scheme for the drainage of the English Fens (Darby, 1940, 1983; Van Veen, 1955: 47). The Electors of Brandenburg called on Dutch and Flemings to drain their peat bogs and 'Hollandries' were also established along broad valleys in Poland and Russia. 'Jan Wind' announced his ambitious plan for draining the formidable Haarlemermeer (16,000 ha.) west of Amsterdam; however, this goal was not to be achieved until the nineteenth century and then with the help of the steam engine.

A range of agricultural specialisms developed in various districts of the northern Netherlands. Reclaimed polders were suited to livestock-rearing and dairying, with the associated production of cheese. Around the fringes of Amsterdam emphasis was placed on supplying fresh milk. In drier districts intensive cultivation systems included hemp, flax, dyestuffs, hops, and tobacco, with related craft industries absorbing surplus labour and raising incomes (Le Roy Ladurie and Goy, 1982: 98). Following the introduction of Turkish tulips from Germany, sandy areas south of Haarlem were devoted to bulb cultivation in the early seventeenth century (Lambert, 1971: 221). Commercial horticulture developed around Amsterdam and other Dutch cities, providing a model that was to be carried to south-east England by Dutch refugees and emulated in an impressive way around London (Van der Woude, 1975: 235). After 1650 the writings of expert Dutch gardeners made distinctive contributions to the diffusion of horticultural and agricultural knowledge throughout north-west Europe (De Vries, 1974: 154).

## England

Movement of people and ideas across the North Sea made a profound contribution to the improvement of English agriculture during the seventeenth and eighteenth centuries. Many contemporaries argued that open-fields and associated communal practices across much of lowland England restricted agricultural progress, which might only be facilitated by remodelling the geometry of fields and farms by enclosures. Descriptions of Flemish farming, such as the famous report of 1644 by Sir Richard Weston about the Waas district between Antwerp and Ghent, were received favourably by many English landowners, who subsequently encouraged their tenants to grow clover and turnips as well as wheat and barley in what came to be known as the Norfolk four-course rotation (Darby, 1976: 15). Fallow was no longer left bare, more livestock were raised, and greater quantities of manure were produced to fertilize the land and increase crop yields. Improved rearing of livestock was possible in hedged, rectangular fields which replaced the open-fields and were well suited to new implements such as seed drills and improved ploughs. In his classic statements on this eighteenth-century 'agricultural revolution' in *Pioneers and Progress in English Farming* (1888), and again in *English Farming Past and Present* (1912), Lord Ernle identified a number of innovating landowner heroes, who insisted that leases incorporate 'improvements', and provided powerful role models for others to emulate (Overton, 1984: 119; Daniels and Seymour, 1990: 487–520).

Such a view has required substantial revision in the light of recent research, which identifies agricultural change in the sixteenth and seventeenth centuries (Yelling, 1978: 151; Overton, 1996). Piecemeal consolidation of holdings, abolition of common rights, and growing demands by London and other cities for regular supplies of food occurred earlier than had once been believed. For example, convertible husbandry dated from before 1560 and spread rapidly during the following hundred years (Mingay, 1977: 22). After 1620 clover and rye-grass seed were being imported regularly from Holland, with turnips and clover being grown in East Anglia as early as 1650. This was long before 'Turnip' Townshend came to manage his estates and, according to Ernle, introduced the cultivation of turnips (Mingay, 1977: 20). A greater degree of flexibility in farming practices was possible in the open-fields than once believed, hence enclosure was not the necessary precondition for some aspects of agricultural innovation. Detailed research has produced evidence which deflates the reputations of many of the 'great men' who allegedly pioneered England's 'agricultural revolution' (Parker, 1955; Overton, 1984: 120). In short, it is no longer possible to believe that an English agricultural revolution began in the eighteenth century. None the less, new plant strains continued to arrive in England during that century, including special varieties of clover, swedes, and mangold-wurzels (Walton, 1978: 247).

Regardless of the degree of revision appropriate to the notion of the agricultural revolution, it is clear that the rapid growth of demand for food in London (whose population rose from 55,000 in 1520 to 200,000 c.1600, 575,000 in 1700, and 960,000 in 1801) provided a powerful incentive for agricultural change across an ever-widening area (Wrigley, 1987b: 167). Indeed, Wrigley (1987b: 190) has argued that the growth of London 'was probably the most important single factor in engendering agricultural improvement' in England. The metropolitan grain market developed rapidly after 1500 and by 1700 called upon supplies from much of the country south of a line between the Wash and the Severn (Gras, 1915). Navigable rivers, notably the Thames and its tributaries, and the opportunities afforded by coastal navigation structured the flow of wheat and barley to market towns around the capital, where grains were processed into meal or malt (Fisher, 1935: 60). Livestock husbandry flourished not only in the western counties but also in parts of East Anglia, where turnips provided a new source of fodder. Cattle and sheep were dispatched in droves from northern counties for fattening in the Midlands or closer to London. The capital was ringed by meadows providing prodigious quantities of hay for metropolitan horses, by orchards, and by a suite of market gardens beyond the urban fringe which were fertilized with night soil, animal manure, and all urban wastes in a style reminiscent of Flanders (Richardson, 1984: 248). Many gardening families were Dutch or French Protestant refugees who had migrated across the North Sea (Thick, 1985: 505; Weinstein, 1990: 79). As the capital's population approached the million mark in the early 1800s, so the opportunity of supplying the metropolitan market transformed an ever-greater share of England's farming, as did the possibility of shipping food to the rapidly urbanizing Netherlands. In broad terms, the concentric pattern of decreasing agricultural intensity away from London evoked the model that J. H. von Thünen was to advance in 1826 from the experience of his estate at Tellow in Mecklenburg-Schwerin (Bull, 1956: 25–30; Chisholm, 1962: 21–32; Pounds, 1985: 252).

## France

By comparison with much of England and the Low Countries, the complex processes contributing to agricultural advance operated slowly across the greater part of France. Relatively self-sufficient family farming survived during the *ancien régime* and remained the anchor of the nation in the nineteenth century, retaining a higher proportion of the workforce in the countryside and in farm-based employment than in neighbouring countries of north-west Europe. This characteristic derived in part from the constraints of the feudal system which operated across the country during the *ancien régime* and also from many collective obligations which restricted individual initiative in areas with open-fields (Bloch, 1966). By contrast, three types of area experienced notable specialization and market orientation, and had done so since medieval times (Duby, 1970: 33–41). Northern districts, which only came into French hands in the seventeenth century, displayed the kind of intensification found in the main territory of Flanders and thrilled travellers such as Arthur Young in the late 1780s (Braudel, 1990: 285). Areas of specialized viticulture in Gascony, the Loire valley, Burgundy, and many smaller districts took advantage of transport by river or sea to supply wine to urban markets not only in France but also in England and the Netherlands (Dion, 1959; Smith, 1978: 496–506). Finally, the relatively loose network of French cities gave rise to a scatter of 'islands' of intensive production which was dominated by Paris (Grantham, 1989: 50).

The capital reached 550,000 inhabitants in 1801, and had been overtaken by London in terms of size. In medieval times, fruit and vegetables were grown within the walls of Paris, especially in the humid Marais. By the end of the *ancien régime* the capital was ringed with market gardens, nurseries, and vines (De Planhol, 1994: 259–63). Horse manure and night soil were much prized commodities and their availability helped explain the location of market gardening around Paris as well as garrison towns nearby (Phlipponneau, 1956: 74; Reid, 1991: 33). Situated near the confluences of the great navigable rivers of the Paris Basin, the capital commanded a very substantial hinterland (Braudel, 1982: 38). It summoned vast quantities of cereals from the extensive granaries of northern France, timber and firewood from the forests of the eastern Paris Basin, dairy products from Brie and Normandy, and wine from Burgundy and the Loire, as well as less-privileged vineyards in northern

France (Braudel, 1988: 258; Pounds, 1979: 153). Daily life in Paris and, indeed, throughout Europe 'turned on the need to procure bread' (Kaplan, 1984: 7). The great majority of Parisians lived arduous and uncertain lives in which subsistence, in the form of cheap wheaten bread, had to be guaranteed if civil disturbance were to be avoided. The city's immediate hinterland, within a 100-kilometre radius, included the large farms and fertile ploughlands of the plains of France, Vexin, Valois, and Beauce, whose cereals were exclusively for the capital (Moriceau, 1994). The second zone included Picardy, Soissonnais, Champagne, and Orléannais, with a third zone embracing the remainder of northern France, whose supplies could be diverted through a network of grain merchants, millers, and tax officials to the capital should the harvests of the first two zones fall short. In extreme circumstances, grain could be brought in at considerable cost from other parts of Europe. The geography of the grain trade in the eighteenth century demonstrated that Paris was truly *the* central place of the whole kingdom.

The notion of agricultural improvement in the English style was advocated in translations of English writings, and in the published proceedings of provincial agricultural societies established after 1750. The physiocrats insisted that farming should be improved since, according to their philosophy, it was the land rather than trade or crafts which provided the essential source of wealth (Blum, 1978: 249). Members of a tiny, educated élite were, indeed, influenced by the principles of agronomy and contributed to a kind of Anglomania (Bourde, 1953; 1967). The Paris Basin, long exposed to market forces, contained estates which displayed their owners' initiative and received praise from English travellers (Clapham, 1961: 17). By contrast, most large landowners in the provinces used their wealth to acquire more property, positions, or residences in the capital rather than to increase the productivity of their land (Forster, 1970: 1600–15). Nor did risk-taking through agricultural innovation enter the collective psychology of family farmers who sought to ensure survival and household cohesion. In any case, feudal obligations and communal traditions inhibited the degree of change that was possible even among the most adventurous.

Abolition of feudal constraints at the Revolution

opened the way for change but, although legally free to work the land as they wished, the majority of French farmers continued with the old, familiar practices used by their fathers and grandfathers (Soboul, 1956: 78–95). The stipulation of the Code Napoléon that property should be divided equally between heirs served to fragment farmland even further. Increases in food production between 1750 and 1850 were more the result of clearing additional land than of increased productivity (Sutton, 1977: 247–300). In the early nineteenth century complex rotations favoured high-yielding production in only a few areas which had responded to market demands for many centuries (Sigault, 1976: 631–43; see Fig. 10.1). The notion of an 'agricultural revolution' in France between 1750 and 1850 may not be sustained in terms of increased productivity or markedly changed practices. The only case might be made with respect to the evolution of ideas among an educated, but not always active, land-owning élite (Slicher van Bath, 1977: 103).

## The Germanic Lands

By contrast with France, agriculture in several parts of the Germanic lands underwent important change during the eighteenth century, with translations of English publications being read by many landowners (Blum, 1978: 250). The rulers of Prussia adhered to mercantilist principles which required considerable state involvement in economic life. In addition, it was widely believed among the German élite that farming was the root of all wealth. The cultivated surface was extended, with heaths and moors being reclaimed (for example, on Lüneburg Heath) and marshes being drained in east Friesland and along the Elbe and the Oder. Indeed, the policies of Frederick the Great were reputed to have

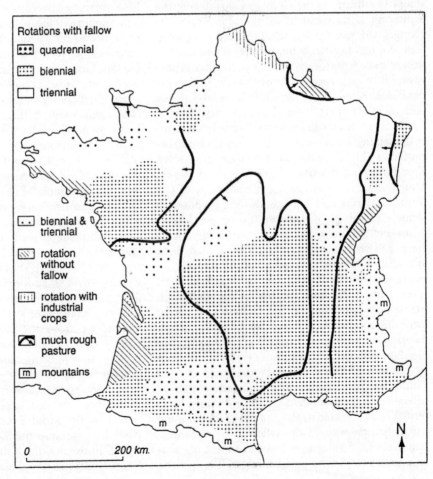

Fig. 10.1. Crop rotations in France c.1820. (Redrawn from Sigaut, 1976.)

Rotations with fallow
- quadrennial
- biennial
- triennial

- biennial & triennial
- rotation without fallow
- rotation with industrial crops
- much rough pasture
- m  mountains

0    200 km.

N

settled 300,000 colonists on reclaimed land (Mayhew, 1973: 121). Farming societies were founded by local aristocrats and their publications included translations of English and Dutch works, as well as local treatises advocating the application of scientific principles to farming in order to enhance wealth (Fussell, 1972: 166). An important example was the Celle agricultural society, which was founded in 1764 with the support of George III of England, who was also Elector of Hanover (Kain and Baigent, 1992: 142). English innovators were invited to visit Hanover and young German farmers were sent to learn from English farms. Publication in 1798 of Albrecht Thaer's *Introduction to the Knowledge of English Agriculture* consolidated agronomic ideas which the author was to teach at the Prussian school of agriculture and the new University of Berlin (Clapham, 1961: 47).

Open-fields with triennial rotations operated across two-thirds of the German lands during the eighteenth century but in many places important changes had been made, including deep ploughing, stall feeding of cattle, better rotations, and cultivation of peas, lentils, turnips, hemp, and flax (Dickinson, 1953: 131). Forward-looking German farmers paid close attention to agricultural book-keeping and in this respect were ahead of their English counterparts (Clapham, 1961: 51). Agrarian systems in remaining areas ranged from primitive infield–outfield to highly complicated rotations in Saxony, East Prussia, and parts of the Rhinelands where Dutch influences were strong (Clapham, 1961: 30). Vegetables, fruit, and dairy goods were produced in densely populated districts of the upper Rhine-Main and around such cities as Berlin and Hamburg (Newman, 1985: 61). Enclosure had been introduced as early as the sixteenth century in Schleswig and south-west Germany (Blum, 1978: 264). During the final decades of the eighteenth century various agricultural changes occurred with differing intensity across the Germanic lands. Political authority was fragmented to the west of the Elbe and, in general terms, enclosure did not accompany liberation of the peasantry from their communal constraints. Open-fields and nucleated villages survived in many areas. By contrast, in Prussia and most territories east of the Elbe peasant emancipation was usually followed by enclosure, removal of old villages, and establishment of modern estates worked by paid labourers. The demand for grain in England and Holland provided a powerful inducement for these structural changes in eastern Germany (Tipton, 1976: 23).

## Southern Scandinavia

By 1800 the imprint of enclosure was widespread in Denmark and southernmost Sweden and was to be followed by profound change in agricultural production after 1850. In the early seventeenth century Denmark had been a land of open-fields and nucleated villages, with large areas of heathland in Jutland (Frandsen, 1988: 117–21). During the 1660s Danish Crown lands were sold to a rising class of landlords who made use of the peasant labour force that was tied to its place of origin (Tracy, 1989: 5). The Royal Danish Agricultural Society (founded in 1769) encouraged innovations and after 1784 Crown Prince Frederick and his reforming government liberated the peasantry and supported the passage of legislation to promote rural improvement (Mead, 1981: 157). Enclosure acts of 1758–60 and 1781 for remodelling landholdings and settlements, and reclaiming heathland, were complemented by laws which made farmers responsible for cultivating their lands and enabled them to obtain low-interest loans to purchase land, construct buildings, and innovate in farming practice (Kampp, 1975: 12). By 1810 most Danish open-fields had been enclosed. The nuclei of pre-enclosure villages remained, with some landholdings radiating from them, but new farmsteads were sited on ring-fenced properties in outer sections of the parishes. In 1814 Denmark made elementary education compulsory and thirty years later the Lutheran Pastor Gruntvig (1783–1872) created the first Folk High School, which provided additional education for people over 18 years of age, mainly from farming families. This combination of structural change, education, and advice from the Royal Danish Agricultural Society would enable Danish farmers to respond rapidly to changing commercial circumstances later in the century (Sivignon, 1992–3: 137).

English experience of enclosure was also echoed in southern Sweden, where innovation was promoted by the Royal Swedish Academy of Science (founded 1739) and the Swedish Economic Society (1767). In the middle of the eighteenth century the Swedish Land Survey Board supported laws of 1749 and 1757 for *storskifte* which consolidated highly fragmented property but did not disperse farmsteads from nucleated villages (Kain and Baigent, 1992: 59). Large landowners, well versed in agricultural literature emanating from England, appreciated the advantages of enclosure for increasing the efficiency of their property (Berglund, 1991: 380). In 1803 legislation was introduced to promote *enskifte*, with property being reassembled into blocks and farmsteads relocated in the midst of new fields, thereby reducing the density of settlement in what remained of the old villages. Subsequent legislation (1827) introduced a less rigorous enclosure system (*laga skifte*) which allowed more than one piece of land per farm (Helmfrid, 1961: 114). The impact of these changes varied considerably across southern Sweden but enclosure enabled farmers to operate without reference to village traditions, and to introduce new rotations and farming practices such as mechanization, under-drainage, and marling (Möller, 1990: 59). In a similar way to their Danish neighbours, many farmers in southern Sweden were ready to change to livestock husbandry after 1850 when European grain markets were challenged by overseas suppliers and animal products commanded preferential prices.

## Changing Spatial Relations

By the dawn of the nineteenth century the transition from a feudal world to one motivated by capitalist imperatives was clearly under way in north-west Europe. Subsistence economies were being replaced by the market economy, with increasing commercialization of production and distribution. Through various brands of enclosure the agricultural landscapes and settlement patterns of some parts of northern Europe had been remodelled. Although traditional landscape features remained in place across much of France and western Germany, communally regulated systems were gradually eliminated and commonlands abolished, thereby allowing flocks, herds, and land to be managed in severalty (Dovring, 1966: 624). Numerous measures turned customary peasants into tenants or freeholders, with farmers now needing to produce surpluses for sale to pay rents and taxes in cash, rather than in kind or by means of labour services.

In the context of continuing population growth, uneven improvements in crop yields, and as yet no major changes in transportation, more land was brought into cultivation, sometimes involving areas which had been farmed previously but had lapsed into waste, as a result of economic decline or war (Dovring, 1966: 618). Thus remaining areas of moor, marsh, and heath were converted into fields in such environments as Jutland, the Ardennes uplands, and Brittany (Olwig, 1984; Clout, 1973–4: 29–60). When sources of food supply were interrupted by war, and cereal prices rose, the margin of cultivation was pushed up-slope into localities that were scarcely suitable for growing grain. But, at the same time, farmers in innovative and market-orientated areas continued to intensify their means of production with the aid of new crops and improved rotations and practices. After a particularly complicated diffusion process and much rejection by country people, whose minds were only changed by sheer hunger, potatoes were cultivated to provide a vital new source of food for humans and livestock alike (Blum, 1978: 271–3). Other innovations included sugar beet, whose cultivation in northern France began during the Napoleonic Wars when supplies of tropical sugar cane were cut off (Clout and Phillips, 1973: 105-19).

By the 1830s the agricultural map of Europe was a complex mosaic reflecting varying degrees of productive intensity and commercial orientation, but also continuity with the past, in terms of both landscape and farming practice (Grantham, 1989: 43). Rural areas close to markets generated high financial yields per unit area which contrasted with low values from isolated and relatively impoverished areas where self-sufficiency remained significant. For example, in 1828 William Jacob noted that fallows planted with fodder crops were 'minute exceptions to the generally established system' over much of France, the Germanic lands, and areas further east (cited in Blum, 1978: 259). Financial data from the great *Statistique* of 1837–40 demonstrated the long-established market orientation of the core of the Paris Basin, from which branches of high-value production extended into Picardy and Flanders and along the Channel coast to northern Brittany (Clout, 1980: 219; see Fig. 10.2.). Two axes of relatively high-value output were shaped by the Rivers Rhône, Rhine, and Moselle and the Atlantic littoral, with intensive production concentrated

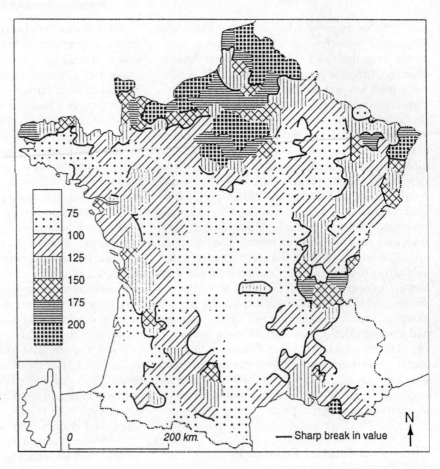

Fig. 10.2. The annual value of agricultural output in France *c.*1840, expressed in francs per hectare. (Redrawn from Clout, 1980, 219.)

Legend:
75
100
125
150
175
200

0    200 km.

— Sharp break in value

N

around large cities. Farmlands in central and southern France were less involved in commercial activities and generated much lower financial yields per unit area. Similar gradations of intensity doubtless existed across other parts of the continent.

Despite agricultural improvements spanning several centuries, the urbanizing core region of northwest Europe required supplementary food supplies to be brought by cart, barge, and sailing ship. Semiperishable goods, such as butter, cheese, and meat, were carried moderate distances, for example from Ireland to London, or from northern Germany to Amsterdam (Peet, 1972: 5). The critical factor was how far and how fast the commodities could be transported before they rotted. Non-perishable goods, grain in particular, were carried westwards from more distant locations around the Baltic and from the valleys of the Elbe, the Vistula, and further east (Maczak, 1972: 676; Kostowicka, 1984: 81; see Fig. 10.3).

These examples responded to a well-established series of spatial constraints which would be overturned by use of the steam engine in transportation by land and by sea. Price (1994: 91) has written of 'a communications revolution creating a world market [which], together with rapid industrialization, urbanization, and the decline of rural population, resulted in a more thorough-going and permanent series of responses' in European farming than ever before. The new railway systems linked increasingly populous urban markets to potential supply areas and opened the way for agricultural specialization, provided that the obsession with self-sufficient polyculture could be dispelled among family farmers. For example, farming activities in parts of south-west England and Normandy were reorientated from mixed husbandry to dairying, with fresh milk being rushed by train to London market and soft cheeses being expedited to eager consumers in Paris (Brunet, 1987: 124). Sections of Languedoc

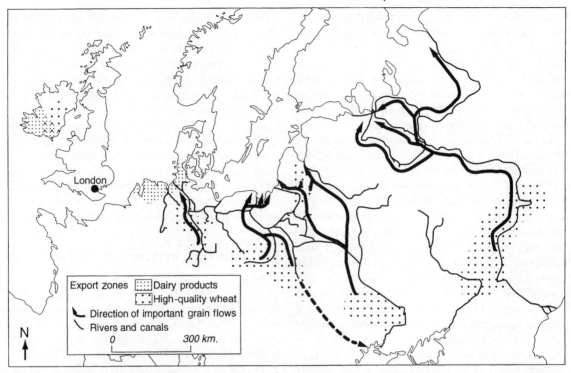

**Fig. 10.3.** Origins of British imports of dairy products and high-quality wheat *c.*1830. (Redrawn from Peet, 1972, 8.)

were converted from arable farming to viticulture in order to produce *vin ordinaire* for the growing mass of consumers in the capital (Clout, 1983: 111–23).

Further afield in eastern Europe large areas had been brought into cultivation during the eighteenth and nineteenth centuries. In the early 1700s much land lay desolate in Hungary after a century and a half of Turkish rule (Blum, 1978: 254–7). This situation changed when the region was restored to Habsburg rule, and settlers from Germany and many surrounding areas streamed into the Great Plain, reclaiming stretches of floodable land and greatly increasing the surface under the plough. After long periods of war, population growth during the eighteenth century encouraged a massive expansion of cereal cultivation in the Danubian lands, Poland, White Russia, Lithuania, and the Ukraine. A similar result was produced in New Russia (the south-western corner of the empire and lands along the Volga) as settlers moved south of the 55th line of latitude to colonize the black earths of the steppe (Pallot and Shaw, 1990). Between 1802 and 1850, the surface planted with cereals increased by half in European Russia,

providing supplies which might in part be directed to north-west Europe. Newly installed railway lines enabled large quantities of grain to be moved from the east European and Russian interiors through ports on the Baltic and Black Sea for shipment to the west. Indeed, at mid-century Russia was the largest grain-exporting country (Tracy, 1989: 15).

By contrast with the railway, which offered the possibility of reshaping regional aspects of European farming, the coming of the steamship offered the potential for change on a much grander scale. During the 1860s territories in the New World began to supply grain to north-west Europe, benefiting from a free-trade interlude which began in Britain with the repeal of the corn laws in 1846 and was emulated in neighbouring countries during the 1850s and 1860s (Tracy, 1989: 17). The impact of cheap overseas grain continued to grow, especially in the late 1870s when North America enjoyed four excellent harvests at a time when yields were miserable in north-west Europe. At the same time steamships laden with frozen meat from the USA, Argentina, and Australia arrived in Europe's ports. These cheap supplies brought down food prices, reduced farm incomes, and offered a powerful sti-mulus for change in west European farming. Large, specialized grain producers in east Prussia, East Anglia, and the Paris Basin were affected most, unlike livestock farmers, whose supplies of meat and milk were in growing demand in urbanizing Europe, or relatively self-sufficient producers who were not greatly involved in commercial activities.

Faced with this 'great depression' governments in Belgium, France, and Germany introduced protec-tive tariffs during the 1880s and 1890s to shield their agricultural sector. By contrast, Britain remained committed to free trade, and the period of Victor-ian high farming, with its large investment in machinery, drainage, fertilization, and livestock-breeding, was followed by a phase of falling prices, collapsing wage levels, rural outmigration to urban workplaces or new opportunities in the Empire, and general neglect of the countryside (Kain and Prince, 1985: 4; Phillips, 1989). Many who remained in farming changed to animal husbandry or market gardening, since demand for meat, milk, and vege-tables continued to grow and prices held up. Dutch farmers also devoted their energies to horticulture and dairying, but the most radical changes were occurring in Denmark, where the farm structure had been reorganized and the farming population educated to adapt to new challenges. Already dur-ing the 1850s Denmark had started to produce dairy goods and meat products for Britain and Germany. The transition was eased by the creation of co-operative dairies and bacon factories, which developed from Danish credit associations and retail co-operatives (Tracy, 1989: 113).

The relationship between town and country remained grounded in food supply, but some farm-ers in north-west Europe were also important con-sumers of urban products. Those with sufficient capital or means of raising loans purchased agri-cultural machinery and 'artificial fertilizers' gener-ated as by-products from industrial processes. Around 1840 Liebig in Germany and Lawes in Eng-land had produced formulas for chemical manures (Dovring, 1966: 654). The production of sugar-beet and other root crops in Germany after 1850 required not only heavy ploughs for deep-working the soil but also abundant supplies of artificial fertilizer to supplement manure from cattle fed on pulpy residues from sugar-making (Tipton, 1976: 62; Perkins, 1981: 84). The earliest artificial fertilizers derived from imported nitrates from Chile but these were superseded by ammonium sulphate and sodium nitrate, which were by-products from gas manufacture and coal-tar distil-lation (Pounds, 1985: 231). In the 1880s German steelworks were able to supply basic slag with a high phosphate content. By 1900 German and Dutch farmers distinguished themselves by heavy applications of artificial fertilizer to their soil, unlike most French farmers who made very little use of it. Ironically, the devastation of the Great War would be most intense in areas of northern France which had adopted something of the German approach to intensive sugar-beet cultivation (Clout, 1994: 138–9).

## Rural Europe on the Eve of the Great War

By 1913 the population of north-west Europe was greater, more urbanized, and less agricultural than ever before. Definitions of rurality are notoriously contentious but it would seem that, despite decades of important city-ward migration, half the popula-tion of Germany and of the Netherlands was living outside cities, and in France and Denmark the proportion exceeded 60 per cent (Pounds, 1985: 505). As the 'first urban nation', England and Wales had only a quarter of its inhabitants living in rural areas. Countryside around Europe's great cities had already become the preferred residence for select groups of commuters travelling by train to urban work each day. The spread of suburbia had begun. In addition, picturesque rural areas were being discovered by urban visitors for recreation and holidays. In 1913 farms and forests still employed two-fifths of the total workforce of Denmark, France, and Germany, with the share for Belgium and the Netherlands standing at one-quarter. In England and Wales it had fallen to 8 per cent.

The volume of agricultural output in Europe was larger than ever, with more people being fed as a result of much smaller inputs of labour than in earlier centuries. Grain production for the whole of Europe (including Russia) was almost three times what it had been c.1815, while the total popu-lation had doubled from 190 million to over 400 million (Pounds, 1985: 118; Dovring, 1966: 604). Despite the impact of the agricultural depression, which had hit Britain particularly severely, much farmland in the innovative rural areas of north-west Europe was better fertilized and drained than before. Patterns of production remained very diverse, with wheat being especially important in northern France, eastern England, central Germany, Hungary, and Romania. By 1913 Europe was importing about one-fifth of the total wheat consumed by its population, but there were great variations among nations as a result of differences in supply and demand and in tariff policies. Thus, France was almost self-sufficient, but neighbouring countries had to import much of the wheat they needed to make white bread (Germany, 20 per cent, Denmark 35 per cent, Belgium 75 per cent). In spite of the growing trend to substitute wheat for secondary cereals, rye cultivation remained very significant in northern Germany and Poland (Collins, 1993: 35). Both potato cultivation and pig-rearing displayed a similar pattern to that of rye, contributing a distinctive combination to the dietary repertoire of modern Europe. More cattle were raised for meat and milk than in earlier times across north-west Europe, but the number of sheep had declined as fallow was eradicated in more advanced rotations, and as moors and heaths were brought into cultivation. Agricultural mechanization and the use of improved implements was extremely uneven, having been favoured on large farms with money

to invest in lowland England, the Paris Basin, and southern Sweden, but being shunned on small, family run holdings (Dovring, 1966: 641–5).

As a result of centuries of agricultural change, combined with innovations in transport and the outworking of recent trade policies, the population of north-west Europe was better fed than before. The dual threats of pestilence and famine identified by Malthus in 1798 had been banished. Plague, smallpox, and cholera had largely disappeared, and severe food shortages were a thing of the past, last occurring in France in 1788, in England in 1794, and disastrously in Ireland in 1845. North-west Europe had been ravaged by the Thirty Years War, the Napoleonic Wars, and countless others, with warfare remaining a continuing threat (Slicher van Bath, 1977: 68). In the summer of 1914 a more devastating but ultimately localized modern war began, which would wreak havoc across highly productive and much-prized farmlands in northern France and Flanders. As many thousands of village war memorials proclaim, its horrific events would distort the population structure of communities throughout rural Europe, since countrymen were sacrificed in vast numbers. When the First World War ended in November 1918 the days of traditional rural Europe were surely over also (Clout, 1996).

REFERENCES

ABEL, W., *Agricultural Fluctuations in Europe, from the Thirteenth to the Twentieth Centuries*, (London, 1980).

BAIROCH, P., 'Agriculture and the Industrial Revolution', in C. M. Cipolla (ed.), *The Fontana Economic History of Europe*, iii. *The Industrial Revolution*, (Glasgow, 1973), 452–506.

BERGLUND, B. E. (ed.), *The Cultural Landscape during 6000 Years in Southern Sweden*, (Copenhagen, 1991).

BERTRAND, G., 'Pour une histoire 'écologique de la France rurale', in G. Duby and A. Wallon (eds.), *Histoire de la France Rurale*, i. *Des origines à 1340* (Paris, 1975), 34–113.

BLOCH, M., *French Rural History*, (London, 1966).

BLOMME, J., *The Economic Development of Belgian Agriculture 1880–1980*, (Leuven, 1993).

BLUM, J., *The End of the Old Order in Rural Europe*, (Princeton, 1978).

BOURDE, A. J., *The Influence of England on the French Agronomes, 1750–1789*, (Cambridge, 1953).

—— *Agromanie et Agronomes en France au XVIIIe siècle*, (3 vols.; Paris, 1967).

BRAUDEL, F., *Civilization and Capitalism*, i. *The Structures of Everyday Life*, (London, 1981).

—— *Civilization and Capitalism*, ii. *The Wheels of Commerce*, (London, 1982).

—— *Civilization and Capitalism*, iii. *The Perspective of the World*, (London, 1984).

—— *The Identity of France*, i. *History and Environment*, (London, 1988).

—— *The Identity of France*, ii. *People and Production*, (London, 1990).

BRUNET, P. (ed.), *Histoire et géographie des fromages*, (Caen, 1987).

BULL, G. B. G., 'Thomas Milne's Land Utilization Map of the London Area in 1800', *Geographical Journal*, 132: 25–30, (1956).

CHISHOLM, M., *Rural Settlement and Land Use: An Essay in Location*, (London, 1962).

CHORLEY, G. P. H., 'The Agricultural Revolution in Northern Europe', *Economic History Review*, 2nd ser. 34: 71–93, (1981).

CLAPHAM, J. H., *The Economic Development of France and Germany, 1815–1914*, 4th edn., (Cambridge, 1961).

CLOUT, H. D., 'Reclamation of Wasteland in Brittany, 1750–1900', *Bulletin de la Société Royale de Géographie d'Anvers*, 84: 29–60, (1973–4).

—— *Agriculture in France on the Eve of the Railway Age*, (London, 1980).

—— *The Land of France, 1815–1914*, (London, 1983).

—— 'Reconstructing the Countryside of the Eastern Somme after the Great War', *Erdkunde*, 48: 136–49, (1994).

—— *After the Ruins: Restoring the Countryside of Northern France after the Great War*, (Exeter, 1996).

—— and PHILLIPS, A. D. M., 'Sugar Beet Production in the Nord Département of France during the Nineteenth Century', *Erdkunde*, 27: 105–19, (1973).

COLLINS, E. J. T., 'Why Wheat? Choice of Food Grains in Europe in the Nineteenth and Twentieth Centuries', *Journal of European Economic History*, 22: 7–38, (1993).

DANIELS, S., and SEYMOUR, S., 'Landscape Design and the Idea of Improvement 1730–1900', in R. A. Dodgshon and R. A. Butlin (eds.), *An Historical Geography of England and Wales*, 2nd edn., (London, 1990), 487–520.

DARBY, H. C., *The Draining of the Fens*, (Cambridge, 1940).

—— 'The Face of Europe on the Eve of the Great Discoveries', in G. R. Potter (ed.), *The New Cambridge Modern History*, i. *The Renaissance 1493–1520*, 20–49, (1957).

—— (ed.), *The New Historical Geography of England after 1600*, (Cambridge, 1976).

—— *The Changing Fenland*, (Cambridge, 1983).

DE MADDALENA, A., 'Rural Europe 1500–1700', in C. M. Cipolla (ed.), *The Fontana Economic History of Europe*, ii. *Sixteenth and Seventeenth Centuries*, (1974).

DE PLANHOL, X., *An Historical Geography of France*, (Cambridge, 1994).

DE VRIES, J., *The Dutch Rural Economy in the Golden Age, 1500–1700*, (New Haven, 1974).

DICKINSON, R. E., *Germany*, (London, 1953).

DION, R., *Histoire de la vigne et du vin en France des origines au XIXe siècle*, (Paris, 1959).

DOVRING, F., 'The Transformation of European Agriculture', in H. J. Habakkuk and M. M. Postan (eds.), *The Cambridge Economic History of Europe*, vi. 2. *The Industrial Revolution and After*, (Cambridge, 1966), 604–72.

DUBY, G., 'The French Countryside at the end of the Thirteenth Century', in R. E. Cameron (ed.), *Essays in French Economic History*, (Homewood, Ill., 1970), 33–41.

—— 'Medieval Agriculture, 900–1500', in C. M. Cipolla (ed.), *The Fontana Economic History of Europe*, i. *The Middle Ages*, (Glasgow, 1972), 175–220.

FISHER, F. J., 'The Development of the London Food Market', *Economic History Review*, (1935), 5: 46–64.

FORSTER, R., 'Obstacles to Agricultural Growth in Eighteenth Century France', *American Historical Review*, (1970), 75: 1600–15.

FRANDSEN, K. E., 'The Field Systems of Southern Scandinavia in the Seventeenth Century', *Geografiska Annaler*, (1988), 70B: 117–21.

FUSSELL, G. E., *The Classical Tradition in West European Farming*, (Newton Abbot, 1972).

GOOSSENS, M., *The Economic Development of Belgian Agriculture: A Regional Perspective 1812–1846*, (Leuven, 1993).

GRANTHAM, G., 'Agricultural Supply during the Industrial Revolution: French Evidence and European Implications', *Journal of Economic History*, (1989), 49: 43–72.

GRAS, N. S. B., *The Evolution of the English Corn Market*, (London, 1915).

GRIGG, D., *Population Growth and Agrarian Change: An Historical Perspective*, (Cambridge, 1980).

HECHTER, M., and BRUSTEIN, W., 'Regional Modes of Production and Patterns of State Formation in Western Europe', *American Journal of Sociology*, (1980), 85: 1061–94.

HELMFRID, S., 'The Storskifte, Enskifte and Laga Skifte in Sweden', *Geografiska Annaler*, (1961), 43: 114–29.

HOHENBERG, P. M., and LEES, L. H., *The Making of Urban Europe, 1000–1950*, (Cambridge, Mass., 1985).

HOUSTON, J. M., *The Western Mediterranean World*, (London, 1964).

JONES, E. L., and WOOLF, S. J. (eds.), *Agrarian Change and Economic Development: The Historical Problems*, (London, 1969).

KAIN, R. J. P., and BAIGENT, E., *The Cadastral Map in the Service of the State: A History of Property Mapping*, (Chicago, 1992).

—— and PRINCE, H. C., *The Tithe Surveys of England and Wales*, (Cambridge, 1985).

KAMPP, A. H., *An Agricultural Geography of Denmark*, (Budapest, 1975).

KAPLAN, S., *Provisioning Paris: Merchants and Millers in the Grain and Flour Trade during the Eighteenth Century*, (Ithaca, NY, 1984).

KOSTROWICKA, I., 'Changes in Agricultural Productivity in the Kingdom of Poland in the Nineteenth and Early Twentieth Centuries', *Journal of European Economic History*, (1984), 13: 75–97.

LAMBERT, A. M., *The Making of the Dutch Landscape*, (London, 1971).

LASLETT, P., *The World We have Lost*, (London, 1965).

LE ROY LADURIE, E., *Paysages, paysans. L'Art et la terre en Europe du Moyen Âge au XXe siècle*, (Paris, 1994).

—— and GOY, J., *Tithe and Agrarian History from the Fourteenth to the Nineteenth Centuries*, (Cambridge, 1982).

MACZAK, A., 'Agricultural and Livestock Production in Poland: Internal and Foreign Markets', *Journal of European Economic History*, (1972) 1: 671–80.

MAYHEW, A., *Rural Settlement and Farming in Germany*, (London, 1973).

MEAD, W. R., *An Historical Geography of Scandinavia*, (London, 1981).

MENDELS, F. F., 'Agriculture and Rural Industry in Eighteenth-Century Flanders', in W. N. Parker and E. L. Jones (eds.), *European Peasants and their Markets*, (Princeton, 1975), 179–204.

MINGAY, G. E. (ed.), *The Agricultural Revolution: Changes in Agriculture 1650–1850*, (London, 1977).

MÖLLER, J., 'Towards Agrarian Capitalism: The Case of Southern Sweden during the Nineteenth Century', *Geografiska Annaler*, (1990), 72B: 59–72.

MORICEAU, J.-M., *Les Fermiers de l'Île-de-France, XVe–XVIIIe siècle*, (Paris, 1994).

NEWMAN, K., 'Hamburg in the European Economy 1660–1750', *Journal of European Economic History*, (1985), 14: 57–93.

OLWIG, K., *Nature's Ideological Landscape: A Literary and Geographic Perspective on its Development and Preservation on Denmark's Jutland Heath*, (London, 1984).

OVERTON, M., 'Agricultural Revolution? Development of the Agrarian Economy in Early Modern Europe', in A. R. H. Baker and D. Gregory (eds.), *Explorations in Historical Geography*, (Cambridge, 1984), 118–39.

—— 'A Critical Century? The Agrarian History of England and Wales, 1750–1850', *Agricultural History Review*, (1990), 38: 185–9.

—— *Agricultural Revolution in England: The Transformation of the Rural Economy, 1500–1850*, (Cambridge, 1996).

—— and CAMPBELL, B. M. S., 'Productivity Change in European Agricultural Development', in B. M. S. Campbell and M. Overton (eds.), *Land, Labour and Livestock*, (Manchester, 1991), 1–50.

PALLOTT, J., and SHAW, D. J. B., *Landscape and Settlement in Romanov Russia 1613–1917*, (Oxford, 1990).

PARKER, R. A. C., 'Coke of Norfolk and the Agrarian Revolution', *Economic History Review*, 2nd ser. (1955), 8: 156–66.

PARKER, W. N., and JONES, E. L. (eds.), *European Peasants and their Markets*, (Princeton, 1975).

PEET, R., 'Influences of the British Market on Agriculture and Related Economic Development in Europe before 1860', *Transactions of the Institute of British Geographers*, (1972), 56: 1–20.

PERKINS, J. A., 'The Agricultural Revolution in Germany, 1850–1914', *Journal of European Economic History*, (1981), 10: 71–118.

PHILIPPONNEAU, M., *La Vie rurale de la banlieue Parisienne*, (Paris, 1956).

PHILLIPS, A. D. M., *The Underdraining of Farmland in England during the Nineteenth Century*, (Cambridge, 1989).

POUNDS, N. J. G., *An Historical Geography of Europe, 1500–1840*, (Cambridge, 1979).

—— *An Historical Geography of Europe, 1800–1914*, (Cambridge, 1985).

—— *An Historical Geography of Europe*, (Cambridge, 1990).

PRICE, R., *The Modernization of Rural France: Communications Networks and Agricultural Market Structures in Nineteenth Century France*, (London, 1983).

—— 'The Transformation of European Agriculture', in D. A. Aldcroft and S. P. Ville (eds.), *The European Economy 1750–1914: A Thematic Approach*, (Manchester, 1994), 72–109.

REID, D., *Paris Sewers and Sewermen*, (Cambridge, Mass., 1991).

RICHARDSON, R. C., 'Metropolitan Counties: Bedfordshire, Hertfordshire and Middlesex', in J. Thirsk (ed.), *The Agrarian History of England and Wales*, v. 1. *1640–1750 Regional Farming Systems*, (Cambridge, 1984), 239–69.

SIGAULT, F., 'Pour une cartographie des assolements en France au début du XIXe siècle', *Annales: Economies, Sociétés, Civilisations*, (1976), 31: 631–43.

SIVIGNON, M., 'La Diffusion des modèles agricoles', *Revue Géographique des Pyrénées et du Sud-Ouest*, (1992–3), 63: 133–54.

SLICHER VAN BATH, B. H., 'The Rise of Intensive Husbandry in the Low Countries', in J. S. Bromley and E. H. Kossman (eds.), *Britain and the Netherlands*, (London, 1960), 130–53.

—— *The Agrarian History of Western Europe, AD 500–1850*, (London, 1963).

—— 'Agriculture in the Vital Revolution', in E. E. Rich and C. H. Wilson (eds.), *The Cambridge Economic History of Europe*, v. *The Economic Organization of Early Modern Europe*, (Cambridge, 1977), 42–132.

SMITH, C. T., *An Historical Geography of Western Europe, 1500–1840*, rev. edn. (London, 1978).

SOBOUL, A., 'The French Rural Community in the Eighteenth and Nineteenth Centuries', *Past and Present*, (1956), 10: 78–95.

SUTTON, K., 'Reclamation of Wasteland during the Eighteenth and Nineteenth Centuries', in H. D. Clout (ed.), *Themes in the Historical Geography of France*, (London, 1977), 247–300.

THICK, M., 'Market Gardening in England and Wales', in J. Thirsk (ed.), *The Agrarian History of England and Wales*, v. 2. *1640–1750 Agrarian Change*, (Cambridge, 1985), 503–32.

TIPTON, F. B., *Regional Variations in the Economic Development of Germany during the Nineteenth Century*, (Middleton, Conn., 1976).

TRACY, M., *Government and Agriculture in Western Europe, 1880–1988*, 3rd edn., (London, 1989).

VAN DER WEE, H., *The Low Countries in the Early Modern World*, (Aldershot, 1993).

VAN DER WOUDE, A., 'The Study of Dutch Rural History', *Journal of European Economic History*, (1975), 4: 215–41.

VAN VEEN, J., *Dredge, Drain, Reclaim: the Art of a Nation*, (The Hague, 1955).

VERHULST, A., *Précis d'histoire rurale de la Belgique*, (Brussels, 1990).

WAGRET, P., *Polderlands*, (London, 1968).

WALLERSTEIN, I., *The Modern World-System i. Capitalist Agriculture and the Origins of the European World Economy in the Sixteenth Century*, (Orlando, Fla., 1974).

WALTON, J. R., 'Agriculture 1730–1900', in R. A. Dodgshon and R. A. Butlin (eds.), *An Historical Geography of England and Wales*, (London, 1978), 239–65.

WEINSTEIN, R., 'London's Market Gardens in the Early Modern Period', in M. Galinou (ed.), *London's Pride: The Glorious History of the Capital's Gardens*, (London, 1990), 80–99.

WRIGLEY, E. A., 'Early Modern Agriculture: A New Harvest Gathered In', *Agricultural History Review*, (1987a), 35: 65–71.

WRIGLEY, E. A., 'Urban Growth and Agricultural Change: England and the Continent in the Early Modern Period', in E. A. Wrigley, *People, Cities and Wealth*, (Oxford, 1987b), 157–93.

YELLING, J. A., 'Agriculture 1500–1730', in R. A. Dodgshon and R. A. Butlin (eds.), *An Historical Geography of England and Wales*, (London, 1978), 151–72.

ZAMAGNI, V., *The Economic History of Italy*, (Oxford, 1993).

# Chapter 11

# Rural Europe since 1500: Areas of Retardation and Tradition

## I. D. Whyte

This chapter considers areas of retardation and tradition in Europe where change occurred belatedly and often incompletely compared with the regions discussed in Chapter 10. Indeed, the belated modernization in many such areas necessitates a different focus from Chapter 10, giving greater consideration to the period after 1914. Although they were disadvantaged in terms of remote locations and harsh, limiting physical environments, defining them is not easy as such criteria were relative rather than absolute (Fig. 11.1). One clear group of regions were those of northern Europe: Iceland, Finland, and most of Norway and Sweden. Another forms the Atlantic periphery of Europe: the Faeroes, the west Highlands and Islands of Scotland, western Ireland, Brittany, and north-western Iberia. Here, climate and topography are often less harsh but maritime conditions generally favour livestock-rearing rather than cereal cultivation. A third category comprises the mountain environments of continental Europe ranging from the Spanish sierras through the Pyrenees and the Alps to the Apennines and the ranges of eastern Europe. Finally, there are the problem regions of the Mediterranean. Some are mountainous such as Corsica, Sardinia, Crete, or most of Greece. Others, including Andalusia, Sicily, and the Mezzogiorno, are areas

of retardation as much due to patterns of land-ownership and social structures as to remoteness.

Between late medieval and modern times European areas of retardation and tradition in Europe have gone through three broad phases. In the first, which lasted until the nineteenth century in many areas, traditional economies and societies continued, though often modified by population growth and the penetration of commercial influences. In the second, spanning the nineteenth and early twentieth centuries, population peaked, to be followed by rapid decline, associated with the opening up of regions to wider political, economic, and social influences. The third phase, since 1945, has been characterized by continued depopulation in many areas and increased government intervention through regional planning policies. The chronology of change has varied between regions: the economies of mountain areas in Albania and of northern Scandinavia continued to expand long after economic collapse and depopulation had transformed many Alpine areas or Mediterranean islands. However, there is sufficient chronological homogeneity to make this three-phase division a useful framework for broad comparison.

Some areas of retardation and tradition were peripheral to the north-west European core at the

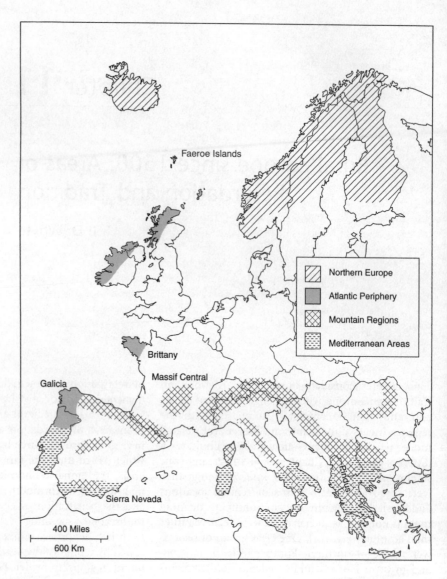

**Fig. 11.1.** Areas of retardation and tradition in Europe.

large scale defined in Wallerstein's world-system (Wallerstein, 1974, 1980). Others were peripheral at a national rather than international scale. By the later sixteenth century, however, such areas were becoming increasingly integrated not merely into a framework of developing nation-states but also into broader economic structures by systems of exchange which transcended local and regional levels of activity. Cores sought to control and exploit peripheral areas as centralized nation-states and national economies developed (Dodgshon, 1987). This process, especially evident with England, France, and Spain, began in medieval times but was

not completed until transport developments in the nineteenth century allowed fuller integration.

## The European Periphery: Pre-industrial Economy and Society c.1500–c.1800

The medieval expansion of population and settlement penetrated many areas remote from core regions of feudalism. Landlords had more freedom to organize relations with their peasants in ways which suited the new, more pastoral environments

(Dodgshon, 1987; Viazzo, 1989). As a result, many areas were only feudalized late, incompletely, and in a different form owing to their lack of potential for demesne exploitation (Dodgshon, 1987). Where serfdom was imposed it often declined early, as in Sweden and Finland (Fullerton and Knowles, 1991; Kellenbenz, 1976). In the Pyrenees, valley communities largely escaped feudalism because of their isolation and only became effectively subject to central authority in the nineteenth century (Gomez-Ibanez, 1975).

Remoteness has been seen as encouraging the continuity of peasant culture over long timespans. In central Finland and south-east Norway it has been suggested that peasant societies changed little from medieval times and only began to alter under the impact of improved communications at the end of the nineteenth century (Millward, 1964; Netting, 1981). On the other hand, studies of communities in such areas by anthropologists have often lacked a convincing time dimension, assuming rather than demonstrating that social structures and farming practices which were evident on the eve of major changes in the late nineteenth or early twentieth centuries had survived with little change from early times. Historical studies like Viazzo's (1989) research on Alpine communities or Dodgshon's (1992, 1993) on west Highland agriculture have revealed hitherto unsuspected changes through time.

The period before the nineteenth century has often been seen as an age of autarky for remote peasant communities but their degree of isolation has frequently been exaggerated. In the Alps, the rise of medieval trade across the major passes involved much local labour and brought the inhabitants into close contact with Europe's great trading centres. In the late fifteenth century the volume of trade dropped and many communities became more closed and self-sufficient, opening up again in the later sixteenth century with economic revival. Just as variations through time were greater than was once believed, recent studies have shown that within particular areas there were also considerable differences between communities (Cole and Wolf, 1974; Netting, 1981). The closed corporate structure of villages like Torbel in the Swiss Valais came close to complete self-sufficiency, with few immigrants, little permanent out-migration, and a carefully balanced, strictly controlled system of exploiting resources (Netting, 1981). Other communities were much more open. Nor were social contrasts between communities always straightforward. In some Alpine valleys literacy levels in the eighteenth and nineteenth centuries were well above those of communities in nearby lowlands, a feature probably linked to emigration, suggesting the possibility that the lowland rural communities were more isolated and autarkic than the mountain ones (Viazzo, 1989).

Nevertheless, relative isolation often perpetuated social structures which were different from those of core areas. It has been widely assumed that while in these areas nuclear families became the norm from medieval times, extended families survived longer in peripheral areas (Flandrin, 1979). The survival of extended rather than nuclear families and the importance of kinship and genealogy was a feature of many areas of retardation, as in the clans of the Scottish Highlands (Dodgshon, 1989) and Corsica (Thompson, 1978) or the *zadruga*, an extended family structure with up to 100 members, in mountain areas of the Balkans (Hammel, 1982; Turnock, 1988). The percentage of extended family households diminished over time in many communities. In Perigord it fell from 36.5 per cent of total households in 1644 to 14.6 per cent in 1836 (Flandrin, 1979). However, the existence of communities with 25–40 per cent of households of the extended type in the nineteenth century did not necessarily mean that such structures had dominated for centuries (Netting, 1981).

Household structure was heavily influenced not only by inheritance customs and environmental constraints but also by systems of pasture management and levels of seasonal migration (Viazzo, 1989). The survival of stem families, where one son remained in his parents' household after marriage and inherited the family holding, was a feature of the Pyrenees, parts of the Massif Central, the Alps, and Corsica in the eighteenth and nineteenth centuries, preventing the subdivision of holdings and reducing pressure on resources. However, the interactions of demography, family structure, and inheritance strategies with cultural and economic variables was highly complex, disproving earlier generalizations that stem family systems were a normal feature of the Alps in contrast to the Mediterranean joint-family system (Burns, 1963). Primogeniture has been seen as an inheritance system better suited to mountain environments and marginal areas because it limited the number of opportunities in communities with restricted resources and controlled population levels by encouraging out-migration, late marriage, and high levels of celibacy. However, Netting's study of Torbel in the Swiss Alps has shown that a system of partible inheritance, combined with high endogamy within the community, could lead to the endless re-combination of inherited shares of land with late marriage, high celibacy, and limited migration which kept population levels in balance with a finite resource base, allowing particular families to survive in the community for centuries (Netting, 1981).

The complexity of the links between family structure, landholding, and inheritance in the Alps shows that it is difficult to generalize concerning patterns of landownership and inheritance over such a wide range of societies. In some areas peasant proprietorship was normal. In Sweden in late medieval times the nobility owned only 8 per cent of the land, in Norway 5 per cent. Even in Iceland, where much land belonged to the Church, the Bonders owned about half the land (Mead, 1981). The owner-occupation of smallholdings, or minifundia, with forests and grazings held in common was a feature of most of the mountain areas of the Mediterranean, contrasting with the large feudal estates, or latifundia, of many lowland areas (McNeill, 1992). In the Alps feudal controls on tenure and inheritance were maintained longer in Austria than in the freer peasant societies of the western Alps. In Portugal there was a marked contrast between the south, where land taken from the Moors was granted out in large estates, and northern areas, where already small peasant holdings were further reduced in size by the effects of partible inheritance. A common feature was that holdings were characteristically small, certainly as far as the amount of cultivated land was concerned, often too small to support a peasant family without an additional source of income from by-employments or seasonal migration. Lack of tight controls over landholding in some marginal upland areas allowed easy colonization and the formation of new holdings, leading to greater population growth. In western Ireland and the Scottish Highlands, small holdings reflected the willingness of estate owners to allow subdivision of tenancies and squatting on areas of waste.

Remoteness from central authority sometimes led to high levels of violence. Bandits and brigands survived in Brittany and parts of the Massif Central into the mid-nineteenth century (Weber, 1977), in the Pindus and the Sierra Nevada into the early twentieth (McNeill, 1992). In Corsica, home of the vendetta, there were c.2,000 murders a year in the late sixteenth century and banditry was only eradicated in the 1930s (Wilson, 1988). Retardation was also reflected in the limited use of money; in societies like the Highlands and western Ireland landowners' incomes were measured in kind and hospitality rather than cash, and society was organized around consumption not capital investment (Graham and Proudfood, 1993). Wealth was measured in livestock in many regions. In remoter villages in the Swiss Valais cash incomes remained the exception into the 1920s and families might go for weeks without spending any money (Netting, 1981).

Peripheral areas were more remote from epidemics and wars but their populations often lived close to the margins of subsistence and were particularly vulnerable to crop failure and famine. Other foods were available when cereals were short. Chestnuts were ground into a flour and used to make bread in Corsica (Perry, 1967), the southern fringes of the Massif Central (Clout, 1979; Le Roy Ladurie, 1976), northern Portugal, northern Italy, and Calabria (Pitte, 1979). Acorns were a major element in the diet of peasants in the Sierra Nevada and the southern Apennines (McNeill, 1992). In Scandinavia the bark of coniferous trees was mixed with straw and reindeer moss to form 'bark bread' (Mead, 1981). Environmental controls on food supply have been seen as major checks on population growth and a stimulus to out-migration. Famine survived into the mid-nineteenth century in many parts of France, Scandinavia, the Scottish Highlands, and, catastrophically, in Ireland. On the other hand, it should not automatically be assumed that because the land was poor the people necessarily were too (Viazzo, 1989).

Population began to rise in many areas during the sixteenth century before levelling out or even falling in the seventeenth, then rising again in the eighteenth and early nineteenth. The limited carrying capacity of many regions meant that population pressure soon became evident. In Sicily rates of population growth between 1500 and 1800 were the highest in Italy owing to a low age of first marriage and a high birth rate. Pressure was eased by migration to towns like Naples and Palermo (Pounds, 1979). Ireland was relatively underpopulated in 1700 with c.2 million inhabitants but by 1841, on the eve of the Famine, this had risen to c.8.5 million, the highest densities, with over 400 people per square mile, being in poorer, marginal areas like Connacht and Connemara, supported by potatoes, intensive subdivision of the existing cultivated area, and an intake of new land (Graham and Proudfoot, 1993).

Braudel (1972) has suggested that fertility in mountain areas like the Alps was high in relation to the available resources, forcing permanent emigration as a safety valve. Many areas do seem to have had high birth rates and where this led to a high rate of natural increase overpopulation and emigration might be expected. On the other hand, if high fertility was accompanied by a high rate of infant and adult mortality, a high-pressure demographic regime with a low rate of natural increase would have resulted. In the Alps, however, there is evidence for a low-pressure system in many areas characterized by a moderate rate of natural increase due to late marriage and high levels of celibacy. In Alagna in the Piedmontese Alps by the end of the sixteenth century the crude birth rate was already low, probably owing to seasonal migration with long periods of separation for married couples, and low levels of nuptiality (Viazzo, 1989). Nuptiality rather than migration may have been an important, even decisive homeostatic mechanism, although seasonal migration itself would have worked to reduce fertility (Netting, 1981; Viazzo, 1989).

Emigration and nuptiality were also related to patterns of inheritance. In theory, impartible inheritance should have been associated with low nuptiality, high rates of out-migration, and limited population growth, partible inheritance with high nuptiality, lower migration rates, and rapid population growth. There is some evidence that such linkages did operate, but interactions between inheritance systems, marriage patterns, and family formations were complex. High levels of celibacy, late age at marriage, and low nuptiality occurred in communities with both partible and impartible inheritance (Cole and Wolf, 1974).

Many regions were characterized by systems of cultivation which were seen by contemporaries from better-endowed areas as backward and primitive. Similar judgements have often been made by modern historical geographers. In Finland, and in areas of Sweden settled by Finns, swidden cultivation was practised in which trees were felled, vegetation burnt, and a crop of rye sown in the ashes. After a second crop the land was used for pasture for a few years and was then allowed to revert to forest. Yields from the first crop could be several times those of conventionally cropped land and as long as population levels were low this was an easy way of obtaining a basic supply of cereals (Millward, 1964). In Corsica the temporary cultivation of land for one or two years before leaving it for ten to twenty years continued in some areas into the 1950s (Houston, 1964). Arable land in Sweden, Scotland, and Ireland, worked on an infield–outfield system, might only produce yields about 3:1 for oats. Such returns, however, were often not much worse than those in prime arable areas where yields were depressed owing to a lack of manure, poor ploughing of the soil, and poor-quality seed (Pounds, 1979).

Although cereals were an important element in diet and, under pressure of population, tended to become more significant (Smout and Gibson, 1989), livestock were significant as sources of food, raw materials, and cash income. Animals marketed on the hoof or their products, long-lasting and valuable in relation to their bulk, helped overcome problems of remoteness. Cattle dominated from the Alps northwards and westwards, sheep in the Mediterranean, but in many areas small-scale mixed livestock farming was normal. Systems of transhumance involving seasonal movements of livestock were common, often operating over considerable distances, providing a fine-tuned response to environmental variations. However, as Cleary (1988) has warned, such systems were more transient and less ubiquitous than has often been believed, less determined by environmental conditions and more influenced by specific cultural, economic, and political contexts. Long-distance transhumance in particular had various preconditions including market demand, political stability, and an élite with sufficient resources to control large herds and flocks (Smith, 1979). In Alpine, Atlantic, and northern areas lowland cattle were moved to the uplands in summer. In the Pyrenees, the southern margins of the Alps, and in the

Apennines inverse transhumance from the mountains to the lowlands in winter to take advantage of the wet season grazing was normal. In Provence both movements occurred together (Smith, 1979).

Despite their isolation few areas of retardation were completely self-sufficient. Between the sixteenth and nineteenth centuries many regions became increasingly drawn into the commercial orbit of Europe's core areas, providing them with the products of pastoral farming and raw materials like timber. In the Massif Central, the Jura, and the Swiss Alps during the sixteenth and seventeenth centuries a subsistence grain-growing economy combined with livestock farming gave way to a concentration on cattle rearing and dairying, supplying surrounding lowlands and their towns (Pounds, 1979). A similar, later trend occurred in the Austrian Alps (Lichtenberger, 1975). For less accessible areas in northern Europe the market was mainly for store cattle. By the end of the seventeenth century livestock were being driven from the western Highlands and islands of Scotland to meet the demands of London (Whyte, 1979a). There was an even greater flow of livestock from southern Scandinavia, eastern Europe, and the Ukraine into north Germany, the Rhineland, and the Low Countries (Pounds, 1979).

Increasing commercialization was reflected by the spread of market centres and towns. In the sixteenth century towns were few and often non-existent in Europe's less advantaged regions (De Vries, 1984). In Norway many small urban centres were established during the seventeenth and eighteenth centuries by landowners promoting the export of primary products. In Sweden and Finland the expansion of urbanization was sponsored by the state (Millward, 1964). Growing commercialization was also reflected in the spread of rural market centres. In seventeenth-century Ireland grants for over 630 markets and 503 fairs were made (O'Flanagan, 1985). These were associated with the growth of export trades, channelling produce to the main port towns, and the increase was especially marked in more remote western areas. A similar trend occurred in the Scottish Highlands to promote the cattle trade (Whyte, 1979b).

A feature of many regions was the availability of new land for occupation, allowing population growth and sometimes encouraging substantial in-migration. Land which had been abandoned owing to population decline in the fourteenth and fifteenth centuries was recolonized, and less attractive environments were occupied for the first time. In Ireland Iar Connacht and Connemara are thought to have been settled only with a mid-seventeenth-century influx of migrants fleeing Cromwellian persecution (Aalen, 1978). In the Alps there were valleys where colonization was limited or non-existent before the later sixteenth century (Lichtenberger, 1975). In the Balkans the extension of estate systems in the seventeenth and eighteenth centuries with formerly free peasants becoming increasingly servile caused migration to mountain areas to escape feudal controls (Turnock, 1988). On the other hand, the climatic deterioration of the Little Ice Age brought colder, wetter conditions to many upland areas between the sixteenth and the nineteenth centuries, forcing cultivation limits downhill, causing glaciers to advance in the Alps and Scandinavia, and leading to settlement abandonment in marginal areas (Grove, 1988). Environmental degradation due to deforestation and the over-exploitation of fragile environments also reduced the potential of some areas, especially in the uplands around the Mediterranean (McNeill, 1992).

There were also frontiers of colonization on a larger scale. In the Pindus mountains it was only in the seventeenth century that scattered nomadic pastoralism gave way to a wave of pioneer agricultural settlement (McNeill, 1992). In Scandinavia there was a northward expansion of settlement from the sixteenth century along the coasts of Norway and Sweden. By the seventeenth century Swedish settlers were pushing inland up the main valleys (Mead, 1981). In Sweden unplanned late-medieval clearances gave way to planned colonization fostered by the state (Millward, 1964). In the eighteenth century the population of Swedish Norrland tripled with an influx of new settlers. In the thickly forested south-west of Norrland the Swedish Crown encouraged the immigration of Finns who, leapfrogging established settlements in the main valleys, penetrated the uncleared forest plateaux between Dalarna and Varmland (Mead, 1981).

Population pressure produced different responses in different regions. Early commentators saw the Scottish West Highlands as a cultural backwater whose agrarian systems preserved many features rooted in the remote past, such as hand cultivation using the spade or cas chrom (foot plough), and the continuing use of hand querns and small horizontal watermills. In fact many of these features were recent developments. Rather than plough cultivation gradually ousting hand tillage, the increasing use of hand cultivation in place of ploughing was a response to population growth allowing the cultivation of marginal land on steep slopes which could never have been ploughed. The system demanded high inputs of labour, but this was available in abundance. As population grew and holding size fell, plough cultivation became less practicable in any case. Increased inputs of manure from dung, turf, thatch, and seaweed were also required. The result was higher yields from small, intensively cultivated plots, an economy which was further sustained by the spread of the potato in the later eighteenth century (Dodgshon, 1992, 1993). The Mediterranean counterpart was the extension of terracing into remote mountain areas, a labour-intensive response only feasible in an economy where excess labour could find no use outside agriculture.

Population pressure was linked with the adoption of two new high-yielding crops, maize and potatoes. Maize was established in northern Portugal and north-west Spain by the end of the sixteenth century but did not reach Greece before the mid-eighteenth century (Pounds, 1979). The widespread introduction of maize into the western Pyrenees in the early eighteenth century provided a greater supply of food for humans and livestock, allowing substantial population growth (Gomez-Ibanez, 1975). The potato had similar results in the central Pyrenees. Potatoes spread into Spain and Italy during the sixteenth century and were introduced into Ireland and western Scotland in the mid-eighteenth century, rapidly establishing themselves as a staple food, allowing further subdivision of land and population growth. In the Alps and Scandinavia the widespread adoption of the potato in many areas in the late eighteenth century reduced the death rate and allowed a more sustained population growth, ending the instability caused by periodic failure of the cereal crop and bringing an end to peaks of crisis mortality (Drake, 1969; Netting, 1981).

Another response to the narrow resource bases of peripheral areas and the limited economic opportunities available for growing populations was migration, both temporary and permanent. In relation to labour requirements many peripheral areas were underpopulated during the summer but overpopulated during the winter, encouraging at least a proportion of the population to engage in non-agricultural activities or to find work elsewhere. Most upland areas from the Cantabrians to the Carpathians exported temporary or permanent labour (Pounds, 1979). Temporary migration to help with lowland harvests allowed families in areas such as western Ireland, the Highlands, or the Massif Central to generate cash income which could buttress a traditional economy or, viewed more negatively, at least allow a proportion of the population to be fed away from their home areas for part of the year, returning home to help with their own, later, harvest (Devine, 1993; Johnson, 1967; Hufton, 1974). In southern France the period of employment was extended after the grain harvest by the grape and olive harvests, woodcutting and the repair of terraces (Hufton, 1974). Such movements increased in importance in many areas as population rose in the eighteenth century. In France in 1810 there were at least 200,000 seasonal migrants, the bulk of them drawn from the Massif Central, the Alps, and the Pyrenees (Hufton, 1974). Not all temporary migrants were agricultural workers. Many had industrial skills or worked as petty traders. Different areas and communities often specialized in particular kinds of work. Limousin provided c.6,000 itinerant masons by the end of the sixteenth century, c.15,000 by the 1760s, most of them working in Paris (Pounds, 1979). In some areas migration was undertaken mainly by women, as with movement from the west Highlands to the lowland harvest or fish-gutting in east-coast Scottish ports (Devine, 1983). In the Massif Central and the Alps it was the men who went, leaving a surplus of females at home (Pounds, 1979).

It has been claimed that emigration was an unwilling response to poverty (Hufton, 1974). The link between poverty and migration has often been assumed rather than proved though. In Alagna, in the Piedmontese Alps, seasonal migration was less a response to overpopulation, indispensable to survival, than a reaction to seasonal underemployment which increased output per head (Viazzo, 1989). In such communities migratory building workers formed an élite group, skilled and well paid.

Large-scale temporary migration did not necessarily make source areas receptive to change, though. Building workers from the Massif Central tended, while in Paris, to keep within their own group, integrating very little with the rest of urban society. Such links may have provided the basis for later permanent out-migration but did not necessarily contribute directly to cultural change (Ogden, 1980).

As an addition, or alternative, to migration industrial work provided a range of by-employments in peripheral areas. Rural industrial production for non-local markets was widespread and, while showing a tendency to locate in areas with a pastoral emphasis, was by no means restricted to peripheral regions. In recent years considerable debate has centred on the concept of proto-industrialization as a way of explaining the distribution and organization of industrial production for non-local markets (Butlin, 1986). Textile production for local use was ubiquitous, but with the development of specialized putting-out systems production became more closely tied to urban markets and more spatially concentrated. Proto-industrialization, such as cotton manufacture in the uplands of Zurich canton, undoubtedly caused demographic expansion in some areas. However, the importance of proto-industrialization in areas like the Alps should not be overestimated. It failed to penetrate the high Alpine valleys and was more a feature of fringe areas like the Zurich uplands and the Vorarlberg, where communal controls were weak compared with lowland communities nearer the towns. The closed corporate structure of many Alpine communities, linked to the regulation of pastures, may have kept proto-industrialization out of the high mountain areas (Viazzo, 1989).

In some districts the mining of iron and non-ferrous metals added another element to the economy. In the Alps mining was long established, but with the development of improved technology there was a boom between the late fifteenth and early seventeenth centuries, especially in the Tyrol, helping to double the population of the region between 1427 and 1600. At a local level waves of immigrants could have marked effects on demographic structures, but these were often short-lived as veins were worked out and the miners moved on (Viazzo, 1989). Areas like the Basque country, the Jura, Savoy, and the eastern Alps possessed high-grade deposits of iron ore which were worked on a commercial scale for distant markets. In Brittany, Ireland, and the West Highlands, on the other hand, remoteness encouraged the smelting of local low-grade ores using primitive bloomeries long after blast furnaces had come into use elsewhere. Such industries survived until they were exposed to competition by improved transport in the nineteenth century.

## Dimensions of Change 1800–1945

During the nineteenth century and the first half of the twentieth the growth of industry and urban populations in core areas produced major changes in the agricultural zones surrounding them (Peet, 1972), impinging increasingly on the societies and economies of peripheral regions as centralized bureaucracies became more powerful. Taxation, military service, and compulsory education spearheaded attacks on language, religion, and culture (Clout, 1987). The spread of education put pressure on languages like Breton and Gaelic (Weber, 1977; Withers, 1984). The more precise demarcation of national boundaries and the greater exercise of central authority in frontier areas began to divide communities in border areas like the Pyrenees, sharpening linguistic differences across frontiers and reorientating inhabitants towards core areas (Gomez-Ibanez, 1975). In Italy and Spain the sale by the state of huge areas of forest and pasture belonging to the Church and local communities led to investment in land by speculators who stripped the forests and impoverished the soils in attempts to gain quick returns on their capital (McNeill, 1992).

A key aspect of change was the spread of transport improvements, providing closer links with core areas and increased penetration of the market into once remote districts. Medieval transport technology continued to serve in some remote areas; in Scandinavia the use of sledges on frozen rivers and lakes continued late into the nineteenth century. In the Apennines and Sierra Nevada roads only reached remoter villages in the 1930s. In many parts of Greece pack-horses or mules remained the principal form of transport into the 1950s (McNeill, 1992).

The development of railways and steamships greatly lessened the friction of distance but road improvements were also vital in opening up many areas. In the later eighteenth and early nineteenth centuries road construction was often promoted by the State for strategic reasons, as in France or Spain. Schemes like these often had little impact on the inhabitants of peripheral areas and carried little traffic (Pollard, 1981). Later road construction reflected economic needs much more closely. In France the construction of networks of secondary roads in the last quarter of the nineteenth century opened up many areas. The process was often initiated by the arrival of the railway, allowing the import of road-building materials (Weber, 1977).

The canal era passed by or barely touched most areas of retardation. Projects like the Caledonian Canal proved to be far more expensive than original estimates and, founded on hopelessly optimistic forecasts, never generated the expected traffic (Youngson, 1973). Coastal shipping had long been vital to areas like Norway, western Scotland, and Brittany, but their rugged coasts made transport slow and often dangerous before the development of steam power. Regular steamship links increased the distance over which semi-perishable goods could be transported to urban centres. Steamships, combined with feeder railways, put Breton dairy produce within reach of London markets in the mid-nineteenth century (Peet, 1972). During the same period steamships improved links down the western coast of Norway while icebreakers lengthened the shipping season in the Gulf of Bothnia from the 1890s (Mead, 1981).

In regions like the Balkans railways were built ahead of their time, out of harmony with the stage of development of local economies (Pollard, 1981). In France, however, the railways were key agents in uniting isolated local economies into a national one, but much of their impact came very late in the nineteenth century (Clout, 1979). Railways began to penetrate Brittany, the Massif Central, and Alpine valleys in the 1860s and 1870s. The railway reached Brest in 1865, but it was the construction of branch lines in the 1880s and narrow-gauge lines in the 1890s which really opened up Brittany, allowing lime to be brought in for agricultural improvement, and encouraging reclamation on a large scale. The coming of the railways led to the rapid demise of livestock-droving. In Languedoc the transhumance of sheep declined as the railways encouraged a concentration on wine production in the lowlands and a shift to cattle production for the Paris market in the Cevennes (Cleary, 1988). In the early twentieth century railways were important in breaking down isolation in Scandinavia. The Bergen to Oslo line was not completed until 1909 and the Oslo to Trondheim line, across the Douvre massif, until the 1920s. The coming of the railways to northern Scandinavia allowed the import of food and led to the abandonment of the swidden system of cultivation (Millward, 1964).

Population continued to rise in many regions despite increasing permanent out-migration. In most areas a peak was reached sometime in the nineteenth century. The expansion of cultivation associated with population growth was achieved with techniques which were essentially those of the pre-industrial era, supported by immense investments in labour. In Brittany population growth was accommodated by the cultivation of potatoes and a major phase of moorland reclamation. By 1836 Brittany had one of the highest rural population densities in France (Clout, 1979). Improved transport encouraged further commercialization in agriculture, with the decline of buckwheat, essentially a subsistence crop, and an expansion of wheat as soils were more intensively fertilized with seaweed, shells, and imported lime. Throughout the Mediterranean population growth in mountain areas led to deforestation and the overcultivation of fragile soils for short-term survival, leading to massive soil erosion and environmental deterioration (McNeill, 1992).

The peak of population was followed by rapid decline in many areas due to out-migration facilitated by faster, cheaper transport, increased literacy, and growing command of national languages among the young people of peripheral areas. Temporary migration to lowland areas tended to be replaced by permanent movement to the towns or overseas. In the Alps greater political barriers to movement after the First World War, followed by the depression of the 1930s, killed seasonal migration abruptly in many areas, causing a period of hardship which led to a more self-sufficient economy than had existed in earlier times (Viazzo, 1989). The process termed 'Hohenflucht' had, by the 1930s, led to the abandonment of a third of the farms in some Alpine areas (Dovring, 1966), a trend paralleled in the Apennines. In Switzerland concern over depopulation led to an official inquiry in the 1920s.

Closer market integration threatened the livelihoods of the inhabitants of peripheral areas with no comparative advantage in anything but labour power. The impact of cheap American grain and meat from Argentina began to affect remote upland areas as well as lowlands by the end of the nineteenth century. Improved transport caused the demise of many local industries. In France traditional artisan industries declined rapidly as they were opened up to competition from mass-produced manufactures. Nail-making by hand survived in the French Alps until *c.*1900 and in the southern Massif Central until the First World War, but by that time a wide range of local crafts formerly protected by isolation had declined, leading to heavy permanent out-migration and population decline (Weber, 1977). In the early nineteenth century in the southern Apennines and the Pindus the collapse of local textile production due to the import of British cotton goods encouraged a switch to silk production, a labour-intensive activity well suited to such areas. However, the import of cheap Japanese silk in the later nineteenth century killed off this industry too (McNeill, 1992).

In the Pyrenees population growth peaked in the first half of the nineteenth century, linked with the uphill expansion of cultivation in mountain valleys and the permanent colonization of shielings. The construction of all-weather roads encouraged farmers on the French side to move from largely subsistence production to commercial livestock farming. The slower pace of road improvement on the Spanish side delayed comparable changes until the mid-twentieth century in many areas (Gomez-Ibanez, 1975). Between 1856 and 1956 the population of the French Pyrenees fell by 56 per cent as out-migration increased. Population decline and the abandonment of arable land encouraged the remaining farmers to move from unspecialized polyculture to commercial livestock production. The cultivation of wheat declined as it became possible to import it, and commercial dairying and cheese production became the mainstay of the economy, especially on the French side, where milk was being supplied to Bayonne and Pau by the late nineteenth century and ewes' milk was used to increase the production of Roquefort cheese.

In Ireland rapid population growth in the early nineteenth century was terminated by the famines of the 1840s and massive emigration to Britain and America. Between the Famine and the First World War the population of Ireland fell by half owing to emigration and demographic factors such as a rise in the age of first marriage and an increase in celibacy. Nevertheless, many western areas continued to hold high populations in relation to the quality of the land. Seasonal harvest migration from Ireland to Britain peaked in the 1860s and then fell steeply. A disproportionately high percentage of migrants in the later nineteenth century came from poorer western areas, temporary migration allowing an archaic pattern of landholding and living to continue for more than a generation after the Famine (Graham and Proudfoot, 1993).

In Corsica, population increased from *c.*110,000 in 1798 to 280,000 in 1880, partly owing to political stability but with no change in basic agricultural systems. The last two decades of the nineteenth century saw a collapse of the traditional economy in the face of imports of cheap food and manufactures due to better communications with the mainland (Thompson, 1978). The cultivated area fell from *c.*200,000 ha. in 1890 to 12,000 by 1960 (Houston, 1964) and population had dropped to nearly half its peak level by 1955. Improved education stimulated out-migration, training many Corsicans beyond the level of opportunities available at home. Many left to pursue careers in the French civil and colonial services, the army, and the police. Within the island there was a large-scale shift of population from the interior mountains to the coast, from the countryside to the towns.

In some areas the shift from small-scale mixed peasant farming to large-scale commercial livestock production was dramatic. A major transformation occurred in the Scottish Highlands where commercial sheep farming, introduced into southern areas in the 1760s, spread to the far north and west in the early years of the nineteenth century. As sheep farming could not be combined with traditional small-scale mixed farming, large numbers of people were cleared from inland glens and settled in new townships on the coast, often on previously waste land. Although most proprietors did not aim initially to remove people from their estate short-distance migration often led to longer-distance moves to the industrial towns of the Lowlands, or to America (Turnock, 1970).

The problems of traditional small-scale agriculture led to official intervention and early attempts at land reform in Ireland and the Scottish Highlands. In Ireland the land reforms of 1881–1900 gave over 328,000 tenants a fair rent and encouraged occupiers to buy their lands with state aid. At a more local scale in western Ireland estate reorganization led to a reduction in the number of tenancies, an increase in average holding size, and the reworking of the landscape into a more regular pattern of enclosures. In areas like West Connacht the traditional landscape of clachans and rundale continued into the late nineteenth century until abolished by the Congested Districts Board, set up in 1891 as a pioneer attempt at comprehensive rural planning (Aalen, 1978). In Scotland the Napier Commission of 1883 also led to fairer rents for crofters but not to peasant proprietorship or larger holdings. The crofting landscape was fossilized by this legislation, preserving an anachronistic labour-intensive system of agriculture lacking economic viability (Turnock, 1970).

In northern Scandinavia during the second half of the nineteenth century a drop in the death rate and the persistence of high birth rates led to a steady growth of population. The population of some northern areas of Sweden more than doubled between 1850 and 1900 (Mead, 1981). Colonization of the northlands continued through the later nineteenth century and was intensified after 1920 by the imposition of immigration quotas by the United States and the growing threat of the USSR. In Norway between 1920 and 1970, 18,000 new farms were created by the State and many more by private initiatives. In Sweden planned colonization continued after the First World War with subsidies to retain existing settlers. Population peaked in South Norrland as late as c.1920 and in North Norrland around 1940.

The economies of some areas were diversified by the development of industries using local raw materials and power sources. In Scandinavia the development of steam-powered sawmills from the 1840s heralded the start of a phase of peripheral industrialization. At the end of the nineteenth century the development of hydroelectric power encouraged a switch from timber to pulp and paper manufacture. Due to the impossibility of transferring electricity over long distances, hydroelectric developments attracted electro-chemical industries like aluminium smelting. Hydroelectric power also made an important contribution to development in

Alpine areas, though in the Pyrenees the associated electro-chemical and metallurgical industries tended to be located in adjacent lowlands (Gomez-Ibanez, 1975).

Tourism also contributed to diversifying the economies of some peripheral areas. In the Alps the first phase of tourism in the mid-nineteenth century, associated with the growing popularity of mountaineering, had only a limited impact. The coming of the railways in the last quarter of the nineteenth century led to the development of exclusive resorts and spas such as Badgastein, Bad Ischl, and St Moritz, in the Alps, Le Mont Dore in the Massif Central, and Bagnères de Luchon in the Pyrenees. Although tourism was highly localized, it produced spectacular changes in some communities. Davos developed as a health and ski resort, its population increasing from c.1,700 in 1860 to 8,000 by 1900 and 11,164 by 1930.

Andalusia and southern Italy preserved feudal regimes with a harsh form of labour exploitation which allowed landowners to offset the cost of shipping grain to northern towns, tying them into a wholly agrarian economy with a colonial-style dependence on the north (Dodgshon, 1987). Agriculture was dominated by large estates or latifundia whose absentee owners spent their incomes outside the region. The rural population enjoyed only a precarious access to land as sharecroppers and insecure tenants. The abundance of cheap labour led to poor techniques, low productivity, high unemployment, and little investment in improvement (Mountjoy, 1973). Population pressure and competition for land pushed rents to high levels. By the 1930s population pressure was continuing to grow, real incomes were falling, and living standards actually declining (Clout, 1987). In Italy per capita incomes of the poorest southern areas were a fifth of those in Milan. Conditions were similar in Andalusia with a population of landless labourers crammed into large villages (Naylon, 1975).

## Regional Development and Change 1945–1995

Since 1945 out-migration from peripheral areas of Europe has continued. At a macro scale migration within what is now the European Union has been dominated at various times by regions of Portugal,

Spain, Italy, Greece, Yugoslavia, and, in the west, Ireland. Northern Sweden and Norway have also begun to lose population. Some migration has been traditional—the migration of harvest workers from Galicia, the southern Apennines, and Thrace—but increasingly movement has been to urban areas and to jobs in industry and the service sector. More recently the net outflow of population has reversed in areas such as Brittany and the Massif Central, owing partly to counter-urbanization (Clout, 1994).

Depopulation has highlighted the continuation of structural problems. The economies of peripheral regions have remained dependent on outside markets, lacking in flexibility, with high proportions of their workforces in primary activities, and vulnerable to the fluctuations of external markets. A consistent trend has been a steady fall in the proportion of the workforce engaged in agriculture as younger people, reluctant to enter farming, have taken up jobs in other sectors or migrated. With the abandonment of high-lying marginal land the upper limit of permanent settlement in parts of the Alps has fallen by 300 m. (Lichtenberger, 1978). The scale of depopulation has been influenced by varying degrees of official concern over the maintenance of agriculture in less well-endowed areas. In parts of the French and Italian Alps the post-war decline of mountain farming has been sharper than in Switzerland and Austria, where governments have tried to support agriculture. In the Balkans losses of young men during the Second World War and the flight of refugees to the towns started the process of depopulation which has continued in Greece and Yugoslavia during the post-war period (Thomas and Vojvoda, 1973).

The sharp decline in rural population in many regions has encouraged structural reform and modernization in agriculture. Recent changes have been especially notable in Spain, Portugal, and Italy. Traditional technology survived late in the eastern Mediterranean. In Greece wooden ploughs were still more numerous than iron ones in 1939 and in Yugoslavia they were still in widespread use in the 1950s (Dovring, 1966). In such areas mean farm size remains low. The EU average is 16 ha. but for Greece the figure is 4.5 ha. and under 5 ha. for most of the Mezzogiorno (Clout, 1987).

The high degree of fragmentation of holdings has increasingly been seen as a barrier to efficient modern farming. Fragmentation has a rational basis in areas like the Alps and the Mediterranean where variations in physical conditions over short distances can be considerable. In traditional Alpine farming the scattering of plots at different altitudes helped to spread out the workload, allowing an optimal use of household labour (Viazzo, 1989). Consolidation of fragmented holdings has made significant progress in some areas. In Brittany the scale of removal of the hedge and talus boundaries of the traditional bocage landscape has given rise to concern about ecological damage. In Spain consolidation programmes began in the 1950s with a peak in the early 1970s when 350,000–400,000 ha. were being consolidated each year (Clout, 1987). There has, however, been a concentration of activity in irrigated areas rather than regions of dry farming. In Portugal, Italy, and Greece little has yet been done about fragmentation. The enlargement of holdings, vital for modernization, has also proceeded slowly.

Land reform has been a feature of some regions, with the division of great estates and the creation of large numbers of new owner-occupied peasant holdings. In Greece since the early 1950s over 205,000 ha. of cultivated land has been taken from large estates. In Italy 680,000 ha. of land was expropriated to provide 113,000 new farms of between 2 and 10 ha. (Clout, 1987). Land reform worked best where it was associated with agricultural improvements, as in the area around the Gulf of Taranto, but it failed in Sicily, where tradition—and the Mafia—proved stronger. In Portugal land reform was concentrated on the large estates of the south with over 1 million ha. being expropriated and turned into large collective farms.

In the 1960s more wide-ranging attempts at rural planning were made in which agricultural improvement was combined with efforts to diversify employment by encouraging tourism, forestry, and industry as well as improving infrastructure, particularly transport. The Mezzogiorno provides a good example of how centralized regional planning policies were applied to peripheral areas. In the immediate post-war period the problems of this region were defined in terms of the backwardness of agriculture and the need to improve conditions on the land. Later, however, measures were extended to the development of industry and tourism. From 1957 the Cassa per il Mezzogiorno provided incentives for the creation of industrial growth poles. Owing to local rivalries, too many sites were developed. There was a concentration on investment by the public rather than the private sector with a limited range of industries. The cost of providing each new industrial job was disproportionally high and most specialist workers moved in from the north. Major investments were made in transport infrastructure, some of it, such as motorways in Sicily, unnecessarily grand. Attempts to encourage tourism were even less successful. By 1980 over 300,000 new jobs had been created, mostly in large capital-intensive units in the chemical, mechanical, and metallurgical industries, a neo-colonial style of development which increased inequalities within the region. Economic recession in the 1970s brought expansion to a halt; some firms closed while others continued to operate at reduced capacity. The net effect was that the south of Italy caught up only a little with the north in terms of living standards and per capita GDP (Clout, 1987).

While many traditional areas have struggled to come to terms with post-war economic conditions, some have moved decisively towards commercial production based on initiatives generated from within rather than being imposed externally. In Brittany farmers have been receptive to market opportunities and government incentives to modernize. In the 1960s and 1970s Breton agriculture changed dramatically from traditional polyculture to intensive livestock production for meat and dairy produce, along with horticulture, developing co-operatives to improve quality control and marketing and backing improvements in infrastructure like the development of port facilities at Roscoff (Ardagh, 1987).

Mass tourism has provided a valuable source of income in many areas such as Brittany, the Alps and, on a grand scale, the coasts of the Mediterranean. Tourism has helped stabilize and even increase population in many Alpine valleys. In Austria, largely owing to tourism, population in the Vorarlberg, Tyrol, and Salzburg Alps increased by between 10 per cent and 20 per cent in 1961–71 while most lowland rural areas experienced slower rates of growth or even decline. The economies of many Alpine communities have changed from predominantly agricultural to near-total dependence on tourism and there is an increasing feeling that the limits of sensible development have been reached or even exceeded in many areas (Kariel and Kariel, 1982).

Tourism has been a mixed blessing, resuscitating declining craft industries but changing the physical appearance of many communities and their demographic and social structures, as well as causing major environmental damage. Its local impact has varied with different planning approaches. In the French Alps the highly centralized planning system has resulted in the creation of large purpose-built resorts, out of sympathy with the landscape and out of touch with local inhabitants. In Austria by contrast there has been a much greater emphasis on smaller-scale local initiatives which have spread the benefits of tourism more widely (Baker, 1982).

In some peripheral areas depopulation and agricultural decline has been a recent phenomenon. Northern Scandinavia remained a pioneering region into the 1950s with colonization still being encouraged, especially in Finland where, after the Second World War, 420,000 refugees from Karelia had to be accommodated (Mead, 1981; Millward, 1964). The period from the mid-1950s to the early 1970s formed a watershed. Long-established government initiatives to create new rural settlement stopped and were reversed. The frontiers of colonization halted and began to retreat followed by heavy out-migration or, more locally, a move from interior rural settlements to coastal industrial ones. Since 1950 over half the arable area has been lost from the interior of northern Sweden as crop yields on the better land have doubled owing to better varieties, chemical fertilizers, and agricultural improvements like drainage.

Regional policy in the 'golden age' of European regional planning from the 1950s to the early 1970s was based on the expectation of continued economic growth and often reflected a top–down, technocratic approach rather than one working from the bottom upwards (Clout, 1987). The economies of peripheral areas nevertheless remained vulnerable, with narrow economic bases, tending to specialize in activities with high labour costs. Down to the mid-1970s regional policies in many areas seemed to be achieving some success, with growing convergence between core and peripheral areas in terms of incomes and living standards. The oil price crisis of the mid-1970s and subsequent economic recession brought to an end this phase of large-scale integrated planning, with less encouragement for peripheries as core areas tried to cope with their own structural problems.

The 1980s saw the number of areas designated for assistance being drastically trimmed. Inequalities increased again, with cuts in regional development spending, while growing unemployment in major urban and industrial areas diminished the inflow of migrants. The pattern of regional development within Europe still reflects variations in resource endowment, regional economies, and the ability of regions to adapt to structural change. The less-developed areas still suffer from low productivity and low incomes associated with production for local markets and the limited value of exports of goods and services. Dunford and Perrons (1994) have identified within the EU an advanced core flanked by a peripheral western Atlantic arc and a southern Mediterranean zone of underdevelopment. Despite over four decades of efforts to reduce disparities there are still sharp variations in average incomes and living standards. The less-developed areas of Europe continue to be characterized by low productivity, low incomes, production for local markets, and a limited range of exports. The areas which are seen as backward and traditional today are, in most cases, the ones which have exhibited such traits throughout the period we have considered. Greater market integration within the European Union may hinder rather than aid the development of these less-advantaged regions in the future (Cole and Cole, 1993).

REFERENCES

AALEN, F. H. A., *Man and the Landscape in Ireland*, (London, 1978).

ARDAGH, J., *France Today*, (London, 1987).

BAKER, M. L., 'Traditional Landscapes and Mass Tourism in the Alps', *Geographical Review*, 72 (1982), 395–415.

BRAUDEL, F., *The Mediterranean and the Mediterranean World in the Age of Philip II*, (2 vols.; London, 1972).

BURNS, R. K., 'The Circum-Alpine Area: A Preliminary View', *Anthropological Quarterly*, 36 (1963), 130–55.

BUTLIN, R. A., 'Early Industrialization in Europe: Concepts and Problems', *Geographical Journal*, 152 (1986), 1–8.

CLEARY, M., 'Transhumance in the Mediterranean World: The Case of Languedoc 1900–40', *Journal of Historical Geography*, 14 (1988), 37–49.

CLOUT, H. D., *The Historical Geography of France on the Eve of the Railways*, (London, 1979).

—— (ed.), *Regional Development in Western Europe*, 3rd edn. (London, 1987).

—— BLACKSELL, M., KING, R., and PINDER, D., *Western Europe: Geographical Perspectives*, 3rd edn. (London, 1994).

COLE, J. P., and COLE, F., *The Geography of the European Community*, (London, 1993).

COLE, J. W., and WOLF, E., *The Hidden Frontier: Ecology and Ethnicity in an Alpine Valley*, (London, 1974).

DEVINE, T. M., 'Highland Migration to Lowland Scotland 1760–1860', *Scottish Historical Review*, 72 (1983), 137–49.

DE VRIES, J., *European Urbanization 1500–1800*, (London, 1984).

DODGSHON, R. A., *The European Past: Social Evolution and Spatial Order*, (London, 1987).

—— 'Pretense of Blude' and 'Place of Thair Dwelling': The Nature of Highland Clans, 1500–1745', in R. A. Houston and I. D. Whyte (eds.), *Scottish Society 1500–1800* (Cambridge, 1989), 169–98.

—— 'Farming Practice in the Western Highlands and Islands before Crofting:

A Study in Cultural Inertia or Opportunity Costs?', *Rural History*, 3 (1992), 173–89.

—— 'Strategies of Farming in the Western Highlands and Islands of Scotland prior to Crofting and the Clearances', *Economic History Review*, 46 (1993), 679–701.

DOVRING, F., The Transformation of European Agriculture, in H. J. Habakkuk and M. Postan (eds.), *The Cambridge Economic History of Europe*, vi. *The Industrial Revolution and After: Incomes, Population and Technological Change*, (Cambridge, 1966).

DRAKE, M., *Population and Society in Norway 1750–1835*, (Cambridge, 1969).

DUNFORD, M., and PERRONS, D., 'Regional Inequality, Regimes of Accumulation and Economic Development in Contemporary Europe', *Transactions of the Institute of British Geographers*, 19 (1994), 163–82.

DUPÂQUIER, J., and JADIN, L., 'Structure of Household and Family in Corsica 1769–71', in P. Laslett and R. Wall (eds.), *Household and Family in Past Time* (Cambridge, 1982) 283–97.

FLANDRIN, J.-L., *Families in Former Times*, (London, 1979).

FULLERTON, B., and KNOWLES, R., *Scandinavia* (London, 1991).

GOMEZ-IBANEZ, D., *The Western Pyrenees* (Oxford, 1975).

GRAHAM, B. J., and PROUDFOOT, L. (eds.), *An Historical Geography of Ireland*, (London, 1993).

GROVE, J. M., *The Little Ice Age*, (London, 1988).

HAMMEL, E. A., 'The Zadruga as Process', in P. Laslett and R. Wall (eds.), *Household and Family in Past Time* (Cambridge, 1982), 335–73.

HOUSTON, J. M., *The West Mediterranean World*, (London, 1964).

HUFTON, O., *The Poor in Eighteenth-Century France 1750–1789*, (Oxford, 1974).

JOHNSON, J. H., 'Harvest Migration from Nineteenth-Century Ireland', *Transactions of the Institute of British Geographers*, 41 (1967), 97–112.

KARIEL, H. G., and KARIEL, P. E., 'Socio-cultural Impacts of Tourism: An Example from the Austrian Alps', *Geografiska Annaler*, B64 (1982), 1–16.

KELLENBENZ, H., *The Rise of the European Economy* (London, 1976).

LE ROY LADURIE, E., *The Peasants of Languedoc*, (London, 1976).

LICHTENBERGER, E., *The Eastern Alps*, (Oxford, 1975).

—— 'The Crisis of Rural Settlement and Farming in the Higher Mountain Regions of Central Europe', *Geographica Polonica*, 38 (1978), 181–7

MCNEILL, J. R., *The Mountains of the Mediterranean: An Environmental History* (Cambridge, 1992).

MEAD, W. R., *The Scandinavian Northlands*, (London, 1974).

—— *An Historical Geography of Scandinavia*, (London, 1981).

MILLWARD, R., *Scandinavian Lands*, (London, 1964).

MOUNTJOY, A. B., *The Mezzogiorno*, (Oxford, 1973).

NAYLON, J., *Andalusia*, (Oxford, 1975).

NETTING, R. M., *Balancing on an Alp: Ecological Change and Continuity in a Swiss Mountain Community*, (Cambridge, 1981).

O'FLANAGAN, P., 'Markets and Fairs in Ireland 1600–1800: Index of Economic Development and Regional Growth', *Journal of Historical Geography*, 11 (1985), 364–78.

OGDEN, P. E., 'Migration, Marriage and the Collapse of Traditional Peasant Society in France', in P. White and R. Woods (eds.), *The Geographical Impact of Migration*, (London, 1980), 152–79.

PARRY, M. L., *Climatic Change, Agriculture and Settlement*, (Folkestone, 1977).

PEET, R., 'Influences of the British Market on Agriculture and Related Economic Development in Europe before 1860', *Transactions of the Institute of British Geographers*, 56 (1972), 1–20.

PERRY, P. J., 'Economy, Landscape and Society in La Castagniccia (Corsica) since the Late Eighteenth Century', *Transactions of the Institute of British Geographers*, 41 (1967), 204–22.

PITTE, J. R., 'L'Homme et le châtaignier en Europe', in P. Flatres (ed.), *Paysages ruraux Européens* (Rennes, 1979), 513–26.

POLLARD, S., *Peaceful Conquest: The Industrialization of Europe 1760–1970* (Oxford, 1981).

POUNDS, N., *An Historical Geography of Europe before 1800*, (London, 1979).

SMITH, C. D., *Western Mediterranean Europe*, (London, 1979).

SMOUT, T. C., and GIBSON, A., 'Scottish Food and Scottish History 1500–1800', in R. A. Houston and I. D. Whyte (eds.), *Scottish Society 1500–1800* (Cambridge, 1989), 59–84.

THOMAS, C., and VOJVODA, M., 'Alpine Communities in Transition: Bohinj, Yugoslavia', *Geography*, 58 (1973), 217–26.

THOMPSON, I. B., *Corsica*, (Newton Abbot, 1971).

—— 'Settlement and Conflict in Corsica', *Transactions of the Institute of British Geographers*, 3 (1978), 259–73.

TURNOCK, D., *Patterns of Highland Development*, (London, 1970).

—— *Scotland's Highlands and Islands*, (London, 1984).

—— *The Making of Eastern Europe: From the Earliest Times to 1815*, (London, 1988).

VIAZZO, P., *Upland Communities: Environment, Population and Social Change in the Alps since the Sixteenth Century* (Cambridge, 1989).

WALLERSTEIN, I., *The Modern World-System*, i. *Capitalist Agriculture and the Origins of the European World-Economy in the Sixteenth Century*, (London, 1974).

—— *The Modern World-System*, ii. *Mercantilism and the Consolidation of the European World-Economy 1600–1750*, (London, 1980).

WEBER, E., *Peasants into Frenchmen*, (London, 1977).

WHYTE, I. D., *Agriculture and Society in Seventeenth-Century Scotland*, (Edinburgh, 1979a).

—— 'The Growth of Periodic Market Centres in Scotland 1600–1707', *Scottish Geographical Magazine*, 95 (1979b), 13–26.

WILSON, S., *Feuding, Conflict and Banditry in Nineteenth-Century Corsica*, (Cambridge, 1988).

WITHERS, C. W. J., *Gaelic in Scotland 1698–1981: The Geographical History of a Language*, (Edinburgh, 1984).

YOUNGSON, A. J., *After the Forty Five*, (Edinburgh, 1973).

# Chapter 12

# Industrial Change, 1500–1740, and the Problem of Proto-industrialization

P. Glennie

## Introduction

Viewed from the twentieth-century's highly specialized industrial regions, industry in early modern Europe appears a poor affair. While some specialized textile or mining districts existed, they were exceptional amidst a proliferation of localities within which diverse arrangements produced non-agricultural goods for local consumption. The rarity of large regional concentrations of industrial producers reflected several factors. Relatively sparse populations and modest living standards meant that markets for the goods produced in any area were comparatively small, except in the few very large cities such as London and Paris. An area's low production costs could easily be swamped by the high transport and transaction costs that accounted for much of goods' prices to consumers. Transaction costs therefore protected local industries, whilst simultaneously limiting their scale. Political fragmentation and the reluctance of governments to become reliant on potential enemies for key goods also supported many local specialisms, via embargoes, tariffs, or other protective restrictions on trade. Numerous factors thus contributed to the very widespread distribution of many basic craft and industrial activities.

Early modern European industries were incorrigibly diverse, arising in environmental and societal settings that varied considerably. Different activities varied substantially in the precise nature of industrial processes; in prevailing technologies (including their use of non-human power); and in forms of organization. Local examples of any one activity were diverse in their scale, organization, and finance; diverse in their relationship to political authority; and diverse in the social and cultural institutions within which individual activity occurred. In accounting for this diversity, and for the later emergence from within it of 'industrial revolutions', historical geographers (and others) have often analysed industrial change through the locational characteristics of 'successful' industrial sectors, and their shifting spatial pattern over time.

National units long provided the staple units for comparison. For much of the twentieth century, debates on industrial development focused on the divergent experiences of England and France, or other comparisons. Early attempts to explain emergent patterns of European industrial specialization placed the greatest emphasis on three factors: the local availability of raw materials, access to international trade routes and ports, and dominance of international politics, shipping, and trade. More recently, comparisons and explanations of economic development have been formulated at both larger and smaller scales.

Increasing work on extra-European areas has made it clear that Europe was not possessed of unique resources that differentiated the continent from parts of Asia, the Americas, or elsewhere (Jones, 1981). Nor was Christian Europe scientifically more advanced than Chinese or Islamic civilizations, and in some respects Europe clearly lagged (Needham *et al.*, 1954–94; Hill, 1993). In certain parts of Europe, however, the deployment of resources does seem to have become increasingly distinctive. This was partly made possible by growing European military, imperial, and trading power, which subordinated societies across the world. This brought greater access to familiar materials, including precious metals, textile fibres, and some dyestuffs, and new access to unfamiliar materials, including such as cochineal dye, foods, beverages, tropical woods, Oriental fabrics and ceramics (Donkin, 1977; Mukerji, 1983). Such access was usually controlled by one or more imperial powers, rather than being available through open commodities markets, and so had important consequences for the relative power of different European states. More generally, European industries were indirectly affected by colonial expansion through changes in domestic consumer tastes (Brewer and Porter, 1993). Several writers stress the connections between European global dominance and a distinctively European orientation to the practical applications of science and mechanical ingenuity. Innovative devices in Europe, in other words, were much more likely to become weapons or machines rather than amusements or toys (Landes, 1983; Swetz, 1987; Livingstone, 1992; Crosby, 1997). Europe was also characterized, notwithstanding government (or other) attempts to maintain secrecy, by the relatively effective diffusion of technological information and innovations. All in all, explicit inter-continental comparisons have helped to highlight the role of facets of European culture which it might otherwise be easy to take for granted.

Important as international comparisons have been, since the early 1980s the most striking debates on early modern industry have focused on a regional scale. Above all, these have involved 'proto-industrialization theories', and these form a central theme of this chapter. The starting point here is the observation that, even in 1750, the regional rather than the national context of industrial activities was crucial (Pollard, 1981: 1–16; Hudson, 1989). The regional theme was being asserted particularly vigorously in the 1980s, in reaction to national-scale econometric modelling which emphasized the slowness of economic growth in European economies restrained by large, static agricultural and artisan sectors (e.g. Floud and McCloskey, 1981).

Explanations of divergences among regional and local industrial trajectories take one or more of three broad forms. The first invokes *intrinsic* features of specific areas, such as the presence of particular resources or raw materials; the form of farming systems; the availability of cheap labour; and the existence of institutions affecting land tenures, common rights, inheritance, family and household structures, any of which could facilitate certain activities. The second invokes advantages that were *acquired* rather than intrinsic, such as the skills and experience of workers and/or entrepreneurs (or 'improved human capital') generated in the performance of industrial activities. Even if such areas lacked or lost intrinsic advantages, the effects of more skilled labour were far-reaching in labour-intensive tasks, and could outweigh the costs of transporting raw materials and products to and from production areas. The third invokes *extrinsic* advantages, where an area benefits from its relative location. Extrinsic advantages would be economic (for example, where processing or refining industries grew at pivotal or gateway locations in trading networks) or political-cum-imperial (as where certain locations benefited from military protection or favourable economic regulation).

Econometric work, whether at national or regional scale, necessarily requires relatively firm quantitative estimates of the economy, numbers that—however frail—can be used in sometimes highly sophisticated models. It must be emphasized, therefore, that all figures for aggregate or sectoral production, productivity, and occupational composition of early modern European populations are frail, and require considerable care. Fundamental problems arise from uneven documentation, and from inappropriate categorizations of households. For no part of Europe is there systematic information on industrial activities at any one time, never mind a series of such sources for the analysis of change over time. Even if they existed, census-type sources with information on 'occupations' would be unhelpful where households survived on a combination of several different activities: for example, the number of householders described as 'weaver' may be a poor guide to the amount of weaving occurring. Most estimates of occupational composition, used to identify national totals of agricultural, craft, and industrial producers, presume a degree of occupational specialization within households that was present in only a few parts of Europe, even by c.1750. In practice, there could be considerable overlap at household level among farming, industrial, trading, and servicing activities. All in all, considerable uncertainty will continue to surround estimates of the exact distribution and scale of different activities, especially in eastern and central Europe.

These empirical obstacles also bedevil the mapping of geographical patterns of industrial activity. Cartographic depictions of industrial distributions are further complicated by a historiography that is chronically, geographically, and thematically uneven. Much more attention has been devoted to events and processes after 1750 than to the preceding centuries, partly owing to better documentation. Parts of England, Spain, France, Italy, the Low Countries, and Germany have received considerable attention, whereas much less is known about events elsewhere before c.1750, except in coarse outline. The uneven work on different areas also reflects a diversity of research priorities, some linked to potent broader contexts of national identities.

Accordingly, it is appropriate to finish this introduction on a cautionary note. These empirical and historiographical difficulties make it impossible to map all areas of Europe on an equivalent basis, either quantitative or qualitative. Maps therefore remain indicative rather than definitive, and caution is required in judging the comparative significance of industries in different places. Simplified continent-wide maps may also omit pockets of potentially specialized activity outside the main concentrations, and understate the internal heterogeneity of 'industrial regions'. Most such regions comprised patchworks of localities with a diversity of product specializations, scales of production, and forms of organization. The most uniform areas

were probably found across eastern Europe, where local economic circumstances were overridden by strong Crown or state influences, and well-defined community structures.

The chapter is divided into three main sections. The first briefly surveys some of the principal changes in industrial processes and in industrial geographies between the start of the sixteenth century and the early decades of the eighteenth century. The second focuses on the often contentious topic of 'proto-industrialization', around which there has been much debate since the mid-1970s. The main aim of that section is to clarify the proliferation of proposals and rebuttals of different elements within proto-industrialization theories. A concluding discussion considers possible responses to recent changes in proto-industrialization debates, and identifies some items for a research agenda.

## Dimensions of Industrial Change

In the course of the sixteenth and seventeenth centuries, Europe became a more industrial continent, in the sense that a greater proportion of its constituent populations relied for their livelihoods on activities other than farming, fishing, or hunting. This guarded form of expression is deliberate, for two reasons. First, an appropriate definition of 'industrial' at this time is very broad, because of the handicraft character of most industrial processes. Secondly, it should be remembered that the rise of specialist income-generating activities was partly offset by declines in such activities for households' own direct comparison, for example, in spinning, weaving, or small woodwork (De Vries, 1993: 108–10).

It is impossible to quantify the total numbers of people involved, but clear that both the proportion of populations engaged in industry (broadly defined), and the absolute numbers of industrial workers, increased. In very round figures, Wrigley suggests that between c.1500 and c.1750 the relative size of the non-agricultural population more than doubled in England (from about 25 to 55 per cent) and approximately doubled in France (from about 20 to 40 per cent), whereas the Dutch economy featured a much more modest increase from about 40 to 45 per cent (Wrigley, 1987: 170–85).

Comparable figures are not available for many areas, though corresponding figures for most of Europe would be considerably lower. Another safe generalization is that the largest relative increases occurred where urban and rural industrial sectors were very small in 1500, whereas scope for rapid growth of industrial employment was limited in areas where it was already high, like the Netherlands or northern Italy.

Statistics are as elusive for output as they are for people's activities, but commodities for which estimates have been made all suggest substantial increases in total production and consumption, although with significant and potentially lengthy regressions. Rising industrial output was based on the exploitation of new resources, both organic and inorganic, and of resources from new areas, both European and colonial. Much was driven by sustained population growth and the geographical expansion of settlements and resource exploitation. Population growth, especially where accompanied by rising living standards, significantly expanded the total market for products. Market expansion produced general benefits through lower transaction costs, comprising lower unit costs for transport, finance, and information, both commercial and technical.

Superimposed on the widespread and very local provision of many everyday requirements (in handicraft forms that would not qualify as 'industry' under modern definitions), late seventeenth-century Europe contained numerous modestly specialized industrial regions. Woollen and linen textiles, and various iron-based industries were prominent among growth sectors. Production was often rural, being co-ordinated at central places from which mercantile or seigneurial power was exercised. Central places from small local markets to the great ports and cities constituted the urban network through which goods were traded inter-regionally and internationally. A number of strikingly new or almost new sectors appeared, for example in relation to printing, papermaking, and clockmaking, but these and other highly technical or specialized refining or manufacturing activities tended to remain the province of a handful of specialized production areas only.

Occasionally, but rarely, individual enterprises attained considerable size. Centralized mines or shipyards might employ a workforce running into hundreds, and foundries and chemical works might employ dozens of (mainly) men. However, most manufacturing activity remained in small domestic production units based around the home or small workshop (these were usually one and the same), and this remained one of the most characteristic features of early modern industry. These small units sometimes formed part of larger networks of producers, co-ordinated by entrepreneurs or administrators. In exceptional cases, hundreds of households might be involved, though most such networks were much smaller. The prominence of such networks has placed them at the core of major theoretical debate in so-called proto-industrialization theories, attempting to explain the nature of household production within such networks, and their highly variable organization. These are discussed in the second section of this chapter.

Although changes in technologies and processes in most industrial sectors between 1500 and 1740 were relatively modest, their geographical distribution changed significantly (Fig. 12.1a and b). The late medieval pattern of core industrial areas in the southern Low Countries, north Italy, and southern Germany was already being modified in the early sixteenth century as parts of England, northern France, Holland, and north-west Spain became settings for important textiles and iron production. In the course of the sixteenth century, further growth and diversification in those areas, and in parts of Sweden, southern France, and Bohemia, 'resulted in the re-drawing of the industrial map of Europe' (Sella, 1974: 417). A more extensive swathe of areas became more involved in industrial activities, and formerly 'core' areas such as northern Italy experienced growing competition from the north, east, and west. The story is not, however, one of inevitable and continual geographical extension. Parts of Spain and Poland emerged as industrial regions in the sixteenth century, only to wither in the seventeenth. The picture naturally varied for different activities as regional economic specialization increased.

The most widely distributed 'industrial' activities were barely present as industries. Carpenters, smiths, tailors, and the like, supplying shelter, tools, and clothing, were particularly widespread, but almost always on a small scale. Grain-processing (including milling, malting, and brewing) was

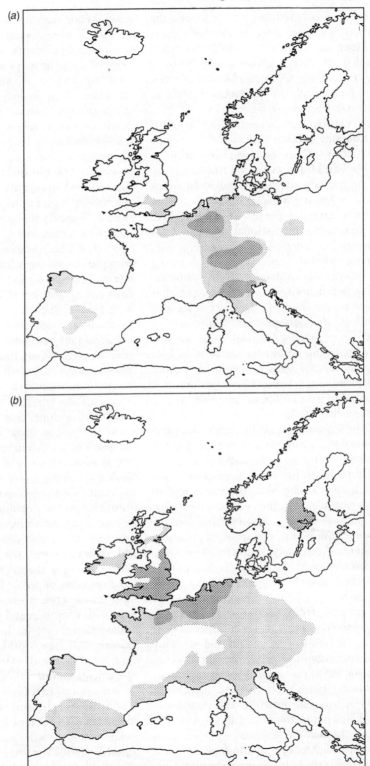

Fig. 12.1. Major areas of industrial activity in Europe: (a) in the early decades of the sixteenth century; (b) in the decades following 1700. Heavier shading indicates areas with larger involvement in industry. It is impossible to provide specific thresholds for the indication of areas: the general and indicative nature character of these maps is discussed on pp. 262, 264. Isolated centres, even if highly specialized, have been excluded.

also very widely distributed and, except around the larger cities, small-scale, since grain's bulk made transport expensive. The greater mobility of livestock enabled butchers, tanners, and leather-workers to concentrate in central places of all sizes, again disproportionately in the largest centres of population and purchasing power. By the later seventeenth century, regional centres serving international (especially colonial and military) markets with leather items such as shoes were appearing, as in the English East Midlands around Northampton.

The major pre-industrialization industrial sector was textiles, embracing a range of fibres and an almost infinite array of grades of fabric. There had been significant international trade in both high-quality and cheaper cloths in medieval Europe, although only the former survived slumping demand from the late thirteenth century (Munro, 1991). Rising populations from 1500 or soon after reversed that trend. The main production areas of 1500 (Flanders, Brabant, northern Italy, eastern Spain, and—increasingly—England) were supplemented by increasing production in areas of Lorraine and Alsace, and later by production from areas of eastern Europe that had hitherto imported large quantities of west European cloth (Fig. 12.2a). At least three distinct channels for the growth of rural cloth production can be identified. In some cases, production spread from urban production centres into less-restricted and cheap-labour countrysides (Lewis, 1994). In others, rural production was a new feature arising from the movement of techniques or people, as in the production of 'New Draperies' in England and the southern Netherlands. Finally, there are examples of peasant-produced wool or (especially) linen fabrics breaking into international trade, and maintaining considerable production growth thereafter (Myska, 1996; Thomson, 1996).

As demand grew, falling transactions costs particularly benefited producers and merchants dealing in lower-quality cloths. At various dates, many proto-industrial regions switched their production from fine woollen fabrics (high quality, high cost, limited markets, extensively dispersed marketing) to coarser woollens and worsteds, with their lower costs, less-skilled production, and greater suitability for dispersed production. The shift of production towards cheaper, lighter fabrics also took the form in many areas of expanding linen production (Fig.

12.2b). While woollen textiles remained the largest element of international trade, the linen sector was of growing significance. Flax and hemp cultivation supplied local linen production in most parts of northern Europe (Evans, 1985). Internationally important linen industries were growing in sixteenth-century Silesia, Poland, and to a lesser extent Ireland, as well as in Flanders.

Other industrial crops were much less ecologically versatile, and spread much less widely. The potentially very profitable cultivation of mulberry trees, to feed silkworms, was extended from ideal hot and dry conditions in Sicily and Calabria to more ecologically marginal areas of Lombardy and Languedoc (with mixed results), as demand and prices rose. The limited scope to extend production prompted increasingly intensive production in suitable areas, with a complementary intensification of silk weaving and finishing in French, Italian, and Swiss areas further north. Dye production was also geographically restricted, much of it in Languedoc and Piedmont or, in the case of indigo and cochineal, the New World. To dye many fabrics required use of a mordant, for which sixteenth-century Europe was almost entirely reliant on the Papal alum works at Tolfa, possibly the largest extractive enterprise in all Europe, employing about 800 men in c.1550 (Delumeau, 1962: 76). After 1600, a limited independent north European supply became available as alum extraction got under way in Yorkshire, Durham, and parts of the Low Countries.

Tanning and other leather-producing industries illustrate how the distribution of an industry could be essentially derivative of another sector, in this case, the supply and demand for meat. The meat and hides trades also exemplify enormously extended supply chains. As a demographically driven expansion of arable farming reduced livestock numbers in 'core' areas of Europe, the supply of hides fell, while demand grew rapidly. Supply networks reached ever further into eastern Europe for animals, and New World colonies for hides (Sella, 1974: 391; Blanchard, 1986). As international cattle movements through Poland and Hungary became the mainstay of European urban meat provisioning, leather-working became more concentrated around major cities and colonial import ports.

Population growth threatened timber production as well as pastoral husbandry. The clearance of woodland for arable cultivation or grazing, and

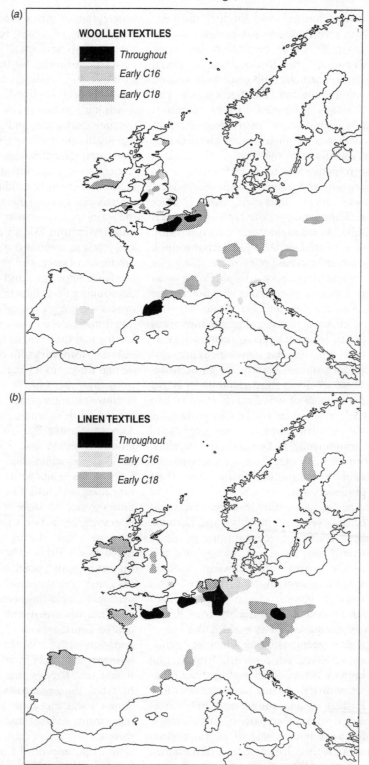

**Fig. 12.2.** Major areas of textile production: (*a*) woollens; (*b*) linens.

the over-consumption of wood for fuel and construction, both contributed to substantial losses of woodland cover. By c.1600 these were becoming severe in Castile and the Biscay coast of Iberia, southern France, northern Italy (especially around Venice), southern England, the Netherlands, and Flanders. Shortages prompted various responses from wood-based industries, partly depending on whether wood was required as a construction material or, sometimes in the form of charcoal, as a fuel. One response was to extend supply areas into Scandinavian and Polish forests where local demand was limited. Imported Baltic timber enabled the Netherlands, a country almost bereft of major timber, to maintain its dominant position in European shipbuilding. Another response took industry to better wood supplies rather than vice versa, as where large-scale shipbuilding was initiated in colonial possessions in Brazil, the Caribbean, New England, and Maryland.

Wood and charcoal served as fuel for numerous industries producing metals, glass, chemicals, ceramics, and other products. Of these, the refining of iron was the largest, and among the most widespread. In these sectors, a third response to actual or potential shortages of wood or charcoal was to substitute coal. In few areas was coal important for industrial rather than domestic uses until after 1700, most prominently in England and Flanders. Earlier commentators may have exaggerated the significance of this response. Finally, more fuel-efficient processes spread, albeit slowly. The mercury-based 'amalgamation process' increased the fuel efficiency of silver extraction and refining, and stimulated Spanish mercury-mining in Almaden and Istria. Major locational changes also followed from the relatively fuel-efficient 'indirect process' of refining iron in blast furnaces. This did not displace bloomery and shaft furnace methods for processing high-quality ores if fuel supplies were adequate, but it enabled the use of lower-grade ores, again assuming a lack of institutional barriers (Short, 1989: 164–6). Iron-mining expanded both in its traditional centres (the Rhineland, Low Countries, Basque country, northern Alps, the English Weald), and in new areas (Sweden, the eastern Alps, other parts of England, and Ireland (Sella, 1974: 391, 489–99). Rising production in the Low Countries, Britain, Galicia, Lombardy, Styria, Dauphiné, and Sweden significantly changed

geographies of iron extraction, refining, and working (Fig. 12.3a). Iron working also witnessed a spread of technological improvements (including powered bellows; tilt-hammers; and stamping, wire-drawing, rolling, and slitting mills) from innovative centres like Nuremberg and Liege.

Mining's vulnerability to cyclical fluctuations, wartime disruption, and political upheavals created a generally poor climate for large-scale investment and enterprise, although warfare could have positive effects through direct demand for weapons, in areas where neither production nor investment was directly disrupted. Thus the Swedish military were the principal customers of iron-working proto-industry around Eskilstuna after 1600 (Magnusson and Isacson, 1987: 263). Others to benefit included producers of guns, gun components around Brescia (Belfanti, 1993: 263), and bullets (Burt, 1995: 31–4). Silver and gold production was exceptional in that profits were high enough to sustain high levels of investment. Output was increased by new mines in Saxony and the Tyrol in the early sixteenth century, before European production was undermined by falling prices following massive bullion imports from the Americas after mid-century. Other non-ferrous mining was also relatively concentrated in the sixteenth century, principally in the Tyrolean Alps, the Banska Byastrica district of Slovakia, and around Manfeld and Schwarz in Thuringia (Fig. 12.3b). Production had been supported by the high profits available from precious metals occurring along with lead, tin, and/or copper, and several mines were both large-scale and deep. When metal prices fell, extraction from deep mines became less economic, and during a prolonged slump from c.1550 much demand was met by recycling lead, tin, and copper objects. Demand rose dramatically after 1600 because of new 'non-recyclable' demands, including copper for coinage, copper and lead for armaments, and tin and copper for plating (Burt, 1995).

As long as precious metal prices remained low, however, old mines faced difficulties in responding. Rising non-ferrous metal prices were insufficient to offset the high costs of deep mining. Instead demand was met largely from new areas (including, from c.1620, imports of Japanese copper through the Dutch East India Company). Production leapt around Falun in Sweden, the most important copper-producing area in Europe by

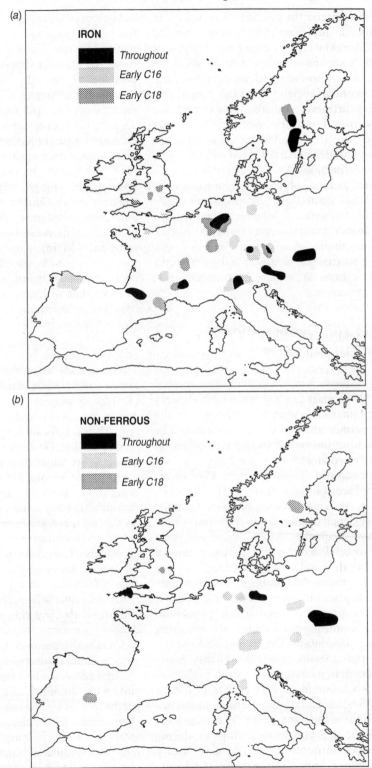

**Fig. 12.3.** The mining and processing of metals: (a) iron; (b) non-ferrous.

1625. British lead and tin production benefited in similar ways. The low silver content of ores had led mines in the Mendips, Derbyshire, north-east Wales, and Northumberland to remain small, so their low-cost, proto-industrial production could be rapidly expanded. The hitherto low output and scale of British mines had also discouraged the kinds of Crown investment and regulation that constrained increased production in some older non-ferrous mining districts (Blanchard, 1976; Kiernan, 1989; Burt, 1995).

If current pictures of industrial processes (as changing rather slowly) and the geographical distribution of industries (changing between 1500 and 1740) largely elaborates and refines the picture described a quarter of a century ago, without overturning it, changes in conceptual frameworks used in their explanation have been much more dramatic.

## Concepts and Geographies of Proto-Industrialization

Concepts of proto-industrialization have proved both a durable and, at times, a remarkably slippery set of ideas since they were first proposed in the 1970s (Mendels, 1972; Kriedte et al., 1981). The emergence of proto-industrialization theories in the 1970s and early 1980s has been reviewed on numerous occasions (Coleman, 1983; Berg et al., 1983: 1–32; Houston and Snell, 1984; Clarkson, 1985; Butlin, 1986; Wrigley, 1988: 91–9; Glennie, 1990: 205–9; Gregory, 1994: 484–6). Although useful commentaries on earlier debates these overviews have been overtaken by major developments in later work (Ogilvie and Cerman, 1996).

A one-sentence definition of proto-industrialization might run: 'the development of rural regions, in which a large part of the population lives largely from household industrial production for eventual sale outside the region, and experiencing internal pressures towards capitalist industrial production.' This may seem relatively straightforward, but in practice such a definition accommodates an enormous variety of situations. Many different elaborations have been suggested, and it is more appropriate to attempt to classify different types of proto-industrialization theory than to derive a pure 'essence'. Any taxonomy-cum-chronology of proto-industrial theories for the period 1500–1740 is rendered somewhat schematic by five factors. First, different proto-industrial theories employ various concepts and perspectives. Secondly, different areas experienced differing chronologies of change, and work on different areas has had very different focal periods. Thirdly, the transfer of ideas among literatures in different languages has often taken several years, so international debates have not developed in a coherent way. Fourthly, some relevant studies have been reluctant to engage with proto-industrialization concepts at all. Finally, the cut-off date of 1740 post-dates substantial proto-industrial developments in some areas, but precedes it in others, in some cases by many decades. For the purposes of this discussion, I will identify three 'rounds' of proto-industrial theorizing, each a diverse body of views, but with certain central concerns, priorities, and orientations.

### 'First-Round' Theories

These early theories were initially based on specific regions: of Flanders (Mendels, 1972) and Germany (Kriedte et al., 1981), also explicitly drawing on earlier studies of 'industries in the countryside' (e.g. Thirsk, 1961, on England). Their central concerns included (i) why proto-industries occurred in certain areas rather than others; (ii) ways in which proto-industries paved the way for factory industrialization; and (iii) generalizing about proto-industrial regions' common features. Most focused on the internal features of proto-industrial regions, either *intrinsic* features of agriculture and local production (especially the low opportunity costs of labour for part-time craftsmen-farmers) or *acquired* features that provided a developmental dynamic (especially accumulated specialist skills and knowledge, growing merchant control of production, or rising fertility). Their characteristic regional scale meant that wider contextual factors, such as social institutions, were often downplayed.

The regions studied were overwhelmingly rural, and with minimal institutional controls over production or marketing. Households mixed proto-industrial textile, metal, or mining activity with part-time farming, but over time growing proto-industrial populations became reliant on increasingly commercial agriculture in neighbouring regions for food supplies. Proto-industries seemed associated with population growth, through earlier marriages, higher marital fertility, and high levels of in-migration (Levine, 1977).

As studies begun to proliferate, they revealed increasingly diverse regional experiences, and the features initially highlighted were called into question as inevitable preconditions for proto-industrial development. Central features of first-round theories apply, at best, only to certain parts of northern Europe. Mendels' assumption that guilds were inimical to proto-industry, Kriedte's argument that feudal breakdown was integral to proto-industrial development, and Mendels' and Medick's connecting of proto-industrial activities to demographic growth were all downgraded from necessary to contingent relationships. That this diversity destroys the deterministic clarity of first-round theories was soon acknowledged (e.g. Mendels, 1982).

Most advocates of proto-industrial concepts abandoned any residual hankering after linear or stage-sequence models of industrial and societal change. Many critics were unable to do so, complaining that the initially highlighted features of proto-industrialization did not consistently 'cause' factory industrialization. If this was to ignore the explicit rejection of such interpretations within the original theories (as witness their discussions of 'de-industrialization'), it was clear that some advocates had claimed too much for proto-industrialization: by applying it to every industrial sector; by extrapolating from one or two study areas to Europe as a whole; or by downplaying earlier and other forms of complex industrial organization. The upshot was that comparisons among areas, and further exploration of the discussion of links among proto-industry, agriculture, and demography, were pushed to the head of research agendas from the early 1980s.

## 'Second-Round' Theories

From the early 1980s, 'second-round' theories focused on distinguishing *necessary* features and processes of proto-industries as focuses for theorizing, from *contingent* features of locally variable settings, which could be jettisoned from proto-industrial theories. Contrasting emphases among first-round theories were recast into underlying structural or processual similarities, still mainly intrinsic economic features. For example, emphases on arable or pastoral settings were recast as more general emphases on time-availability not tied to specific ecotypes. Other factors that likewise became articulated in more flexible, less deterministic terms included extra-regional market orientation; connections with commercial agriculture; and accumulating capital and skills (e g. Mendels, 1984).

As further studies in the late 1980s and early 1990s produced yet more heterogeneous findings, three topics received particular attention: namely urban proto-industries; diverse agricultural contexts for rural proto-industrial households (including full-time producers); and relationships between proto-industries and demographic change.

Significant urban activities spanned a spectrum from large workshops to individual households, and mixtures of part-time and full-time work. They particularly characterized textile manufactures, as in Krefeld (Kriedte, 1983, 1986), Nîmes, Carcassone, and Rouen (Deyon, 1996: 41). The blurring of the distinction between urban and rural producers has had several consequences. The influential argument of Jan de Vries (1985) that expanding rural industrial production accounts for sluggish European urban growth between 1600 and 1750 now looks too simple, for several reasons. At a national scale, some of the most heavily proto-industrialized economies, like England, were also the most urbanizing (Wrigley, 1987). Almost all regions of rural industry were organized around urban central places, in which finance, trading, provision of materials, and the most highly skilled production stages (especially if economies of scale applied) were all concentrated. The division between rural and urban places was also blurred when rapid growth transformed once-rural settlements into highly specialized urban economies.

In a similar way, both regional- and household-level relationships between proto-industry and agriculture came to be recognized as particular examples of more complex possibilities. The general idea of symbiotic specialisms between commercial agrarian production and proto-industries survived, but not necessarily at the regional level originally discussed. In areas such as the Pays de Caux, for example, proto-industrialists and large, commercial farms coexisted at every level

(Gullickson, 1983, 1986). At a household level, the classic 'dual economy' proto-industrial household, combining small-scale farming and industrial activity, was only sometimes found. Elsewhere, weaving or mining households might be better described as *peasant* (dovetailing industry with mainly subsistence agriculture on a 'traditional' family holding); *landless* (dovetailing industry with labouring, semi- or unskilled artisan work); *commercial* (dovetailing industry with market-orientated grain, livestock, or industrial crop production); or *artisan-trading* (dovetailing weaving or mining with work in wood, metal, or leather, or the running of a small retail shop). Trends over time have proved similarly diverse. In some areas dual-economy households retreated under pressure of intensifying proto-industrial work and population growth, so that production became full-time, but in others they remained a stable element of enduring household economies and structures (Hudson, 1989; Belfanti, 1993).

The demographic consequences of proto-industries were highly variable. There was no simple dichotomy between 'traditional' and 'proto-industrial' fertility regimes. Among proto-industrial regions and localities cases of earlier marriage and higher fertility (Levine, 1977) have been outnumbered by cases of delayed marriage and lower marital fertility (Gullickson, 1986; Gutman, 1989; Terrier, 1994; Vandenbroeke, 1996). This undermines viewing proto-industry as a unique or special case of economic-demographic interaction. Ten years after his original study, Levine (1987) argued that both proto-industry and commercial agriculture were instances of a more general process of proletarianization. These studies point to the multitude of factors that affected nuptiality and demographic behaviour directly, or via family and household structures, or via life-cycle patterns of household and family formation. Generally, in other words, demographic change depended more on wider institutional trajectories than on specific developments within proto-industrialization.

The plurality of relationships among proto-industry, demographic behaviour, and social institutions has become much debated. A key theme of 1990s proto-industrialization literature has been the influence of social institutions on proto-industry. This has been seen as of considerable general importance. On this new account, that different proto-industrial regions developed in such radically different ways is revealing not only about processes of industrial change, but especially about the effects of societal contexts on industrial organization and production.

## 'Third-Round' Theories

In a radical departure, 'third-round' theories place social institutions at centre stage, rather than as peripheral elements of 'locally variable settings'. This involves two major changes in how institutions have been viewed.

First, market restrictions are no longer seen as necessarily obstructive to proto-industrial developments, but as potentially constructive ways of protecting them or channelling them into certain areas. It had been axiomatic for first-round theories that proto-industries both depended on, and promoted, the erosion of restrictive institutions by free markets in materials, labour, and land. A late medieval dissolution of feudal controls, including restrictions on production and marketing in some areas, seemed self-evidently to have benefited proto-industries, whereas much of Europe was affected by an intensification of feudal social controls (Brenner, 1976, 1982; De Vries, 1974; Ogilvie, 1993; Braun, 1978).

This characterization of feudal institutions was first contested by work on eastern Europe and central Europe, recognizing that feudal regulatory powers could encourage proto-industry (Myska, 1979; Rudolph, 1985; Klima, 1991). Seigneurial lords could exclude proto-industry by direct prohibition, or through insisting on intense agricultural labour services, or by restricting settlement rights. But they could also exploit their position as monopoly employers, or as monopoly channels between producers and foreign merchants. Rigid institutions could protect markets and lower labour costs, as well as opening markets and raising labour costs. Fixing wages, exacting industrial labour services, or proscribing other activities were all used to lower labour costs used in Bohemia (Myska, 1996), early eighteenth-century Russia (Parker, 1968; Rudolph, 1980), and by the Swedish Crown in iron production (Floren and Ryden, 1993). Over much of Europe, then, structures of monopoly rights lay at the centre of successful proto-industrial districts, even towards 1750. In short, proto-industry was less influenced by feudal institutions in themselves than by their specific effects on industrial costs (Ogilvie, 1996: 29), effects that varied over space and time.

It was clearly not the case that free-market areas could proto-industrialize whereas feudal areas could not, the mere existence of certain regulatory institutions being the determining factor. Both free-market and heavily regulated areas contained districts of proto-industry, and other districts without them. Both free-market and heavily regulated areas provide examples of proto-industries that grew over long timespans, some that were short-lived, and some that withered fairly quickly. English, Flemish, and Dutch proto-industrial production and exporting were controlled by quasi-capitalist entrepreneurs, and this was occasionally the case elsewhere. But this was not the typical European pattern. Control was exercised by guilds, merchant companies, or groups of product finishers, and it rested not on market power, but on state privileges, production or trading monopolies, or power to enforce below-market wage levels. Where the two groups of areas did vary more systematically was in exhibiting significantly different proto-industrial forms. The differences lay in the structures of economic and political control, and the locations of power over co-ordination, decision-making, and profits.

A second change has been the use of broader definitions of 'institutions', going beyond marketing rights and arrangements, and the privileges of corporate groups such as guilds and merchant companies. Social institutions are now interpreted broadly, as 'the sets of established rules and practices through which people organized their economic, social, demographic, political and cultural activities' (Ogilvie, 1996: 23–4). They include frameworks of settlement and landholding; of community, family, and household organization; and of work, wages, and saving. All these could shape the costs and availability of proto-industrial labour, and how they were apprehended by households. Social and familial institutions became prominent in work on southern Europe (Mitterauer, 1986; Belfanti, 1993). Although certain institutional factors had been discussed in north-western Europe (Hudson, 1981, 1983, on Yorkshire), broader reinterpretations of northern European proto-industry in institutional terms have appeared relatively recently (Ogilvie, 1990, 1993, 1996a; Pfister, 1992).

For such 'third-round' theories, 'profound and enduring differences in social institutions among European societies' account for much (though not all) of the divergence in proto-industrial trajectories and experiences. Institutions 'profoundly affected the costs of different economic, social and political decisions', through their impacts on the availability and costs of raw materials, labour, transport, exchange, information, and security (Ogilvie, 1996a: 23, 37). Political and social institutions displayed considerable geographical variation, and those affecting community and work organization could vary over very short distances (as in affecting land and settlement rights, or gendered work patterns). Their effect was to promote or inhibit proto-industrial activities, or to channel them into particular forms that could enhance or diminish their vitality, competitiveness, and adaptability.

Whether particular institutions inhibited or promoted proto-industries depended on local circumstances. Assuming that seigneurial or community institutions did not altogether exclude proto-industrial work, the form of such work was still flexible. For example, institutions in peasant or other rural societies might restrict landholding to certain hereditary groups, excluding non-rights holders from any access to land. This could lead peasant households to engage in part-time proto-industrial activity at little opportunity cost (especially if households needed to remain on the land to retain access rights), whereas landless households could not do so. The capacity of landed households to move easily into and out of weaving or mining as market conditions changed could create an extremely flexible low-cost labour force, as in south-west England (Burt, 1995). However, in some situations of restricted access to land, it might be the landless rather than the landed households that entered proto-industrial work, maybe on a full-time basis, as around Cento in northern Italy (Belfanti, 1993).

The importance of symbiosis and mutual support between states and other local institutions, both political and cultural, looms large in many studies (Ogilvie, 1996b; Tilly and Blockmans, 1994). Besides the direct organization of production, states were profoundly important in many indirect ways: they could support or undermine other institutions (such as guilds or regional landholding customs), and they might indirectly affect industrial activity through *unin*tended consequences of other state policies. Few proto-industrial guilds or merchant companies survived in the long run without state support. The use of guild powers to regulate and co-ordinate proto-industries was particularly important within small city-state polities in the territorially fragmented areas of Italy, Germany, and Austria (Belfanti, 1993, 1996; Ogilvie, 1992; Tilly and Blockmans, 1994). And where states did attack monopoly structures and restrictions, they often sought to replace them with more congenial monopolies rather than to establish free-market conditions (Ogilvie, 1996a: 36). Even mainly unregulated areas, like England, saw Elizabethan governments attempt to promote certain metallurgical activities by granting royal monopolies (Pallister, 1983: 318–25). Proto-industry could be a means to broader strategic ends. Many states protected import-substituting production to promote national self-sufficiency in particular products. Across much of north and central Italy, states fostered proto-industries in remote and frontier areas for a mixture of strategic and anti-guild reasons (Belfanti, 1993: 260). Increasingly, too, new types of state apparatus were emerging, with far-reaching potential effects (Tilly and Blockmans, 1994: 218–50).

The recent upsurge in writing on social institutions and proto-industrialization opens up some interesting new topics, on which little research currently exists. As yet, it is rather early to gauge the long-run significance of this 'institutional turn'; its appeal is likely to vary among researchers within different disciplines. With regard to future work, three cautionary comments may be appropriate. First, the broadening of definitions of 'social institutions' creates some scope to make everything a facet of social institutions (after all, societies must have institutions of some kind). However, to treat social institutions as a new, independent determining variable would encounter the same problems that made first-round theories ineffective. Secondly, many writers, whilst not ignoring the potentially encouraging roles of feudal or guild institutions, continue to emphasize that breakdowns of state, seigneurial, or guild restrictions were significant factors in many 'successful' proto-industrial areas, as in recent studies of northern Italy (Belfanti, 1993: 255, 1996: 156) and

Switzerland (Pfister, 1992). The general outcome of state-sponsored proto-industry was often unsuccessful, as in the case of Spanish calico printing (Thomson, 1991). Even where states, seigneurs, or guilds were successful in establishing activities, the resultant proto-industries tended to need continued protection, rather than developing an effective independent dynamic. Thirdly, there are dangers of biasing empirical work to understate the inhibiting effects of institutions. Most studies of, for example, guilds and proto-industries analyse proto-industrial regions, rather than areas with guilds but not proto-industry. Comparably detailed studies of both types of areas are required to clarify contingent relations between institutions and industrial change.

## Conclusions

As a greater volume of research on a widening array of places has been published over the last quarter-century, the findings have borne heavily on the bold and schematic theories proposed in the 1970s. The situations and contexts of (broadly) proto-industrial activities have proved far more variable than first envisaged. Likewise the trajectories of established proto-industries varied much more than the 'intensify or de-industrialize' schemes allowed. And the consequences of proto-industrialization for demographic regimes and for market and other institutions have proved highly varied among proto-industrial regions, and not distinctively proto-industrial.

References earlier in this chapter to the chronological, spatial, and thematic unevenness of proto-industrial literature point to at least four components of a research agenda. The first follows from the observation that 'the centre of gravity of most empirical studies on proto-industrialization still lies in the eighteenth or even in the nineteenth century' (Schlumbohm, 1996: 18). This underlies continuing over-emphasis on transitions to factory industry as a criterion for evaluating proto-industrialization theories. Studies of sixteenth-century industrial change are particularly needed, although some have appeared (Rollison, 1992, on Gloucestershire; Stromer, 1986; Kiessling, 1989; and Zorn, 1988, for parts of Germany). Secondly, substantial empirical explorations of the distributions of ecotypes, agricultural systems, land ownership, community, or family cultures would be welcome, for independent comparison with distributions of proto-industrial activity. Previously, 'given' regions have functioned both as 'containers' for the accumulation and synthesis of information, and as explanations of patterns of proto-industrial activity, and have been prone to circular argument. Thirdly, the precise forms and effectiveness of regulation in terms of proto-industrial development have yet to be clarified for many parts of Europe: 'how any given corporate group affected costs and prices in any given proto-industry therefore requires detailed investigation, of a sort which has, in most cases, not yet been carried out' (Ogilvie, 1996a: 32). Fourthly, the time is also ripe for deeper explorations of physical geographical influences on geographies of agricultural and industrial activities, which have remained marginal and/or very generalized, within regionally focused accounts, to match the increasing attention to social institutions.

Given the proliferation and diversification of industrial contexts and trajectories, how might we respond? Of a variety of possible responses, four will be discussed here. One reaction, evident since shortly after proto-industrialization theories were first proposed, would be to abandon them altogether, on the grounds that they over-generalize about courses of industrial change, do not consistently 'predict' where industrialization would occur, and not all industrialization was preceded by proto-industries in the same areas (Coleman, 1983; Clarkson, 1985)—although not all these claims were ever made by proto-industrialization advocates. For those interested in proto-industrialization theories in a narrow sense as predicting factory industrialization, this view remains more widely held (e.g. Darnton, 1995: 169). Where interest is less factory-focused, the abandonment view is less widespread.

A second response is to define proto-industry much more broadly. Some commentators suggest that large urban manufactories, urban craftsmen, full-time rural industrial workers, and craftsmen serving purely local markets point towards 'abandoning the original demarcations and fully integrating urban production centres into the concept of proto-industrialization' (Schlumbohm, 1996: 19, see also Hohnenberg, 1991; Cerman, 1993). Such groups are neglected by narrow definitions of proto-industry, and the boundaries between

proto-industrial and other households are blurred. Much recent work has in practice incorporated a broader range of household types. Each broadening, however, makes 'proto-industrial' less acute as an analytical category (and to critics, smacks of *ad hoc* adjustments to rationalize any situation). Views on this response partly depend on how far definitions are broadened. For example, would a less-restricted definition than 'production in small, separate units, oriented to extra-regional markets, and related to agricultural change (although not necessarily at one particular scale)' retain worthwhile meaning? If additional variations in location, scale, and organization are incorporated, we risk emptying the concept of any meaning beyond 'not factory'. The analytical utility of such a vague definition is minimal and would hardly justify its retention (whilst leaving a need for terminology for differing forms of industrial organization).

A third response draws the opposite lesson from proto-industrialization theories' neglect of other organizational forms. This is to view proto-industrialization as one of a number of coexisting processes, rather than an entire industrial system, and as 'only one path toward the industrial revolution . . . [if] one of the more privileged forms of access to it' (Deyon, 1996: 43). For proponents of this response, diverse early modern industrial forms do not necessarily require resolution *within* proto-industrialization theories. What is striking, they suggest, is not so much the problematic status of proto-industrialization concepts, as the scarcity of other concepts for analysing industrial forms. For example, Jan de Vries (1974, 1993, 1994) excepted, what concepts are available for analysing artisan or workshop production? There is therefore a case for restricting the forms and trajectories that qualify as proto-industrial, whilst seeking comparable analytical concepts appropriate to change in artisan, workshop, or extractive plant. There are two main objections to this response. First, a series of separate theories risks the downgrading of key comparative questions on which proto-industrialization concepts have sought to integrate industrial, demographic, agricultural, and cultural history. Secondly, the boundaries between different industrial forms may be very blurred: much mining was proto-industrial, for example, and locally orientated crafts may merge into export-orientated proto-industries (Ogilvie and Cerman, 1996: 7).

Those desiring to maintain proto-industries' distinct status as 'heading-for-modernity' will find this response unacceptable. There might also be some danger of exacerbating the preoccupation of proto-industrial concepts with textiles.

Finally, a rather different kind of response seeks to recast proto-industrialization theories as networks of contingent relationships and processes, rather than as templates of necessary and sufficient conditions for industrial change. The goal of theorizing becomes less a predictive model than a framework for comprehending how various conjunctions of circumstances could direct industrial change in particular directions, according to different combinations of environmental, economic, and social institutional factors. In the terms recently used by Schlumbohm (1996: 12–22), proto-industrialization becomes a research and conceptual strategy applicable to economic and social circumstances that varied hugely across Europe, rather than a generalization about courses of events. This approach retains concern with the links among different dimensions of economic, cultural, and political life, and links among macro-, meso-, and micro-scale changes, but abandons portraying those links as anything but highly contingent and equifinal. That is, different proto-industrial areas derived cost advantages from very different configurations of environmental, economic and institutional factors. No one factor was necessary and sufficient for successful proto-industry. Many combinations of resources, agricultural systems, social settings, trading, and administrative frameworks, and demographic regimes that could provide an accommodating context. While various arrangements might 'succeed' in different contexts, in any particular context industrial configurations are limited 'path-dependent' contingent relations. Here, notions of path-dependence and non-deterministic models may be unacceptable to some commentators.

Not all these responses are mutually exclusive. Several commentators combine elements of various responses (e.g. Schlumbohm, 1996). The last three responses illustrate how debates on proto-industrialization continue and how, in some ways, they have recently received considerable stimulation from 'third-round' theories. The current liveliness of discussion justifies continued—though not unreflective—use of proto-industry concepts. Were their use to be banned tomorrow, the proto-industrial literature of the last twenty-five years would leave three important legacies for historical geographical work. First, they have played a pivotal role in comparative studies across (and beyond) Europe, drawing attention to some profound differences in national and regional circumstances, processes and outcomes. Secondly, theories of proto-industrialization have raised issues around the connections among economic change, cultural change, and political power, touching

upon almost every aspect of pre-industrial society: people's thoughts and motivations, their sexual and family behaviour, their use of time in work and play, their ownership of land and equipment, their standard of living and nutrition, their inequalities and conflicts, the social institutions which they used (but were also constrained by) in the attempt to survive and the mechanisms by which economy and society gradually changed between c.1500 and c.1800. (Ogilvie and Cerman, 1996: 11)

And thirdly, proto-industrial concepts have been associated with a decisive shift to focusing on the spatial dynamics of industrial processes. Not the least outcome of this has been the maintenance of links with work in contemporary economic geography, dealing with its own issues of economic convergence and divergence in economic restructuring, as Europe and the world move way from, rather than towards, the Fordist factory era.

REFERENCES

ALLEN, R., *Enclosure and the Yeoman: Agricultural Development of the South Midlands 1450–1850*, (Oxford, 1992).

ARACIL, R., and BONAFE, M. GARCIA, 'La protoindustrialitzacio i la industria rural espanyola al segle XVIII'. *Recerques*, 13 (1983), 83–1102.

ASTON, T., and PHILPIN, C. E. (eds.), *The Brenner Debate*, (Cambridge, 1985).

BELFANTI, C. M., 'Rural Manufactures and Rural Proto-industries in the "Italy of the Cities" from the Sixteenth through to the Eighteenth Century', *Continuity and Change*, 8 (1993), 252–80.

—— 'The Proto-industrial Heritage: Forms of Rural Proto-industry in Northern Italy in the Eighteenth and Nineteenth Centuries', in Ogilvie and Cerman (1996), 155–170.

BERG, M. (ed.), *Markets and Manufacturers in Early Industrial Europe* (London, 1991).

—— HUDSON, P., and SONENSCHER, M., *Manufacture in Town and Country before the Factory* (Cambridge, 1983).

BLANCHARD, I., 'English Lead and the International Bullion Crisis of the 1550s', in D. C. Coleman and A. H. John (eds.), *Trade, Government and the Economy in Pre-industrial England* (Newton Abbot, 1976).

—— 'The Continental European Cattle Trades, 1400–1600', *Economic History Review*, 39 (1986), 427–60.

BRAUN, R., 'Protoindustrialization and Demographic Changes in the Canton of Zurich', in C. Tilly (ed.), *Historical Studies of Changing Fertility*, (Princeton, 1978).

BRENNER, R., 'Agrarian Class Structure and Economic Development in Pre-industrial Europe', *Past and Present*, 70: 30–75.

—— 'The Agrarian Roots of European Capitalism', *Past and Present*, 97 (1982), 16–113.

BREWER, J., and PORTER, R. (eds.), *Consumption and the World of Goods*, (London, 1993).

BURT, R. 'The Transformation of the Non-ferrous Metals Industries in the Seventeenth and Eighteenth Centuries', *Economic History Review*, 48 (1995), 23–45.

BUTLIN, R. A., 'Early Industrialization in Europe: Concepts and Problems', *Geographical Journal*, 152 (1986), 1–8.

CERMAN, M., 'Proto-industrialization in an Urban Environment: Vienna 1750–1857', *Continuity and Change*, 8 (1993), 281–320.

CLARKSON, L., *Proto-industrialization: The First Phase of Industrialization?* (London, 1985).

COLEMAN, D., 'Protoindustrialization: A Concept Too Many?', *Economic History Review*, 36 (1983), 435–48.

CROSBY, A. W., *The Measure of Reality: Quantification and Western Society, 1250–1600*, (Cambridge, 1997).

DAHLMAN, C., *The Open Field System and Beyond: A Property Rights Analysis of an Economic Institution* (Cambridge, 1980).

DARNTON, M., *Progress and Poverty: An Economic History of Britain 1700–1850*, (Oxford, 1995).

DELUMEAU, P., *L'Alun de Rome: XVe–XIXe siecle*, (Paris, 1962).

DE VRIES, J., *The Dutch Rural Economy in the Golden Age*, (New Haven, 1974).

—— *The Economy of Europe in an Age of Crisis, 1600–1750*, (Cambridge, 1976).

—— *European Industrialization, 1500–1800*, (London, 1984).

—— 'Between Purchasing Power and the World of Goods: Understanding the Household Economy in Early Modern Europe', in J. Brewer and R. Porter (eds.), *Consumption and the World of Goods* (London, 1993), 84–132.

—— 'The Industrial Revolution and the Industrious Revolution', *Journal of Economic History*, 59 (1994), 249–70.

DEYON, R., 'Proto-industrialization in France', in S. Ogilvie and M. Cerman (eds.), *European Proto-industrialization*, (Cambridge, 1996), 38–48.

DONKIN, R. A., *Spanish Red: An Ethnogeographic Study of Cochineal and the Opuntia Cactus*, (Washington, DC, 1977).

EVANS, N., *The East Anglian Linen Industry: Rural Industry and Local Economy*, (Aldershot, 1985).

FLINN, M., *Origins of the Industrial Revolution*, (London, 1977).

FLOREN, A., and RYDEN, G., 'Arbete, husall och region: Tankar om industrialiserings processor och den Svenska jarnhanteringen', *Uppsala Papers in Economic History*, 29 (1993).

FLOUD, R., and MCCLOSKEY, D., *The Economic History of Britain Since 1700*, (2 vols.; Cambridge, 1981).

GLENNIE, P. D., 'Industry and Towns', in R. Dodgshon and R. Butlin (eds.), *An Historical Geography of England and Wales*, 2nd edn., (London, 1990), 199–222.

GREGORY, D. J., 'Protoindustrialization', in R. J. Johnston *et al.* (eds.), *The Dictionary of Human Geography*, 3rd edn., (Oxford, 1994), 484–6.

GULLICKSON, G., 'Agriculture and Cottage Industry: Redefining the Causes of Protoindustrialization', *Journal of Economic History*, 43 (1983), 831–50.

—— *Spinners and Weavers of Auffay*, (Cambridge, 1986).

GUTMAN, M. P., *Toward the Modern Economy: Early Industry in Europe 1500–1800*, (Philadelpia, 1989).

HILL, D. R., *Islamic Science and Engineering*, (Edinburgh, 1993).

HOHENBERG, P. M., 'Urban Manufacturers in the Protoindustrial Economy: Culture versus Commerce', in M. Berg (ed.), *Markets and Manufacturers in Early Industrial Europe*, (London, 1991), 159–72.

HOUSTON, R., and SNELL, K., 'Protoindustrialization? Cottage Industry, Social Change and the Industrial Revolution', *Historical Journal*, 27 (1984), 483–92.

HUDSON, P., 'Proto-industrialisation: The Case of the West Riding Wool Textile Industry in the Eighteenth and Early Nineteenth Centuries', *History Workshop Journal*, 12 (1981), 34–61.

—— 'From Manor to Mill: The West Riding in Transition', in M. Berg *et al.* (eds.), *Manufacture in Town and Country before the Factory*, (Cambridge, 1983), 124–46.

—— 'The Regional Perspective', in P. Hudson (ed.), *Regions and Industries: A Perspective on the Industrial Revolution in Britain*, (Cambridge, 1989), 5–38.

JONES, E. L., *The European Miracle: Environments, Economics and Geopolitics in the History of Europe and Asia*, 2nd edn., 1987 (Cambridge, 1981).

KIERNAN, D., *The Derbyshire Lead Industry in the Sixteenth Century*, Derbyshire Records Society, 14 (1989).

KIESSLING, R., *Die Stadt und ihr Land: Umlandpolitik, Bürgerbesitz und Wirtschaftsgefüge in Ostschwaben vom 14 bis ins 16. Jahrhundert* (Cologne, 1989).

KLIMA, A., *Economy, Industry and Society in Bohemia in the Seventeenth to Nineteenth Centuries*, (Prague, 1991).

KRIEDTE, P., 'Proto-industrialisierung und grosses Kapital: Das Seidengewerbe in Krefeld und sienem Umland bis zum Ende des Ancien Regime', *Archiv fur Sozialgeschichte*, 23: 219–66.

—— 'Demographic and Economic Rhythms: The Rise of the Silk Industry in Krefeld', *Journal of European Economic History*, 15 (1986), 259–89.

—— MEDICK, H., and SCHLUMBOHM, J., *Industrialization before Industrialization: Rural Economy in the Genesis of Capitalism* (Cambridge, 1981); originally published in German, 1977.

—— —— —— 'Protoindustrialization on Test with the Guild of Historians: Response to Some Critics', *Economy and Society*, 15 (1986), 254–72.

—— —— —— 'Proto-industrialization Revisited: Demography, Social Structure, and Modern Domestic Industry', *Continuity and Change*, 8 (1993), 217–52.

LANDES, D. S., *Revolution in Time: Clocks and the Making of the Modern World*, (Cambridge, Mass., 1983).

LEVINE, D., *Family Formation in an Age of Nascent Capitalism*, (London, 1977).

—— *Reproducing Families: The Political Economy of English Population History*, (Cambridge, 1987).

LEWIS, G., 'Proto-industrialization in France', *Economic History Review*, 48 (1994), 150–64.

LIVINGSTONE, D., *The Geographical Tradition: Episodes in the History of a Contested Enterprise*, (Oxford, 1992).

MAGNUSSON, L., and ISACSON, M., *Protoindustrialization in Scandinavia: Craft Skills in the Industrial Revolution*, (Leamington Spa, 1987).

MENDELS, F, 'Proto-industrialization: The First Phase of the Industrialization Process', *Journal of Economic History*, 32 (1972), 241–61.

—— 'Protoindustrialization: Theory and Reality: General Report', in F. Mendels and R. Deyon (eds.), *Eighth International Economic History Society Conference: A Themes, Protoindustrialization* (Budapest, 1982), 69–107.

—— 'Des industries rurales à la proto-industrialization: historique d'un changement de perspective', *Annales Economies, Societes, Cultures*, 39 (1984), 911–1008.

MITTERAUER, M., 'Peasant and Non-peasant Family Forms in Relation to the Physical Environment and the Local Economy', *Journal of Family History*, 17 (1986), 139–59.

MUKERJI, C., *From Graven Images: Patterns of Modern Materialism*, (New York, 1983).

MUNRO, J. H., 'Industrial Transformation in the North-West European Textile Trades, *c.*1290–*c.*1340: Economic Progress or Economic Crisis?', in B. M. S. Campbell (ed.), *Before the Black Death: Studies in the 'Crisis' of the Early Fourteenth Century*, (Manchester, 1991), 110–48.

MYSKA, M., 'Pre-industrial Iron-Making in the Czech lands: The Labour Force and Production Relations *c.*1350–*c.*1840', *Past and Present*, 82 (1979), 44–72.

—— 'Proto-industrialization in Bohemia, Moravia and Silesia', in S. Ogilvie and M. Cerman (eds.), *European Proto-industrialization*, (Cambridge, 1996), 188–207.

NEEDHAM, J. *et al.*, *Science and Civilization in China* (16 vols.; Cambridge, 1954–94).

OGILVIE, S., 'Women and Proto-industrialization in a Corporate Society: Württemberg Woollen Weaving 1590–1760', in P. Hudson and W. R. Lee (eds.), *Women's Work and the Family Economy in Historical Perspective* (Manchester, 1990), 76–103.

—— 'Germany and the Seventeenth-Century crisis', *Historical Journal*, 35 (1992), 417–41.

OGILVIE, S., 'Proto-industrialization in Europe', *Continuity and Change*, 8 (1993), 159–79.

—— 'Social Institutions and Proto-industrialization', in Ogilvie and Cerman (1996a), 23–37.

—— 'Proto-industrialization in Germany', in Ogilvie and Cerman (1996b), 118–36.

—— and CERMAN, M. (eds.), *European Proto-industrialization*, (Cambridge, 1996).

PALLISTER, D., *The Age of Elizabeth: England under the later Tudors 1547–1603*, (London, 1983).

PARKER, W. H., *An Historical Geography of Russia*, (London, 1968).

PFISTER, U., 'The Protoindustrial Family Economy: Towards a Formal Analysis', *Journal of Familiy History*, 17 (1992), 201–32.

—— 'Proto-industrialization in Switzerland', in Ogilvie and Cerman (1996), 137–54.

POLLARD, S., *Peaceful Conquest: The Industrialization of Europe 1760–1970*, (Oxford, 1981).

RIGBY, S., *English Society in the Later Middle Ages: Class, Status and Gender*, (Basingstoke, 1995).

ROLLISON, D., *The Local Origins of Modern Society: Gloucestershire 1500–1800*, (London, 1992).

RUDOLPH, R., 'Family Structure and Protoindustrialization in Russia', *Journal of Economic History*, 40 (1980), 111–18.

—— 'Agricultural Structure and Protoindustrialization in Russia: Economic Development with Unfree Labour', *Journal of Economic History*, 45 (1985), 47–69.

SCHLUMBOHM, J., 'Protoindustrialization as a research strategy and a historical period: a balance sheet', in Ogilvie and Cerman, (1996), 12–22.

SELLA, D., 'European Industries, 1500–1700', in C. Cipolla (ed.), *Fontana Economic Short History of Europe*, ii. *The Sixteenth and Seventeenth Centuries*, (London, 1974).

SHORT, B., 'The Deindustrialization Process: A Case-Study of the Weald', in P. Hudson (ed.), *Regions and Industries: A Perspective on the Industrial Revolution in Britain*, (Cambridge, 1989), 156–74.

STROMER, W., 'Gewerbereviere und Protoindustrien in Spätmittelalter und Frühneuzeit', in H. Pohl (ed.), *Gwerbe- und Industrielandschaften vom Spätmittelalter bis ins 20. Jahrhundert*, (Stuttgart, 1986).

SWETZ, F. J., *Capitalism and Arithmetic: The New Math of the Fifteenth Century*, (La Salle, Ill., 1987).

TERRIER, D., *Les Deux Âges de la Protoindustrialisation* (Paris, 1994).

THIRSK, J., 'Industries in the countryside', in F. J. Fisher (ed.), *Essays in the Economic History of Tudor and Stuart England in Honour of R. H. Tawney*, (Cambridge, 1961), 70–88.

THOMSON, J. K. J., 'State Intervention in the Catalan Calico Printing Industry in the Eighteenth Century', in M. Berg (ed.), *Markets and Manufacturers in Early Industrial Europe*, (London, 1991), 57–89.

—— 'Proto-industrialization in Spain', in Ogilvie and Cerman (1996), 85–101.

TILLY, C., and BLOCKMANS, W. (eds.), *Cities and the Rise of States in Europe*, AD 1000 to 1800 (Oxford, 1994).

VANDENBROEKE, C., 'Proto-industry in Flanders: A Critical Review', in Ogilvie and Cerman (1996), 102–17.

VARDIE, L., *The Lord and the Loom: Peasants and Profit in Northern France, 1680–1800,* (Durham and London, 1993).

WHYTE, I., 'Proto-industrialization in Scotland', in P. Hudson (ed.), *Regions and Industries: A Perspective on the Industrial Revolution in Britain,* (Cambridge, 1989), 228–51.

WRIGLEY, E. A., 'Urban Growth and Agricultural Change: England and the Continent in the Early Modern Period', *Journal of Interdisciplinary History,* 15 (1985), 683–728.

—— *People, Cities and Wealth: The Transformation of Traditional Society,* (Oxford, 1987).

—— *Continuity, Chance and Change: The Character of the Industrial Revolution in England,* (Cambridge, 1988).

ZORN, W., 'Ein neues Bild der Struktur der ostschwabischen Gewerbelandschaft im 16. Jahrhundert', *Vierteljahrschrift für Sozial- und Wirtschaftgeschichte,* 75 (1988), 153–87.

# Chapter **13**

# Industrialization, 1740 to the Present

## S. Pollard

## The Meaning and Sequence of European Industrialization

Industrialization should be understood as a technical term, carrying a particular meaning; it should not be taken to imply that before its onset there was no industry in Europe. The preceding chapter will have shown how much industry in fact there was. By general consent, the term has come to be applied to particular changes in industrial structure and technology, together with changes in other aspects of social life associated with them.

In the manufacturing sector itself, 'industrialization' implies a change from workshops to factories or other large enterprises, including shipyards or coal mines, using technologies superior to those that had gone before. These, in turn, require concentrated inanimate power sources, such as steam engines. They require substantial capitals and large markets, with appropriate means of transport to reach them. The provision of capital, in turn, does not depend merely on the accumulation of sufficient wealth in society, it also requires that those who own it are willing to invest it in economic enterprises of the appropriate kind, rather than waste it in conspicuous consumption or use it in trading only; that is to say, it requires a change in attitudes of wealth holders, as also, usually, of wage workers. The social framework is also important: at the least, security of property and the rule of law should obtain.

All these, it will be evident, are interdependent and have to advance together. A broad social movement, therefore, is involved, beyond changes in big industry. Among other consequences, there will also be technical progress in small workshops and even in agriculture: paradoxically, industrialization frequently implies the transformation of agriculture, as it did conspicuously in Britain and Denmark.

What set the whole process going? From one point of view, there was no special departure point, merely a continuation of preceding developments in new directions and with new centres of gravity: recent literature has stressed the slow rate of growth of supplies in the early phases of the British Industrial Revolution in the eighteenth century, labour productivity growing by only 0.8 per cent and total factor productivity by a mere 0.15 per cent a year between 1760 and 1801, according to one calculation (Harley, 1982; Crafts, 1985; Crafts and Harley, 1992: 718; Hawke, 1993; Berg and Hudson, 1994). Other historians have tended to point to the growth of demand as the major factor (Eversley, 1967; Gilboy,

1932; but see Mokyr, 1985). However, any population will accept additional incomes, goods, and services if they can afford them because of their own rising productivity, while productivity will not necessarily rise simply because people want more goods; it is therefore difficult to give demand the primary role. In any case, even if we take the opportunities provided by growing markets as the key, our interest still lies in determining where, how, and by whom they were taken.

Most studies, therefore, begin with the changes in supply. One common starting point is to see the driving force behind the industrialization process as being essentially technological and economic: by using new machinery and novel processes to produce more efficiently and more cheaply, the new inevitably drives out the old, which then has to adapt or go under. Industrialization, therefore, is expansionary, and invention and innovation, accompanied by associated or independent progress in science, become part of the system. Output goes up, and with it national income and the spending power of individuals. Once the system 'takes off', expansion may therefore be said to be irreversible and unstoppable. Herein lies the justification for the widespread belief that industrialization opened up an entirely new chapter in the history of mankind.

Given the widespread distribution of domestic or 'proto-industry' in the mid-eighteenth century in Europe, as described in Chapter 12 above, with its associated skilled labour, its commercial capital, and its links to distant markets, and given that these were the apparent preconditions for the 'take off', one might have expected new forms of technology and organization to have arisen simultaneously in many places out of the pre-existing workshops. But such was not the case. The breakthrough occurred in one country only, Great Britain, and there, perhaps even more surprisingly, not in one location only, which would point to a single originating initiative, but in several districts at once. Thereafter, for about fifty years, to the end of the Napoleonic Wars, British industrialization proceeded on a broad front on its own while Continental countries scarcely even made a beginning.

It was only after 1815 that there were clear signs of progress also in parts of the Continent of Europe, especially in its western half. By the last third of the nineteenth century those parts were becoming fully

'industrialized', while the technology implied by that term was itself constantly advancing. Their progress now created a new gap, between the successful advanced areas in the West as a whole and the remainder of Europe, which even by the end of the century had hardly got going. The dividing line between these two regions which cut Europe in half ran in part through the middle of countries, political boundaries being not necessarily decisive in this context. It corresponded roughly to the course of the River Elbe, leaving the eastern provinces of Imperial Germany in the retarded part; it separated the Czech- and German-speaking provinces of the Habsburg Monarchy from its eastern territory; reaching the Adriatic, it swung westward, separating northern Italy from the centre and south, and then, crossing a stretch of the Mediterranean, it divided the north-eastern corner of Spain from the rest. One might even think of it as turning northward thereafter, entering the Irish Channel and, swinging west again, separating the southern countries of Ireland from the rest of the United Kingdom.

It is significant that eastern and southern Europe, the stagnating regions, were not merely behind the north-western countries in an economic sense in the nineteenth century, industrializing only in the later twentieth century, if at all; they remained in other respects also, such as the rise of liberal democratic institutions, several generations behind the west. The Scandinavian countries form a conspicuous exception to this: among the poorest and least industrialized before 1850, they had caught up with the leaders, and even overtaken some of them, by the 1920s. It is evident from Fig. 13.1 that the gap between the United Kingdom and the Continent had become widest by the mid-nineteenth century, though there were signs of growing industrialization also in some western continental countries by then; by 1913 these had almost caught up, and the drive to convergence was to continue in the twentieth century. At least as striking as the national differences is the extent of the increase everywhere between 1800 and 1913 and again in 1913–80. Fig. 13.2 presents the same evidence, with the absolute size of countries taken into account.

It remains to be added that both halves of Europe should really be subdivided again, for both contained concentrated advanced regions which industrialized well ahead of the rest of their territories.

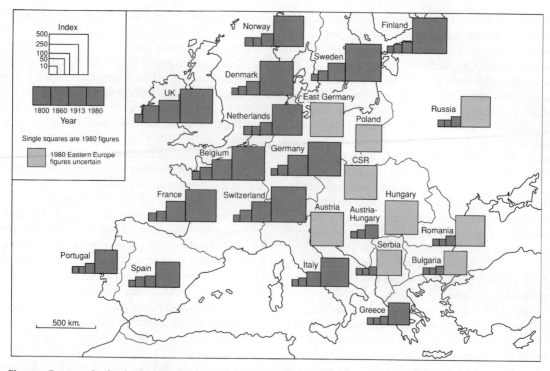

**Fig. 13.1.** European levels of industrialization 1800–1980—Index: UK 1900 = 100. (*Source*: Bairoch, 1982: 281, 302.)

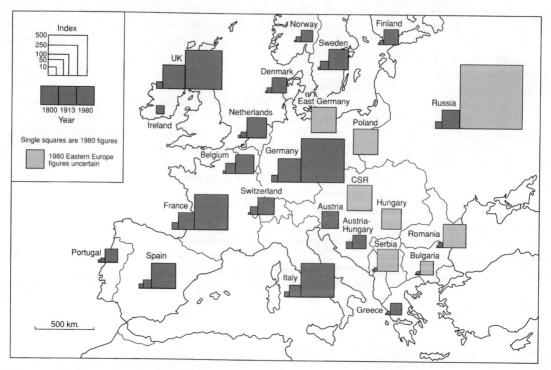

**Fig. 13.2.** Total industrial capacity of European countries 1800–1980—Index: UK 1913 = 100. (*Source*: Bairoch, 1982: 292, 299.)

There were islands of modernized economic activity in seas of traditionalism both in the west and in the east, though appropriate to their different absolute levels.

There is thus a distinct geographic dimension to the industrialization process in Europe. It first affected Great Britain, hardly disturbing the rest; it next, in a second phase, spread outward into north-western Europe; and only after almost a century's delay did it begin to affect substantially the European periphery in the east and the south of the continent, operating by then in different circumstances and under different technical and world economic conditions. This chapter will follow the same sequence.

## The Industrial Revolution in Great Britain

Because of its unique nature and early start, the industrialization process in Britain is frequently referred to as an industrial revolution, or perhaps even as *the* Industrial Revolution, or its prototype which other countries then followed—though the term has recently come in for some criticism (O'Brien, 1993; Pollard, 1992). This leads immediately to the question: Why was Britain first?

Leaving aside the possibility that the breakthrough in Britain occurred by mere chance, widened by cumulative advantages arising from that lead (Crafts, 1977; Landes, 1993), several explanations have been offered. The likelihood is that all contain an element of truth and that it is the conjunction of all of them which favoured Britain in such remarkable fashion. Some favourable preconditions which were conducive to economic growth in Britain existed in other parts of Europe also, especially in France, the most likely competitor (Crouzet, 1990: 12–43). These included a certain standard of income and wealth, of industrial skills and scientific knowledge, a reasonably favourable legal framework and internal security, and expanding foreign trade links. Britain, additionally, enjoyed, among other factors, the advantages of freedom from foreign invasions, a secure monetary and tax base, the virtual abolition of guild restrictions, a long tradition of individualism (Macfarlane, 1978), and, more concretely, a good internal transport network and excellent harbours, a situation athwart the world's major sea routes, and some valuable resources, including iron, copper, tin, and coal.

Coal was perhaps the single most important factor in ensuring a British lead. Found in several regions and in part outcropping or easily mined, coal clearly determined the locational distribution of British industry in the period of industrialization (Flinn, 1984; and Fig. 13.3). Providing cheap, concentrated energy, it was a main component of that key aspect of the industrialization process, the substitution of relatively limited organic sources of energy and raw materials by almost unlimited mineral resources (Wrigley, 1988). With an output of over 2.6 million tons in 1700 and 15 million tons in 1800 (Hatcher, 1993: 68; Flinn, 1984: 26), Britain produced at each stage several times more coal than the rest of Europe put together.

Possibly even more important than a supply of energy was the role played by coal mining in furthering the advance of technology. Coal had allowed Britain to take the lead in some earlier innovations, such as metal-smelting by reverberatory furnaces, salt-boiling, brewing, and glass-making. In the Industrial Revolution itself, the needs of mining and the possibilities created by it were responsible for two of the most important innovations of the age: the steam engine, generally considered to be at the heart of the industrialization process in Britain, and the railways, at its heart in much of the continent. Both James Watt's steam engine and its predecessor, the Newcomen engine, were originally conceived as methods of pumping water out of mines, and later also for haulage work about the pits. Of 2,191 engines known to have been built in Britain by 1800—possibly ten times as many as in the whole of the rest of Europe—no fewer than 828 were erected in coal mines (Kanefsky and Robey, 1980: 181).

Transport also owed much to the collieries. Something like a transport revolution was a necessary accompaniment to industrialization. Road improvements were largely the work of turnpike trusts, which charged tolls to road users: by 1750, 143 trusts had been established responsible for 3,000 miles of roads. The next stage was the building of canals, mostly in two bouts, 1759–74 and the 1790s: by 1850, the system possessed 7,200 km. of them (Ville, 1990: 14, 31). Most of them were built for the purpose of transporting coal. Coal

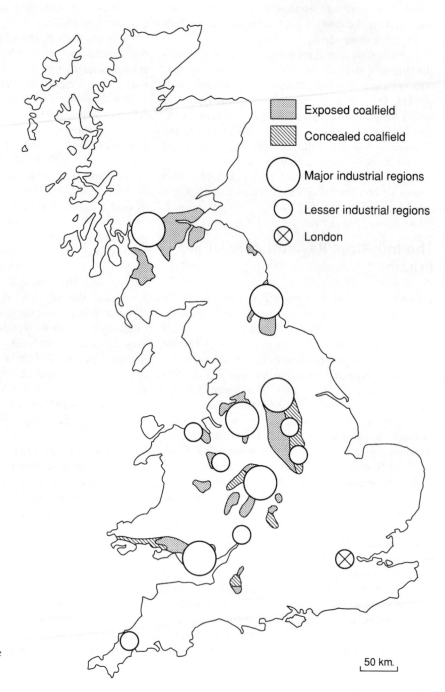

Exposed coalfield

Concealed coalfield

Major industrial regions

Lesser industrial regions

London

**Fig. 13.3.** The location of the major industrial regions in relation to coal supplies in Britain, eighteenth and nineteenth centuries.

50 km.

mining was responsible more directly still for the development of steam railways until their emergence into a complete system in the Stockton and Darlington Railway, itself a coal line, in 1825. After an unexampled expenditure of capital, raised mainly in two 'manias', 1835–6 and 1845–7, 9,797 km. of line were open in 1850 (Ville, 1990: 115). It should be noted that, while in the rest of Europe railway building was part of the industrialization process, or even preceded it, in Britain, the pioneer country, it took place at its end and, in a sense, as its culmination.

Next to coal mining, it was the textile mills which were the most extensive users of steam power (Tunzelmann, 1978). Textiles were, with coal, somewhere near the centre of the British industrialization process. Cotton spinning, a relatively new industry, became the 'leading sector' in Rostow's theory (Rostow, 1960: 52–4) after the inventions of the 1760s and 1770s by Hargreaves, Arkwright, and Crompton, and it may be said to have originated the factory system. Early water-driven mills were established in remote valleys in the southern Pennines (Chapman, 1967) and in Scotland, where land and power were cheap, but the industry soon moved to the towns of Lancashire and the adjoining portions of Derbyshire and Cheshire where coal for steam power and a larger labour supply were available. With its cotton mills, its engineering, chemical, and other works, this became the most highly concentrated industrial region in Britain. A second concentration was established along the lower Clyde. From its beginnings, cotton manufacture became the fastest growing and leading export industry (Chapman, 1987; Farnie, 1979). The woollen and worsted industries, an old manufacture with earlier concentrations in the West Country and East Anglia, settled on the eastern slopes of the Pennines, in Yorkshire, and applied mechanization some thirty years after cotton (Heaton, 1965; Jenkins, 1975). The linen industry, regionalized in Scotland, Northern Ireland, and North Yorkshire, followed in the 1830s.

Iron making was the third significant innovating sector. Following the earlier discovery of the coke smelting process, iron puddling, which permitted the mass production of refined forge or bar iron, was perfected in the early 1780s. Again highly concentrated in a few regions, above all the West Midlands, South Wales, and Scotland, the industry's annual output rose from 27,000 tons of pig iron in 1720–4 to 250,000 tons in 1805 and 2,700,000 in 1852 (Riden, 1977: 448, 455), a hundredfold increase in little more than a century. The output of bar iron, which became in many ways the basic material of the Industrial Revolution, grew at about the same rate.

While these four sectors—coal, engineering, textiles, and iron—are generally considered to have been the key industries of the British industrialization process, there was hardly a manufacturing industry that was not also affected. Thus in the metal trades, numerous processes were mechanized in workshops (Berg, 1991), chemicals underwent revolutionary changes, especially in the production of sulphuric acid, and in the dyeing and bleaching processes; printing, paper, pottery, grain milling, and brewing were among others that were transformed, in many cases out of recognition (Musson, 1978). The Great London Exhibition of 1851, which celebrated the triumph of the new industrialism, drew most of its exhibits from the less spectacular, but equally progressive, non-factory sectors.

It remains to be emphasized that most of these changes were concentrated in a small number of provincial industrial regions and in London. Among the former, Lancashire, the West Riding of Yorkshire, the West Midlands, the North-East of England, South Wales, and South-West Scotland stand out (see Fig. 13.3). Much of the rest of the country remained rural and agricultural, supplying food, raw materials, and labour power to the industrial conurbations. There are no reliable occupational census data before 1851, but the census of 1811, despite its numerous weaknesses, may be used with caution to illustrate this uneven regional development. In Fig. 13.4 the numbers of 'families chiefly employed in trade, manufactures, or handicraft' are compared with those chiefly employed in agriculture, on a county basis: the ratios varied from 0.11 in Sutherland to 8.64 in Renfrew. In England (leaving out London), the extremes were 0.40 for Hereford and 4.91 in Lancashire, a differential of 12:1.

## The Industrializers on the Continent

The industrial developments in the British Isles were observed with interest on the Continent, and governments as well as private individuals

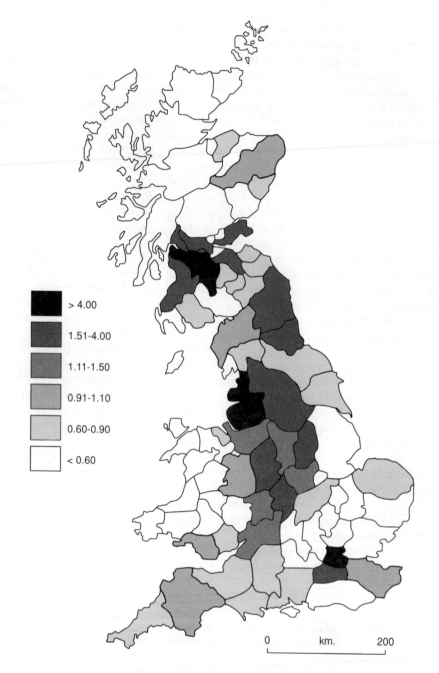

**Fig. 13.4.** Employment in trade, manufactures, and handicrafts in Great Britain, 1811, by counties. Ratios represent the number of families listed as chiefly employed in trade, manufactures, and handicrafts, compared with those listed as chiefly employed in agriculture. (*Source*: Census of Great Britain, 1811.)

> 4.00

1.51–4.00

1.11–1.50

0.91–1.10

0.60–0.90

< 0.60

0    km.    200

made deliberate attempts to copy them. Technical details were available from scientific journals and from the reports of industrial spies, and numerous British emigrants, entrepreneurs and skilled workers alike, found their way abroad in attempts to introduce British methods there (Henderson, 1954). With rare exceptions, these had no lasting success before 1815. There were several causes for this. Some were 'natural': lack of coal or iron ore, or their inaccessible location; distance from the sea or from navigable waterways; high mountains or severe climate. Others were social or historical:

governments and class systems hostile to middle-class entrepreneurs; poverty and illiteracy; the absence of industrial traditions and of established trading links; and, not least, the destruction and diversion of resources caused by war. There were also the sheer practical difficulties of mastering complex new techniques, such as steam-engine building and maintenance (Tann and Breckin, 1978), and the problems faced by pioneers when associated services and skills, taken for granted in Britain, were missing in Continental locations (Harris, 1986).

When industrialization finally did take place on the Continent, it followed, in its first phases, until c.1870, the British model: steam power, iron and coal technology, textile mechanization, the factory, and in due course the railways were the hallmarks of industrialization there, too. Only rarely did Continental technicians strike out on their own, one well-known example of this being the development of water power in France. In consequence, the regions which industrialized first were generally those with endowments closest to those of Britain (Pollard, 1981: 85–6).

Belgium's resources matched those of Britain best: coal and iron in the east and south, a long-established textile industry in the west as well as the east of the country, good water communications, and an excellent location in relation to other advanced areas. Belgium was therefore not merely the first Continental country to become industrialized in the modern sense; she also took over some of Britain's role in spreading industrialization further, by the export of capital, and of coal, iron, machinery, and know-how to neighbouring countries, in turn. In the boom year of 1837, that small country had a larger coal output (3.2 million tons) than France or Germany (3.0 million tons each). Her output of pig iron, at 150,000 tons was only slightly below that of Germany, which registered 175,000 tons and just under half of that of France, where 332,000 tons were produced in that year (Mitchell, 1975: 360, 391–2).

Politics played an important part in the Belgian story. During the revolutionary and Napoleonic period, her provinces benefited by having privileged access to the huge protected French market. From 1815, something of the same benefit arose by being linked to the Netherlands. When, after 1830, Belgium was on her own, her industrial progress was sustained by what was then a unique form of support given by two giant investment banks, the Société Générale (established in 1822) and the Banque de Belgique (1835) (Caron, 1978: 495–7). There was also the advantage of being a 'small country', in which the industrial sector was not overwhelmed numerically and politically by a powerful landed class which opposed legislation favouring industry; instead, on the contrary, Belgium possessed a farming community dependent on sales to the manufacturing population and therefore willing to encourage it (Milward and Saul, 1973: 451–3). Government support of railway building, designed to create a system that would draw to itself north–south and east–west trade between Belgium's larger neighbours, was particularly notable (Laffut, 1983: 203).

While Belgium benefited from being situated in the middle of the north-western European coalfield (Wrigley, 1961; see Fig. 13.5), the other small country among the very first industrializers, Switzerland, had no such advantage.

In fact, her landlocked position and high mountains, and the absence of coal or iron resources, added to the lack of political unity before 1848, which weakened her hand in trade negotiations, seemed to have destined her for relative backwardness. Yet Switzerland was among the first to mechanize her textile industry, to which were added engineering and chemicals, while the watch industry, centred on Geneva and the Jura mountains, came to dominate world markets in the first half of the nineteenth century.

One explanation for the Swiss success is that the mountains turned out not to be entirely a disadvantage. For one thing, they provided water power and cheap land for the industrial population which began to settle in the foothills. Secondly, by forcing the population to depend on meat and dairy farming from the Middle Ages onward, they encouraged early trading links with the Italian plains in the south and the grain of Swabia and the Rhine valley in the north, fostering the growth of a class of trading and transport entrepreneurs.

Moreover, the upland poverty, which forced so many young men to hire themselves out abroad as soldiers between the fifteenth and nineteenth centuries, not only channelled capital into the Swiss cities that clinched these lucrative mercenary deals, but brought trading concessions to Swiss cantons

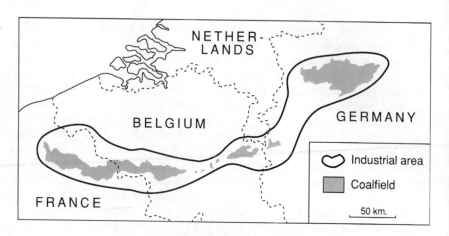

Fig. 13.5. The coalfields of
north-western Europe.

from the leading hirers of troops, above all the kingdom of France (Pfister, 1992; Bergier, 1983: 52). In the critical early decades of the nineteenth century, the Swiss also contrived to benefit by the trade with Britain in a way which was later used successfully also by the German states: they imported cheap machine-produced yarn, which their low-paid workers then worked up by weaving and embroidery at a time when no machines existed as yet for those stages in the production process. Later, the Swiss were not only to build their own textile machinery, but to pioneer the development of machine embroidery.

France, much the largest economy in western Europe in the early nineteenth century, included regions of widely differing character. While her overall slower advance has often been compared unfavourably with that of the United Kingdom (Crouzet, 1990), she had some regions that were among the most advanced in Europe. These included the north, an old textile region, but becoming, after the development of the concealed coalfield discovered in 1847, a major European centre for heavy industry; Alsace, France's most important cotton textile region, an innovative centre of mechanization and applied chemistry; Lyons, the centre of the silk industry, together with the nearby upper Loire valley around St Etienne, with its coal, its iron industry, and silk-ribbon weaving; and, not least, Paris and suburbs, where paper and printing, engineering, fashion goods production, and many other flourishing consumer goods industries were to be found.

It was mainly in regard to the key industries of textiles, iron making, and steam engineering that the industrialization of France has traditionally been regarded as disappointing and laggard (Landes, 1969: 164). One problem was the disadvantage of the scattered distribution of the country's natural resources. It was counteracted in part in the first half of the nineteenth century by constructing, mainly between 1818 and 1848, an extensive system of navigable canals linking the main rivers with each other. After the mid-century, as Chapter 15 shows, a substantial railway network was built up following the outlines of two major state-backed plans, that of Legrand (1842) and that of Freycinet (1879). The start was slow, and in 1850 rather less than 3,000 km. of line were open, but by 1910 this had risen to 40,000 km. (Ville, 1990: 115; 119–20).

Coal resources were not only scattered, they were also less in quantity than those available to Britain and Germany. Other causes for the slow growth of the key industries have been said to include the disturbances due to political upheavals, high tariffs allowing poor entrepreneurship to survive, and a strongly traditional peasantry firmly anchored to its soil as a result of the revolutionary settlement, reluctant to leave agriculture for the towns and industrial employment. There was also the tendency to invest too much capital abroad, especially in European railways (Cameron, 1961), starving French home industry in the process.

However, recent work which has shown that the French economy, including the manufacturing sector, achieved quite respectable growth rates (O'Brien and Keyder, 1978: 57–61; Lévy-Leboyer and Lescure, 1991: 153–5; but see Crouzet, 1990: 348–9) has led to a major reappraisal of the French industrialization process. For one thing, the country's slow population growth hides a rather better per capita performance than the low absolute growth rate would lead one to believe: on a per capita basis, French output growth for 1815–1914, at 1.2 per cent a year, was only marginally below the British annual rate of 1.3 per cent (Pollard, 1990: 30). Secondly, given her different resource endowment and the inherited dominance of world markets by Britain, it would have been the wrong policy for France to have followed the British path and create the pattern of a classical 'industrial revolution' (Trebilcock, 1981: 139 f.). Instead, by using the asset of a skilled labour force and of gifted designers, the French maintained their momentum by 'upstream industrialization', expanding their fashion and other consumer industries first, rather than the crude early stages of production (Cameron and Freedeman, 1983), Even in those, however, there were always some French firms that were among the leading European concerns. French technology never lost contact with the latest available technical improvements and not infrequently took the lead, and the standard of living of the French population remained among the highest in continental Europe. Having thus travelled by a quite different route, the French economy had by 1914 nevertheless converged on the structure achieved by the other leading countries, though the national figures were dragged down by a larger, more traditional agricultural sector.

Germany, by contrast, has always been considered to be the Continent's most successful industrializing country. This view is based on at least three factors: her backwardness at the beginning of the nineteenth century which made her later achievements all the more striking; she followed the British model closely, thus conforming to the expected pattern of developing coal mining, iron making, textiles, and engineering; and by 1914 she had caught up with and in some cases even overtaken Britain in absolute terms: thus in 1913, her output of crude steel, generally considered to be a good guide to industrial prowess, was 17.6 million tons, thus greatly exceeding that of Britain's 6.9 million tons (Mitchell, 1975: 400–1). Her overtaking of France is clearly depicted in Figs. 13.1 and 13.2. Although overall German output and incomes per head were still only around 70 per cent of the British in 1913, Germany's GDP growth rate, at 1.6

per cent a year per head and 2.8 per cent overall in 1870–1913 which greatly exceeded the British rates of 1.0 and 1.9 per cent respectively (Maddison, 1982: 44–5; Tilly, 1991: 180) threatened to catch up with and overtake the United Kingdom in the foreseeable future. This threat was emphasized by Germany's successes in some of the more modern industries, such as chemicals and electrical engineering. By that time she had become much the most formidable continental economy.

Her natural resources certainly helped Germany to achieve a rapid rate of industrialization: thus her leap forward would have been unthinkable without such assets as the Ruhr coal mines. But there were favourable social factors also. These included a skilled, well-educated, and disciplined labour force, widely credited to compulsory schooling introduced in Prussia and in other German states earlier than elsewhere in Europe, and by universal military service. Critics, however, maintain that schooling was still defective in the villages where a large part of the urban working class had its origin, while 42.6 per cent of the employed population still actually worked in agriculture in 1882 and 28.4 per cent in 1907 (Mitchell, 1975: 156). Similarly, compulsory military service does not seem to have aided the discipline of either the Austrian or the Russian workforce. Among other factors, it cannot be doubted that German society threw up some vigorous entrepreneurs, aided by a tradition of forming cartels and associations which were encouraged by the law rather than being hampered, as in Britain.

The State helped development in Germany, first by engineering the economic unification which began with the Customs Union of 1834, and latterly by a carefully balanced 'scientific' tariff which favoured both manufacturing and grain producers. Germany was thus able to use her central European position to acquire, in the first half of the century, cheap and advanced machines and semi-manufactures from Britain, and sell finished manufactures to eastern European, overseas, and even western European markets. Finally, the role of the large 'universal' German banks should be mentioned, which provided ample finance for industrial companies in growth periods and nursed them during slumps.

German industrialization was possibly even more concentrated regionally than that of France. Much the most important industrial region of Germany, and the most powerful concentration of industry in Europe in 1913, developed along the Ruhr and Rhine rivers. As a latecomer to the industrial scene, the rise of the 'Ruhr' was meteoric. It was based on coal, of which 1.6 million tons were raised in 1850 and 88.3 million in 1908–12 (Pounds, 1985), a more than fiftyfold increase. The largest and most modern concentrations of steel making, armaments, and heavy engineering in Europe were added, as well as chemicals and other industries. Older textile regions to the southern and western rims of the coalfield may be added to the complex. Saxony, the second most important region, was an old industrial area, with textiles in the plains and mining and miscellaneous industries in the mountains. It was the most densely populated country in Europe and in the second half of the nineteenth century the proportion of its working population engaged in industry and mining was as high as that of the Ruhr area, being over 60 per cent (Trebilcock, 1981: 51). Upper Silesia, rich in coal and mineral ores, was another industrial concentration, and so was the capital city of Berlin.

Germany was among those countries (the USA was another) in which the building of railways was a critical element in the industrialization process (Pollard, 1990: 51; Fremdling, 1983: 121; Fremdling, 1985). In 1850 some 5,856 km. of lines were open, forming much of the largest network on the continent at the time; by 1913 this had grown to 63,379 km. (see Fig. 15.2). The backward linkages were particularly significant in the stimulation of iron, later steel, and coal production. The forward linkages are best represented by the increase in ton-kilometres of freight carried, from 303 million in 1850 to 61,744 million in 1913, a 200-fold increase (Fremdling, 1983: 132). Railways linked the formerly sleepy valleys, the eastern agricultural hinterland, and the industrial cities of Germany with world markets more effectively than in any other part of Europe, with the possible exception of Belgium.

## The Industrialization of the Periphery, 1870–1939

From the 1860s onward, industrialization took a new turn, sometimes referred to as the second industrial revolution. Technical breakthroughs of a novel kind occurred in both traditional and new sectors: steel making, electrical engineering, motor-car production, chemicals, and precision instruments, among others. They might be described as 'high-tech', requiring much sophisticated technology; they were also, with few exceptions, carried forward by large, dominant, even monopolistic firms. At the same time, there was also a rapid development in mass-produced consumer goods industries, such as the production of cigarettes, chocolates, patent medicines, and newspapers. While Britain still led in the latter, Germany now led in the former. Very soon the development of electricity technology was beginning to benefit regions well supplied with water power, including Switzerland, northern Italy, eastern France, and Norway, as well as initiating a move away from the coalfields elsewhere. What all this meant was that the later industrializing countries of the European 'periphery' faced a different and harder task, since they had to jump a wider technological gap to become up-to-date, while at the same time being less well endowed for industrialization, as proved by their relative backwardness. They therefore required different, more deliberate industrialization policies than the early industrializers—a point associated particularly with concepts made familiar by Alexander Gerschenkron (1962).

The term 'periphery' is somewhat misleading in this context: it is meant to refer to the remainder of Europe outside the north-western countries discussed so far. Some of the countries concerned were by no means peripheral in European history: they included the Netherlands, which managed to modernize its economy without difficulty from an originally high standard, in the later nineteenth century, and the four Scandinavian countries. The latter, starting from a low level of incomes around 1850, made exceptional progress to 1913, as shown in Figs. 13.1 and 13.2, and have kept up their momentum since then. Among the causes of their success, compared, for example, with the Mediterranean countries starting from similar levels, must be included an established rule of law, a high level of education, and a reasonable distribution of property and political power. The three northern countries additionally possessed more valuable natural resources, such as timber and minerals, which they managed to develop with the help of western capital, thus refuting the common notion that suppliers of raw materials are always exploited victims of the advanced economies. Denmark, without rich raw materials, had to break through into modernity by revolutionizing her agriculture.

In the rest of Europe, industrialization was introduced only after much greater delay. Among the physical causes were lack of resources, above all coal and iron ore, and distance made worse by the delay in creating modern transport networks, an important factor in the Russian Empire. Of at least equal significance were social causes: a poverty-stricken, ill-educated population; harsh, even feudal conditions in the countryside, including true serfdom in Russia to 1861; corrupt bureaucracies and incompetent and irresponsible governments (Berend and Ranki, 1977).

Not all the 'periphery' exhibited all these features. There was, in fact, a transition zone between the north-west and the regions of the most serious underdevelopment. Typical for this transition was the Habsburg Empire. Parts of it had always been considered as lying within the core of Europe, and those parts were also among the early industrializers, but were swamped, within their political boundaries, by backward agrarian provinces. The differences between the advanced and laggard parts of Europe are thus in a curious way mirrored within the smaller compass of Austria-Hungary (Ashworth, 1977: 148; and see Fig. 13.6).

Upland Bohemia, possessing old-established textile, glass, and mining industries, now added coal, iron, and engineering to form a modern industrial complex. Vienna with its suburbs in Lower Austria, as well as Budapest, and Prague contained other concentrations of modern industry and transport networks, and had relatively high incomes, while Vorarlberg, a surprisingly precocious factory

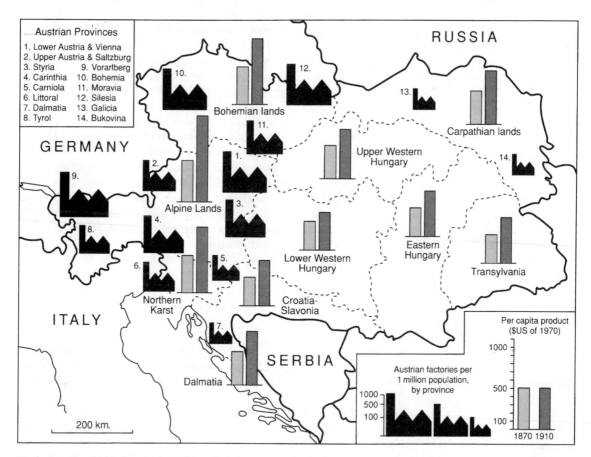

**Fig. 13.6.** Regional variations in the Habsburg Empire 1870–1910: real per capita product and numbers of factories. (*Sources*: Good, 1991: 228; Matis and Bachinger, 1973: 232.)

district, benefited by an overflow from neighbouring Switzerland.

Up to 1850, Austria had kept pace with Germany and even France in some key sectors and fell markedly behind only thereafter. She had a highly developed capital market and, regionally, a good railway system. What held her back was the deadweight agrarian population in the east. After the breakup of the empire, neither the rump German-speaking Austria nor the newly formed Czechoslovakia had any difficulty in approaching western industrialization levels in the inter-war years.

A similar kind of 'dual' economy also developed, slightly later, in Italy after the country's unification, which combined the go-ahead, almost central European north with a group of southern provinces that in part were little above North African levels. The differential between them pre-dated the unification and cannot fairly be blamed on the government, as it often has been; but it is undoubtedly true that new investment, especially by the banks which played a crucial role as in Germany, tended to go to the regions which were already advanced. Northern Italian towns had a long-established manufacturing, trading, and banking tradition. In the absence of any substantial coal or iron deposits, the particular strengths of the Italian manufacturing sector were, on the one hand, engineering in the 'industrial triangle' between Turin, Milan, and Genoa, and textiles along the Alpine slopes in mills that were originally water-driven, but later linked to the hydroelectric network. Motor car production, which was to become one of Italy's most important successes, and after the Second World War was to take her to the forefront of industrial advance, was one consequence of early Italian engineering progress, as were mopeds, domestic electric equipment, and office machinery.

As a relative latecomer, Italy depended more than the other large countries considered so far on foreign investment, as well as on government initiatives, which included railway building, an active tariff policy, and a catching-up effort in the educational field. After the great 'spurt' c.1897–1913 (Cafagna, 1973: 297) it was clear that Italy had joined the industrial part of Europe, though the absolute level of per capita incomes and output was still well below that of north-western Europe (see Figs. 13.1 and 13.7).

Spain, also, had a long tradition of urban wealth and international trade, but her reaction to the pressures exerted by the advancing European core were slower even than those of Italy, and her industrializing regions formed a smaller part of the whole. Catalonia, a highly industrialized textile and engineering region, found it difficult to keep up with the changes of the later nineteenth century, and only the Basque country was fully integrated with the world economy, even if it was as an exporter of raw material, her valuable iron ore, rather than as a semi-finished steel producer. The proportion working in agriculture, 56.3 per cent in 1910, was comparable to that of Italy and Austria, but much of the agricultural sector was poor and backward. As in southern Italy, the productivity even of the large estates or latifundia, which represented 17.6 per cent of the cultivated land in 1900 (Trebilcock, 1981: 323, 328) was low and did nothing to advance national progress.

Leaving out the Balkan states, much the most backward part of the European economy was to be found in Russia: it had the lowest incomes, lowest agrarian yields, and poorest communications. Yet the Russian Empire, though handicapped by vast distances and a severe climate, was rich in natural resources and, given the large population, even a relatively small per capita output would soon show up in European totals (see Fig. 13.2). Much of Russia's traditional manufacturing output had been produced in domestic, peasant *kustarny* workshops, but there were also more modern units to be found from the first half of the nineteenth century onwards, including textiles and metal factories around Moscow, various industries, including later railway engineering in St Petersburg, and mining in the Urals. The discovery and development of coal in the Donets Basin in the late nineteenth century, which helped to raise the empire's total output from 1 million tons in 1872 to 36 million tons in 1913, heralded a new era in conjunction with the ironfield discovered at a convenient distance in Krivoy Rog in 1882. Together they were largely responsible for the increase of steel production from 307,000 tons in 1880 to 4.9 million tons in 1913 (Mitchell, 1975: 399–40; Portal, 1966: 828). To these has to be added the huge oil production in the Baku field, which represented one-third of the world total in 1900 and still one-sixth in 1913. Industry and mining grew overall at a rate of 5

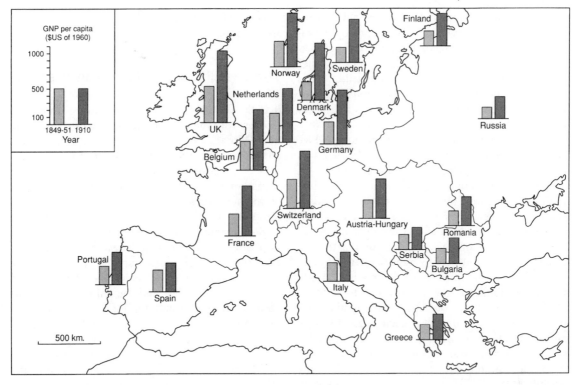

**Fig. 13.7.** Europe's per capita Gross National Product 1849–1919 ($US of 1960). (*Sources*: Based on Bairoch, 1976: 286; Crafts, 1983: 389, 394.)

per cent a year between 1860 and 1913, at a time when population increased by 1.5 per cent: continued over such a long period this growth rate was unprecedented (Grossman, 1973: 489).

In view of the general poverty and low rate of saving in the country, much of the capital came from foreign investors whom the government consciously attracted, particularly under finance minister Witte (1892–1905). Much of this foreign capital was channelled either through the banks which played a large part in financing modern industry, or by investment in government loans which were then transferred to building up the railway system. By 1914 foreign investments represented one-third of the capital of the 2,000 or so companies, but the proportion was higher in certain key sectors, being 90 per cent in mining and 50 per cent in chemicals (Portal, 1966: 851). As the last of the latecomers, Russia would have not only to make use of government and banking initiatives to overcome the enormous gap in development but, according to Gerschenkron, required also an ideological drive

to break down the traditionalist barriers, and possibly also to make palatable the inevitable hardships of the transition period. This clearly applied before 1914, but took on a particular shape after the Bolshevik revolution of 1917 thanks to the strongly ideological motivation of the new rulers, who turned what remained of the Russian Empire after the war into the Soviet Union.

When internal peace had at last been established in 1921, the Soviets were faced, not merely by the devastation of war and civil war, but also by the loss of most of the educated professional classes and by what amounted to a world boycott. A first phase lasting to 1928, the 'New Economic Policy', restored the economy to some kind of equilibrium; it was followed by a massive and revolutionary transformation, designed to carry through industrialization in the shortest possible time in order to lay foundations for the building of 'Socialism in one country', and protect it from foreign aggression. Quite apart from the immense capital investments required, there had to be also a massive transfer of labour from the villages into the towns while keeping up the food supply for a growing population and preventing the rise of a capitalistic agrarian class, the kulaks, in the process. As this transfer could not be carried through voluntarily, it was enforced at enormous cost and with appalling brutality: possibly 10 million people died as a result of hunger and persecution.

In the end, by means of a set of consecutive Five Year Plans, and by concentrating on capital goods industries while holding down the production of consumer goods, something akin to a classic industrialization process, appropriate to the twentieth century, was accomplished by 1940. With an output of 166 million tons of coal, 31.1 million tons of oil, 18.3 million tons of steel, and 5.7 million tons of cement, to cite but a few representative figures (Pollard, 1981: 299) the Soviet Union had clearly joined the ranks of the industrialized world. If available figures are to be believed, her total output index, with 1913 = 100, had reached 210 by 1939, and even output per head stood at the respectable index of 161.5 in 1939 (Maddison, 1969: 155, 159). Partly for sound strategic reasons, and partly because of the opportunities opened up by modern technology, the Soviet industrial revolution also involved a major shift eastward, into the Asiatic provinces. Thus the share of coal accounted for by the Donets Basin fell from 87 per cent in 1913 to 60 per cent in 1937, while Siberia, Kazakhstan, and the Urals now delivered 28 per cent (Dobb, 1948: 390, 393); Magnitogorsk in the southern Urals became the new metallurgical centre; and an increasing share of hydroelectric power was developed east of the Urals, where also many of the most valuable minerals were to be found.

Although private consumption hardly improved at all in the Soviet Union between 1928 and 1940, the State had carried through an immense educational programme to eliminate the widespread illiteracy and produce the scientists and technologists required by modern industry, and it had also built up a medical service of European standards almost from scratch. It should be remembered that this advance was accomplished while western Europe was caught in the Great Depression of mass unemployment and the destruction of assets. Whatever the total costs and needless sufferings, in the end her industrialization effort enabled the Soviet Union to withstand the onslaught of the most

powerful war machine of the age, the German army, in 1941.

## The Last Phase: 1945 to the Present

It took some years for the European economy to recover from the worst of the destruction caused by war. When it had done so, from the late 1940s onwards, it set out on a course of quite unprecedented economic growth. Some observers have termed it the third industrial revolution. Among its main features, apart from the growth in consumption and the extension of leisure which it made possible, were the shift from coal and iron to plastics, non-ferrous metals, and electricity; the enormous rise of motor-car and air traffic; the expansion of the products of the electrical and electronics industries; and the shift from agricultural to industrial and service employment, and from blue-collar to white-collar jobs. The transition from coal to oil and electrical power also allowed locations to be chosen more freely: the result was a significant geographical shift from the old, crowded, and polluted industrial centres to more desirable locations, usually involving a move to the south: to the French Mediterranean coast, to Baden-Württemberg and Bavaria, to the Thames valley and the Bristol area. The industrial regions which once carried the industrialization process, and had the highest incomes, now became problem areas of unemployment and inner-city deprivation.

Ultimately, it was the growth in productivity which was the foundation of the third industrial revolution (see Fig. 13.8). It will be seen that while productivity tended to double between 1870 and 1913, it has increased around fivefold since then in almost all countries, most of the improvement having taken place since 1950. Moreover, there has been a strong tendency to convergence among the western European economies. The basis for both these tendencies lies in the astonishing technical advances which have become available to all.

In one way, this has made the task of the countries which still had to pass their industrialization stage all the harder, since they had an even wider chasm to bridge. But in another sense it has eased their progress, for they could now skip some costly earlier stages, and choose among a variety of technologies and a wider range of energy sources. They

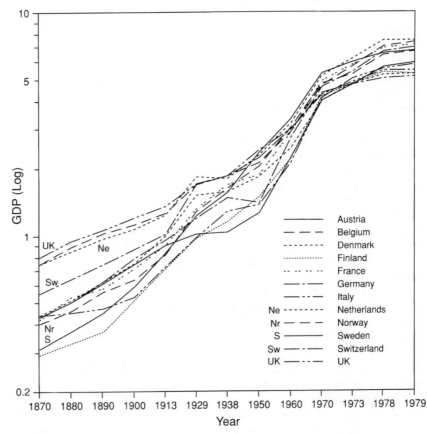

GDP (Log)

| | Austria |
|---|---|
| | Belgium |
| | Denmark |
| | Finland |
| | France |
| | Germany |
| | Italy |
| Ne | Netherlands |
| Nr | Norway |
| S | Sweden |
| Sw | Switzerland |
| UK | UK |

1870  1880  1890  1900  1913  1929  1938  1950  1960  1970  1973  1978  1979

Year

**Fig. 13.8.** Productivity per work-hour in Europe 1870–1979. (*Source*: Maddison, 1982: 212.)

also benefited from the fact that many of the new technical innovations were capital-saving, and thus reduced the costs of modernization.

The country which made the most successful use of these opportunities was Spain. With a per capita income in 1913 of $367 at 1960 prices (Bairoch, 1976: 307), or between one-quarter and one-third of the leading countries, and scarcely a modern industrial base even in 1960, Spain has since then undergone a remarkable phase of modernization and industrialization, including the foundation of that symbol of the third industrial revolution, a sizeable motor-car industry. Her GDP per head, at $11,738 in 1993 at purchasing power parity according to the OECD, was still well below that of the leading European countries; but her industrialized regions, such as Navarre ($15,236), Catalonia ($14,273), and the Basque country ($13,921) were by then only marginally below the United Kingdom level at $15,882. One factor which has helped Spain in recent years was her membership of the European Union. The same

could be said of Ireland, which should be included in the same category.

Quite different was the path taken by the countries in eastern Europe and the Balkans which were forced into the Soviet orbit. Despite the great differences between the economies of Poland, Hungary, Romania, Bulgaria, Albania, and Yugoslavia—Czechoslovakia and East Germany being already industrialized from the outset—they adopted remarkably similar policies. These involved the public ownership of all major industrial and commercial enterprises, total economic planning, a rapid shift of employment from agriculture to industry and the services, and the building up of capital goods industries while keeping back the manufacture of consumer goods. All these were to be expected from governments looking to the Soviet Union as their model. Rather more surprising was the decision of all of them to aim for autarky, that is, to become self-sufficient, with foreign trade playing a marginal role only. What had been a reasonable policy for the Soviet Union with her vast extent and natural resources, surrounded by enemies, autarky was a less obvious solution for the smaller countries with limited resources, which one would have thought might have gained a great deal by planned co-operation on the basis of an international division of labour.

The policies actually pursued were remarkably successful for a time, at least on their own terms. Thus employment in agriculture fell from 82.1 per cent in 1950 to 41.8 per cent in 1976 in Bulgaria, from 49.8 to 30.0 per cent in Hungary and from 53.5 to 39.9 per cent in Poland, and in the other countries in proportion (Pollard, 1981: 331). Employment in industry and the services rose correspondingly. Basic transport networks and heavy industry were created and the pressures of urbanization met by rapid, if shoddy, residential building.

Among the individual countries, Poland built up her most important industrial concentration round the older mining and metallurgical region of upper Silesia, extending it westward into former German Silesia and eastward as far as Cracow, and adding chemical and engineering plants. Textiles followed tradition in being concentrated around Lodz, and Warsaw was the centre of a third major industrial region (Pounds, 1969: 374–81). Hungary, short of minerals, has relied largely on expanding her electricity-generating capacity. Iron and steel making depended mainly on imported raw materials and was concentrated, as was much other industry, on Budapest and the north-east of the country. Yugoslavia, also, has few fuel resources apart from widely scattered brown coal. Her iron and steel industry is concentrated in the Zenica area, other works being in Kiksic and Skopje. Romania is the only European country with substantial oil

supplies; the rest of eastern Europe came to be supplied by pipeline from the USSR. The expansion of industry in a formerly largely agricultural country has been particularly rapid in Romania, and possibly even more so in Bulgaria. Only Albania remained largely untouched by industrialization.

One feature of this forced industrializing pace to strike the eye has been its sheer wastefulness (Aldcroft, 1978: 220), hidden for a time by the absence of consistent pricing. Managers went for output at any cost, ignoring long-term interests and alternatives. In consequence, the dynamic of rapid growth could not be maintained after the first twenty years or so, nor could eastern manufactured goods be sold in world markets: in due course, all the eastern countries became desperately short of hard currency, which blocked their access to western technology, and forced them to rely on that of the Soviets. Overall growth also slowed down, though at present it is not possible to be certain by how much. What is clear is that the output figures published and supplied to international organizations were unreliable (see Figs. 13.1 and 13.2), invented to hide the failure of the system. Now that something of the truth has been permitted to emerge, it appears that planning mistakes and management incompetence have meant that the sacrifices made by the past generation have largely been in vain. These countries are still the poor relations of Europe.

Nevertheless, the revolutions of eastern Europe have at least ensured that industrialization has reached that region also. Only the far corners, Portugal and possibly Greece, remain pre-industrial, and in their case, too, it is to be expected that, with membership of the European Union to help them, they will take the final steps in the near future.

REFERENCES

ALDCROFT, D. H., *The European Economy* (London, 1978), 1914–1970.

ASHWORTH, W., 'Typologies and Evidence: Has Nineteenth-Century Europe a Guide to Economic Growth?', *Economic History Review*, 2nd ser. 30 (1977), 140–58.

BAIROCH, P., 'Europe's Gross National Product: 1800–1975', *Journal of European Economic History*, 5 (1976), 273–340.

—— 'International Industrialization Levels from 1750 to 1980', *Journal of European Economic History*, 11 (1982), 269–333.

BEREND, I. T., and RANKI, G., *The European Periphery and Industrialization, 1780–1914* (Cambridge and Paris, 1977).

BERG, M., 'Commerce and Creativity in Eighteenth-Century Birmingham', in M. Berg (ed.), *Markets and Manufacture in Early Industrial Europe* (London, 1991), 173–201.

—— 'Factories, Workshops and Industrial Organization', in R. Floud and D. N. McCloskey (eds.), *The Economic History of Britain since 1700*, i (1994), 123–50.

—— and HUDSON, P., 'Growth and Change: A Comment on the Crafts–Harley View of the Industrial Revolution', *Economic History Review*, 2nd ser. 47 (1994), 147–9.

BERGIER, J. F., *Die Wirtschaftsgeschichte der Schweiz* (Zurich and Cologne, 1983).

CAFAGNA, L., 'Italy 1830–1914', in Cipolla (1973), 279–328.

CAMERON, R., *France and the Economic Development of Europe 1800–1914*, (Princeton, 1961).

—— and FREEDEMAN, C. E., 'French Economic Growth: A Radical Revision', *Social Science History*, 7 (1983), 3–30.

CARON, F., 'Les "Pays suiveurs": France et Belgique', in P. Léon (ed.), *Histoire économique et sociale du monde*, iii (1978), 453–503.

CHAPMAN, S. D., *The Early Factory Masters* (Newton Abbot, 1967).

—— *The Cotton Industry in the Industrial Revolution* (London, 1987).

CIPOLLA, C. (ed.), *Fontana Economic History of Europe*, iv (London, 1973).

CRAFTS, N. F. R., 'Industrial Revolution in England and France: Some Thoughts on the Question: "Why was England First?" ', *Economic History Review*, 2nd ser. 30 (1977), 429–41.

—— 'Gross National Product in Europe 1870–1910: Some New Estimates', *Explorations in Economic History*, 20 (1983), 387–401.

—— *British Economic Growth during the Industrial Revolution* (Oxford, 1985).

—— and HARLEY, C. K., 'Output Growth and the British Industrial Revolution: A Restatement of the Crafts–Harley View', *Economic History Review*, 2nd ser. 45 (1992), 703–30.

CROUZET, F., *Britain Ascendant: Comparative Studies in Franco-British Economic History* (Cambridge and Paris, 1990).

DOBB, M., *Soviet Economic Development since 1917* (London, 1948).

EVERSLEY, D. E. C., 'The Home Market and Economic Growth in England, 1750–1780' in E. L. Jones and G. E. Mingay (eds.), *Land, Labour and Population in the Industrial Revolution* (London, 1967), 206–59.

FARNIE, D. A., *The Cotton Industry and the World Market 1815–96* (Oxford, 1979).

FLINN, M. W., *The History of the British Coal Industry*, ii. 1700–1830: *The Industrial Revolution* (Oxford, 1984).

FREMDLING, R., 'Germany', in O'Brien, (1983), 121–47.

—— *Eisenbahnen und deutsches Wirtschaftswachstum 1840–1879* (Dortmund, 1985).

GERSCHENKRON, A., *Economic Backwardness in Historical Perspective* (Cambridge, Mass., 1962).

GILBOY, E. W., 'Demand as a Factor in the Industrial Revolution', in A. H. Cole (ed.), *Facts and Factors in Economic History* (Cambridge, Mass., 1932).

GOOD, D. F., 'Austria-Hungary' in Sylla and Toniolo, (1991), 218–47.

GROSSMAN, G., 'Russia and the Soviet Union', in Cipolla, (1973), 486–531.

HARLEY, C. K., 'British Industrialization before 1841', *Journal of Economic History*, 42 (1982), 267–89.

HARRIS, J. R., 'Michael Alcok and the Transfer of Birmingham Technology in France before the Revolution', *Journal of European Economic History*, 15 (1986), 7–57.

HATCHER, J., *The History of the British Coal Industry*, ii. *Before 1700* (Oxford, 1993).

HAWKE, G., 'Reinterpretation of the Industrial Revolution' in O'Brien and Quinault (1993), 54–78.

HEATON, H., *The Yorkshire Woollen and Worsted Industries* (Oxford, 1965).

HENDERSON, W. O., *Britain and Industrial Europe 1850–1970* (Liverpool, 1954).

JENKINS, D. T., *The West Riding Wool Textile Industry 1770–1815* (Edington, Wilts., 1975).

KANEFSKY, J., and ROBEY, J., 'Steam Engines in Eighteenth-Century Britain: A Quantitative Assessment', *Technology and Culture*, 21 (1980), 161–86.

LAFFUT, M., 'Belgium' in O'Brien (1983), 203–26.

LANDES, D. S., *The Unbound Prometheus* (Cambridge, 1969).

—— 'The Fable of the Dead Horse, or, The Industrial Revolution Revisited', in J. Mokyr (ed.), *The British Industrial Revolution: An Economic Perspective* (Boulder, San Francisco, Oxford, 1993), 132–70.

LÉVY-LEBOYER, M., and LESCURE, M., 'France' in Sylla and Toniolo, (1991), 153–74.

MACFARLANE, A., *The Origins of English Individualism* (Oxford, 1978).

MADDISON, A., *Economic Growth in Japan and the USSR* (London, 1969).

—— *Phases of Capitalist Development* (Oxford, 1982).

MATIS, H., and BACHINGER, K., 'Österreichs industrielle Entwicklung', in A. Brusatti (ed.), *Die Habsburgermonarchie 1848–1918*, i. *Die wirtschaftliche Entwicklung* (Vienna, 1973).

MILWARD, A. S., and SAUL, S. B., *The Economic Development of Continental Europe 1780–1870* (London, 1973).

MITCHELL, B. R., *European Historical Statistics 1750–1970* (London, 1975).

—— with DEANE, P., *Abstract of British Historical Statistics* (Cambridge, 1967).

MOKYR, J., 'Demand versus Supply in the Industrial Revolution', in J. Mokyr (ed.), *The Economics of the Industrial Revolution* (Totowa, NJ, 1985).

MUSSON, A. E., *The Growth of British Industry* (London, 1978).

O'BRIEN, P. (ed.), *Railways and the Economic Development of Western Europe 1830–1914* (London, 1983).

—— 'Introduction: Modern Conceptions of the Industrial Revolution', in O'Brien and Quinault (1993), 1–30.

—— and KEYDER, C., *Economic Growth in Britain and France 1780–1914* (London, 1978).

—— and QUINAULT, R. (eds.), *The Industrial Revolution and British Society* (Cambridge, 1993).

PFISTER, U., *Die Zürcher Fabriques* (Zurich, 1992).

POLLARD, S., *Peaceful Conquest: the Industrialization of Europe 1760–1970* (Oxford, 1981).

—— *Typology of Industrialization Processes in the Nineteenth Century* (Chur and London, 1990).

—— 'The Concept of the Industrial Revolution', in G. Doso *et al.* (eds.), *Technology and Enterprise in a Historical Perspective* (Oxford, 1992), 29–62.

PORTAL, R., 'The Industrialization of Russia', in H. J. Habbakuk and M. M. Postan (eds.), *The Cambridge Economic History of Europe*, vi. (Cambridge, 1966), 801–72.

POUNDS, N. J. G., *Eastern Europe* (London, 1969).

—— *An Historical Geography of Europe 1800–1914* (Cambridge, 1985).

RIDEN, P., 'The Output of the British Iron Industry before 1870', *Economic History Review*, 2nd ser. 30 (1977), 442–59.

ROSTOW, W. W., *The Stages of Economic Growth* (Cambridge, 1960).

SYLLA, R., AND TONIOKO, G. (eds.), *Patterns of European Industrialization* (London and New York, 1991).

TANN, J., and BRECKIN, M. J., 'The International Diffusion of the Watt Engine, 1775–1825', *Economic History Review*, 2nd ser. 31 (1978), 542–64.

TILLY, R., 'Germany', in Sylla and Toniolo (1991), 175–96.

TREBILCOCK, C., *The Industrialization of the Continental Powers 1780–1914* (London, 1981).

TUNZELMANN, G. N. von, *Steam Power and British Industrialization to 1860* (Oxford, 1978).

VILLE, S. P., *Transport and the Development of the European Economy 1750–1918* (London, 1990).

WRIGLEY, E. A., *Industrial Growth and Population Change: A Regional Study of the Coalfield Area of North-West Europe in the Later Nineteenth Century* (Cambridge, 1961).

—— *Continuity and Change: The Character of the Industrial Revolution in England* (Cambridge, 1988).

# Chapter 14

# The Urbanization of Europe since 1500

## A. Sutcliffe

## The European Urban Network in 1500

Thanks to the central role of the town in European development in the Middle Ages, almost all of the western part of the continent was integrated into a mature urban system by the sixteenth century (see e.g. De Vries, 1984: 9–10). Town foundations were very rare in western Europe after 1500. In eastern Europe, and especially in Russia, on the other hand, the very low level of economic development sustained a sparse distribution in which towns could grow, contract, or even disappear in just a few years of fickle fortune or tragedy (Bairoch, 1988: 170–2). In Russia, no functioning urban network existed. Most Russian towns were virtually isolated, with Bairoch estimating a national urbanization level of 3–6 per cent in 1500 (Bairoch, 1988: 179). Even the capital, Moscow, had a population of only 36,000 in 1500, and the largest cities in Russia were the Black Sea port of Feodosia and the Dnieper river port of Smolensk, both with about 50,000 inhabitants (Bairoch, et al., 1988: 60–5).

The level of economic development, even in western Europe, was however too low to sustain more than a marginal level of urbanization. In Europe as defined by De Vries (lying roughly west of a line drawn from Kaliningrad to Trieste), the level of urbanization, again as defined by De Vries (towns of over 10,000 inhabitants at some time between 1500 and 1800) was only 5.6 per cent of the total population in 1500 (De Vries, 1984: 20–1, 38–9). Bairoch, enumerating the population of towns of 5,000 or more people in the whole of Europe except Russia, arrives at a figure of 10.7 per cent urbanized in 1500 (Bairoch, 1988: 177).

In western Europe, urbanization was highly concentrated in a number of regions. The two main urbanized regions were northern Italy and the Low Countries (see e.g. Prevenier et al., 1992). Food and migration were generated largely within the regions, with population moving to the large towns via the settlement network. Additional supplies could be readily brought in by sea, and distributed by river. Both these regions had an important industrial production, notably of textiles, and much of it was carried on in the towns, which could dominate the countryside more easily here than in less urbanized areas. They also enjoyed a broader influence based on trade and finance, and they were closely linked with each other. Their largest towns, such as Florence (55,000 inhabitants in 1500), Bruges (35,000), Venice (100,000), and Ghent (55,000), had a great political and cultural influence (Bairoch *et al.*, 1988: 11–12, 40–9). Long-distance trade generated sea-based networks of towns outside the immediate region, with Venice controlling a number of ports in the eastern Mediterranean, while in the Baltic a similar network of ports was formally co-ordinated by the institutions of the North German Hanse, based on Lübeck (25,000 inhabitants in 1500). Functionally it was held together by the rich commerce between the corn-producing regions of northern Germany, Poland, and the Baltic states and the industrial towns and cities of the Low Countries.

These local, regional, and long-distance commercial links generated very high levels of urbanization, with Bairoch estimating 30–45 per cent in 'Belgium', 20–26 per cent in the Netherlands, and 15–20 per cent in Italy as early as 1500 (Bairoch, 1988: 179). Outside these regions, however, most towns were dominated by their links with the nearby countryside, and their existence was partially determined by rural life.

# Urban Growth after 1500

Seaborne trade expanded after 1500, particularly in northern Europe as the influence of Atlantic and Baltic commerce grew. Inland water transport also expanded and European urbanization tended to concentrate even more on the seaports and their hinterlands (Bairoch 1977: 18). Exchange continued to be the prime function of towns, and faster population growth in most of the continent during the sixteenth and early seventeenth centuries tended to boost urban populations, especially in western Europe (see De Vries, 1984: 28–9). A series of demographic and political setbacks thereafter slowed the rate of population growth, but urbanization continued, especially in the larger towns. The level of urbanization in the Europe of De Vries increased to 7.6 per cent by 1600 and to 9.2 per cent in 1700 (De Vries, 1984: 39). Bairoch's definitions produce percentages of 11.5 per cent and 12.3 per cent respectively (Bairoch, 1988: 177). In a new estimate based on towns of more than 2,000 inhabitants, Lepetit puts the urban population of France at 18 per cent around 1740, slightly higher than Bairoch's estimate of 12–16 per cent (Lepetit, 1994: 435; Bairoch, 1988: 215). Meanwhile, much of Germany was still recovering from the Thirty Years War (1618–48) and

subsequent discouragements to urban growth. Its level of urbanization, already 7–9 per cent in 1500, was still no more than 8–10 per cent in 1750 (Bairoch, 1988: 179, 215).

The distribution of towns in Europe scarcely altered, and the Low Countries and northern Italy saw their urbanization levels increase, with the Netherlands growing faster than any European region (Hohenberg and Lees, 1985: 47–55). In the Netherlands as seen by Paul Klep, the level of urbanization rose from 8 per cent in 1350 to 34 per cent in 1700 (Klep, 1992: 239). Bairoch suggests an even higher level in 1700, with 38–49 per cent urbanized, but both estimates make the Netherlands the most heavily urbanized territory in Europe (Bairoch, 1988: 179).

However, the reinforcement of the large, territorial states which had emerged in the Middle Ages generated a number of big capital cities such as Paris and London. These capitals were normally larger than any other city in their territories. Some were founded from scratch, such as Madrid (1561) and St Petersburg (1703). Others were enhanced medieval regional centres such as Vienna and Prague, and former county capitals, now promoted on the strength of territorial expansion, such as Berlin.

The rise of these national capitals, together with the big seaports, began to affect the structure of European urbanization after 1700. While the population of much of Europe remained stagnant into the eighteenth century and many medium-sized and small towns declined, the national capitals and seaports accounted for the bulk of Europe's net urban growth between 1600 and 1750 (De Vries, 1984: 141). Most of the growing seaports were part of the Atlantic system, and their ability to import bulk food over very long distances gave them a clear advantage over more remote inland towns, as De Vries has noted of Amsterdam in the later seventeenth century (Bairoch, 1988: 200). Inland trading and manufacturing towns were especially vulnerable to stagnation or decline and Europe's urban network moved towards domination by a number of large centres (De Vries, 1984: 139–42).

Capital cities had a multiple influence. For instance, the presence of a royal court in a capital city boosted élite demand and encouraged consumption styles which could influence a whole country, as Wrigley (1967) has shown for London.

They also led to the creation of permanent trading links over very long distances, which reinforced the role of the capital and of other large cities as a regional or territorial entrepôt. Amsterdam and later London both benefited in this way, rising to become leading European cities in the years c.1580–c.1800 and after c.1690 respectively.

European urban growth thus progressed from the sixteenth century onwards on the basis of the late medieval urban network, reinforced by the emergence of some giant cities (Sutcliffe, 1984: 2–7). The main pattern of growth was the spatial extension of the existing heavily urbanized regions (Schwarzer, 1993: 259–60, 264). In the sixteenth and early seventeenth centuries southern Europe still predominated, as it had in the Middle Ages, in terms of the number and size of its large cities and its general urbanization level (De Vries, 1984: 28–9). However, the beginnings of a shift in the balance of urbanization towards northern Europe were already visible. The sixteenth century saw the clear emergence of a 'Rhine axis' of trade routes and staging-points such as Cologne linking the Low Countries and northern Italy, and stimulating economic growth in a wide band along its length (Schwarzer, 1993: 263).

Northern Italy meanwhile went into a relative decline. Its trade with the eastern Mediterranean was harmed by the Turks and by competition from the direct sea route to the Orient. Its financial techniques were emulated and improved in the big northern cities. Worst of all, nearly all its cities lost the political vitality and independence which their strenuous efforts had won in the Middle Ages. They remained Europe's key centres of Renaissance culture, but their populations stagnated or even declined. Italy's biggest cities by 1700 were the great centres of authority, Rome (135,000) and Naples (300,000), together with a dwindling Venice (138,000) (Bairoch, 1988: 182–3; Bairoch et al., 1988: 278).

Meanwhile, Spain moved from an argentine prosperity in the sixteenth century to chronic economic decline in the seventeenth (Bairoch, 1988: 180–1). Once thriving towns lived on into the twentieth century within their medieval or sixteenth-century fortifications (see Gutkind, 1967). The most important cities were the large seaports, of which Seville (72,000) was still the biggest in 1700 after a century of chronic decline, but a few inland

cities such as Granada (70,000) and Valencia (50,000), together with the new capital of Madrid (140,000), had important regional functions (Bairoch et al., 1988: 15–21). As in Russia, a barren countryside was overlooked by a small number of very large cities, sustained by administrative functions carried out for a centralized Crown.

It would be misleading, however, to emphasize this northward movement to the exclusion of other developments. For instance, the Turkish threat in south-eastern Europe was not entirely negative in its effects. In inner Austria a group of cities around Graz was heavily fortified to resist Turkish expansion in the seventeenth century. Their commercial functions benefited from this investment and the resulting protection. Manufacturing, including that of arms, provided much employment (Valentinitsch, H., 1988: 194). In eastern Europe, where only 6–7 per cent of the population were urbanized in 1700 (Bairoch, 1988: 182), a number of large cities prospered in the service of underpopulated but very large regions which could not sustain intermediate service centres. Regensburg (20,000), Leipzig (20,000), České Budějovice (16,000), Skopje (45,000), and Cracow (30,000) were flourishing regional capitals by 1700 (Schwarzer, 1993: 262; Bairoch et al., 1988: 3–69).

The main exception to the picture presented so far was Russia. Here, in a great territory which accounted for over half Europe in both population and area, the level of urbanization was only about 5 per cent (Bairoch, 1988: 182). The main trade routes followed the big rivers which flowed roughly north and south across an underdeveloped expanse. Much trade was carried on by colporteurs. The best-located region was the Ukraine, and here the urban network bore some resemblance to western Europe. Kiev, however, had only 15,000 inhabitants in 1700. Elsewhere, there were a number of larger cities at key crossings or administrative locations, and a scattering of small market towns, but the medium-sized towns which formed the backbone of the western European urban network were virtually absent. Only Moscow (130,000) had more than a hundred thousand inhabitants in 1700, and the largest inland town. Smolensk, had only 40,000 (Bairoch et al., 1988: 60–5). In any case, many urban dwellers engaged in agriculture, making the towns partly self-sufficient like the 'agro-towns' of Sicily (Bairoch, 1988: 182).

# The Urban Fabric

A new wave of urban rebuilding occurred in the sixteenth century and the early part of the seventeenth, generally using stronger materials than in the Middle Ages, and fires and conflagrations became less common. Houses tended to be built taller as growing populations were confined within costly artillery fortifications. From the seventeenth century, building regulations were strengthened in many of the larger cities to reduce the risk of fire and to ease movement through the streets. London's post-Fire building regulations of 1667 and later years were outstandingly ambitious and they were influential in Paris and elsewhere (Sutcliffe, 1993: 45).

Buildings became increasingly tall and solid in the eighteenth century, at any rate in the largest cities (Meyer, 1983: 163–75). In Paris, large, purpose-built apartment houses appeared around 1750 and the authorities imposed stricter height restrictions in 1783 and 1784 (Sutcliffe, 1993: 65–6). By now, Europe's towns were environmentally far superior to those of the later Middle Ages and many, such as London, Paris, and the new English spas, had become places of aristocratic resort. This new role had a big effect on urban design.

## The Planned Town

In the sixteenth century the ideal of the planned town revived for the first time in Europe since the Roman Empire (Hohenberg and Lees, 1985: 151–9). Some of the new towns of the Middle Ages had been given a grid layout, but the Renaissance generated a universal interest in planned cities using perspectives and vistas, with monuments standing in piazzas (Lavedan, 1959: 9–34). From the fifteenth century the building of new *palazzi* in the cities of northern Italy was often combined with street regularization and the reinforcement of the façade as a townscape element (Friedman, 1992: 100–6). Planned urban extensions and improvements were often linked to the building of artillery fortifications after about 1500 (Stoob, 1988: 42–54; White, 1984: 22–7). Layouts based on streets radiating from plazas became the ideal, if rarely the practice.

At first these concepts reflected a new rationalism and idealism, but by the seventeenth century they were associated more with royal and princely Absolutism. It was mainly in this context that they spread from their Italian origins to the whole of Europe after about 1600 (Benevolo, 1993: 138–51). However, the maturity of the urban network meant that there was little scope for new towns, and the improvement of existing cities raised massive financial and legal problems. The popes made some important and influential changes in Rome in the later sixteenth century (Lavedan, 1959: 35–70). Later, in France and in Germany, urban developments on Renaissance lines were largely symbolic expressions of Absolutism inspired mainly by Louis XIV's palace and town of Versailles (1671 onwards). Private housing was subordinate in these projects and at Versailles and Karlsruhe (from 1715) the grandiose radiating avenues remained largely undeveloped into the nineteenth century. At St Petersburg (1703) the foundation of a new city and port by Peter the Great to link Russia more closely to western Europe made much use of Absolutist planning principles and forms (Bater, 1980: 11–13). The example of St Petersburg went on to influence urban planning in Moscow and elsewhere in Russia from the later eighteenth century (Bater, 1980: 13–15).

Paradoxically, the most impressive application of the new aesthetic was in eighteenth-century Britain where the weakness of the Crown allowed aristocratic landowners, municipalities, and syndicates of developers to create Europe's most impressive classical cityscapes in towns such as Bath, Edinburgh, and the West End of London (Benevolo, 1993: 151–9).

# Industrialization and a New Urban Pattern, 1800–1900

The emergence of the great trading network of north-western Europe and its growing links with other continents had encouraged enterprise in a once-remote country—England. During the eighteenth century it overhauled the Netherlands in economic dynamism and France in military power. By 1700 its capital, London, had a population of 575,000 and had overtaken Paris (500,000) to become the largest city in the western world (Barker, 1993: 43; Bairoch *et al.*, 1988: 28, 33).

England's trade, agriculture, and institutions helped it during the second half of the eighteenth century to move towards a system of manufacturing using powered machinery and processes, such as iron and glass manufacture, based on coal (see pp. 283–5). By 1820 the coalfield districts of the Midlands and North were studded by new urban development linked to manufacturing and mining. This process of development, now known universally as industrialization, lifted English productivity well above the Dutch, which had been the highest in Europe in the seventeenth century. English living standards surpassed those of continental Europe by the early nineteenth century. The English urbanization level rose, on Bairoch's criteria, from 13–16 per cent in 1700 to 22–24 per cent in 1800. It would go on to 45 per cent in 1850 and to 75 per cent in 1910 (Bairoch, 1988: 215, 221).

Industrialization could, however, be emulated elsewhere. Sidney Pollard has shown how mechanized manufacture spread across Europe along the major routes of exchange from the early nineteenth century until the process was completed in eastern and southern Europe in the 1960s (Pollard, 1981; see also, above, Chapter 13). First to be absorbed by the process were other coalfield regions of Britain, notably southern Scotland and South Wales. After the Napoleonic Wars, Belgium (from c.1820), France (from c.1830), and Germany (from 1840) were affected. More distant regions were almost untouched by industrialization until the end of the nineteenth century—northern Scandinavia, Iberia apart from Barcelona, Bilbao, and regions, central and southern Italy, south-east Europe, and Russia. Nordic Europe was drawn into the industrial urbanization process at about the same time as Germany, but progress was not as rapid. The level of urbanization in Sweden started to rise rapidly in the 1840s after forty years of very slow growth, though industrialization did not get under way until the 1870s (Öhngren, 1981: 183, 200).

The effect of this diffusion was that the level of urbanization in Europe (excluding Russia) rose from 12.1 per cent in 1800 (a level slightly lower than in 1700) to 18.9 per cent in 1850 and to 40.8 per cent in 1910 (Bairoch, 1988: 216). This urban revolution, which would continue into the twentieth century, was the biggest turning-point in the urban history of Europe since the Roman Empire.

Until the end of the nineteenth century, when electric power became available, industrialization was largely restricted to the coalfields and their vicinity. A secondary localizing factor was the previous accumulation, in some places, of a manufacturing workforce and bulk methods which has come to be known as proto-industrialization. These factors produced big concentrations of mechanized, large-scale manufacture in a small number of regions, mainly in north-western Europe.

Despite the use of highly productive methods, industrialization needed a large labour force to handle greatly increased volumes of production. Consequently, the new manufactures were best carried on in towns. The coalfield regions had not been highly urbanized before the eighteenth century, and industrialization not only multiplied the size of existing towns within a few decades, but it converted many villages and hamlets into urban centres. Neither the State nor the entrepreneurs founded new towns, and private capital poured into housing as well as into industry. The canals and railways which provided the main transport links were also largely the product of private capital, but even when the State influenced or subsidized the new routes, as in France, parts of Germany, and later Russia, the arterial lines linked existing major cities and generated further growth there.

Owing to the scale economies and better transport now available, large towns were consistently able to grow faster than smaller ones, and the national capitals and the seaports grew almost as fast as the large manufacturing centres (De Vries, 1984: 101). The result was increasingly a Europe of large cities, many of which had already been prominent before 1750 (De Vries, 1984: 44, 101). By 1800 the proportion of the urban population of Europe (without Russia) living in towns of more than 100,000 people had reached 42.5 per cent, compared to 33.5 per cent in 1600. By 1910, it had leapt to 58.4 per cent (Bairoch, 1977: 44).

In the Ruhr, in Belgium, and northern France, and in the English Midlands and North, cities of 100,000 or more people could coexist just ten or twenty miles apart by the end of the nineteenth century, some of them forming giant clusters of a type unknown in pre-industrial Europe (Freeman, 1959: 136–41). Smaller towns tended to multiply in the vicinity and the result was the urbanized regions or 'conurbation' as defined by the Scottish urban scientist Patrick Geddes at the turn of the century. By 1900 Europe possessed six cities with more than a million people: London, Paris, Berlin, Vienna, St Petersburg, and Moscow. Overall, there was a shift in the balance of urbanization towards the coalfields during the century, but the new transport systems ensured access to the existing larger towns away from the coalfields and the pre-1750 urban network was not seriously disturbed. Above all, the growth of the national capitals, always at the focus of the national railway networks, was reinforced.

Industrialization split Europe into three big zones. The first, northern Europe, prospered mightily and completed its gradual replacement of southern Europe as the continent's leading region. The Mediterranean region's share of Europe's urban population fell from one-half in 1600 to one-third in 1800 (De Vries, 1984: 31). Meanwhile, the urban population of northern Europe had risen to half De Vries's European total. Southern Europe, however, had a traditional structure of settlement and production which could support industrialization at a later date. The third zone, much the biggest, did not have this advantage. As in the past, Russia lagged strikingly behind the changes which were transforming northern Europe. In the absence of a strong pre-industrial urban network, Russia lacked a basis for industrialization. St Petersburg, Russia's main link with the Baltic and by now the largest port in Russia, was typical of the large cities in that it overlooked a very bleak region. Its population rose from 220,208 in 1800 to 1,637,111 in 1910, allowing it to rank with Vienna and Berlin as one of the leading capitals of eastern Europe. At this time, however, a large part of its unskilled workforce was still made up of rough *okhtodniki*, seasonal peasant labourers (Engman, 1993: 76–7). In Russia as a whole, peasants were the largest social category in the urban population (Bater, 1980: 10). This and other symptoms of backwardness separated Russian urbanization from that of western Europe even when, as at St Petersburg above all, up-to-date industrial methods were in use in some sectors by 1910.

## The Spatial Structure of the Town

The spatial structure of European towns was affected by two factors above all. First, their increased area tended to promote the emergence of districts with distinct functions. Second, the industrial towns, seaports, and capital cities, in particular, had to accommodate a greatly increased proportion of manual workers on low earnings. In most respects the working of the land market ensured that the pre-industrial structure remained largely in place, with business creating dense development at the centre and densities declining outwards along the main roads. In country towns the spatial structure remained entirely that of pre-industrial times down into the twentieth century. However, one important change upset the pre-industrial pattern. The masses of manual workers came to reside in the inner districts of the town, in contrast to the poorer residents of the pre-industrial town, who had lived towards the outskirts. As for the middle classes, they began to look towards the periphery. Only the rich could afford to continue living in the central areas (Conzen, 1978: 1–48).

Manual workers and their families accounted for up to 70–80 per cent of the population in industrial towns and seaports. Their low effective demand for housing could not, however, generate the building of anything other than rudimentary accommodation, and the poorer among them lived in rooms in older structures. At first, much building was on a small but congested scale, filling in small sites in the inner districts. Overcrowding created very poor health conditions, and at first the shortage of pure water and poor drainage made the environment very dangerous.

Housing was especially poor on the fringes of industrializing Europe. Here, the larger cities tended to attract impoverished migrants from the countryside who went on to suffer severely from lack of employment and very low wages in what, despite their size, remained very backward urban economies. Dublin's degraded, multi-occupied houses and Lisbon's single-storey *ilhas* in crowded courts and rows are representative of this fringe phenomenon (Aalen, 1992: 149–52; Teixeira, 1992: 268–77). In Russia, a similar effect was visible from the 1860s, with crude wooden hutments housing the bulk of the urban population.

Some of Europe's coal and ore fields generated straggling, fragmented settlements which housed large numbers of workers without generating urban characteristics. In the Ruhr, this non-urban form of concentration was linked to the recruitment of rural migrants from East Prussia and elsewhere who valued the low-density housing provided by many of the employers (White, 1984: 16–18). In Belgium, whose large Wallonia region industrialized from the 1820s, an intermediate form emerged with limited industrial growth in the main cities but much development in the smaller places around them, and along the linking roads (Van den Eeckhout, 1992: 190). This was associated with a coal/textile/metal economy which had much in common with the English Midlands, and which, in the Black Country, produced similar results. In the Netherlands, urbanization between 1800 and 1930 produced an annular network of cities (the *Randstad*) without a single dominating centre but including Amsterdam, Rotterdam, and The Hague. This pattern was the product of trade and communications before industrialization, which did not begin in earnest until about 1890.

In industrializing north-western Europe from as early as the 1780s, towns tended to spread at low densities with individual houses built in villa or terrace formations. This pattern was probably the result of high average earnings in the regions of early industrialization, proto-industrial traditions, and the early removal of fortifications, especially in England. As industrialization spread into Germany, central France, and northern Italy between the 1830s and the 1880s, it moved into areas of lower wages and high urban land values. The result was a pattern of housing based on tenement houses which could reach a height of four–six storeys in the bigger cities (Reulecke, 1985: 40–9). Even the middle and upper classes lived in these taller buildings, and only the very rich escaped to villas on the edge of the city or in the nearby countryside. Paradoxically, the growing separation of the social classes throughout urban Europe did not generate radically divergent housing forms in individual localities.

The rapid growth of towns generated congestion, reflected in rising rents, and private enterprise began to create systems of public transport from the later 1820s. The first Paris horse omnibus lines of 1828 were emulated by a London entrepreneur, George Shillibeer, in 1829. Omnibus companies spread to provincial cities in both England and France in the 1830s and 1840s. Horse tramways, an American innovation, came to Britain in 1859,

and a number of experiments in Europe's larger cities in the 1860s led to a general adoption in the 1870s. Experiments with mechanized traction had some success, but a general mechanization did not come about until the 1890s and the early 1900s on the basis of electric traction. Electric trams, generally twice as fast and twice as cheap as their horse-drawn predecessors, allowed even the working class to become commuters, and the outward spread of the city accelerated throughout Europe after 1890.

At first, however, the local structure of land values and the inertia of the housing market merely produced an extension of multi-storey development in high-rise urban Europe. In the north-west, meanwhile, the electric trams led to the spread of low-density suburbs but, here too, existing patterns of terraced and villa housing were not replaced by something new. This paradox was unwelcome to urban reformers and it helped prompt a new departure in which civic authority would play a bigger part in the creation of the urban environment.

## Regulation and Planning

The cities of the industrial era were so complex that they required much more public regulation than those of pre-industrial times. Conditions in the working-class districts began to cause concern as early as the 1830s in Britain. Building regulations and other public controls were complemented from the 1840s by public works concentrating on streets, water supplies, and sewerage. Prefect Haussmann's modernization programme in Paris (1853–70) set an example for the whole of Europe, with streets, squares, and parks applying Renaissance principles on an unprecedented scale, and the crowded inner districts partially reconstructed (Olsen, 1986: 44–53; Benevolo, 1993: 169–88). From the 1850s the term 'Haussmannization' spread through Europe, and not a single capital city was untouched by modernization works (Hall, 1986). Even many of the provincial cities were influenced by the French example, as were Birmingham with its new Corporation Street and Nottingham with its boulevards.

In Germany, Austria, Italy, and Spain, the removal of artillery fortifications, which took place mainly in mid-century, was often accompanied by municipal extension plans which guided peripheral growth. Their geometrical layouts of broad streets eased traffic movement and improved health, but they encouraged high land values, and the related building regulations led to an accentuation of the multi-storey building form (White, 1984: 22–7). In Germany, peripheral growth was so rapid in most towns that extension plans were almost universal, whether or not they had been fortified (Fehl and Rodriguez-Lores, 1983).

As early as the 1850s housing reformers, especially in Germany and France, sought to promote cottage housing for urban workers, but little had been achieved by 1900. More effective was the allotment garden movement. In Germany, Scandinavia, and northern France, especially, workers' allotment gardens proliferated from the 1880s. The land was often provided by the local authorities but in France the Catholic Church played the leading role (Hyldtoft, 1992: 53).

By the 1890s the growth of building regulation and the town extension procedure, combined with the emergence in England of a low-density ideal for workers' housing, generated the idea of comprehensive town planning. Germany was the leader, making a smooth transition from town extensions to comprehensive planning incorporating city-wide zoning on the lines pioneered by Mayor Adickes at Frankfurt in 1891 (Rodriguez-Lores and Fehl, 1985). By the turn of the century the way lay open in Germany to the creation of low-density housing on the periphery. English influence was reinforced by Ebenezer Howard's Garden City idea after its publication in 1898. Utterly independent garden cities proved hard to establish, even in England, but the ideal of city-wide planning designed to encourage small houses in the suburbs for both middle and working classes was widespread in Europe by 1910. The development of other planning skills, such as traffic engineering, had helped create a general expertise by 1914.

# The Twentieth Century: Urbanization Completed

After 1914, industrialization continued its diffusion from its nineteenth-century core in northern Europe. Europe's urbanization level, according to Bairoch, rose from 36.0 per cent in 1910 to 39.1 per cent in 1930. By 1950 it had reached 46.0 per cent and by 1970, 59.7 per cent (Bairoch, 1977, 17). In 1990, according to United Nations estimates, the level of urbanization in Europe (excluding the USSR) was 73.1 per cent. In the USSR (notwithstanding the inclusion of its Asian republics) the level was 67.5 per cent (United Nations, 1989: 5). Hall and Hay (1980: 225), using a generous definition of urbanization based on their extensive 'metropolitan areas', found that 86 per cent of the population of western Europe was urbanized in 1950, and over 88 per cent in 1970.

The main areas of continuing urbanization were now southern and eastern Europe, Russia, the Nordic countries, and France. In north-western Europe, meanwhile, the larger cities experienced a deconcentration of population which favoured nearby towns above all. Population tended to stagnate or decline in the older coalfield districts and the large seaports. Growth took place in or near the capital cities and other areas of high demand. These developments were backed by per capita incomes which rose faster after 1918 than ever before, and especially so after 1945. The leader in these tendencies was Britain, but from the 1950s most of urbanized northern Europe was moving in the same direction (Hall and Hay, 1980: 226).

In the industrialized regions of Europe the largest urban areas continued to grow fastest, at any rate until the 1960s. The number of cities with populations of more than one million, as defined by Mitchell, reached twenty-six by 1971. Seven of them were in the European Soviet Union, and four more were in other European Comecon countries. Six were in Italy and Spain (Mitchell, 1978: 12–14). This meant that the great majority of Europe's large cities were in the once-backward east and south. This transition was, however, due partly to the post-war decentralization of population in the richer urban north. Here, by the 1960s, the largest cities as defined by Bairoch (more than two and a half million inhabitants) were beginning to stabilize or decline as people and employment moved to outlying towns. The fastest growth was among the medium-sized and smaller towns (Bairoch, 1977: 29–30, 47). Decentralization from the very largest cities continued during the 1970s and the 1980s, when its effects were accentuated by the economic troubles which began in 1969–73 (Berry, 1981: 182–3). This did not mean, however, that a major flight to the country took place. The great majority of western Europeans continued to live in large urbanized areas which, in the north, were now spreading and disaggregating on an unprecedented scale.

In southern Europe the process of rapid urbanization from a low level meant that population transferred quickly from rural areas into cities that were growing in a concentrated fashion, very much like the cities of nineteenth-century industrial Europe (Hall and Hay, 1980: 227–8; Berry, 1981: 197). Government encouragement of rapid free-market industrialization meant that planning constraints were few, and serious congestion effects built up very quickly. Milan and Barcelona were prime examples of the clogged environment generated by rapid growth.

In eastern Europe and Russia industrialization proceeded rapidly. Poland, Czechoslovakia, and adjacent countries began to benefit, more than before 1914, from the diffusion of industrialization from Germany. In Russia, State efforts to introduce industrialization did not begin in earnest until the 1880s and little had been achieved by 1914. Stalin put this right in the most brutal fashion between 1928 and his death in 1953. Industrialization continued under his successors, but industrial location policies, the virtually uncontrolled movements of public enterprises, and continuing restrictions on the movement of labour slowed the growth of the largest cities after 1945 and produced a more even spread of medium-sized towns (Bater, 1980: 30).

Urbanized Russia thus came to contain a hierarchy of medium-sized and large cities not greatly different from western Europe's, except that the distances between them were generally very large. In 1850 only one Russian in twenty had lived in towns, and in 1917 still only one in six did so after sixty years of rapid urbanization (Bater, 1980: 1, 15). In 1850 a mere 1.6 per cent of the population of Russia had lived in towns of more than 100,000 people. By 1930 this proportion had risen to 8.9 per cent, compared to 22.6 per cent in the rest of Europe. In 1970 the differential had almost disappeared, with the Russian proportion being 31.7 per cent and that of the rest of Europe, 38.8 per cent (Bairoch, 1977: 40–1). Moscow's population rose from just over one million at the turn of the nineteenth century to 2,026,000 in 1931. Despite a horde of State controls, its population had risen to nearly nine million at the end of the 1980s, when it was nearly as large as Europe's biggest city, London (Vodarsky, 1993: 87).

## War and the City

Twice in the twentieth century most of Europe was affected by modern industrial war. The towns were affected from as early as 1915 when widespread rent controls distorted the housing market and led on to the construction of public housing after the war. Much of this new building reinforced the separation of the working and middle classes, and provided a distinctive physical and aesthetic form for the habitat of the workers. Indeed, public urban development on this scale, combined with enhanced town planning, diverged strikingly from the nineteenth-century city. Most of the new housing was sited on the city fringe where large sites were available. Even when flats were built, as was generally the case outside England, densities were reduced and a planned environment including open space and social facilities was created. There were exceptions, however. In neutral Sweden, the State and local government did not become involved in house provision until 1939–45 (Strömberg, 1992: 19). Outside some cities, notably on the Paris lotissements, the homeless built their own shanties with chaotic results. In England, local authorities built great swathes of 'cottages' with very few amenities, and the standard of their planning fell below that achieved in many German cities.

The country worst damaged in the First World War was Belgium, and urban reconstruction was helped by an international effort led by France. The towns were rebuilt largely on their original lines in order to retain a pre-industrial heritage (Smets, 1985). This policy might appear paradoxical for so early an industrializer, but it reflected pride in the country's urban prominence in the sixteenth century.

In the Second World War extensive urban destruction, especially in Germany, led to planned reconstruction on a scale not previously seen in Europe, except perhaps in Haussmann's Paris (see e.g. Forshaw and Abercrombie, 1943). Much of this reconstruction, as at Rotterdam, followed the tenets of the Modern Movement in architecture and its Athens Charter of 1933 (Benevolo, 1993: 197–202). The combination of space and height contributed to the almost universal ideal of the 'city of towers' from the mid-1950s (Relph, 1987: 119–210). The momentum of reconstruction led on to the demolition and reconstruction of dilapidated slum areas, whether or not they had been affected by bombing.

## High Incomes and the Spreading City after 1945

In northern Europe the growth of urban employment took place mainly in the tertiary sector after 1945. Central business districts expanded both outwards and upwards. Commuting into the central city increased from spreading suburbs and nearby towns. From the later 1950s, beginning in Britain and Germany, urban road-building on the lines pioneered in the United States since the 1930s sought to cater for these growing flows (Tetlow and Goss, 1965). Meanwhile, some of the older industries in the inner city began to decline from the 1960s, and new manufactures set up in the suburbs.

More affluent citizens demanded more space, and with their multiplying cars they were able to secure access to it further and further out. Suburbs spread and, from the 1960s, more and more second homes were acquired or built outside the big cities, especially in France. In southern and eastern Europe, on the other hand, rapid urbanization with low incomes produced high-density peripheral districts of which the *barrios neuvos* of Madrid were extreme examples.

British governments chose a novel means of resolving these problems. They pursued a big programme of planned 'new towns' between 1946 and 1977 (Aldridge, 1979). Other European countries were happy for the most part to tolerate peripheral spread, provided that it was efficiently planned. In southern Europe, much of it was not, and in eastern Europe and the Soviet Union planning, to say the least, lacked sensitivity and coherence. France, northern Europe's late urbanizer, developed a practical solution. Anticipated rapid growth in the Paris region from around 1960 prompted the French government to adopt the most ambitious strategy of planned peripheral development in Europe. This, the 'ville nouvelle' programme was based on commuting along improved roads and railways, and the decentralization of some employment to peripheral 'towns' such as Évry and Cergy-Pontoise. Public sector building provided the initial impetus, but by the mid-1970s the lower middle classes were moving into the 'new towns' in large numbers. The same was true of planned developments near Lyons such as L'Îsle d'Abeau.

Soviet planning and urban reconstruction continued on modernistic lines, though the pompous public structures built in honour of Stalin in the late 1940s and early 1950s used ornate, historicist styles. Socialist town planning principles were extended to the territories of eastern Europe which were brought within the socialist system after 1945. They were designed to replace the bourgeois city, but in practice utilitarian needs produced cities which were not greatly different in functions and layout from those of western Europe. Limited funds and an incomplete planning machinery often produced inefficiency and a poor quality of environment (Benevolo, 1993: 210–11). The survival of the older cores of most of these cities, moreover, prevented a complete break with the past. In the German Democratic Republic the older areas were left to rot so that people would eventually move into the new high-rise housing estates where a socialist lifestyle could be encouraged. Conservation did not begin until after the union with the German Federal Republic in 1989. In Warsaw and in Riga, on the other hand, the restoration of devastated city centres began in the 1950s.

## The Inner City

The spreading city of post-war prosperity left room for a very different 'inner city' which emerged in the older districts of the larger cities, beginning with Britain in the 1950s and following in the industrialized regions of the Continent from the 1960s. This was the zone of non-European immigrants and their descendants. Sometimes known, imprecisely, as the 'ghetto', this area was made up of older housing, dating mainly from the nineteenth century. Abandoned by the natives, it housed ill-paid newcomers who created there a version of their traditional, non-European lifestyle. Unlike their European predecessors, they did not merge readily into the established culture of their city (Robinson, 1981). The Muslim religion was especially resistant to European culture, and the Arab residents of urban France and the Turks in Germany were well on the way to creating their own urban society in the inner city at the end of the twentieth century.

The creation of space in the inner city by the departure of population and industry sometimes led to the creation of a more open, attractive cityscape emphasizing historic buildings and landscaped areas. Inner Glasgow was a prime example of this transformation by the 1980s. This marked the end of the more destructive process of slum clearance and 'urban renewal' which had marched through north European cities in the 1950s and 1960s (Grebler, 1964). Almost everywhere, a revival of the centres of the older cities was visible in the 1980s and 1990s as carefully directed conservation and renewal programmes, pedestrianization, higher living standards, and the gradual decline in the numbers of the poorer residents of the inner districts created the spread of middle-class tertiary activity. This formed part of a general stabilization of the larger, older cities, which often went with a deliberate 'return to the past' as the qualities of the traditional city were at last recognized (Parkinson *et al.*, 1994: 1).

As the European city approached the end of the millennium, its appearance and atmosphere confirmed its established position as a basic component of European society. Deconcentration since the 1950s had not destroyed it and the city centre remained the focus of urban life. The traditional city stood at the heart of a spreading, motorized network, but it still remained the central place which it had been in the Middle Ages.

REFERENCES

AALEN, F., 'Ireland', in C. G. Pooley (ed.), *Housing Strategies in Europe, 1880–1930* (Leicester, 1992), 132–63.

ALDRIDGE, M., *The British New Towns: A Programme without a Policy* (London, 1979).

AMMANN, H., 'Wie gross war die mittelalterliche Stadt?', *Studium Generale*, 9 (1956), 503–6.

BAIROCH, P., *Taille des villes, condition de vie et développement économique* (Paris, 1977).

—— *Cities and Economic Development: From the Dawn of History to the Present* (Chicago, 1988).

—— *et al.*, *The Population of European Cities: Data Bank and Short Summary of Results, 800–1850* (Geneva, 1988).

BARKER, T., 'London: A Unique Megalopolis?', in T. Barker and A. Sutcliffe (eds.), *Megalopolis: The Giant City in History* (London, 1993), 43–60.

BATER, J., *St Petersburg: Industrialization and Change* (London, 1976).

—— *The Soviet City: Ideal and Reality* (London, 1980).

BENEVOLO, L., *The European City* (Oxford, 1993).

BERRY, B., *Comparative Urbanization: Divergent Paths in the Twentieth Century* (London, 1981).

BLOTEVOGEL, H. (ed.), *Kommunale Leistungsverwaltung und Stadtentwicklung vom Vormärz bis zur Weimarer Republik* (Cologne, 1990).

BOLLEREY, F. *et al.* (eds.), *Im Grünen Wohnen—im Blauen Planen: ein Lesebuch zur Gartenstadt* (Hamburg, 1990).

—— and HARTMANN, K., *Wohnen im Revier* (Munich, 1975).

BORDE, J. *et al.*, *Les Villes Françaises* (Paris, 1980).

CERASI, M., and FERRARESI, G., *La residenza operaia a Milano* (Rome, 1974).

CLIFTON-TAYLOR, A., *Six English Towns* (London, 1978).

CONZEN, M., 'Zur Morphologie der englischen Stadt im Industriezeitalter', in H. Jäger (ed.), *Probleme des Städtewesens im industriellen Zeitalter* (Cologne, 1978), 1–48.

CORFIELD, P., *The Impact of English Towns, 1700–1800* (Oxford, 1982).

DAUNTON, M. (ed.), *Housing the Workers, 1850–1914: A Comparative Perspective* (London, 1990).

DE VRIES, J., *European Urbanization, 1500–1800* (London, 1984).

DICKINSON, R., *The City Region in Western Europe* (London, 1967).

DYOS, H., *Victorian Suburb: A Study of the Growth of Camberwell* (Leicester, 1961).

ENGMAN, M., 'An Imperial Amsterdam? The St Petersburg Age in Northern Europe', in T. Barker and A. Sutcliffe (eds.), *Megalopolis: The Giant City in History* (London, 1993), 73–85.

FEHL, G., and RODRIGUEZ-LORES, J. (eds.), *Stadterweiterungen, 1800–1875: von den Anfängen des modernen Städtebaues in Deutschland* (Hamburg, 1983).

FORSHAW, J., and ABERCROMBIE, P., *County of London Plan* (London, 1943).

FREEMAN, T., *The Conurbations of Great Britain* (Manchester, 1959).

—— 'Palaces and the Street in Late-Medieval and Renaissance Italy', in J. Whitehand and P. Larkham (eds.), *Urban Landscapes: International Perspectives* (London, 1992), 69–113.

GREBLER, L., *Urban Renewal in European Countries: Its Emergence and Potentials* (Philadelphia, 1964).

GUTKIND, E., *Urban Development in Central Europe*, i (London, 1964).

—— *Urban Development in Southern Europe: Spain and Portugal*, iii (New York, 1967).

HALL, P., 'Introduction: Defining the Problem', in P. Hall (ed.), *The Inner City in Context* (London, 1981), 1–8.

—— and HAY, D., *Growth Centres in the European Urban System* (London, 1980).

HALL, T., *Planung europäischer Hauptstädte: zur Entwicklung des Städtebaues im 19. Jahrhundert* (Stockholm, 1986).

HARPER, R., *Victorian Building Regulations* (London, 1985).

HOHENBERG, P., and LEES, L., *The Making of Urban Europe, 1000–1950* (Cambridge, Mass., 1985).

HYLDTOFT, O., 'Denmark', in C. G. Pooley (ed.), *Housing Strategies in Europe, 1880–1930* (Leicester, 1992), 40–72.

JOHNSON-MARSHALL, P., *Rebuilding Cities* (Edinburgh, 1966).

KLEP, P., 'Long-term Developments in the Urban Sector of the Netherlands (1350–1870)', in *Le Réseau urbain en Belgique dans une perspective historique (1350–1850): une approche statistique et dynamique* (Brussels, 1992), 201–42.

LAVEDAN, P., *Histoire de l'urbanisme: Renaissance et temps modernes* (Paris, 1959).

LAWLESS, P., *Britain's Inner Cities: Problems and Policies* (London, 1981).

LEPETIT, B., *The Pre-industrial Urban System: France, 1740–1840* (Cambridge, 1994).

MCKAY, J., *Tramways and Trolleys: The Rise of Urban Mass Transport in Europe* (Princeton, 1976).

MARMARAS, M., *E Astike Polikatoikia tes Mesopolemikes Athenas* (Athens, 1991).

MATZERATH, H., 'Städtewachstum und Eingemeindungen im 19. Jahrhundert', in J. Reulecke (ed.), *Die deutsche Stadt im Industriezeitalter: Beiträge zu modernen deutschen Stadtgeschichte* (Wuppertal, 1978), 67–89.

MERLIN, P., *New Towns: Regional Planning and Development* (London, 1971).

—— *Les Villes nouvelles de France* (Paris, 1991).

Meyer, J., *Études sur les villes en Europe occidentale (milieu du XVIIe siècle à la veille de la Révolution française): generalités—France* (Paris, 1983).

Mioni, A. (ed.), *Urbanistica fascista: ricerche e saggi sulla città e il territorio e sulle politiche urbane in Italia tra le due guerre* (Milan, 1980).

Mitchell, B., *European Historical Statistics, 1750–1970* (London, 1978).

Öhngren, B., 'Urbanisation in Sweden, 1840–1920', in H. Schmal (ed.), *Patterns of European Urbanisation since 1500* (London, 1981), 183.

Olsen, D., *The City as a Work of Art: London, Paris, Vienna* (New Haven, 1986).

Parkinson, M. *et al.*, 'Introduction: The Changing Face of Europe', in A. Harding *et al.* (eds.), *European Cities towards 2000: Profiles, Policies and Prospects* (Manchester, 1994), 1–15.

Plowden, S., *Towns against Traffic* (London, 1972).

Pollard, S., *Peaceful Conquest: The Industrialization of Europe, 1760–1970* (Oxford, 1981).

Pooley, C. (ed.), *Housing Strategies in Europe, 1880–1930* (Leicester, 1992).

Prevenier, W. *et al.*, 'Le Réseau urbain en Flander (XIIIe–XIXe siècle): composantes et dynamique', in *Le Réseau urbain en Belgique dans une perspective historique (1350–1850): une approche statistique et dynamique* (Brussels, 1992), 157–200.

Relph, E., *The Modern Urban Landscape* (London, 1987).

Reulecke, J., *Geschichte der Urbanisierung in Deutschland* (Frankfurt am Main, 1985).

Ringrose, D., *Madrid and the Spanish Economy, 1560–1850* (Berkeley, 1983).

Robinson, V., 'The Development of South Asian Settlement in Britain and the Myth of Return', in C. Peach *et al.* (eds.), *Ethnic Segregation in Cities* (London, 1981), 149–69.

Rodriguez-Lores, J., and Fehl, G. (eds.), *Städtebaureform, 1865–1900: von Licht, Luft und Ordnung in der Stadt der Gründerzeit* (Hamburg, 1985).

Schöller, P., *Die deutschen Städte* (Wiesbaden, 1967).

Schwarzer, O., 'Die Stellung "Westeuropas" in der Weltwirtschaft, 1750–1950', in M. North (ed.), *Northwestern Europe in the World Economy, 1750–1950* (Stuttgart, 1993), 257–90.

Smets, M., *L'Avènement de la cité-jardin en Belgique: histoire de l'habitat social en Belgique de 1830 à 1930* (Brussels, 1977).

—— (ed.), *Resurgam: la reconstruction en Belgique après 1914* (Brussels, 1985).

Stoob, H., 'Die Stadtbefestigung: Vergleichende Überlegungen zur bürgerlichen Siedlungs—und Baugeschichte, besonders der frühen Neuzeit', in K. Krüger (ed.), *Europäische Städte im Zeitalter des Barock* (Cologne, 1988), 25–56.

Strömberg, T., 'Sweden', in C. G. Pooley (ed.), *Housing Strategies in Europe, 1880–1930* (Leicester, 1992), 11–39.

Sutcliffe, A., 'Introduction: Urbanization, Planning and the Giant City', in A. Sutcliffe (ed.), *Metropolis 1890–1940* (London, 1984), 1–18.

—— *Paris: An Architectural History* (New Haven, 1993).

Teixeira, M., 'Portugal', in C. G. Pooley (ed.), *Housing Strategies in Europe, 1880–1930* (Leicester, 1992), 268–96.

Tetlow, J., and Goss, A., *Homes, Towns and Traffic* (London, 1965).

United Nations (Department of International Economic and Social Affairs), *Prospects of World Urbanization, 1988* (New York, 1989).

VALENTINITSCH, H., 'Die innerösterreichischen Städte und die Türkenabwehr im 17. Jahrhundert', in K. Krüger (ed.), *Europäische Städte im Zeitalter des Barock* (Cologne, 1988), 169–94.

VAN DEN EECKHOUT, P., 'Belgium', in C. G. Pooley (ed.), *Housing Strategies in Europe, 1880–1930* (Leicester, 1992), 190–220.

VAN ENGELSDORP GASTELAARS, R., and WAGENAAR, M., 'The Rise of the "Randstad", 1815–1930', in H. Schmal (ed.), *Patterns of European Urbanisation since 1500* (London, 1981), 229–46.

VODARSKY, Y., 'The Impact of Moscow on the Development of Russia', in T. Barker and A. Sutcliffe (eds.), *Megalopolis: The Giant City in History* (London, 1993), 86–95.

WEIS, U., 'Zentralisation und Dezentralisation: von der englischen Gartenstadt zur Frankfurter "Gross-Siedlung" ', in F. Bollerey *et al.* (eds.), *Im Grünen wohnen—im Blauen planen: ein Lesebuch zur Gartenstadt* (Hamburg, 1990), 228–46.

WHITE, P., *The West European City: A Social Geography* (London, 1984).

WRIGLEY, E., 'A Simple Model of London's Importance in Changing English Society and Economy, 1650–1750', *Past and Present*, 37 (1967), 44–70.

# Chapter **15**

# Changing Patterns of Trade and Interaction since 1500

## J. R. Walton

## An Outline History of European Trade

In 1500 European trade was about to undergo a sea-change, both literally and metaphorically. Between 1460, when the death of Prince Henry closed a chapter in Portugal's tentative exploration of the West African coast, and Magellan's achievement of the first global circumnavigation during the years 1519–22, European seafarers discovered the existence of continuous sea passages from ocean to ocean around the world, that 'all the seas of the world are one' (Parry, 1981; Boorstin, 1983; Livingstone, 1992). The discovery of the sea, largely by Iberians or by Iberian-sponsored expeditions, over such a short period of time is a measure of the extent of Europe's existing level of development. Financial and business institutions, shipbuilding and military technology, and navigation science had already progressed, within a European frame of reference, to the point where they were capable of supporting European adventures overseas (Cipolla, 1965; Parker, 1988). The major lure was the prospect of profit from trade. Unlike explorers of the late nineteenth and early twentieth centuries, Europeans of the age of reconnaisance did not seek unknown lands, but new and more profitable sea routes to known markets and known sources of tradable commodities. The period had need of new routes to the East. The Ottoman conquest of the east Mediterranean littoral following the fall of Constantinople in 1453 threatened an end to profitable trading via traditional overland routes (Peet, 1991: 117). Occasionally, the search for another way round threw up the unknown and unexpected, most significantly in the course of Columbus's westward quest for Asia in 1492 (Parry, 1981; Childs, 1995).

The European discovery of the sea had two sets of long-term consequences for trade. First, it gave impetus to the expansion and diversification of trade. Secondly, it helped reorientate the focus of European trading activity away from its traditional Mediterranean channels to the Atlantic seaboard. Neither amounted to an overnight transformation. The roots of north-west Europe's trading history were already deep (Hodges, 1988), while the strategic and trading importance of the Mediterranean never faded completely, and was rekindled when the loss of the American colonies intensified British interest in India and other eastern possessions (Farnie, 1969: 1–31). Furthermore, even after three centuries of European contact with the Americas, the absolute size of the consequent commerce was still modest. In the second half of the eighteenth century sugar and tobacco were carried from Jamaica and Virginia to London in vessels averaging little more than 150 tonnes and making no more than one return vogage per year (French, 1987). It deserves emphasis that the impacts of commercial exchanges were not a simple function of their absolute size. Indeed, some of the most significant impacts in terms of their consequential effects for commerce were not in the first instance commercial at all.

The most devastating was Europe's microbial and biological invasion of the New World, the effects of which have been explored by numerous authors (for instance, Crosby, 1986, 1994; McNeill, 1979; Stannard, 1992). Disease began with contact, and the death toll of indigenous populations mounted vertiginously shortly thereafter. A significant European presence was not a prerequisite for contact mortality. But contact mortality and the thinning of indigenous populations lent impetus to different European commercial activities including, ultimately, those involving large-scale European settlement of the temperate zones. Settler capitalism was preceded by several centuries of mercantile exploitation of various kinds. Spanish conquistadores, having pillaged Meso- and South America of gold religious artefacts, set about the single-minded exploitation of unmined resources of silver and gold (Vilar, 1976). Exploitation of the region's agrarian resource initially focused on densely populated tributary societies, later shifting to less-populated areas suitable for plantation agriculture using slave labour (Peet, 1991: 118). Meanwhile, persistent fashion demand for furs drove not only the rapid eastward movement of Russians into Siberia, but a parallel westward movement of European fur traders into North America. By the early eighteenth century, the two movements had encircled the globe, and were beginning to contend the rich sea-otter resources of the American northwest coast (Wolf, 1982: 158–94). Despite constraints of technology, institutions, and ambition, the global periphery was increasingly drawn into Europe's trading orbit during the mercantile period.

The effects on Europe of modest traded volumes were also profound, none more so than the biological counterpart to the westward spread of European microbes, the eastward movement of American flora and fauna. Starkly contrasting with the destructiveness of the former, American crop plants and American domesticated animals and birds provided new food sources, including some (manioc, maize, and potatoes, white and sweet) rich in carbohydrate. Thanks to these novel sources of sustenance, Old World populations grew as those of the New World came under increasing pressure. Manioc (cassava) became a staple of African subsistence, while maize, already growing in Castilé as early as 1498, had reached the eastern Mediterranean by the end of the sixteenth century. Northern Europe took to the potato more slowly. However, by the second half of the nineteenth century, the crop's ability to sustain (and on occasion fail) the rural masses had been demonstrated across cool-temperate Europe from Ireland and Scotland to Russia (Crosby, 1996; Denecke, 1976). In essence, Africa's slaves and Europe's mass migrants of the late nineteenth and early twentieth centuries were drawn, either directly or indirectly, from populations of the rural poor whose numbers had been inflated by food crops originating in the continent of their ultimate destination. A transfer of botanical material almost negligible in scale was absorbed into what were essentially low-technology peasant agricultures to produce, by propagation, a large volume of biomass and, in turn, large numbers of people.

The food crops which fuelled much of Europe's population growth were of American origin. America was also implicated as a cause of the European mass migrants' displacement, albeit obliquely (settler capitalists ultimately undercut European food producers in their own markets), and as only one aspect of the shift from small- to large-scale trading, from mercantilist to capitalist exploitation of the non-European parts of the globe. Different scholars have focused on different features of this process. The application to transport of a succession of technological innovations, culminating in the widespread adoption of iron and steam on both sea and land, brought cost reductions which made possible large-scale transfers of people and goods (Vance, 1986; Ville, 1990). However, for many, the most important changes are not the technological, which merely allowed latent pressures to be released, but the structural, which caused such pressures to develop in the first place. Wallerstein's essentially Marxian world-system model has become the focus of thought about the structural underpinnings of world capitalist development and its inequalities (Wallerstein, 1974, 1979; and critically Giddens, 1981). Under this model, Europe's sixteenth-century expansion into the world periphery marked the beginnings of a capitalist accumulation at a world scale, the essential features of which were threefold (Peet, 1991: 49–52; Taylor, 1994). First, economic activity and economic decision-making were increasingly informed by the recognition that there was a single global market and a single global division of labour. Secondly, a single state might gain hegemony within the world system, as did Britain after 1760, but the essence of the system was its multi-state character. Interstate economic rivalry among European core states ultimately ensured that crises of capital accumulation did not lead to the total paralysis of economic activity, albeit such rivalry also resulted in a succession of interstate conflicts, notably the First World War (Offer, 1989). Third, Wallerstein's model divides the world into core, periphery, and semi-periphery. These three elements are a persistent feature of the world-system, but not geographically static. Areas formerly part of the periphery, for instance the United States, became by turn semi-periphery and core, while former core areas, such as northern Italy, the nerve-centre of the medieval Mediterranean trading economy, slipped down the hierarchy into the semi-periphery during the Atlantic centuries.

World-system thinking naturally invites reflection on global patterns of advantage and disadvantage at the macro scale, and in particular on the persistently favoured position of Europe *vis-à-vis* the rest. At one level, European privilege presents a simple contrast to peripheral exploitation. Thus the health benefits to Europeans of washable cotton clothing, and the economic multiplier effects associated with the factory production and export of cotton in eighteenth- and early nineteenth-century Britain contrast with both the oppressive production regime under which the crop was grown in America, and the deliberate suppression of household cotton spinning and weaving in eighteenth-century India as a necessary prelude to the invasion of that market and others by the British factory product. Similar contrasts may be claimed for sugar, tea, and coffee, which became items of European mass consumption as costs fell during the late eighteenth and early nineteenth centuries,

contributing to the diversification of the popular diet (Shammas, 1993; Mintz, 1993). This in itself did little to change their status as instruments of mass oppression in the regions of production. Likewise, there is merit in the argument that even the least privileged of Europe's emigrants to the United States, southern and eastern Europe's 'second wave' of the period 1880 to 1910, though targeted by discriminatory exclusionist legislation from 1925, still derived advantage from being European. At least, exclusionist pressures on them were much less draconian than those directed against Asians, and civil rights were accorded far more readily to European-Americans, whatever their origin within Europe, than to native- or African-Americans.

The notion that Europe enjoyed multiple and uniform advantage derives from global comparisons which contrast an exploitative Europe and an exploited residue. For Blaut (1989, 1992), Europe's fortuitous proximity to the newly discovered Americas is itself sufficient to explain why Europe should have gained relative to those parts of Africa or Asia which in 1492 enjoyed levels of development similar or superior to those of Europe itself. However, a Europe which appears uniformly advantaged in the light of global comparison presents a much more textured picture when scrutinized closely, without reference to conditions elsewhere. Even in regions where trade brought rapid economic growth, the mass of the population only began to obtain a just share of the gains when successful agitation resulted in a fairer measure of electoral representation during the second half of the nineteenth century. Even then, the economic cycle remained, leading to recurrent crises which, among other things, prompted demands for protective tariffs and the suppression of free trade, widely thought, in these circumstances, to be destructive of well-being. Social life at the heart of the core was not invariably well served by the regions' position of economic privilege. Trade between Europe's core and non-core regions ensured that the latter also shared the experience of instability, often much magnified.

Europe's trading history was therefore multi-layered, each layer interacting with the layers above and below it. At the top of this hierarchy were a succession of leading centres, the core locations of the emergent world-system. Each rose to prominence and then declined according to its particular circumstance. Spain and Portugal, 'Europe's street corner', offered a more convenient stepping-off point for transatlantic commerce than the north Italian city-states which were the source of much of its initial funding. However, primacy was a transient experience for Iberia, lasting no more than a century, a brief interlude between the Mediterranean and North Sea eras. In part, this may be attributed to the destructive inflationary impacts of the American gold and silver which flowed there. This sapped productive activity and energy within the Iberian peninsula rather than stimulating it (Vilar, 1976: 166–7). Madrid grew rapidly on the basis of its imperial administrative functions to become primate city of Castile. But as a centre of luxury and conspicuous consumption, unable by virtue of its landlocked location to draw upon the resources of a wider area, it made unbalanced and parasitic demands upon its immediate hinterland, a major contribution to Castile's decay (Ringrose 1970, 1989).

The southern North Sea region, focused successively on the leading centres of Bruges, Antwerp, Amsterdam, and London, came to prominence on the back of more durable trading activities as Spain declined. The clockwise migration of primacy round the coastline of the west European peninsula marked the advent of new trading relationships, as regions which had previously participated fitfully or not at all in continent-wide commerce were connected to the European core. While these developments were influenced by immediate historical events (the struggle of the Netherlands for independence from Spain, the dislocations of the Thirty Years War, the Anglo-Dutch conflicts of the second half of the seventeenth century), they were also shaped, as was the history of Braudel's Mediterranean (1972), by deeper and more slowly changing structural forces.

The new leading centres of European trade had common structural features. In each, the political environment had developed in ways sympathetic to merchants, especially foreign merchants, and trading activity. The Dutch Republic of the seventeenth century was shaped and controlled by the urban merchant classes. The English Civil War of the mid-seventeenth century marked an important step in the ascendancy of urban, merchant interests over the rural and aristocratic (Fox, 1989: 339). Both contrast with France, where, as Fox (1971) has shown, merchant classes long deferred to an entrenched aristocracy in a polity which exhibited increasingly pronounced symptoms of schizophrenia. Each successful trading centre was home to innovations in the management of trade, offering markets which traded not just physical goods, but finance, shares in trading enterprises, and futures contracts. Each successful trading centre became not just a place of importation, but a centre for redistribution, processing, manufacture, and consumption. The Dutch 'golden age' of the seventeenth century was distinguished, above all, by the profusion and quality of its material culture of domesticity (Schama, 1987). Each successful trading centre was supported by transport connections of high quality. In the context of the mercantile age, this meant that each centre had not merely access to waterborne transport, but command of an extensive waterborne transport network capable of generating large trading profits.

In the case of Amsterdam, this comprised the transoceanic routes to the East Indies and the Americas, the coastal routes through the narrow seas connecting Baltic grain and Mediterranean consumer, and the Rhine river system, the latter linking Amsterdam and German cities, formerly active traders with north Italian towns via the Alpine mountain passes. London enjoyed a similar if differently weighted combination of oceanic, coastal, and riverine transport connections. Unlike most European cities, neither Amsterdam nor London needed to draw fuel wood from land in their immediate hinterlands which might otherwise have been used to grow food. Amsterdam burned peat, brought by canal and river barge from nearby deposits and, as these were exhausted, from the eastern provinces of Groningen, Friesland, Drente, and Overijssel (De Zeeuw, 1978). London's 'seacoal' was shipped coastwise from the collieries of Northumberland and Durham. Each trade had important multiplier effects, although with contrasting long-term consequences. The Amsterdam peat trade influenced the development of the Dutch canal system, and made an early contribution to the evolution of the unique and remarkable 'trekvaart' network of dedicated passenger-carrying canals (De Vries, 1978: 56). The 'trekschuit' underwent no significant change over the three centuries of its operation, and represented a technological dead end. By contrast, the Northumbrian coal trade turned out to be not merely a 'nursery of seamen' but a nursery of railway and steam technology, seamlessly connecting the transport and trading activities of the mercantile and capitalist eras.

The remainder of this chapter looks in greater depth at four aspects of the five-century history of European trade and interaction under review. First, it will consider the developing role of traders, especially foreign traders, in the evolution of trading. Secondly, it will examine the effects of developing urban provisioning needs on the geography of the trade in foodstuffs. Thirdly, it will explore innovation in transport and communication, focusing especially on the role of state intervention during the major innovation phase of the eighteenth and nineteenth centuries. And fourthly, it will look at the arguments advanced for and against free trade, as perceptions of national and factional interests became increasingly confused and ambivalent.

## Traders and the Evolution of Trading

By its nature, trade reaches to and beyond the margins of single polities and cultures. The associated problems have become less marked through time as the volume of international trade has grown. Present-day international trade is generally conducted within a culturally homogeneous and internationally regulated environment, the world of international business, by nationals of the trading partners, generally during short-term overseas visits. In the past, the key figure in cross-cultural trade was the resident expatriate, whose role might take one of a number of forms. In some cases, expatriates operated within a trading enclave which retained close ties with and was sometimes under the sovereignty of their place of origin. The foundations of the Hanseatic League of independent trading towns were laid in the medieval period when merchants from Cologne established trading enclaves along the Rhine and the Baltic and North Sea coastlines (Dollinger, 1970). Until the mid-sixteenth century, Calais was an English enclave from which raw wool originating in Britain was distributed to continental processors (Curtin, 1984: 3–4). Extra-territorial enclaves of this kind, established by European polities elsewhere in Europe, faded from the scene during the mercantile period, at the same time as the creation of similar

enclaves along the shorelines of Asia was fostering the growth of European trading world-wide (Murphey, 1975; Basu, 1985; Broeze, 1989).

More commonly, expatriates operated within polities over which they exercised no control. Any major trading city was host to expatriate trader communities, their number generally proportionate to the complexity of that city's trading links. Whatever their accustomed vernaculars, traders all understood the lingua franca of commerce, which increasingly took place in the city's own exchanges, gradually displacing Europe's great periodic fairs (Braudel, 1982: 81–114). Many German cities of the sixteenth century had a significant Italian presence. Braudel (1972: 316) refers to Augsburg as a city 'half-German, half-Italian'. In 1664, a writer observed in Amsterdam representatives of all the European trader polities, plus 'Armenians, Turks and Hindus' (Curtin, 1984: 203). The increasing use of commission agents meant that over time the growth of trade was not associated with a proportionate growth in numbers of resident expatriates (Curtin, 1984: 4). Even so, what Curtin (1984) calls 'trading diasporas', spreading across the map of Europe as trade itself expanded, were integral to both the economic and cultural evolution of the continent.

Two groups, the Armenians and the Jews, were cross-cultural traders *par excellence*. Neither in any sense served the interests of their weak or non-existent polities of origin. Both to an extent, but the Jews especially, suffered, by virtue of the many dimensions of their 'otherness', constant discrimination, recurrently severe, at the hands of their European hosts. Sustained tolerance of a large Jewish presence was one of the attributes of a successful trading polity.

The Armenians' trading credentials were forged over a long period working the caravan routes which linked the eastern Mediterranean and east Asia. By the early years of the sixteenth century, Armenian trading diasporas already extended eastward across Asia and northward into eastern and central Europe, where Armenians rivalled Tatars and Russians as traders in Persian silks and Caspian fish along the Caspian–Volga axis. Towards the end of the sixteenth century, Armenian traders began to reappear in western Mediterranean cities which were linked by seaborne trade to the Levant. They were responsible, among other things, for spreading the coffee-drinking habit in France, where they were proprietors of most of the early cafés. This brought Armenian culture a fleeting voguishness, but did little to improve the Armenians' standing with French merchants and traders. Better tolerated in the Netherlands, their fortunes there failed to survive the collapse of the Levant silk trade in the second half of the eighteenth century. Consequently, west European Armenians started to serve other trading diasporas, with some becoming fully integrated members of west European societies (Curtin, 1984: 179–206).

More numerous than the Armenians, dispersed, to quote Braudel (1972: 804), 'like tiny drops of oil over the deep waters of other civilizations, never truly blending with them yet always dependent on them', the Jews had a more fractured and complex experience which touches many facets of European trading history. It was the Jew's recurrent lot to be cast as scapegoat. As the economic climate in any location worsened, so the probability of persecution, massacre, expulsion, or forced conversion increased. For Braudel (1972: 816), this imparted a simple dynamic to Jewish migrations, a drift towards regions of growth from regions of crisis. Thanks to its economic underpinnings, persecution operated as a kind of self-correcting mechanism which saw Jews constantly relocating towards areas of opportunity.

However, there was in fact no simple correlation between economic vitality and Jewish advancement. As Israel (1985) has shown, the late eighteenth and nineteenth centuries represented for central and west European Jews not only an era of unparalleled economic opportunity, but a period of political and legal emancipation. Yet, over the span of four centuries to 1900, it was precisely during this period that west European Jewry suffered its most marked numerical, cultural, and institutional decline. Moreover, the wave of expulsions which hit western and central European Jewry during the century after 1470 resulted in an eastward movement which ensured survival, but at the cost of immediate large-scale participation in the economic life of Europe's most successful trading centres in the west. Admittedly readmission to west and central European urban life began tentatively during the last quarter of the sixteenth century. By the early seventeenth century, Jewish communities across Europe were sufficiently numerous,

sufficiently wisespread, and sufficiently integrated by shared cultural values as to be a crucial influence in the circulation of precious metals and loans within Europe, east and west. But Jewish involvement in finance was partly a response to continuing discriminatory legislation which barred participation in crafts and trades. And the largest Jewish populations, those of eastern Europe, the Balkans, and the Levant, took their origins from forced migrations of German and Spanish Jews, the Ashkenazim and Sephardim respectively.

At one level, the Jews of eastern Europe were both promoters and beneficiaries of the region's increasing involvement in Europe-wide trading activity, occupying several niches within the emergent east European trading economy. In newly colonized regions of eastern Poland and Lithuania, which supplied grain and timber to western Europe, Jews became managers and leaseholders, effectively running estates for the Polish nobility who owned them. Lower down the social hierarchy, itinerant Jews are estimated to have commanded about one-half of Poland's peddling trade by the seventeenth century (Braudel, 1982: 77). However, in eastern Europe rural development intensified and reinforced the rigidities and inequalities of the feudal order rather than undermined them. This meant that its eventual dissolution, in the second half of the nineteenth century, was cataclysmic rather than gentle. Jews were again cast as scapegoats, again becoming forced migrants. Curtin (1984: 3) observes that 'trading diasporas tend to work themselves out of business'. In the case of the Jews, prejudice ensured a long-lived trading diaspora, its participants recurrently denied even the opportunity to trade.

## Urban Provisioning and Evolving Agricultural Geographies

Trade was ordered and controlled from towns, and wealthier urban populations took the lead in consuming traded goods and commodities. 'Consumer revolutions', which occurred when the masses began to imitate the consumption habits of the wealthy minority, were also urban events, at least initially (McKendrick et al., 1982: 20–1; Fairchilds, 1993). Even in relatively developed consumer societies, like eighteenth-century England, where wealth appeared in both town and country, certain new and decorative goods became almost commonplace in urban inventories before they were recorded in those of the countryside (Weatherill, 1988, 1993). Thus the urban orientation of trade may be illustrated by reference to the consumption geography of any one of the goods which trade put into circulation. However, the historical geography of food production, trading, and consumption commands attention as the essential foundation both of urbanism itself and of all other forms of urban consumption.

The localized dimension of urban provisioning was of enduring importance, and this deserves emphasis. In the early sixteenth century, according to Braudel (1972: 386), all south European cities not enjoying access to waterborne commerce had to take essential foodstuffs from within a radius of thirty kilometres. Until cheap and rapid railway transport undermined the cost and quality advantages of milk, fruit, and vegetables drawn from intra- or sub-urban locations, significant quantities of these foodstuffs continued to be produced within or on the fringes of cities (Atkins, 1978, 1987). Former market-gardening specialization within the Paris suburban belt is still commemorated in the adjectival attachment of place-names 'Clamart' and 'Saint-Germain' to pea dishes, while 'Montmorency' indicates a recipe involving cherries (Grigson, 1978; Montagné, 1961).

Before the industrial age began to impose its conformities, the provisions available in any town were highly variable with respect to quantity, quality, variety, and price. This lent an extra and important dimension to the individuality of place, and was likely to be commented on, favourably, or otherwise, by travellers. In general, a combination of good local soils, varied local farming systems, excellent market provision, and good transport connections, especially by water, meant praise for the urban food regime. At Shrewsbury in the early eighteenth century, Defoe found 'the greatest market, the greatest plenty of good provisions, and the cheapest that is to be met with in all the western part of England' (Defoe, 1962: 76). However, travellers also recognized that urban provisioning was competitive, and that a town's local supplies might be pre-empted or outbid by rivals. Thus Celia Fiennes noted at Penzance in the late seventeenth century an 'abundance of very good fish, though

they are . . . ill supplied [locally] because they carry it all up the country east and southward' (Morris, 1982: 208).

In any given polity, the largest city commanded the best of everything. In part, this was simply a function of the quality of its transport links. No major city rose to prominence without the benefit of a coastal or at least a riverine site. The tentacles of the great cities of northern Italy and Spain explored the Mediterranean world, bringing essential sustenance from some of its remoter backwaters. In the case of Venice, canalized waterways brought the produce of the Po lowlands, supplementing the grain, oil, wine, fish, and firewood conveyed by Venetian ships from places as far distant as Crete and the Levant, and meat on the hoof brought overland from as far away as Hungary. Braudel (1972: 579) likens wheat-exporting Sicily to a sixteenth-century Canada or Argentina. Capitals were not merely the leading centres of consumption in their own right but also the places where provisioning for armies, navies, and merchant marines was organized and collected. Their inhabitants therefore benefited from scale economies in trading unavailable elsewhere. The urban consumer was a larger consumer than his rural cousin: it has been estimated that average per capita meat consumption in late eighteenth-century Paris was about three times that of the rest of France (Braudel, 1981: 196). Further, major cities were inclined to retain the best that came their way and trade the residue. Genoans ate expensive quality grain from Romagna and re-exported cheap inferior grain brought in from the Levant. Marseilles imported grain for distribution to the lesser cities of Provence, but itself preferred to consume the superior local crop. Polish peasants subsisted on a diet of oats and barley, while Dutch town-dwellers ate best Polish wheat (Braudel, 1981: 125).

Responses to the consumption behaviour of major cities were ambivalent. Any ill-fed population was likely to be troublesome to the established political order, but the populace of an ill-fed capital city most troublesome of all. The regulation of provisioning was therefore guided by an instinctive reluctance to compromise the subsistence privileges of major cities in times of dearth (Tilly, 1975; Chartres, 1977: 60–2). This naturally led to accusations of metropolitan 'food imperialism' when dearth threatened, as in northern France in 1793 (Cobb,

1970: 285). Rural food rioters typically demanded that the produce of the locality should be made available to the locality at a just or fair price (Tilly, 1985: 142). The authorities' stock response was stricter control of hated grain-trade middlemen, even though their activities were more a symptom of the problem than its cause.

The notion of urban parasitism, intrinsic to mercantilist thought, was slow to fade. Urban opulence provided an enduring challenge to moral sensibilities and constant visual confirmation of parasitism in action. With the benefit of hindsight, we can appreciate that pre-industrial technology offered slender and insecure support to urbanism. Focusing on what was achieved and accentuating the positive, Wrigley (1986) uses urban population estimates to measure comparative rates of agricultural output growth in England, France, and the Netherlands before 1800. Emphasizing agricultural constraints to urban growth and thus the negative, De Vries (1984: 242–9) shows that south and central Europe's prevailing yield : seed ratios of between 4:1 and 5:1 were barely sufficient to sustain a significant non-agricultural population. Three circumstances made sustained urban growth possible. The first was the sizeable rural non-farm population, who were on hand to assist during seasonal labour peaks in agriculture, and were therefore less parasitic than true urbanites. The second was the development of regional specialization in agriculture, and hence the localized pursuit of comparative advantage, prompted by the stimulus of urban demand. And the third was the tendency of city dwellers to invest in ways which improved transport, lowered production costs, and encouraged regional agricultural and manufacturing specialization across expanding hinterlands. Network expansion, which, for instance, brought Brittany into the food supply hinterland of Paris by the mid-seventeenth century (De Vries, 1976: 162), was promoted by urban merchants who saw possibilities for profit, wealth, and further personal consumption.

In the countryside, urban provisioning presented many challenges, not least because its demands rarely remained stable over extended periods of time. Neither von Thünen's distance-decay model of agricultural land-use zonation, formulated in the east Prussian grain-exporting region in the early nineteenth century, nor Christaller's market-based central place model of a century later give much recognition to change through time, even though the forces responsible for the distributions analysed were dynamic and destabilizing. Over the very long term, increasing numbers of consumers and higher standards of living meant increases in the general demand for food, and an increase in the demand for livestock products relative to the demand for cereals. In the shorter term, demand shifts were often erratic, unpredictable, and uncomfortable for source regions of traded agricultural commodities.

During the period under survey, autarky became increasingly uncommon in rural Europe, though pockets persisted. For instance, as late as the mid-nineteenth century, peasant agriculture in the remoter parts of France was a byword for self-sufficiency (Weber 1977: 30–49; Crone, 1989: 23–33). Instinctively, historians have fallen in with the consensus that such regions were 'backward' and their incorporation into the wider market economy therefore a form of deliverance. However, deliverance carried a price, at the very least the subservience of local to extra-local market forces (Chartres, 1986: 169), at worst subservience to the needs of a remote and unconcerned polity. As a result, there developed in peripheral Europe (Wallerstein's 'semi-periphery') symptoms of dependence and exploitation reminiscent of those found more extensively in the colonized global periphery.

Consider the consequences of England's developing control of the Celtic periphery of the British Isles. These areas gained access to English urban markets in consequence, but not on their own terms (Hechter 1975). First, the nature of the market and the logic of their location demanded of them cattle and sheep, and this necessitated painful adjustments for labour-intensive peasant systems geared to subsistence grain production. Secondly, when market participation threatened English interests, then it was peremptorily suppressed if circumstance allowed, witness the various Irish Cattle Acts. And thirdly, the effects of market instability were always likely to be magnified in the periphery. Ireland quickly became England's granary when the Napoleonic Wars cut off other overseas sources of supply (Thomas, 1982). Harris and Ross (1987, 129–36) see the consequential short-term boost to labour demand as the underlying cause of the rapid population growth which the Irish potato famine subsequently reversed.

Some of the same effects appeared on a much larger scale across vast swathes of eastern Europe during the fifteenth and sixteenth centuries. Increasing demand for meat in the cities of Germany and the Netherlands could no longer be satisfied from indigenous sources of supply, which were under pressure to produce grain for an expanding population. A long-distance trade developed (see Fig. 15.1), with cattle driven to fattening pastures in the lower Rhineland from rearing areas in Hungary, southern Sweden, Denmark, Schleswig Holstein, and, ultimately, from low-cost production zones in Poland, the Ukraine, and Russia (Glamann, 1974: 443–5; Blanchard, 1986). By 1570, the trade flourished only as long as demand was sustained. Further, because merchants responsible for the trade sought economies of scale, the cheapest cattle were entirely absorbed by metropolitan markets while markets in the smaller centres saw little which was affordable. Consequently, this international trade enhanced the privileges of major urban centres but did a great deal less for everywhere else.

## The Means of Trade: Innovation in Transport and Communication

Communication and transport technology before the industrial age was no more static than the trade it served. The growth of European trade therefore both helped to foster and was made possible by technological changes of increasing significance. Even before 1500, pressures, partly linked to trade partly to the need of centralizing states for improved communication, led to a search for improvement and innovation. For instance, the transfer to bridge construction of building methods perfected in the erection of France's great Gothic cathedrals and the churches of Byzantium brought a bridge-building renaissance to southern France and to the emergent Turkish empire in the Balkans (Vance, 1986: 9–10). Yet, according to Vance, it was only in the early sixteenth century that there began a 300-year transport revolution during the course of which, for the first time, major innovations in transport, navigation, and communication originated exclusively in western Europe and its diasporas, not elsewhere in the world. Again, the driving forces were the long-term growth of trade,

with its tendency to place existing transport facilities and technologies under stress, and the state-building instincts of European polities, which saw improvements in transport and communication as key to more effective command of existing state space, and, prospectively, as a means of extending it.

Everywhere in Europe, the parallel demands of trading and governance fostered developments in transport. But the balance between these two forces was not everywhere identical. In this respect, Britain and France were bipolar opposites, with the rest of Europe occupying an indeterminate middle ground generally somewhat closer to the French than to the British model. In Britain, the role of the state was confined largely to the development of an appropriate legislative framework for controlling innovations in transport after they had appeared. An island polity had little need to create road systems in the interests of state-building: the military roads constructed in northern England and Scotland in the aftermath of the eighteenth-century Jacobite rebellions were the exception. Coastwise shipping, requiring little in the way of either fixed infrastructural investment or official intervention, carried a large proportion of total trade. Innovation in transport and communication responded to the pressures arising from an expanding national economy; it was not an attempt to construct a spatial framework within which such an economy could develop. In France, Europe's most populous and centralized polity, the hexagon of state space was an achievement of the ambitions of Paris, imposed upon a discordant topography by a web of transport routes radiating from the capital. These were more the creation of the state than of the commerce they carried.

Each development in transport followed a different course in Britain and France. As elsewhere, the maintenance of roads remained the exclusive responsibility of localities until it became clear that such arrangements served long-distance travel poorly. In Britain, individuals interested in improved road travel formed turnpike trusts, each sanctioned by Act of Parliament. These took over responsibility for the maintenance of defined stretches of road in return for the right to levy tolls from users (Albert, 1972: 23–5; Pawson, 1977). By contrast, the French monarchy actively promoted road improvement, first the medieval *routes royales*, then, following the creation in 1713 of a central road

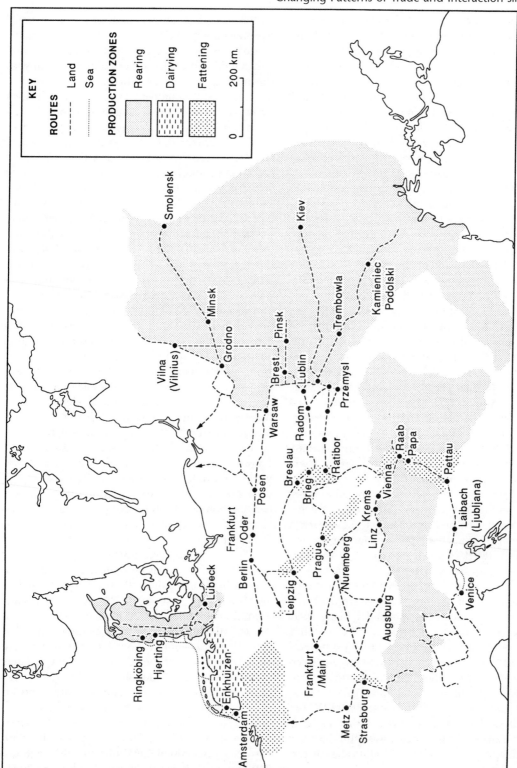

Fig. 15.1. The Continental cattle trade c.1595. (*Source*: Blanchard, 1986: 445.)

agency, the Corps des Ponts-et-Chaussées, a national system of *grandes routes* supported by central taxation (Vance, 1986: 154). Whereas the British canal network developed piecemeal funded by and to serve the specific transport needs of emergent industrial regions, the French system of canals and canalized waterways was conceived and developed as a coherent network, its primary purpose to establish connections between the river systems which disobligingly fragmented French national space (Ward, 1974; Vance, 1986: 53, 83). Likewise, Britain's first railways served the specific and highly localized needs of mining and manufacturing interests requiring outlets to ports and water carriage. Network development occurred rapidly if chaotically as private capital pursued the highest investment returns, parliament in effect sanctioning whatever was proposed. France, lacking an equivalent pattern of provincial industrial activity, followed her accustomed routines. Following private-enterprise construction of a number of early experimental or demonstration lines, Corps des Ponts-et-Chaussées engineers prepared a plan for a national rail network centred on Paris, which became law in 1842. The State undertook to provide the basic infrastructure of trackbed, bridges, and tunnels, while the railway companies were responsible for the superstructure of rails, ballast, stations, rolling stock, and working capital (Cameron, 1961: 204–8; Price, 1983: 208–9; Vance, 1986: 236). A second law in 1859 authorized State guarantees of dividend payments to companies prepared to invest in less profitable secondary lines, while further legislation in 1865 and again under the Freycinet plan of 1878 extended support to developers of a third-tier network of purely local lines, often of the narrow-gauge or tramway variety (Price, 1975: 23–5). Indeed, only about half of the Freycinet funds, which totalled nine billion francs, were dedicated to railway investment, the remainder being spent on other rural transport projects. The intention and the outcome was twofold: to improve the economic lot of remoter regions by opening up access to national markets, and to cement national unity by creating an unbreakable bond of sentiment between these regions and the central state (Weber, 1977: 208–11; Fox, 1971: 33). Even at this late date, French transport investment was still centrally directed, and as much aimed at political as at economic goals.

Elsewhere in Europe, statehood and internal communication shared a symbiotic developmental relationship reminiscent of that found in France, with the difference that other polities were in general neither as large nor as centralized nor such early achievers of modern statehood. Rulers were always attracted by the revenue-raising possibilities of trade. In areas where the political map was fractured or fragmented, trade over any distance therefore faced recurrent customs exactions and tolls. Any simplification of the political map was in itself a stimulus to trade. When the French Revolution and its aftermath reduced the number of nominally independent German states from more than 300 to 38, trading across German-speaking territory was considerably eased (Cameron, 1961: 9). Trade then provided part of the impetus to further simplification of political structures, the customs union or *Zollverein* of 1834 foreshadowing German unification in much the same way that, at a larger scale, the European Common Market has served as foundation for the construction of European political institutions (Henderson, 1939; Hahn, 1984).

In one European country after another, railway networks were shaped by a combination of State direction, private finance, and foreign expertise. Britain contributed about one-third of the capital invested in French railways prior to 1848, but thereafter invested in the United States and the British Empire, leaving the Continent to the French. Companies controlled by French capital were responsible for the greater part of Continental railway construction everywhere except Scandinavia and Germany. On occasion, their activities served the foreign policy interests of France. For instance, France saw in the development of Spain's railways possibilities for increasing French economic influence in Iberia and prising Portugal away from Britain (Cameron, 1961: 249). But in general states administered railway development within their borders in what they perceived as their own interests, whether economic, political, or strategic. Thus the government of newly independent Belgium undertook the construction of a basic grid of two mainlines, one east–west and the other north–south, with two major branches, then invited private enterprise to fill out the network (Cameron, 1961: 208; Milward and Saul, 1977: 171–3). In Switzerland, constitutional revisions in both 1848 and 1873 provided a greater degree of federal authority, partly because of the cantons' inability to handle railway development. The French consul in Geneva likened the economic effects of the resulting Alp-piercing network to the complete removal of trade barriers (Cameron, 1961: 302–4). Confronting the problem of providing rail connections over vast and lightly populated distances promising only modest traffic revenues, Russia eventually settled in 1881 for a system of state ownership and con-struction, but only after a basic network for Euro-pean Russia had been built essentially by French enterprise and capital underwritten by the financial guarantees of the Russian government (Cameron, 1961: 275–83). The emergent European railway net-work (Fig. 15.2) was thus a child not only of a new age of industrial technology, but of financial and governmental institutions and attitudes which bear an ever-closer resemblance to their present-day counterparts.

While the history of transport and trade was one of continuous development within institutional frameworks which were themselves slowly evolving, the nature, scale, and effects of the discontinuities in that historical process were highly significant. The history of transport technology, like that of the economy it served, divides into organic and industrial segments. Organic conveyance meant construction technologies which used expensive iron very sparingly and copious quantities of wood. Locomotion was supplied either by animate energy, or by wind or river flow. Innovation within the constraints of this technology was certainly possible. For instance, the advent of wheeled car-riage in Europe's remote places marked an impor-tant step in their incorporation into larger trading areas. The period between 1550 and 1650 saw larger and heavier wagons introduced into the English carrying trade, with consequential increases in the system's overall capacity (Chartres, 1977: 40–1). The Dutch *fluyt*, developed in the late sixteenth century, proved a cost-effective coastal vessel, adaptable to many different cargoes (Parry, 1967: 210–15). But growth sooner or later ran up against one ceiling or another. The size of tree timber, especially for masts, was a fundamental constraint on the size of ships. The depletion of the best local trees meant that, even to build to established standards, west European shipbuilders had increasingly to source timber in distant Scandinavia and north-east America (Malone, 1964; Astrom, 1970). As the

number of horses grew, concern was expressed about the acreages their feeding required (Braudel, 1982: 349; Gomez Mendoza, 1983; Thompson, 1983). With encouragement, cattle obligingly walked towards their consumers from distant places of breeding, but numbers were constrained by the quantity of feeding which could be procured along the way and by the capacity of fattening pastures close to the ultimate point of slaughter. Under organic transport technologies, the diverse resources of the European peninsula were exploited and traded, but not to the levels which market-expanding industrial technologies subsequently allowed.

The effect of the industrial transition on trade volumes is clear from the available statistics. Som-bart's estimates for late eighteenth-century Ger-many suggest 500 million kilometre-tons per year by road and 80 to 90 million kilometre-tons per year by river. By 1913 the railways alone were hand-ling each year 130 times the road traffic of the earlier period (quoted in Braudel, 1982: 350). Across Europe as a whole, Bairoch (1989: 1) estimates a 44-fold growth in trade volumes between 1815 and 1914 compared with an increase of no more than two- or threefold during the previous century. Overall, transport costs during the nineteenth century declined by an estimated 90 per cent, the highest reductions being for land and the lowest for river transport (Bairoch, 1990: 142). Reducing transport costs and increasing network capacity made large-scale urbanization possible. Bairoch (1990: 149) estimates that, under pre-industrial transport and agricultural technologies, the world could not sus-tain more than one city of a million people. By the end of the nineteenth century, London, with a population approaching ten million and requiring for its own subsistence cultivated land equivalent to four times the area of France, was merely the pin-nacle of a developed European urban hierarchy which simply could not have existed without cheap long-distance transport.

## Free Trade versus Protection

As the recent history of negotiated tariff reduction has shown, the concept of international compara-tive advantage remains problematic for policy makers intent on securing the best deals for

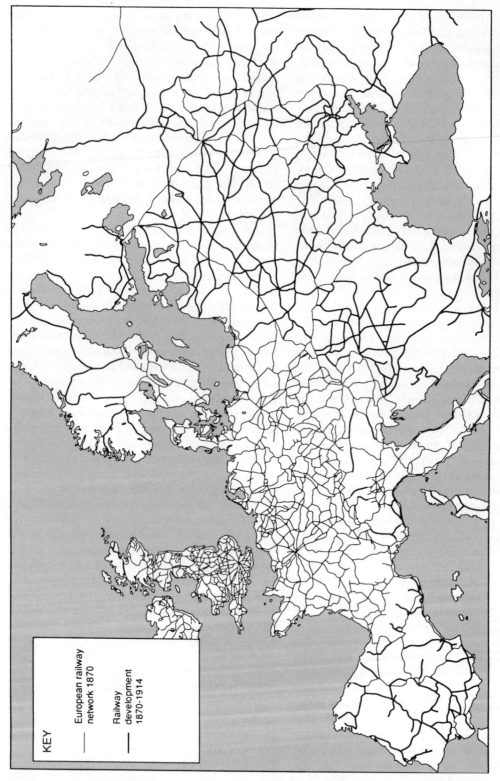

**Fig. 15.2.** Principal European railways to 1914. (*Sources*: Barraclough, 1978: 212–13, 257; Carter, 1959: 402–44; Cameron, 1961: 204–5; Westwood, 1964: 60, 168; Millward, 1964: 270–3; Sömme, 1968: 283, 340–1; Pounds, 1985: 449–61; A. Moyes (personal communication).)

individual nation-states. The argument that unrestricted international free trade generates the highest aggregate levels of consumption by releasing goods to markets at the lowest possible price would be more acceptable if many consumed but few produced. As it is, unrestricted free trade threatens too many producers ever to be an acceptable mode of global economic organization. Tariff barriers are therefore not an incidental feature of the current international trading system, despite a consensus in favour of continued tariff reduction and a tendency, at least among mainstream economists, to regard tariffs as the anachronistic residue of a less enlightened age.

Contemplating the entire industrial era from the wreckage of the Second World War, Karl Polanyi, an Austrian refugee, wrote a damning indictment of free trade (Polanyi, 1944). Polanyi acknowledged that it was the collapse of nineteenth-century free trade which through protectionism, acrimony, and depression, had created the drift towards war. But for him the real problem lay less in the circumstances or the outcome of that collapse than in the ill-considered adoption of the 'self-regulating' or free market which had preceded it. For Polanyi, the hidden hand of the market carried more sinister implications than Adam Smith had ever envisaged: 'The congenital weakness of nineteenth century society was not that it was industrial but that it was a market society. Industrial civilization will continue to exist when the utopian experiment of a self-regulating market will be no more than a memory' (Polanyi, 1944: 250).

To our age, inclined to regard economic liberalism as self-evident truth, Polanyi's analysis seems odd. But it does help counter a teleological tendency to read the past from the standpoint of the present, and, in particular, helps us to perceive that the economic history of Europe was not synonymous with the history of *laissez-faire*. Three points deserve emphasis.

First, Europe's historical free-trade experiment was relatively short-lived, occupying little more than the third quarter of the nineteenth century. Peel's liberal tariff reforms, culminating in corn law repeal in 1846, marked its onset; the Anglo-French treaty of 1860 and the brief period of mutual tariff reduction which ensued was its apogee; to be followed, from the 1870s, by depression and decline (Bairoch, 1989). Prior to the free-trade era,

economic management had been guided by mercantilist principles. The term 'mercantilism' was first used by Adam Smith as a means of conceptualizing the economic policy of the early modern period. No agreed definition exists, but mercantilism is generally associated with the idea that the economic welfare of the State can only be secured by regulation of an essentially nationalistic character, involving the creation and protection of monopolies (for instance, the English Navigation Acts, the East India Company), or the payment of bounties to encourage production (for instance, under the English corn laws of 1663, 1670, and 1689) (Coleman, 1969; Wilson, 1967; Ormrod, 1985). The policy differed from its medieval predecessor by virtue of its emphasis on the creation of surplus rather than the management of scarcity, on production rather than provision or distribution. Wilson (1965: 237–8) suggests that mercantilist policies assisted the early development of the English industrial economy. The same could be argued for the late nineteenth-century reversion to protectionism, which was associated with relatively rapid growth in some of Europe's less-developed countries (Bairoch, 1989: 60–78). Britain and other imperial powers were left to construct systems of colonial preference in place of free trade.

Secondly, even in the free-trade era, significant state intervention occurred. The scale of government involvement in the development of railway infrastructure has already been noted. Ironically, the threat posed by cheap railway transport to indigenous production, especially marked in peripheral Europe, was one of the major causes of the death of free trade. We might also note that the cotton industry, Britain's lead industrial sector, major exporter, and leading advocate of corn law repeal, prospered on freedom from regulation in the sphere of production, but long held out against freedom from regulation in the sphere of exchange (Polanyi, 1944: 136). For instance, the industry's lobbying ensured that a ban on the export of textile machinery remained in place until 1842 (Bairoch, 1989: 12).

Thirdly, agriculture proved to be the European economic sector most at threat from international free trade. During the first quarter of the nineteenth century, the supply of food was so inelastic as to compromise the essential subsistence of Europe (Post, 1977). By the last quarter of the nineteenth century, the supply was so copious as to threaten the livelihoods of European food producers, who, everywhere except Britain, were still sufficiently numerous to constitute an influential political lobby (Pollard, 1981: 261). French peasants, newly connected to markets by the railways of the Freycinet plan, were readily persuaded of the merits of protectionism as they found those markets submerged by imports from America or the Ukraine (Fox 1971: 143). An agricultural sector could not simply be written off merely because it lacked international comparative advantage; there is, in any case, good evidence that the most successful economies have not remorselessly pursued whatever comparative advantage they possessed (Ferleger 1995). At the height of the nineteenth-century industrial age, and not for the last time, European agriculture became a fundamental influence on the shaping of European policies, institutions, and politics.

## Concluding Observations

Consumption habits changed as trade developed. By the end of the nineteenth century, not only were the traditional staples of European subsistence as likely as not to have been traded across international borders, but novel goods or goods previously consumed only by the élite had also become items of mass consumption. Transport, the means of trade, had itself become a mass consumption good as, increasingly, had interaction or the transmission of information without personal mobility, via newspapers, magazines, and cheap, reliable postal systems. In the twentieth century, what was initiated by the railway, the steam engine, iron ocean vessels, the roller press, and mass primary education has been extended by the internal combustion engine, the fan jet, containerization, bulk carriers, telecommunications, the mass media, and mass secondary and tertiary education. Cost reductions have been pronounced (Fig. 15.3). The shrinking world thus has two dimensions. One is rooted in material goods and commodities, traded worldwide and physically available world-wide. The other is represented by the exchange of information and experiences, through travel, print, radio waves, and electrical and electronic impulses. The democratization of consumption has occurred at the same time as the diversification of what is consumed.

At each stage in this evolving process, the élite has commanded more than the masses. Distance has not shrunk at the same rate for everyone. The information available to rulers has remained superior to the information available to the ruled. Of the sixteenth century, Braudel observed 'a state had to wage not one but many struggles against distance'. Some of these struggles were epic and unsuccessful: 'the Turkish empire was a sum of delays laid end to end'. The most powerful in the world of the blind were the one-eyed: 'to look over Philip II's shoulder as he deals with his papers means constantly being aware of the dimensions of France; it means becoming familiar with the postal services, knowing which routes have regular stages and which have not; noting the delays in the mails caused here and there by civil wars' (Braudel, 1972: 372). In the late nineteenth century, the commercial and political power of Britain was cemented by the construction of the first global telegraph network, completed at a time when rival powers were still preoccupied with filling in their domestic communication systems (Hugill, 1993: 320–2; Headrick, 1991). Today, the steady, distinctive motion of a satellite crossing the clear night sky causes one to wonder about the nature of the information being collected and who will enjoy access to it. In all likelihood, the technology of the powerful is being deployed in securing and extending power.

Since the end of the Cold War, governments have gathered more commercial or industrial and less military or political intelligence. The high margins claimed by the first to market a successful new technology tempt subterfuge. More generally, a free-trade rhetoric can be a screen for protectionist stratagems designed to serve national interests or appease national interest groups. Few countries are confident that the competitive advantages which their particular factor endowments currently confer will prove enduring. Trade seems set to remain a paradox, the foundation of economic well-being but also a source of tension and dissonance.

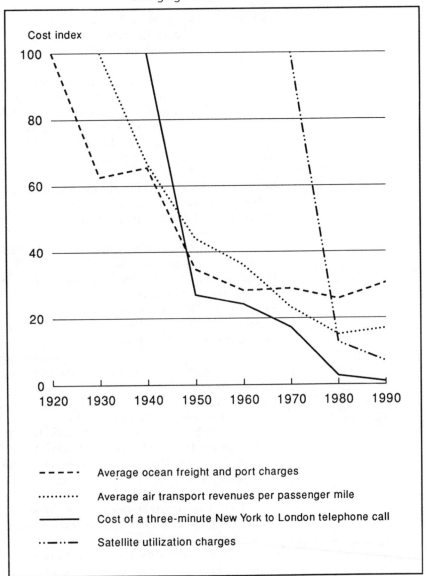

Fig. 15.3. International transport and communication costs, 1920–1990. (*Source:* World Bank, 1995: 51.)

REFERENCES

ALBERT, W., *The Turnpike Road System in England 1663–1840*, (Cambridge, 1972).

ASTROM, S. E., 'English Timber Imports from Northern Europe in the Eighteenth Century', *Scandinavian Economic History Review*, 18 (1970), 12–32.

ATKINS, P. J., 'The Growth of London's Railway Milk Trade, c.1845–1914', *Journal of Transport History*, 4 (1978), 208–26.

—— 'The Charmed Circle: Von Thünen and Agriculture around Nineteenth-Century London', *Geography*, 77 (1987), 129–39.

BAIROCH, P., 'European Trade Policy, 1815–1914', in P. Mathias and S. Pollard (eds.), *The Cambridge Economic History of Europe*, vii. *The Industrial Economies: The Development of Economic and Social Policies*, (Cambridge, 1989), 1–160.

—— 'The Impact of Crop Yields, Agricultural Productivity, and Transport Costs on Urban Growth between 1800 and 1910', in A. D. van der Woude, A. Hayami, and J. de Vries (eds.), *Urbanization in History: A Process of Dynamic Interactions*, (Oxford, 1990), 134–51.

BARRACLOUGH, G. (ed.), *The Times Atlas of World History*, (London, 1978).

BASU, D. K. (ed.), *The Rise and Growth of the Colonial Port Cities in Asia*, (Lanham, Md., 1985).

BLANCHARD, I., 'The Continental European Cattle Trade, 1400–1600', *Economic History Review*, 2nd ser. 39 (1986), 427–60.

BLAUT, J. M., 'Colonialism and the Rise of Capitalism', *Science and Society*, 53 (1989), 260–96.

—— 'Political Geography Debates 3: On the Significance of 1492', *Political Geography*, 11 (1992), 355–85.

BOORSTIN, D J., *The Discoverers: A History of Man's Search to Know his World and Himself*, (New York, 1983).

BRAUDEL, F., *The Mediterranean and the Mediterranean World in the Age of Philip II*, (2 vols.; London, 1972).

—— *Civilization and Capitalism 15th–18th Century*. i. *The Structures of Everyday Life: The Limits of the Possible*, (London, 1981).

—— *Civilization and Capitalism 15th–18th Century*. ii. *The Wheels of Commerce*, (London, 1982).

BROEZE, F. (ed.), *Brides of the Sea: Port Cities of Asia from the Sixteenth to the Twentieth Centuries*, (Honolulu, 1989).

CAMERON, R. E., *France and the Economic Development of Europe 1800–1914: Conquests of Peace and Seeds of War*, (Princeton, 1961).

CARTER, E. F., *An Historical Geography of the Railways of the British Isles*, (London, 1959).

CHARTRES, J. A., *Internal Trade in England 1500–1700*, (London, 1977).

—— 'Food Consumption and Internal Trade', in A. L. Beier and R. Finlay (eds.), *The Making of the Metropolis: London 1500–1700*, (London, 1986), 168–96.

CHILDS, W. R., '1492–1494: Columbus and the Discovery of America', *Economic History Review*, 2nd ser. 48 (1995), 754–68.

CIPOLLA, C., *Guns and Sails in the Early Phase of European Expansion, 1400–1700*, (London, 1965).

COBB, R., *The Police and the People: French Popular Protest, 1789–1820*, (London, 1970).

COLEMAN, D. C., Introduction, in D. C. Coleman (ed.), *Revisions in Mercantilism*, (London, 1969), 1–18.

CRONE, P., *Pre-Industrial Societies*, (Oxford, 1989).

CROSBY, A. W., *Ecological Imperialism: The Biological Expansion of Europe 900–1900* (Cambridge, 1986).

—— *Germs, Seeds and Animals: Studies in Ecological History*, (Armonk, NY 1994).

—— 'The Potato Connection', *World History Bulletin*, 17 (1996), 1–5.

CURTIN, P. D., *Cross-Cultural Trade in World History*, (Cambridge, 1984).

DEFOE, D., *A Tour through the Whole Island of Great Britain*, ii. (London, 1962 Everyman edn.).

DENECKE, D., 'Innovation and Diffusion of the Potato in Central Europe in the Seventeenth and Eighteenth Centuries', in R. H. Buchanan, R. A. Butlin, and D. McCourt (eds.), *Fields, Farms and Settlement in Europe*, (Hollywood, Co. Down 1976), 60–96.

DE VRIES, J., *The Economy of Europe in an Age of Crisis, 1600–1750*, (Cambridge, 1976).

—— 'Barges and Capitalism: Passenger Transportation in the Dutch Economy, 1632–1839', *A.A.G. Bijdragen*, 21 (1978), 33–398.

—— *European Urbanization 1500–1800*, (London, 1984).

DE ZEEUW, J. W., 'Peat and the Dutch Golden Age: The Historical Meaning of Energy Attainability', *A.A.G. Bijdragen*, 21 (1978), 3–31.

DOLLINGER, P., *The German Hansa*, (London, 1970).

FAIRCHILDS, C., 'The Production and Marketing of Populuxe Goods in Eighteenth-Century Paris', in J. Brewer and R. Porter (eds.), *Consumption and the World of Goods*, (London, 1993), 228–48.

FARNIE, D. A., *East and West of Suez: The Suez Canal in History, 1854–1956*, (Oxford, 1969).

FERLEGER, L., 'Comparative Advantage and Crop Specialisation in the Nineteenth Century', in M. A. Havinden and E. J. T. Collins (eds.), *Agriculture in the Industrial State*, (Reading, 1995), 33–43.

FOX, E. W., *History in Geographic Perspective: The Other France*, (New York, 1971).

—— 'The Argument: Some Reinforcements and Projections', in E. D. Genovese and L. Hochberg (eds.), *Geographic Perspectives in History*, (Oxford, 1989), 331–42.

FRENCH, C. J., 'Productivity in the Atlantic Shipping Industry: A Quantitative Study', *Journal of Interdisciplinary History*, 27 (1987), 613–38.

GIDDENS, A., *A Contemporary Critique of Historical Materialism*, (London, 1981).

GLAMANN, K., 'European Trade 1500–1750', in C. M. Cipolla (ed.), *The Fontana Economic History of Europe: The Sixteenth and Seventeenth Centuries*, (London, 1974), 427–526.

GOMEZ MENDOZA, A., 'The Role of Horses in Transport in a Backward Economy: Spain in the Nineteenth Century', in F. M. L. Thompson (ed.), *Horses in European Economic History: A Preliminary Canter*, (Reading, 1983), 143–55.

GRIGSON, J., *Jane Grigson's Vegetable Book*, (London, 1978).

HAHN, H. W., *Geschichte des Deutschen Zollvereins*, (Göttingen, 1984).

HARRIS, M., and ROSS, E. B., *Death, Sex and Fertility: Population Regulation in Preindustrial and Developing Societies*, (New York, 1987).

HEADRICK, D. R., *The Invisible Weapon: Telecommunications and International Politics, 1851–1945*, (New York, 1991).

HECHTER, M., *Internal Colonialism: The Celtic Fringe in British National Development, 1536–1966*, (London, 1975).

HENDERSON, W. O., *The Zollverein*, (Cambridge, 1939).

HODGES, R., *Primitive and Peasant Markets*, (Oxford, 1988).

HUGILL, P., *World Trade since 1431: Geography, Technology and Capitalism*, (Baltimore, 1993).

ISRAEL, J. I., *European Jewry in the Age of Mercantilism, 1550–1750*, (Oxford, 1985).

LIVINGSTONE, D. N., *The Geographical Tradition: Episodes in the History of a Contested Enterprise*, (Oxford, 1992).

MCKENDRICK, N., BREWER, J., and PLUMB, J. H., *The Birth of a Consumer Society: The Commercialization of Eighteenth-Century England*, (London, 1982).

MCNEILL, W. H., *Plagues and Peoples*, (Harmondsworth, 1979).

MALONE, J. J., *Pine Trees and Politics: The Naval Stores and Forest Policy in Colonial New England 1691–1795*, (London, 1964).

MILLWARD, R., *Scandinavian Lands*, (London, 1964).

MILWARD, A., and SAUL, S. B., *The Development of the Economies of Continental Europe, 1850–1914*, (London, 1977).

MINTZ, S. W., 'The Changing Role of Food in the Study of Consumption', in J. Brewer and R. Porter (eds.), *Consumption and the World of Goods*, (London, 1993), 261–73.

MONTAGNÉ, P., *Larousse Gastronomique*, (London, 1961).

MORRIS, C. (ed.), *The Illustrated Journeys of Celia Fiennes 1685–c.1712*, (London, 1982).

MURPHEY, R., *China Meets the West*, (New York, 1975).

OFFER, A., *The First World War: An Agrarian Interpretation*, (Oxford, 1989).

ORMROD, D., *English Grain Exports and the Structure of Agrarian Capitalism 1700–1760*, Occasional Papers in Economic and Social History, 12 (Hull, 1985).

PARKER, G., *The Military Revolution: Military Innovation and the Rise of the West, 1500–1800*, (Cambridge, 1988).

PARRY, J. H., 'Transport and Trade Routes', in E. E. Rich and C. H. Wilson (eds.), *The Cambridge Economic History of Europe* iv. *The Economy of Expanding Europe in the Sixteenth and Seventeenth Centuries*, (Cambridge, 1967), 155–219.

—— *The Discovery of the Sea*, (Berkeley, 1981).

PAWSON, E., *Transport and Economy: The Turnpike Roads of Britain*, (London, 1977).

PEET, R., *Global Capitalism: Theories of Societal Development*, (London, 1991).

POLANYI, K., *The Great Transformation*, (Boston, 1944).

POLLARD, S., *Peaceful Conquest: The Industrialization of Europe 1760–1970*, (Oxford, 1981).

POST, J. D., *The Last Great Subsistence Crisis in the Western World*, (Baltimore, 1977).

POUNDS, N. J. G., *An Historical Geography of Europe 1800–1914*, (Cambridge, 1985).

PRICE, R., *The Economic Modernization of France 1730–1880*, (London, 1975).

—— *The Modernization of Rural France: Communication Networks and Agricultural Market Structures in Nineteenth-Century France*, (London, 1983).

RINGROSE, D. R., *Transport and Economic Stagnation in Spain, 1750–1850*, (Durham, NC, 1970).

—— 'Town, Transport and Crown: Geography and the Decline of Spain', in E. D. Genovese and L. Hochberg (eds.), *Geographic Perspectives in History*, (Oxford, 1989), 57–80.

SCHAMA, S., *The Embarrassment of Riches: An Interpretation of Dutch Culture in the Golden Age*, (London, 1987).

SHAMMAS, C., 'Changes in English and Anglo-American Consumption from 1550 to 1800', in J. Brewer and R. Porter (eds.), *Consumption and the World of Goods*, (London, 1993), 177–205.

SÖMME, A. (ed.), *A Geography of Norden*, (London, 1968).

STANNARD, D. E., *American Holocaust: Columbus and the Conquest of the New World*, (New York, 1992).

TAYLOR, P. J., 'World-systems Analysis', in R. J. Johnston, D. Gregory, and D. M. Smith (eds.), *The Dictionary of Human Geography*, 3rd edn., (Oxford, 1994), 677–9.

THOMAS, B., 'Food Supply in the United Kingdom during the Industrial Revolution', *Agricultural History*, 56 (1982), 328–42.

THOMPSON, F. M. L., 'Horses and Hay in Britain, 1830–1918', in F. M. L. Thompson (ed.), *Horses in European Economic History: A Preliminary Canter*, (Reading, 1983), 50–72.

TILLY, C., 'Food Supply and Public Order in Modern Europe', in C. Tilly (ed.), *The Formation of National States in Western Europe*, (Princeton, 1975), 380–455.

TILLY, L. A., 'Food Entitlement, Famine and Conflict', in R. I. Rotberg and T. K. Rabb (eds.), *Hunger and History: The Impact of Changing Food Production and Consumption Patterns on Society*, (Cambridge, 1985).

VANCE, J. E., *Capturing the Horizon: The Historical Geography of Transportation since the Transportation Revolution of the Sixteenth Century*, (New York, 1986).

VILAR, P., *A History of Gold and Money 1450–1920*, (London, 1976).

VILLE, S. P., *Transport and the Development of the European Economy 1750–1918*, (London, 1990).

WALLERSTEIN, I., *The Modern World-System: Capitalist Agriculture and the Origins of the European World Economy in the Sixteenth Century*, (New York, 1974).

—— *The Capitalist World-Economy*, (Cambridge, 1979).

WARD, J. R., *The Finance of Canal Building in Eighteenth-Century England*, (Oxford, 1974).

WEATHERILL, L., *Consumer Behaviour and Material Culture, 1660–1760*, (London, 1988).

—— 'The Meaning of Consumer Behaviour in Late Seventeenth- and Early Eighteenth-Century England', in J. Brewer and R. Porter (eds.), *Consumption and the World of Goods*, (London, 1993), 206–27.

WEBER, E., *Peasants into Frenchmen: The Modernization of Rural France 1870–1914*, (London, 1977).

WESTWOOD, J. N., *A History of Russian Railways*, (London, 1964).

WILSON, C. H., *England's Apprenticeship 1603–1763*, (London, 1965).

—— 'Trade, Society and State', in E. E. Rich and C. H. Wilson (eds.), *The Cambridge Economic History of Europe*, iv. *The Economy of Expanding Europe in the Sixteenth and Seventeenth Centuries*, (Cambridge, 1967), 487–575.

WOLF, E. R., *Europe and the People without History*, (Berkeley, 1982).

World Bank, *World Development Report 1995*, (New York, 1995).

WRIGLEY, E. A., 'Urban Growth and Agricultural Change: England and the Continent in the Early Modern Period', in R. I. Rotberg and T. K. Rabb (eds.), *Population and Economy: Population and History from the Traditional to the Modern World*, (Cambridge, 1986), 123–68.

# Chapter 16

# Towards an Environmental History of Europe

## I. G. Simmons

## Introduction

To attempt an account of the last 10,000 years of the history of the European environment from Scandinavia to the Mediterranean and from the Dingle peninsula to the Urals, some broad generalizations of both time and space must be made. For the former, time will be divided according to the predominant human economies which have occupied the European space and which are set out in Table 16.1. Each phase has left some traces of its existence in today's landscapes but only in pockets are there functioning examples. For example, remnants of non-agricultural economies persisted until very recently in reindeer-herding cultures such as the Saami. In the case of solar-based agriculture, a few pockets exist in the poorer parts of Europe, but all have to some extent been affected by access to industrial products and knowledge. In the post-industrial period, electricity becomes a key form of delivery of energy. Manufacturing remains important but is increasingly supplanted by electronics-based information flows as sources of wealth. Not all the lands discussed here have yet entered on this phase. For the division of space, the categories adopted are shown in Table 16.2 and Fig. 16.1. Iceland is not included; neither is Russia east of the

Urals, nor the European-related islands such as Madeira and the Canaries.

Some intellectual distinctions must also be introduced. We have long moved away from the crudities of environmental determinism that derived national and individual traits from features of the natural environment, in which southerners were lazy and feckless owing to the heat and northerners were hard-working and puritanical. Yet, since escape from the reality of the land and the sea cannot be complete, they may pose some risks to inhabitants, which are usually categorized as environmental hazards; here is one place where culture meets environment. In this context, Europe is not a particularly hazardous region of the world in environmental terms, though it can be argued that it may well get more so.

In practical terms, therefore, we can structure this examination of humanity–nature relations in Europe around some matrices of regions against economic types: one each for environmental opportunities, constraints, environmental impact and environmental hazards. Forcing all the diversity of European history and culture into such cells (Tables 16.3 and 16.4 respectively) is a Procrustean exercise and only undertaken while keeping in mind what happened to him: today's book

I. G. Simmons

**Table 16.1.** Chronology of main periods used for the last 10,000 years

| Economy | Energy sources | Dates |
|---|---|---|
| Hunter-gatherer | Food *collection* rather than *production*. Uses recent solar energy from biota and older solar energy from wood. | Predates the last 10,000 years but is in sole occupancy of Europe as the ice wanes; disappears rapidly as agriculture succeeds. |
| Agricultural | Dependent upon solar energy from plants, animals, humans, wind and water. | Spread from east to west across Europe, starting *c.*8000 BC and coming to Britain *c.*3500 BC. |
| Industrial | Use of fossil fuels (coal, oil, and natural gas). Use of coal before 19th century, but major development of industrialism in 19th and 20th centuries. | Smelting of iron using coke in early 1700s; steam locomotives in 1820s; internal-combustion engine 1890s. |
| Post-industrial | Oil becomes more important than coal; nuclear power added to repertoire. Energy for pleasure becomes important, e.g. in tourism. | Becomes apparent in the 1950s in e.g. Britain, France. Not yet much in evidence in some regions of Europe. |

**Table 16.2.** Regional division for environmental history

| Region | Areas included | Label on Fig. 16.1 |
|---|---|---|
| Scandinavia | The Nordic group (Norway, Sweden, and Finland) but minus Denmark, which belongs more to western Europe. Iceland is not discussed. | A |
| Russia and eastern Europe | For this purpose, confined to Russia west of the Ural mountains, but including some other republics of a basically European character, such as Georgia, Ukraine, and Belarus. To this is added the countries of the former Warsaw Pact but excluding the Balkan lands, which are placed with the Mediterranean. | B |
| Alps and mountain Europe | The Alps are the centre of this category, but also encompassed are several other ranges of importance such as the Carpathians, Caucasus, the Pyrenees, Apennines, and the ranges of Cantabria and the Sierra Nevada in Spain. | C |
| Western Europe | The central lands from Germany to Ireland leaving out the mountainous regions of Switzerland and Austria, but including Denmark, which is, however, culturally close to the other Scandinavian nations. | D |
| The Mediterranean | The lands bordering the sea to the north, including the Balkans and, on the grounds of some cultural contiguity, Portugal. Turkey is omitted from consideration. The major islands of the Mediterranean are also in this category, including Malta. | E |

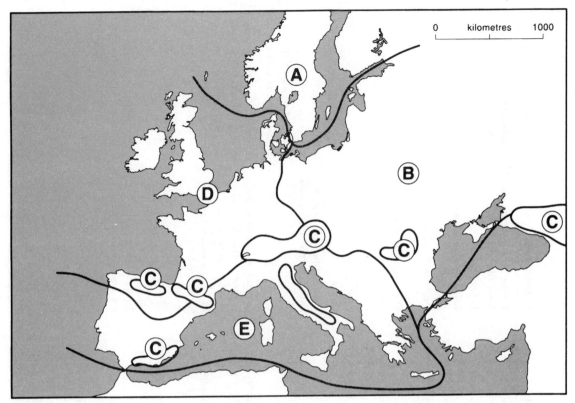

**Fig. 16.1.** A sketch map of Europe, with divisions corresponding to the regional sections of the text and to Tables 16.1–6: A = Scandinavia; B = Russia and Eastern Europe; C = Alps and mountain Europe; D = Western Europe; E = Mediterranean region.

reviewers often see themselves as lineal descendants of Theseus. Tables 16.3 and 16.4 contain a summary of some of the various constraints and openings experienced at different times in European regions. The following text takes us through the rows, highlighting some of the characteristic regional features.

## Environmental Constraints and Opportunities

### Scandinavia

In the beginning, 10,000 years ago, there was ice. In these northerly latitudes, the effects of the last Pleistocene glaciation lingered visibly until the demise of the Fenno-Scandinavian ice-sheet *c*.8500 years BP.[1] A further geophysical effect has been the isostatic recovery of the land surface when released from under the weight of the ice. The rate of uplift in the Baltic during the period 9000–7000

BP was between 2.0 and 2.7 cm./yr. and this kept pace with the eustatic rise of sea level from the melting ice; this was so of the Norwegian coast as well. Rising sea levels, however, invaded the freshwater lake that preceded the Baltic and salinified that water body after *c*.7500 BP. Hunter-gatherers in Scandinavia were therefore faced with the cold climate of a periglacial zone, with rapid changes in the topography and biological resources of the marine zone. The deciduous forests typical of lands further south were kept at bay until the mid-Holocene and then were found only in southern Norway, Sweden, and Finland; further north belts of coniferous forest, tundra, and birch scrub

[1] Following the conventions of radiocarbon dating, radiocarbon years Before Present (where present = AD 1950) are used for the early part of the Holocene; after about 7000 BP, I shall use calendrical years BC/AD, assuming that where these are derived from carbon-14 dating that they have been calibrated from radiocarbon years to calendar years. Dates with no BP/BC/AD annotation are assumed to be AD.

Table 16.3. Constraints

| | Hunter-gatherers | Agriculturists | Industrialism | Post-industrial |
|---|---|---|---|---|
| Scandinavia | Climate restricts flora. Rising sea level diminishes area of land. Late glacial–Postglacial transformation to forest in south. | Climatic limits on agriculture. Soils often v. peaty or v. acid—lots of inputs needed (Sweden, Denmark). Climatic limits on forestry also. | Limited energy resources of fossil fuel kind. Lots of water. Baltic poor processor of pollutants. Easily acidified waters. | Limitations of location for P-I economy, except tourism. |
| Russia and European republics | Eventual transformation from forest unhelpful. | Climate Erodability of soils Exhaustibility of forests | Distances and climate means transport expensive. | Not there yet. |
| Eastern Europe and Alps | Forest necessitated management for open areas. | Mountains: altitudinal limit on cultivation | Coal, water only energy sources. Conflicts between scenery and industrial development. | Limited capacity of air, water to process indust. wastes. Not enough snow in many years; not enough slopes. Avalanches related to sports. |
| Western Europe | Rising sea level forces hunter-gatherers into smaller area. | Northern limit to some desirable crops, e.g. vine, wheat. Rainfall + cropping = nutrient loss from agroecosystems. Slow growth of dominant deciduous trees. | Dependence on fossil fuels v. strong—rush for nuclear in 1950s onwards. Incapacity of fish stocks to withstand heavy exploitation. | Limitations on waste levels in air, gd water, rivers, seas. Need to devote land to conservation. |
| Mediterranean | Early influence of agriculture. Changes in hunter-gatherer ways in postglacial. | Dry summer before irrigation. Fast soil loss from high slopes. Heavy grazing produces poor swards quickly. Fire destructive of forest. Volcanism occasionally big hazard. | Separated from north-western Europe by Alps. | Ability of food agroecosystems to withstand intensification. Limits on supply of irrigation water |

covered the recently deglaciated land. In these eco-systems, an economy which depended upon hunting and gathering of land and marine resources persisted well after the establishment of agriculture in southern Scandinavia. The sea was so important to a number of communities that a case can be made for the adoption of agriculture as a response to the decrease in salinification of the Baltic (Rowley-Conwy, 1984). Elsewhere there was seasonal mammal-hunting, especially of reindeer. It was the practice to follow the animals on their seasonal round, and with the Saami this became the reindeer-herding which has until very recently been their economic base (Beach, 1990). The reindeer were left to browse for much of the year in a quasi-wild state but were corralled for the purposes of establishing ownership. The disruption of feeding by land-use change, the greater attraction of catering to tourism as a way of life, and the devastation to the economy caused by the compulsory mass slaughter of reindeer after the Chernobyl incident (1986), have made this remnant of the hunting-gathering era even more precarious. These environment-related changes took place in the setting of local socio-technological metamorphoses in which the advent of the snowmobile meant fewer families owned more reindeer but concentrated them on a smaller area of land so as to tend them on a daily trip basis. A form of 'overgrazing' began to appear in the 1950s.

In northern Scandinavia arable agriculture has never become established, though grassland (to feed winter-stalled cattle) is found surprisingly far north: in Kuusamo (Finland) this economy removed the Saami in the seventeenth century. At the northern limits of cereal cropping, oats were more cold- and wet-tolerant than wheat and rye (barley and buckwheat might be sown as spring crops in bad years), and the potato was a useful addition in the later part of the eighteenth century. Soils hitherto under conifers are acid (though capable of shifting cultivation, which lasted in Finland until the nineteenth century), less so under the oak woods of Skåne and southern Norway. Climate also placed limits on the northerly growth of trees, with spruce forest giving way to Scots pine as the pole-ward belt of exploitable trees; in northern Finland and Norway, there is also a belt of scrub birch altitudinally below the tundra heath, and this zone reaches as far as the relatively warm Varanger

Fjord. Scandinavian forests were an early (four-teenth-century) source of export of timber for wood-hungry nations like France and England. The conifers (especially the spruces) yielded straight timbers for masts and spars as well as resin products. Sweden and Russia were major sources, along with Finland, not then a nation-state. The forests were, however, sufficiently vast that no management was deemed to be necessary to safeguard the resource. The same was true of the sea at that time: it provided a cheap and easily preserved source of protein for local populations, and this characteristic was even more useful in the export trade, with salt cod and herring. As fishermen took to the open oceans, so Norway became involved in whaling on a large scale, to provide for export the whale oil that was the precursor of paraffin in domestic and commercial lighting.

The entry of Scandinavia into the industrial age had its precursors in the large mining enterprises of, for example, copper in Dalarna (Sweden) in the eighteenth century. The development of iron ore deposits in northern Sweden in the nineteenth–twentieth centuries is a continuation of this geological opportunity. Apart from those non-renewable deposits, the materials for European-style industrialization were thin: there was very little coal, although hydro-power could easily be exploited when capital became available in the twentieth century. Before that, entry into the burgeoning timber trade and, still more important, the provision of woodpulp, paper, and paper products for the reading- and wrapping-hungry richer world to the south has an air of almost determinist inevitability. The industrialization of fisheries, however, depended on only small amounts of coal and later, oil, and so was complete by the 1890s except for small artisanal enterprises. Fish exports grew and so contributed to the 'ghost acreage' by which other nations supplied protein to growing populations of the continent's industrial core. Manufacturing industry was never established in Scandinavia on a mass scale, though environment was clearly not the major determinant unless via the route of low population densities. The primary industries have, however, exerted considerable environmental impact, discussed below. Given, though, the low population densities of these countries, the urban areas are small and the cities generally spread-out: both Stockholm and Helsinki, for example, seem

**Table 16.4.** Opportunities

|  | Hunter-gatherers | Agriculturists |
|---|---|---|
| Scandinavia | Variety of environments in short distance—coastal forest, waters in Sweden, Norway.<br><br>Coastal biological production.<br><br>Concentration of reindeer. | Development of cereals like rye not grown further south.<br><br>Reservoir of forest for conversion to grassland for domesticated animals.<br><br>Timber resources for export, e.g. for ship-building and naval stores.<br><br>Fish exports. |
| Russia and Eastern Europe | Later continuance of late-glacial conditions in higher latitudes—mammal herds persisting.<br><br>Reindeer herding possible. | Grassland soils a big reserve of soil nutrients<br><br>Forest resources for export<br><br>Fuel for industrialization |
| Eastern Europe and Alps | Big stretches of deciduous forest with big game (European bison, aurochs) as well as deer. Dry periods in Continental climate good for fire management. | Open terrain—big agricultural units feasible.<br><br>Easy minerals to work, e.g. salt in Alps.<br><br>Mountain pastures.<br><br>Reservoir of wetlands for conversion to agriculture or fish production. |
| Western Europe | Access to coasts for some. Interspersed uplands give resource variety. Wetlands an additional resource. Movement along ocean fringes and coastal resources. | Brown Earth forest soils for conversion to agriculture: good fertility levels.<br>Year-round rainfall. Warm enough for vine to England, Rhine.<br><br>Forest resources.<br><br>Wetland reserve. |
| Mediterranean | Dry season for fire management. Access to coast plus adjoining mountains: resource variety. | High temperatures year-round, esp. summer ripening.<br><br>Suite of exclusive crops, e.g. olive.<br><br>Salt easy to make and export. |

stretched across a series of islands so that their immediate environmental relations seem almost accidental, as if many another site would have been equally good and cultural factors were paramount in their rise.

What of the Scandinavian environment in a post-industrial age? The opportunities it presents are mostly those of benefit to local people, in the form of ample space for summer cottages and also perhaps winter-sport second homes as well: cattle shielings (*faböd*) in Sweden have converted well to weekend ski retreats; the many islands of the Baltic are havens for summer sailors, and lakes provide one essential component of the sauna experience,

along with birch twigs, and mosquitoes in summer. Tourism is limited by the climate and the perception of long distances and high prices, though the Saami areas attract cultural tourists and wilderness-seekers. Since Scandinavia never experienced the full development of Fordist production methods (Volvo being a pioneer in team production rather than line assembly), it is difficult to see environmental considerations being directly relevant to the current industrial situation except in so far as the Nordic nations are generally amongst the leaders in 'green' thinking, a record in the case of Norway both promoted by the global involvement of Gro Bruntland and diminished by the recent insistence

| Industrialism | Post-industrial |
|---|---|
| Ores in large quantities, e.g. iron in Dalarna, Sweden. | Water power for high value-added industry |
| Wood for initial phases, water power for later. Wood for paper and products. Fisheries in Baltic. | Thin population for tourism and indigenous outdoor recreation (summer homes). |
| Wood for initial fuel coal, oil, natural gas, HEP. Huge energy resources from rivers. Open landscape good for mechanical agriculture/collectivization. | 'Cheap' energy for export, esp. natural gas. But not there yet? |
| Open landscape good for mechanized agriculture/collectivization. Brown coal easy for surface mining. Early tourism for climbing, walking in summer. | Cheap tourism, winter and summer. Reserve of little industrialized agriculture, e.g. Romania. Mass winter tourism in Alps. |
| Coal. Low relief = low transport costs. Coast = ports = trade. | Cultural tourism: natural and cultural resources. |
| Water power near to valleys, ports. Early interest in Mediterranean history. | Mass sun/sand/sea tourism. |

*Notes*: Many countries had a 'ghost acreage' in whaling, fishing, and empires. 'Eastern Europe' includes European republics of the former USSR.

on the resumption of whaling in the face of the International Wildlife Commission moratorium. Sweden announced in 1997, however, that its only nuclear power station was to be closed.

## Russia and Eastern Europe

Hunting and gathering in these lands was largely determined by the vegetation zone. In the tundra lands to the north, a pattern similar to that of Scandinavia developed and in the forests the yearly resource cycle was akin to that of western Europe. Thus there were coastal groups of hunter-fishers who hunted reindeer and also relied upon the resources of the sea, both fish and mammals, pursuing an economy not very different from the coastal Inuit of North America. The Inuit, however, never developed a herding relationship with the caribou, as happened with the Saami and equivalent Russian (and Siberian) peoples and reindeer. In the mixed deciduous–coniferous forests, mesolithic

cultures similar to those found further west hunted mammals such as red and roe deer, wild pig, wild cattle, and a variety of smaller mammals as well as fish. Plant foods were also important in the diet, though their precise importance is difficult to adduce from the evidence. A further element in the east European diet was the European bison or wisent, a few of which remain in protected status in the forests of the Poland–Belarus boundary such as Bialowieza.

Agriculture in this region provides examples of both constraints and opportunities, though it had reached the Urals by about 2000 BC. Pre-industrial agriculture, as always, was reined in by climate; in parts of Russia, for instance, rye was the ubiquitous cereal, though wheat and barley are both feasible crops in more southerly parts of the region, to the great benefit of Bohemian and Moravian beer production, just as Slovakia and Hungary are wine regions. But the loess soils are fertile, even if easily eroded under conditions of heavy cropping. Large areas of forest and grassland, too, always gave the

341

impression of an unlimited land bank waiting for immediate conversion to arable land. The climate combined with soils and native flora to make eastern Europe the cradle area of sugar-beet production. This started in the 1780s in Silesia and Bohemia, with a factory as early as 1804 in Russia (Pounds, 1979); however, the catalysing event was the cessation of imports from the West Indies during the Napoleonic Wars, not a climatic change. Low population densities combined with environment to make much of Moldavia, Ukraine, Wallachia and the Hungarian Plain, and the Great Alföld into cattle ranges, with the animals being driven into central and western Europe: in the eighteenth century the burgers of Frankfurt am Main were eating meat from as far away as Transylvania. The forest reserves of this region were, however, further east and north, so that Russia was the main exploiter of trees for export as well as local use, for example potash production (Berdnikov, 1995). Russia also had a flourishing iron export trade between 1775 and 1815, smelting for which depended upon charcoal rather than coke. The latter came only slowly to eastern Europe, with a successful plant in Silesia at Malapenew in the 1790s. Thus Russia and eastern Europe had a reservoir of experience in industrial enterprises and the resource uses needed thereby when the full impact of the industrial revolutions further west were experienced.

This full impact was to see Russia grow to be one of the great world powers, though as part of the USSR, which federation brought considerable resources of manpower as well as of nature. As in many such agglomerations, the centre profited at the expense of the periphery and so core-Russia experienced full exploitation of her natural resources even as outlying republics remained rather undeveloped under the Soviet regime after 1917. Energy in the USSR was enhanced by the damming of the great rivers for hydropower (celebrated in much of the art of Socialist Realism) and, eventually, by access to the oil and natural gas reserves of Siberia. Further west, brown coal was worked in great pits in Czechoslovakia and Silesia. A revolutionary agricultural response to environmental conditions was perhaps less spectacular, partly owing to social conditions. In Poland, for example, steel ploughshares were in use by 1900 but on a mixture of cereals and fallow that had been unchanged for centuries, with the potato the

only new element. No fertilizers were used and no machines. The common land, especially forest, still belonged to the gentry, even after the Tsar's edict of 1864 which gave peasants full possession of their holdings. So the *szlachta* turned to the forest as a source of wealth, clearing 35 per cent of it in the nineteenth century (Pounds, 1985). Much later the industrial development of this region was disrupted to an immense extent by the Second World War, for the flat terrains of the north European plain might have been made for *blitzkrieg*. Not only were rapid advances (and retreats) possible across such terrains but large areas of them were mined, to the point where Poland still has areas of land unfit for agricultural production because of unexploded munitions of that type. Further, unfashionable though such direct linkages may be, Generals January and February played a large part in the defeat of the Nazis in front of Moscow in 1941, just as they had in the débâcle confronting Napoleon in 1812.

It is difficult to talk of a post-industrial era in the case of Russia and the former members of the Warsaw Pact. Since the late 1980s there is an increasing public awareness of the costs of pollution, which is particularly widespread in the atmosphere and in ground-water. The Czech Republic (Jeleček, 1995) and Poland are perhaps the most advanced in thinking about these effects if we except the remerged eastern part of Germany. The non-renewability of the forest resource under present regimes of exploitation is now more clearly understood in Russia, though the chance to sell large quantities to earn currency to underpin other economic changes is a conservative force, as it is with sales of oil and natural gas. Tourism is increasing as fewer areas are closed to foreigners, though political instability and urban crime will retard its growth. Provided political transitions can eventually be made, the types of constraints and opportunities affecting this region need be no different from those further west, which is not to say they will not exist.

## The Alps and Mountain Europe

Attested evidence for the postglacial period in the Alps is confined to a scatter of find-sites of which rock shelters and overhangs are the commonest. It seems likely that the climate enforced a seasonal occupation mostly centred on the valleys but extending upwards from time to time: tools have been found on the Reiterjoch (Italy) at 2,214 m, and there is a definite settlement site in the Sud-Tirol (Italy) at 6000–5300 BC, shortly before the coming of agriculture (Pauli, 1984). The fringes of the Bavarian Alps feature in models by M. Jochim (1976) of the mesolithic occupation of the upper Danube as summer hunting grounds which are little accessible in winter: terrain and climate shut out humans from such regions at that time. Use of the summer hunting grounds would, of course, have been subject to any vagaries of climatic fluctuation.

The constraints of climate and land-form upon agriculture are obvious in mountain areas, but this did not result in the preservation of wildernesses. Tree species like beech spread at human hands (Watson, 1996) and the natural grasslands and cleared woodlands became the focus of pastoralism, most often of sheep and cattle. As an adaptation to the seasons, transhumance was practised, either on a local scale from the valley to the higher pastures in summer, or on a regional scale from the lowlands of Provence to the Hautes-Alpes, for example. The animal products were traded for the bread grains needed to support the relatively sparse population involved. The stability of mountain systems can be easily interrupted, nevertheless. In some instances this was due to natural hazards such as landslides, avalanches, and floods; in others social factors weighed against the maintenance of a balance between using pastures for immediate milk production or for hay to feed the animals in winter (Netting, 1981). The result, however, has often been a set of cultural landscapes with a high scenic and biological diversity which was very attractive for the tourism that opened up in the nineteenth century.

Since those woods that were not cleared had often to be maintained as slope protection (*Bannwald*) between the valley floor and the upper slopes, a fuel hunger developed in most European mountain systems. The coming of the railways in the nineteenth century allowed this to be satisfied in the Alps by coal and coke (Pfister, 1990); this was not so in the Mediterranean and is one of the causes of contrast in quantities of forest present in those regions today. Another feature has been the lack of demand for forest products from the mountains once cheaper timber could be imported; many of these woodlands are now single-age stands

that are becoming senescent. All this dates from about 1850 in the Alps and the succeeding twenty years saw the beginnings of emigration on a large scale, which undermined local industries and handicraft production. The latter saw some revival in response to tourism, which was a restricted phenomenon until the 1950s. Likewise, a few communes developed industry based on hydropower after 1880, but this suffered from isolation and in the face of recessions most of these branch plants have closed. So, in some senses, the industrial era has reinforced the environmental constraints: an economy dependent on snow, for example, has bad times in warm years; isolation makes for the lopping of branch plants during hard times.

The post-industrial period starts with mass summer tourism in the 1950s, followed by the winter season after 1965: 25 per cent of annual world-wide tourist turnover is generated in the Alps. There are strong national differences in policy to add to natural regional differences: the government of Slovenia, for example, inhibited tourism in favour of industry, with consequent air and water pollution problems. Overall, the mountain environment has not inhibited urbanization and, where a halt has been called to large-scale development for recreation (as in the Tyrol), government policy (Viazzo, 1990) and not the natural environment is the immediate cause. In the Swiss Alps and some parts of Austria, communes have developed their own plans to limit the friction between a more traditional culture and the incursion of mass-market travel. A few governments have been persuaded to subsidize more traditional (though not unchanged) ways of life. Such developments were pioneered in, for example, the Krkonoše and Tatra mountains of Czechoslovakia as early as the 1960s by socialist regimes. Their effect has been to allow the ecological basis of montane life to show through the current developments which ignore so many environmental as well as social impacts (Bätzing, 1991; Stone, 1992).

## Western Europe

Into the postglacial world of the European peninsula and islands there emerged a marine-orientated set of hunting, fishing, and gathering groups known as mesolithic cultures. Their population densities probably never exceeded 1.0 per square kilo but they coped well with the changes in climate, vegetation, and sea level that attended their tenure of the land until c.5500 BC when agriculture finally became the dominant way of life. The period is often divided into two: the earlier groups were hunters of the tundras and open scrubs, and took a variety of marine resources. The later group had a warmer climate in which to live, but that brought two obvious environmental disadvantages. The first was the way in which eustatic rises of sea level outstripped isostatic recovery (and in some areas exacerbated isostatic downwarping) so that areas of land were lost to the mesolithic folk, such as the whole of the area now covered by the southern North Sea, finally severing Britain in about 7800 BP. The second was that the large, cohesive herds of tundra animals such as reindeer and wild horse were replaced by the mammals of the deciduous forests which live in small groups. The predictability of meeting thousands of reindeer in one valley every October was replaced by the need to react quickly to an encounter with three or four red deer in a woodland glade. As will be recounted below, improving the odds led to some early examples of environmental impact. The richest and most predictable sources of food lay at the coasts, and the most developed mesolithic societies seem to have been those of Denmark and the extreme south of Sweden. Careful tuning to seasonal potential (especially the finding of a dependable lean-season food such as sessile shellfish) may have removed the need for nomadism, and so territories could be permanently seen as belonging to a cultural group which might have a characteristic style of hand-axe, and symbolize its occupation by means of having a cemetery, for instance.

The coming of agriculture to western Europe has an interesting ambiguity related to the question of whether an indigenous group saw that the new lifeways might flourish in their existing environments, whether an immigrant cultural group displaced the hunter-gatherers to marginal areas and, eventually, to extinction, or whether we are dealing with a fusion of both kinds of process (see Chapter 1). Debate continues but this transition can be seen essentially as the beginning of an era which lasted until the cheapness of fossil fuels allowed a more mechanized and industrial set of rural land uses to take over (i.e. between c.6000 BC and AD 1800). In this region, the deciduous forests were always at once a land-bank for future cultivation and a source of valuable materials, food, and fuel, and so the replacement of woodland by arable and grassland is one of the major impacts of the era. Opportunities were provided by a climate which in general lacked extremes, though not significant shifts such as the Little Ice Age (1550–1850) but which certainly lacked a dry season like the Mediterranean and a frozen-soil season like the far north. Equally, the soils released from under deciduous forest were high in natural levels of nutrients, though cropping regimes had then to mimic natural cycles to replenish them, a task most easily accomplished with the aid of domestic beasts folded upon enclosed fields. Constraints were as always climatic in providing northern limits to desirable crops such as the vine, though any of the major cereals might ripen successfully within this zone (Herlihy, 1974; Jones, 1981). In general, wheat was favoured since it has a gluten content which makes for a better-textured bread than barley and rye. Barley, however, is the source of malt and hence of beer. The forests provided many materials and full use was made of the resource: timber, poles and wattles, fuel-wood, pannage for swine, leaf-fodder for cattle, bark for tanning, fruits and roots, and a plentiful supply of legends and fairy-tales all came from them. Many forests also provided food for animals, especially cattle among the domestic and deer among the wild. The latter might be preserved for hunting by the privileged and many measures, both legal and practical, were taken to ensure the success of a day's sport, especially in lands under Norman authority. In due course, the relatively gentle relief of much of this region, coupled to a tolerant climate, made the construction of landscape parks quite easy, whether these were of the apparently gentle modification of the natural scene as in the English landscape gardens of the eighteenth century or the regimented microcosm of an authoritarian society as at Versailles in the seventeenth century. In amongst all these bucolic scenes, industry flourished though on a small scale compared with what was to come: coal was mined from bell-pits in fourteenth-century Northumbria, iron smelted with charcoal from 1000 BC to AD 1800 wherever there was a workable ironstone, and manufactories such as tanneries and glass-making were widespread, as was the winning of precious and less-common metals like silver, lead, and copper. Salt, too, was produced from underground (*Salz*burg) or by the sea (*Salt*fleet). The seas, too, were harvested well into the deep oceans: cod and whales were fished from vessels harboured in Brittany and the Netherlands among many other home ports. In general we can say that all the opportunities afforded by the landscapes and the available technology were taken up: nothing, for example, was set aside as too sacred to use—another facet of the story in its own right.

A major turning point in the resource history of Europe (and thence of the world) was the ability to smelt iron with coke rather than charcoal. Pioneered in the Ironbridge Gorge district of England in the early seventeenth century (Palmer and Neaverson, 1994), this is sometimes claimed as a result of the depletion of the woodland resource, though more complex explanations involving labour costs are currently more popular. But the shift from wood to coal as a basic resource caused immense shifts in the geography of Europe and changed environmental relations for ever. The alterations involved the growth of industrial complexes based on major coalfields and led to an enormous nineteenth-century growth in urban area at the expense of less intensively used land. Around these centres of urban-industrial economy was a periphery of lower impact but which was nevertheless contributory to the centre or receptive of effects from it. The resource opportunities afforded by the new cheap power included the use of leaner ores of most metals, so that less attractive areas could be worked, and cheap transport of low-value products. The conquest of space was thus to some extent a major opportunity to which entrepreneurs responded enthusiastically. Fast travel by rail and steamship not only facilitated economic development within Europe, but in a peripheral imperium as well; it also helped with pleasure, making the Alps and the French Riviera accessible to the rest of Europe's well-off, or allowing the Parisian *haute-monde* to get to the Normandy coast by taking the 1.22 p.m. even in the nineteenth century; in Trollope's novels, MPs seem to be able to manage a day's hunting with fashionable Midland packs on a day-return ticket. More important, however, was the lifting of the constraint upon food supplies that had affected European populations during the agricul-

tural era: nutrient cycles were and are subsidized with fossil fuels in the form of machines, fertilizers, and biocides so that productivity is high. Cheap transport means the import of foodstuffs from anywhere in the world at prices which a great number of people can afford. To talk of constraints is not very fashionable: access to energy enabled western Europe, it seemed, to escape the checks of nature and of Thomas Malthus, and 'economic development' has ever since tried to follow the same path. Weather, relief, the Channel: none shall interrupt the flow of goods and people. The success of industrialization in Europe set the tone for the world economy and hence for large portions of its present-day ecology.

Only very recently have signs of reaction to this all-embracing modern era become apparent. It has come, for example, in attitudes to nuclear power, which in both its generation and waste reprocessing phases has gained little in public acceptance after a number of accidents. But many of the elements of the industrial era persist and indeed are magnified by the cheap transport and technology of post-industrial Europe. Mass tourism is the most obvious of these, with the cheap-to-run jet aircraft a crucial feature of post-1950s travel patterns. Both sun/snow-seekers and culture-seekers appear to be getting to the saturation level, however, and other outlets will form part of Europe's coming vacation experiences: fantasy fuelled by developments like Virtual Reality may not be very many steps behind the fabricated world (except for the prices) of Disneyland Paris and other similar theme-parks. In these it may be said that human experience is very largely cut loose from the rest of the environment altogether. This seems to parallel the post-industrial economy's dependence upon services, which are exemplified by the financial houses that move large amounts of capital electronically without regard to space and with very little regard for real time.

## The Mediterranean

In many respects, the mesolithic occupation of this basin conforms to expectations: the sea was used to support a fish-based economy and the land for hunting and gathering. The fish included tunny, grouper, wrasse, and seabream, and the plant list contained almond, pistachio, pears, and vetch.

Hunted animals included wild sheep and goats and the ibex. There seems every reason to suppose that this was a viable economy which would be destabilized only by massive climatic shifts or other geophysical traumas, or by runaway population growth.

Nevertheless, after a period of unknown length, when a husbandry of certain wild species was practised but which came below the horizon of full domestication, the Mediterranean was the first European region to receive the full impact of the new way of life, in the millennia after 8000 BC. This became quickly established and was stable, especially where animals could tap the inedible-to-humans vegetation of the mountains which so closely surround the basin on the north. Food from goats and sheep especially could supplement the yields from the land, which were restrained by the lack of summer rainfall. Hence the coming of widespread irrigation, usually credited to the expansion of Islamic cultures into the region, allowed a degree of intensification hitherto impossible to attain (Watson, 1983). After that, and given that population levels were only diminished by the Bubonic Plague (AD 1350–1450), anywhere that cultivation could flourish became densely populated and human actions shaped the spaces. This has led to a considerable environmental impact in both those places and in the peripheries from which they drew resources, especially the mountain lands which were converted to sclerophyll scrub by centuries of grazing and fire management. One environmental opportunity not wasted was the combination of unirrigable land in such a warm climate to produce valuable crops such as the olive and the vine. Both became staples rather than luxuries and provided calories as well as, apparently, components of a healthy diet. All the resources of the region, including those of the sea, were intensively utilized even before the coming of industrialization.

The industrialization of the European Mediterranean was slow in coming and never so thorough that it hid many of the traces of past periods (including those of classical antiquity) which were so appealing to the visitor from northern regions. Fast rivers provided hydropower (La Scala, Milan, was the first opera house to have electric lighting, in this case powered by hydroelectricity) and eventually local deposits of iron ore founded modern steel industries, as at Terni in Italy. Cultural factors such as the economics and national policy of the new nation were always likely to have been dominant over the natural environment in such developments. Technology, however, did not save the Mediterranean region from the devastation of the vine caused by Phylloxera in the second half of the nineteenth century: only regrafting onto Californian stock revived European wine production, though we might note that few of the traditionally famous vintages came from the Mediterranean regions, with perhaps the exceptions of the lower Rhône (Châteauneuf-du-Pape) and Tuscany (Vino Nobile de Montepulciano) although, extending the concept of Mediterranean a little, we can include the fortified wines of Jerez and Oporto, that is, sherry and port.

The post-industrial era of the Mediterranean has two outstanding characteristics which are environmentally related. The first is the explosion of mass tourism in the 1960s, especially into mainland Spain and Greece and their islands, along with Cyprus and Malta. This is based largely on sun, sand, and sea together with the importation of eating and drinking habits from the source nations like the UK. The second is a retreat from cultivation and pastoralism in rural areas and their replacement with scrubland and with plantations of exotic conifers and eucalypts. This latter exacerbates the summer fire hazard; the former has been much excoriated on aesthetic grounds but is environmentally little different from much urban development; indeed, to bring in the tourists, levels of some environmental pollutants have been kept relatively low; as with the Alps, a few communities have introduced restrictions on development and its spread both outwards and upwards. The sum total, according to Lepart and Debussche (1993), is that 'Landscape patterns that used to be maintained by humans are now changing as the natural dynamics of the vegetation is [*sic*] becoming more important.'

## Environmental Impact

In responding to both the opportunities and the constraints of the natural environment, European societies have from early times altered those surroundings, to the point where these landscapes are as comprehensively reflective of culture and technology outlooks as anywhere in the world. Table 16.5 pulls out some of the many strands in the story.

**Table 16.5.** Environmental change by humans

| | Hunter-gatherers | Agriculturalists | Industrialism | Post-industrial |
|---|---|---|---|---|
| Scandinavia | In mesolithic forests: management for wild animal fodder, wood production. | Clearance of forests → increased soil erosion. Grassland and arable created therefrom. Urban growth. Mining, e.g. of copper. | Reindeer herding by ski-mobile. Mining. Recreation: winter, lakeside cottages. Urban growth. | Avoidance or phasing-out of nuclear power. Attention to 'green' ideas in cities. |
| Russia and eastern Europe | Clearings opened and maintained in lowland forests for attracting animals. | Full suite of alterations from agricultural era (see text). Mining. | Energy production—and consumption-related changes. Over-use of forests. Ploughing of dry grasslands. Urban and industrial growth. | As yet little evidence. Nuclear accidents. Beginnings of 'clear-up'. |
| Alps and mountain Europe | None known. | Mining. Grasslands from woodlands; forests managed for wood and fodder production. Soils more prone to slip and rivers to flood. | Retreat of 'traditional' populations. Tourism: housing, communications, for winter and summer use. Forest area stabilization | Subsidization of 'traditional' way of life. Some mountain towns ban motor vehicles |
| Western Europe | As for eastern Europe: much of the woodland managed in part. | Full suite of alterations from agricultural era (see text). Industrial pollution. | Energy consumption-related changes. Conifers replace deciduous woodlands. Pollution of environment. Transport net places demands on environment. Massive urban growth. | Intensification of farming → soil erosion → overproduction → set-aside. Nuclear accidents. Patchy adherence to 'green' ideas on materials. Attrition of conservation areas. |
| Mediterranean | Probable use of fire to maintain open areas with e.g. harvestable wild grasses | Forests → grassland → *garrigue*. Reclamation of marshlands. Ploughing produces soil erosion, coastal lagoons form. Forests managed for wood, pastoralism. Quarrying, mining. | Slow to arrive, based on HEP initially: rivers altered. Early popularity of winter tourism, e.g. French Riviera. Some reforestation with pines and eucalyptus Tourism in cities, at coasts. | Intensification of agriculture → impacts on soils. More understanding of fire ecology of forest and *garrigue*. Some restriction of tourist-related urbanization. |

## Scandinavia

The first evidence from this region of human-induced environmental change comes from the mesolithic period of southern Sweden, in the years between 7000 BC and the advent of agriculture. Evidence from deposits in lakes and peat bogs suggests that the hunter-gatherer communities manipulated the woodlands with the intention of attracting large mammals and also to provide wood of the sizes most useful in their economy. This meant curtailing the area of high forest and making sure instead that there was plenty of edge where the more shrubby species (rowan, birch, hazel, wild cherry, for example) might flourish; ring-barking was probably used to kill large forest trees. Once openings were made they could be kept thus by the use of controlled fire, possibly allied to high densities of grazing and browsing animals attracted to them (Göransson, 1986). So agriculture was not introduced into a pristine ecology, but one in which the hand of humans was already quite apparent. Finland provides a different possibility of impact: Siiriäinen (1981) has proposed that over-hunting of elk and bear led to a later mesolithic (towards 4000 BC) concentration on seals. Reindeer populations proved to be most sustainable under a herding and ownership system which avoided over-using the lichen forage necessary for the animals until recently (1950 onwards) when the diversion of forest land to other uses (construction, dams, roads) has meant increased felling of trees to provide epiphytic lichens for the animals. Two old spruce trees per day are needed for each reindeer under such a feeding regime.

Agriculture in Scandinavia has often been similar in its impact to other parts of Europe. The initial inroads were made into forests but at very different times: cereals such as wheat are found in southern Sweden as early as 5400 BC but in Kuusamo, Lapland, the Saami occupied the spruce forests of the region until the AD 1670s. There then came immigrants who practised slash-and-burn agriculture in the forests, with a three-year use cycle (growing one crop of rye in that period) followed by forty to fifty years of regeneration. Additional ecological impact came from the use of pine phloem to make bread, at a rate of 500 trees per family per year. So today's forests are the secondary successors of seventeenth–nineteenth-century areas which were heavily used

and altered (Vasari, 1990). Changed though they were, they predominated in much of Scandinavia south of the tree limit and were cleared permanently in small quantities compared with lands further south. They were replaced with a mixture of grasslands and cultivated areas predominantly growing rye and barley, with the former land-use the more common. To help produce a better hay crop, irrigation was used in Norway as elsewhere in mountain Europe (Michelsen, 1987). Wherever possible, as along the coastal plains and strandflats of Norway, the exiguous produce of small cornfields and patches of grassland was supplemented with the more reliable harvest of the sea, a pattern established by the time of the Sagas AD 400–800) and essentially continued until the industrialization of the fisheries.

The impact of industrialization upon the environment of Scandinavia includes as elsewhere a rise in population, which helped to increase the area of cultivated land at the expense of forest: in the case of Sweden from 1 million ha. in 1800 to 2.5 million by 1860, after which it declined again as productivity rose. But any map (e.g. Pounds, 1985) of crop production in say 1913 shows that Scandinavia was far less arable than the rest of Europe except the mountains, and the environmental impacts were correspondingly lower. Pollution from industry has been, in fact, one of the major noticeable effects upon environments, with the pulp and paper industry leading the concentrations, especially into air and freshwater, of compounds of chlorine (used in bleaching), sulphur, and nitrogen. Fungicides used in the industry contained mercury, and levels of this potential poison of the nervous system were for a time in the 1960s and 1970s very high in the Baltic; run-off of untreated sewage with nitrogen and phosphorus levels contributed to the formation of an abiotic zone at depth in this sea. The northern areas were also highly susceptible to fallout from the atmospheric testing of atomic weapons before the 1962 treaty and the slow-growing lichens concentrated isotopes with long half-lives. These then got into the reindeer long-bone marrows which were a favourite Saami food, to the point where excess body burdens of Sr-90 and Ce-137 have been carried for some time. After the Chernobyl incident in 1986, moves to slaughter large numbers of reindeer were swift. The northern lands, especially of Sweden, have been host to development for minerals, hydropower, and tourism, with all the ecological changes (mostly local) that go with such growths.

Since the 1950s, Scandinavia has been the recipient of other unwelcome environmental impacts. In particular, the effects of acid precipitation have been severe, with many trees killed or injured and many freshwaters acidified to the point of losing fish populations more or less entirely. Fallout from atmospheric sulphur compounds released in Britain and east-central Europe seems to have been the main cause, made worse by the low buffering capacity of the young soils of these, the most recently glaciated regions. Oslo, as one of the denser-built of northern cities, has been a pioneer site for showing that nitrous emissions from vehicles are associated with rapid rises in asthma, especially among children. However, nowhere in Europe has the environmentalist message been taken so seriously, especially in terms of materials recycling, to the point where Swedish car manufacturers have been world leaders in making cars whose parts are reclaimable when the vehicle's life is over. All this attention to environmental concerns seems easier when there are small populations without huge disparities in wealth. 'National Parks', I once heard a Finn say, 'are what we designate so as to destroy the nature only in the one place.'

## Russia and Eastern Europe

The mesolithic of this region shares many characteristics with that of southern Scandinavia and with western Europe, so that to describe it in detail would be repetitious. Suffice it to say that management of woodlands by hunter-gatherers has been detected in Poland, among other places, and that although the wild cattle (*Bos primigenus*) became extinct during prehistory, as in the rest of Europe, the European bison (*Bison bonasus*) did not, and small populations survive in eastern Poland and the adjacent area of Belorus to this day.

The early stages of agriculture were not singular to the region, and the development of that way of life during the centuries AD has allowed the full suite of environmental impacts associated with a solar-based agriculture. A few of these are special to this region. The expulsion of the Turks from the Hungarian plain in the eighteenth century allowed a massive change from a localized subsistence economy to a vast grassland area which exported millions of head of cattle westwards, especially to Buttstadt in Thuringia, reversing any general trend from grassland and forest to arable. One particular characteristic of these countries is the progress of the abolition of serfdom (mostly achieved only in the early nineteenth century), which often left the landowners in sole charge of the resources of forest and waste, and these were often exploited heavily in terms of timber sales and reclamation in order to raise revenues for the gentry. In European Russia the years 1700–1800 saw a reduction in forest area from 60 per cent to 53 per cent, which then declined to 37 per cent by 1914.

Even where the nineteenth century saw little overt industrialization, changes in environmental impact proceeded from population growth and from the opening up of markets. In Poland, for example, the land under forest in 1816–20 was 30 per cent, whereas in 1909 it was 18 per cent. In Hungary the equivalent area declined from 30 per cent in 1846 to 24 per cent in 1895, and similar changes occurred in Bulgaria, Romania, and Serbia. Since Hungary was virtually alone in being a country of large estates, it seems as if leaving the landowners alone also led to deforestation. The major interest, however, is the growth of industrial areas with all the environmental changes thus implied: as Pounds (1985: 356) puts it, 'every river that flowed through an industrial region was beginning to acquire the same fetid and lifeless character, for which there was no satisfactory remedy before 1914'. To this might be added that not a great deal has changed in some parts of this region. What was different in areas like Upper Silesia and around Łódź was that there had been no proto-industry in the eighteenth century and so the impact was the more startling. Indeed, Łódź was probably the fastest-growing city in eighteenth-century Europe, sharing many of the features of Manchester as described by Engels. Russia's environment changed the fastest under Bolshevik rule (after 1917) since the collectivization of agriculture in the 1930s was also the mode of introduction of machinery; in both the rural and the urban-industrial sectors production targets had to be met no matter at what environmental cost, and a whole catalogue of ecological problems has come to light as a result: poor forest management; the drying-out of the Aral Sea; air and water pollution with all manner of gross pollutants, as well as insidious release of radioactivity from military plants; and the disastrous ploughing-up of the semi-arid grasslands of Khazakstan in the 1950s (Alayev *et al.*, 1990; Artobolebskiy, 1991; Pryde, 1991).

The greatest engineering achievements are probably the series of dams on the Volga and Dnieper producing electricity and irrigation water on a huge scale; plans to divert the Ob and Yenesei southwards were prevalent in the 1970s but seem for the present to have disappeared with the Communist Party. Indeed, possibly the greatest change in post-industrial times in these lands has been the public recognition of the magnitude of environmental change during the years of Communist Party dominance and the perception of much of this as problematic, something not officially permitted before the 1980s. The great symbol of this was the 1986 nuclear accident at Chernobyl in the Ukraine, which resulted in very high levels of radiation locally, sterilizing a large area for settlement and agriculture (Park, 1989). It also doused various parts of western and northern Europe with enough radioactivity for there to be restrictions on the movement of hill sheep from parts of upland Britain even in 1997. So a great clean-up of eastern Europe and Russia could be planned, using the best western knowledge and techniques. Money and the institutional structure are both obstacles.

## The Alps and Mountain Europe

The impact of hunter-gatherer cultures in the Alps and analogous areas is not well known. In deciduous forests it is generally thought that there may have been manipulation, as in Scandinavia; in the conifer belts and above the tree-line this seems less likely. Evidence from the Pyrenees shows that the hunted fauna in the mesolithic may be 90–99 per cent of the faunal assemblage (in this case the ibex, *Capra pyrenaica*) and we may wonder if this may not betoken some impact upon ibex populations, although these were in any case replaced as Holocene warming produced a fauna of more temperate woodlands, along with trout (Geddes *et al.*, 1989). Detailed work on the fringes of the Alps in Bavaria shows a decrease in the number of sites of late mesolithic age, and one interpretation is that of population decline, which would presumably decrease environmental impact,

though direct palaeoecological confirmation is lacking (Jochim, 1990). So a broad generalization for the later mesolithic might be that some openings may have been maintained in deciduous woodlands, but at the higher altitudes environmental impact is likely to have been restricted to the possible local extirpation of the less common mammals.

Agriculture spread to the central mountain chain of Europe about 5500 BC and transhumant pastoralism some 1,500 years later. Thus the basis of the traditional land-use system of the Alps was laid early and survived long enough for it to have considerable environmental effects. A major example is the weakened ability of a forest to regenerate when grazed by domesticated animals, combined with the need for wood for space heating, construction, and cheese-making. Hence the upper edge of many woodlands moved downwards, creating good-quality pasture by way of compensation. The ratio of artificial grassland to natural areas is about four to one. Clearance of forest was also the way of creating cultivable land on lower slopes, at first on the alluvial cones of tributary streams. These flooded less than the main valleys and shed some of the pools of cold air which fill Alpine valleys. As this process spread during the Middle Ages, so forest retreated upward and a protective zone has to be enforced by legislation, the *Bannwald*. The reclamation of the flood plains of the main valleys is more a process of the nineteenth–twentieth centuries than of the agriculturalist era. The land-use system thus achieved, while still vulnerable to flood, landslide, and avalanche, is generally held to have been stable and sustainable until the nineteenth century even though all Alpine ecosystems except rocky areas and glaciers were altered by the traditional patterns of land occupancy and use. Such customary patterns included mining, and contemporary illustrations show well how this denuded areas of trees, created heaps of waste, and polluted watercourses.

When it became expedient to reclaim valley land in recent centuries, rivers were straightened and deepened and so land hitherto little used was divided into plots, with leisure uses such as golf courses taking their place. Only after about 1960 were houses built on these terrains. The theme of leisure bulks large in any discussion of environmental impact since after the 1960s winter recreation was added in large measure to the hitherto summer-dominated pattern. The latter had brought forth settlement expansion and a transport network (including millions of signed paths) but the former has seen the import of large quantities of machinery to all parts of the mountains and the introduction of pylons and overhead lines to most slopes, especially on north-facing inclines. As the attraction of the isolated areas' traditional way of life has faded, depopulation and conservation have combined to stabilize the area of forest, though it will not regenerate wherever a ski piste is established. These pistes and downhill runs also pack the snow and reduce its insulation value; frost penetrates deeper into the soil and so allows it to slip more readily in the spring. Shallower soils mean faster run-off, and so skiing means more landslides and more floods, as well as more avalanches in winter.

The overall impact has been such that some Alpine and other mountain communities have enacted self-limiting ordinances on development but given the demand, especially in winter, it is always difficult to resist the pressures which will change the local environment 'only a little' and not very differently from everywhere else.

## Western Europe

Evidence is now strong for the hypothesis that mesolithic folk managed parts of the deciduous forests of Europe. They seem to have created openings (with their small lithic implements, ring-barking seems the most likely method), maintained natural and human-created openings using fire, and gathered branches to attract wild animals. In many places, such actions were commensurate with the scale and nature of natural ecological change; if the people ceased their actions, then the forest resumed its own successional change. Removal of trees, however, especially in the wetter parts of Europe, encourages the waterlogging of the soil and in some places (even on water-shedding sites), paludification set in, with consequent accumulation of peat. On the wet uplands of the British Isles, peat depths of 4–6 metres resulted eventually from mesolithic manipulation of the vegetation (Simmons, 1996). Still, the density of the population (not known accurately but probably between 0.1 and 1.0 per square kilometre) was low and some

of the people were quasi-nomadic, so that overall impact was slight compared with what was to follow.

After 6000 BC, agriculture came to western and central Europe. Shifting agriculture has often been reckoned to have been the dominant mode of production, though some later evidence now points in the direction of more permanent clearings in forest whose fertility was maintained by cattle which may have been stalled and foddered. Detailed evidence of the gathering and feeding to animals of the leafy branches of hazel, alder, ash, and lime has been published by Rasmussen (1988), who also makes the point that this practice continued into historic times in Denmark and other European countries. After landholding patterns of some cultural variety were established in the Middle Ages, the environmental impact exhibited some regular lineaments. The relationship between settled agriculture and the woodlands was a continuing tension. The ways in which the forest was used for these purposes have been detailed by Rackham (1976, 1980): here it suffices to say that the result was frequently a wood-pasture which was distanced from the woodlands of prehistoric times, in both physiognomy and floristics. One significant signal of such changes was the construction of boundary banks and ditches, to signify management units. These might be surrounded by a fence if the preservation of deer was involved, and this protective attitude must have led to the eventual demise of the wolf in this region in the eighteenth century. Pleasure, too, led to the development of gardens associated with the great houses. Such pleasures were not only visual but gustatory since these gardens, like their predecessors in southern Europe (and way back beyond that), yielded both beauty and utility. The latter came from orchards and kitchen gardens, herb and medicine beds; the former from formal flower beds and raised grassy enclosures with ornamental trees and shrubs (the import of the rose was a significant addition to these pleasures) and water features. Formal gardens of whatever age obliterate the original ecology, and an interesting development in eighteenth-century England was the 'landscape garden' which, while being no less artificial than formal gardens, was designed to look 'natural', though perhaps a rather tidied-up type of nature and one in which structures had classical allusions.

Industry was rarely absent from anywhere that

had both wood and a mineral resource. Iron was the most sought, and a great deal of charcoal was needed to fire the smelting process: this in turn led to the conversion of much woodland to coppice which yielded poles for the furnaces; there seems no reason why smelting should have been an enemy of the woodland area when coppicing was a sustainable-yield practice. Other metals, glass, and salt were treated in the same way and oak bark was much in demand for tanning, itself a process which (like metal-smelting) polluted the streams of its neighbourhood. Towns and cities inevitably exerted an environmental impact, though the pre-industrial urban area did so on a scale nowhere near that of its successors. The city itself obliterated some ecosystems but provided the structure for others, in which scavengers and parasites (including several human diseases) flourished: the red kite in the streets and the cholera bacterium in the bloodstream. The outreach of the town may well have extended to a zone of fruit- and vegetable-growing just outside the walls, worked by the citizens themselves. It went further for grain and meat, and often depended upon river transport for a dependable supply. In turn this encouraged the early regulation of rivers by means of weirs and sluices, something also advantageous to the many watermills used in processing both food and other materials. A complex interaction between climate, water relations, and land use is seen in the rise of the biting midge in the Scottish Highlands after the eighteenth century (Roberts, 1995).

If charcoal production preserved woodlands, then it is a curious irony that the coming of the Industrial Revolution in Europe removed many of them: in Europe as a whole the period 1860–1919 saw 27 million ha. of land move into crops, of which half at least was probably woodland, and the other grassland and wetland. In the succeeding years of 1920–78, only 14 million ha. moved into crops and 12 million moved out, much of which was into forest, as well as into urban-industrial use (Richards, 1986). Demands for timber and other wood products have meant large shifts to the faster-growing conifers at the expense of the ancient deciduous woodlands, which survive now only as protected remnants. Many things were made possible by access to cheap coal and then oil and natural gas: the symbols of environmental change might perhaps be first the railway locomotive and

then the bulldozer. So all the processes in train in the eighteenth century could be extended and intensified and the greatest of all has been the urban-industrial area with its outreach for materials and energy and their storage as structures and organisms, followed by the emission of wastes (Douglas, 1983; Clapp, 1994). No part of this region within Europe has been free from one or more of such processes. The mycelial web of pipes for water leading to wells, reservoirs, and rivers, followed by the impact of untreated industrial wastes and sewage on rivers and the sea, is one obvious example that could be multiplied thousands of times. The industrialization of the food system, with fossil fuel-derived energy important at every stage from the machines in the field (and the arrival of the vet) to the household kitchen, is another. On the seas, the European fisheries were industrialized very early and so the exhaustion of stocks can be seen by the 1880s. This phase of over-use was caused by the efficiency of the steam trawler; a second phase happened after the Second World War with the advent of nylon nets, which enable the ships' steam winches to pull up larger loads from deeper waters. The sea has literally retreated since large cranes make easier the task of building sea-walls and so the safety of the coast of the Netherlands has been more or less guaranteed for the first time in history, though not perhaps in perpetuity. This might be a useful summary phase for the industrial era: it has been based on non-renewable resources and increasingly on foreign supplies, with all the military and political entanglements thus implied. If the world oceans have been a 'ghost acreage' supplying Europe with animal protein, then the oilfields of the Middle East have been a set of 'ghost mines' supplying equally life-enhancing matter: but not for ever.

Along with Japan, Hong Kong, Singapore, and North America, this part of Europe is reckoned as one of the leading centres of post-industrial economies, in which services are as important as manufactures and in which the latter are flexible and rapidly changing as they follow small market segments. This means that energy intensities can fall a little, as indeed they have (as measured per unit value of product), but that overall demand for energy remains high, prompting first the development in the 1950s of civilian nuclear power, whose environmental impact is measured perceptually by the number of accidents that have caused the unplanned release of radionuclides, most famously at Windscale (now Sellafield) in 1957. A second development is that of 'alternative' energy sources, of which the most popular with any environmental impact is the hi-tech windmill, usually in a group called a wind farm. Their impact is both visual and, more significantly, aural, for they emit a constant loud hum to add to the pervasive noise which is a feature of an electronicized society, with it being difficult to travel without overhearing somebody's phone conversation or buy a sweater without being assailed by the chart-toppers. The success of energy-subsidized agriculture and fisheries has led inevitably to those groups being paid not to farm or fish, and so they are adding to the leisure resources which are so central to a post-industrial economy. Most famously, these are facilitated by twin-engine jet planes like the Boeing 737 and 757, which can haul millions of people to holiday destinations within Europe, notably to the Mediterranean shores and islands. In western and central Europe, the leisure is often more 'cultural', urban areas taking their share, with the consequent tidying-up of towns so that the hanging basket becomes a major visual feature, as does the peripheral camp-site. But even the chillier shores of the North Sea basin have their environmental manipulations: resorts dependent upon sandy beaches invest in groynes to trap that sand, at the expense of the next place down-drift. A new environmental consciousness has pervaded such places, with EU standards of coliforms and visible sewage being publicized, for example, and there is throughout western and central Europe a patchy adherence to 'Green' ideas on matters like materials recycling and public transport: Germany is generally in the lead and the UK some considerable distance behind, along with southern Europe.

Perhaps the most important environmental impact since the 1950s is the sheer quantity of knowledge available in every medium. Just how this produces changes in environmental impact is uncertain: in the negative sense it has prevented some transformations from coming about, so that its impact is in what we do not see.

## The Mediterranean

Knowledge of the environmental impact of hunter-gatherers in this region is still sporadic but there is little doubt that they were masters of controlled fire, and that in the pine forests and sclerophyll woodlands considerable change could have been wrought. It seems likely that one aim was to keep the forests at bay in favour of grasslands which supported game. It has also been suggested that some of the wild grasses ancestral to domesticated cereals were husbanded in such open patches. But of all the regions under consideration here, that of the hunters was the shortest in the Holocene since it received the products of western Asian domestication first of all.

The impact of agriculture in the Mediterranean was profound. All the main crops (wheat, vine, and olives) required cleared land, much of which might be on slopes that eroded easily. Only olive groves might maintain an accompanying grass cover to keep the soil and its moisture in place (Sallares, 1991). Thus a major impact in all ways was the adoption of the terrace. Itself one of the most widespread of all human artefacts in pre-industrial times, the terrace altered the type of soil and its water relations, as well as the slope itself. Where terracing was not deemed necessary, the division of the land, for example in Roman times, into plots is still visible in the landscape and even more so from the air. As population grew, the extension of the cropped area might come about through irrigation, with its impress of managed watercourses and salinified soils, or through the reclamation of wetlands. In Italy in particular, many coastal marshes and deltaic zones (that of the Po being the largest) were subject to drying-out from classical times onwards, to provide *bonifica*. Reducing the breeding-grounds of malarial mosquitoes was no mean collateral advance. An equally high profile has been generated by the impact of pastoralism in the hills. Classical sources onwards refer to the conversion of forest to grassland and *garrigue* (and its Spanish, Italian, and Greek equivalents), with references to the use of fire and the appearance of the bones of the landscape as soil was lost. Later work suggests that there may have been a selectivity of description and that until the nineteenth century forests were relatively abundant in all the countries of the Mediterranean basin. Certainly many of the medieval and early modern republics of Italy had forest management policies: that of Venice is perhaps the best known, with its direct aim of providing wood for the galleys constructed in the Arsenal.

This policy extended beyond the Veneto, to Dalmatia and Istria, for example (Susmel, 1956). Intensive management of woodlands has been traced for Tuscany in the time of the Florentine republic and after (Piussi and Stiavelli, 1988). In spite of such management, the mountainous nature of much of the Mediterranean Basin meant that soils were indeed translocated from the upper slopes to the valleys and the coasts. The usual other instances of rural impact are also present in the Mediterranean. Gardens, for example, have been widespread features of large country houses, and the formal Italian garden provided a model for most of the rest of Europe until the eighteenth century; its relative the botanical garden nurtured introductions from overseas. The garden at Padua was the first scientific establishment of its kind in Europe (dating from the early sixteenth century) and housed the first lilacs and carnations, for example, to be brought to Europe. In Spain, many large pleasure gardens remained from the period of Islamic rule (those of Granada are the best remaining examples), and in them the role of water was paramount since it reflected the Islamic view of heaven as well as a source of cool air in the very hot summers. Water supply to towns from distant rivers was a Roman specialty but one that was generally lost until the nineteenth century, so it and the regulation of rivers were virtually absent from the Mediterranean compared with regions further north. Hence the Mediterranean was a relatively wild place until the nineteenth century, undergoing only slow change (a feature much aided by the imposition of Ottoman rule at the eastern end of the basin) but certainly not one in which there were natural landscapes. The slow evolution of the *dehesa* of Spain is an example of dry forest management for tree and animal products (Oviedo, 1989; Stevenson and Harrison, 1992).

The industrialization of the Mediterranean Basin came late but led to the customary developments of the twentieth century, including transport networks, river regulation, coastal development for recreation and tourism, and a high level of basin-wide pollution. The Mediterranean has also been over-fished. Tourism began its apparently inexorable rise before the great boom of the 1960s, since the Italian and French Rivieras in particular had attracted the rich and their followers in winter: the building of hotels, harbours, and marinas is not therefore solely a response to the post-industrial mass market. The railway that had been so influential in starting off tourism also allowed the exploitation of other resources: the forests of the Sila region of Calabria were subject to only local demand until the late nineteenth century but were nearly all gone by 1915 and replaced by sclerophyll scrub. The nineteenth century even saw great changes in the subsistence use of a tree like the sweet chestnut (Moreno, 1990), so thorough was its eventual impact.

In the last thirty years, agriculture in the region has been intensified by the pressures of EU policies: many slopes are being denuded of their soil-holding tree crops and planted instead to early ripening varieties of luxuries like avocados and strawberries (Mannion, 1995). As a few farmers prosper, so the marginal holdings disappear into the holiday homes market to join the 'self-catering' boom in the holiday market fuelled by motorway access across Europe and the fly-drive option. The impacts of development of mass urbanities in Spain, the Balearic Islands, Greece and its islands, the former Yugoslavia, and Cyprus is obvious: more sewage in the Mediterranean (often untreated), more sand trucked in to form a beach at almost any price, and more demand to be connected to a motorway network. Any environmentalist reaction to this is slow in coming, apart from a few architectural concessions and the 'Blue Plan' for cleaning up the Mediterranean, which is proceeding with all the verve of an Italian city debating the opening of its museums for more than three hours per day, feast days excepted (Pearce, 1995).

## General: Opportunities, Constraints, and Impacts

If we venture upon a few high-level generalizations, for debate if not for immediate acceptance, then they might be of the following kind:

1. The environmental constraints of Europe have often been ameliorated by imports (food, timber, oil, silver and gold, and lilac trees are examples) from trading partners, from the open oceans, and from empires. There has been a consistent 'ghost acreage' supplementing home production.

2. Many characteristically European processes bear no obvious relation to environmental factors:

there is a strong case against environmental determinism for all except the most obvious cases, but even then technology might allow the growing of kiwi fruits in Spitsbergen; it is the highly culture-laden neo-classical economics which militates against it.

3. If we accept the dominant role of technology and the authority of an environmentally permissive world-view, then in general as time progressed the constraints became fewer than the opportunities and so the impact increased: there are practically no pristine ecosystems in Europe.

## Environmental Hazards

The past may have been fraught with environmental hazards, but for reasons of space, Table 16.6 concentrates on the period since AD 1800. Two kinds of hazard are distinguished: those which result from the unmediated processes of nature, as with earthquakes and volcanic eruptions for example, and those in which extreme natural events wreak greater harm because of human activity, as when housing is built on river flood plains or when ground-water is contaminated with nitrates.

The natural hazards are dominated by the weather: in spite of all the resources of technology, periods of extreme weather are still destructive of life, livelihood, and property. In northern regions, it is usually the winter which can cause disruption to travel and still trap unwary travellers, usually those on affected roads. Winter fog normally pushes up the road accident figures. Secondary effects may be the trapping of air pollutants under an unusually long-lasting anticyclone, with extra morbidity and mortality. In the south, drought is more important, with fire at the interface with human-directed systems. Most Mediterranean forests receive heavy fire suppression rather than controlled burning and so, when they are set alight in especially dry periods and fanned by the often strong afternoon winds, devastation of tree resources, wildlife and, often, rural homes is widespread (Lekakis, 1995; Lourencao, 1988, 1992). Between 1987 and 1993, European wind damage took up 31 per cent of the world's insurance payouts for major environmental catastrophes involving windstorms. The British Isles has received occasional and poorly predicted hurricanes, with loss of buildings, trees, and life. Droughts, too, affect mainland Britain, as in 1976 and 1995. The

Table 16.6. Environmental hazards

| | Natural hazards | Environmental hazards |
|---|---|---|
| Scandinavia | Severe winter weather. | Acid precipitation from sources to west and south. |
| Russia and European republics of former USSR | Severe winter weather; drought in south. | Soil erosion; contamination of soil and water from effluents including radioactivity; latter includes Arctic Ocean. |
| Eastern Europe and Alps | Soil loss. Avalanches in mountains. | Contamination of soil, air, and water by effluents, especially acid precipitation. Mines from Second World War. In Alps and other mountains: mud-slides. |
| Western Europe | Occasional earth tremors; onshore storms; very high winds; drought. | Groundwater contamination; air contamination especially from vehicle exhausts; radioactivity in Irish Sea; sea-level change; flooding. |
| Mediterranean | Earthquakes; volcanism; sea-level changes. | Floods; fires in forest and sclerophyll scrub; soil loss; lowering of water-tables. Air contamination from vehicles. |

main obvious environmental consequences are in the browning off of amenity grass in parks and gardens, in fires on moorlands and in conifer forests, and in the outbreaks of flea populations on domestic pets.

The Mediterranean Basin is probably the most volatile, however, with active faults giving earthquakes, especially in the Balkans and in Italy. The volcanoes of Italy are the most active, with regular eruptions of Etna, for example, threatening villages and communications on the lower slopes. Such is the faith in technology now, that the lava flows are the scenes of heroic engineering efforts to divert them. Helicopters, for example, were used in the 1990s to drop concrete and metal structures into the path of molten lava to try and divert it away from buildings and crops. Earthquakes are relatively frequent in Greece, Italy, Albania, and former Yugoslavia but few devastate as thoroughly as the 1953 impacts on the Ionian islands of Kefallinia and Zákynthos. Here, very few structures (some of which dated from the islands' time as part of the Venetian Republic) survived and many people emigrated to the mainland or to the USA and Australia.

Human activity has produced its own increased level of risk, where an environmental process acts as the delivery mechanism for a human-directed process or is exacerbated by human presence or actions. Acid deposition is a good example: the economy puts gases into the air but their effect depends upon climatic and meteorological patterns, upon the buffering effects of soil type, and on the vegetation. Fire and flood have worse effects where settlement and other installations become dense: industrial and housing investment on flood plains is often carried out without a prior appreciation of the statistics of flood hazard. Not many places have the detailed record of floods from 1500 onwards possessed by the Gotthard Pass region of Switzerland (Hächler, 1992). Sea-level change in northern latitudes may be an entirely natural phenomenon where isostatic recovery takes place after the withdrawal of ice; *if* it is due to global warming, then it interacts with human-affected systems. Like flooding, it has impacts upon low-lying installations of all kinds and the Netherlands are especially vulnerable, as are parts of eastern England and Scotland and nations like Estonia (Kont *et al.*, 1996).

Nevertheless, Europe is probably one of the less

hazardous parts of the globe to inhabit. The magnitude of unpredictable natural hazard is not very great compared, for example, with South Asia, and the technological capability of dealing with the aftermath is high. Given also that extreme events are relatively rare, national governments are usually ready with financial compensation for the worst-affected, though this rarely extends to rebuilding in a less precarious location.

# Overview

By way of ending this essay, let us first try to put the information into the framework which is currently so pervasive, that of trying to see whether the systems created are stable at 'present' (that is, in a short time frame) or likely to be 'sustainable' in a longer-term future. Secondly, a brief look back will remind us of the validity of the past as an element in our present lives.

## Stability and Sustainability

Any consideration of stability and sustainability must engage with the reasons for the adoption of agriculture in the period 8000–5500 BP. The features of this new economy (cattle together with domesticated grasses from western Asia) seem often to have had a trial period but after then to have dominated economy and culture completely (see Chapter 1). The original model of an immigrant culture and people has been supplemented with that of adoption of new ways of indigenous folk and this begs the question even further: rundown of animal populations, and human population growth are possible material explanations. Direct forcing by environmental change is not, however, adduced as a reason for this transition although, in a negative sense, hunter-gatherer economies lasted longest in the northern regions owing to a sort of 'climatic protectionism'.

The stability of pre-industrial (solar-powered) agriculture has been much debated. In general, it seems to have met the needs of growing populations, with only occasional Malthusian checks. Improvements in crop breeding and in farming practices seem to have allowed intensification without violating the fertility of the land, recognizing the need to maintain the nutrient status of the soils.

This was all accomplished without recourse to substantial imports of carbohydrates, though it did necessitate the extension of the cropped area at the expense of forests, freshwater wetlands, and salt-flats. Equally, some land was still devoted to pleasure rather than productivity and much lay beyond the margins of cultivation, especially during periods of cooler climate. So an interim judgement on agriculture in Europe to AD 1800 might be that it was sustainable without the subsidies of fossil fuel that were to follow, provided that population growth was relatively slow and that trade was unimpaired by war. Some periods of instability were manifest in particular regions, for instance where there were calorie or fuel deficits. An equally interesting but much more complex question asks whether there were environmental factors in the transition to full industrialism (Jones, 1977).

What of the two later eras? The fast answer is that anything based on a non-renewable resource is not sustainable. There are two possible softenings of that position. The first is that even if a resource is non-renewable it may with careful husbandry last a long time; the second is that if a resource is converted into knowledge of how to do without it, its exhaustion does not matter a great deal. Pearce and Atkinson (1992) measured whether national economies save more than the depreciation on their natural and man-made (*sic*) capital; the sustainability index is 0 for marginally sustainable nations and negative for the unsustainable. Not all European countries have been measured but some examples are:

| | |
|---|---|
| Czechoslovakia (*sic*) + 13 | Finland + 11 |
| Germany (pre-1990) + 10 | Hungary + 10 |
| The Netherlands + 14 | Poland + 16 |
| UK 0 | |

These may be compared with the extreme values of + 15 for Costa Rica and + 17 for Japan and −7 for Ethiopia and −14 for Mali. So that, extrapolating and interpolating from these examples, it looks as if Europe is in general on a sustainable path as measured by economic indices of this kind, with exceptions like the UK which reflects the value of pollution damage and the lack of capital accumulation gained from the exploitation of North Sea hydrocarbon reserves. Uncertainties abound: will global warming in fact result in colder winters in Europe, as well as increased aridity in the southern tier?

## Looking Back

Some of the matters that may be considered under this heading are given in Table 16.7. Readers may think that it is scarcely defensible to think of there being already relict landscapes of the post-industrial era, but huge structures which are losing money badly might, unless they pick up in an economic upturn, be regarded as monuments to the electronic fantasy age where business and pleasure are scarcely distinguishable: the Canary Wharf development in London's Docklands and Euro-Disney in Paris have both faltered during the 1990s, though both looked more viable at the time of writing (1995). They must, however, be large projects: the financial future of the Channel Tunnel is still rather uncertain. The plan of a busy modern city may in all its essentials be medieval and the environmental impact of pre-industrial agriculture may still dominate the landscape in parts of southern and eastern Europe, though it is fast disappearing: the Minho of Portugal still retains much of it. But the landscape of the hunters has mostly gone except perhaps on the English moorlands in their bleak peatiness, where an observer can directly be in touch with a landscape initiated back in the mid-Holocene.

Relics of the environmental *Zeitgeist* of the past also survive. These may be tangible, as in the visual examples suggested in Table 16.7: the cave painting, for example, depicts a human figure as it was stopping a large mammal in its tracks: a paradigm of what was soon to come even if not at that instant. Hobbema's landscape from the Netherlands of the seventeenth century shows the capabilities of humans even in pre-industrial times in producing straight lines out of nature, something carried forward by the railway, which forges through the weather without much regard for this feature of the natural environment, though Turner did leave in a small hare or rabbit, not visible in most reproductions. The final choice of Mondrian can easily be challenged, but in this final abstraction we seem to have the essential modernism of the office tower of Frankfurt or Birmingham. Relics may also be intangible. For medieval Europeans, nature might be an epiphany but it might also be the haunt of fearsome creatures as well, and the clearing of forest by the Cistercians was accompanied by holy water, incense, and psalms, just in case. Needing

**Table 16.7.** Looking back

| Economy | Relict landscapes | Non-material processes | A symbolic visual encapsulation | Places to visit |
|---|---|---|---|---|
| Hunter-gatherers | Mostly gone except for remnants of ancient woodlands and some areas of blanket bog in upland Britain; tundra and *tunturi* in far north. | 99% of human history as hunter-gatherers: some genetic traces even now? | Cave-silhouette from caves of France. | Glaisdale Moor, North Yorks (UK).[a] |
| Agriculturalists | City plans; rural road layouts and settlement patterns; ancient woodland; structures. | Attitudes to the wild; valuation of the non-human. | Mindert Hobbema: *Landscape, Middleharnis.* | Northern Portugal. |
| Industrialists | Much of present landscape; but many large installations disappeared, e.g. Consett, Ruhr. | Western world-view of 'growth' as a major aim; the 'modern'. | J. W. M. Turner: *Rain, Steam and Speed.* | Nova Huta, Poland. |
| Post-industrialist | Unused office blocks, e.g. Canary Wharf, London; fantasy lands. | Too soon to say? Loss of faith in the 'modern'? | P. Mondrian: increasing abstractions. | Open-air industrial museums; Euro-Disney. |

[a] See I. G. Simmons, 'The Earliest Cultural Landscapes of England', *Environmental Review*, 12, (1988), 105–16; I. G. Simmons and J. B. Innes, 'The Later Mesolithic Period (6000–5000 BP) on Glaisdale Moor, North York Moors', *Archaeol. J.* 145, (1988), 1–12.

to tame this otherness has certainly persisted: see how every mountain peak has a cairn if not a chapel or a café (Schama, 1994). Equally unspoken is the world-view of the nineteenth century and beyond, that growth is a sign of correct behaviour and that more is beautiful, whether of population (hence the encouragement to large families in authoritarian regimes) or goods. Only with the neo-Malthusian environmentalist movement of the 1960–72 period was this challenged seriously. Little fundamental notice was taken, and recently there has been zero growth of both economies and population in some (though not all) European countries. The scramble is to get out of the latter at least, and so a truly postmodern consciousness which incorporates environment into the automatic field of concern of humans and their own being seems a long way away. But possibly one lesson of the past, from the changes in the glacial/interglacial cycle, through the acceptance of agriculture to the burgeoning of manufacturing in the nineteenth century, let alone the rapidity of change with which we have all grown up, is that there can be rapid metamorphoses of thought and action (Simmons, 1993). In all of them, though, there is gain and there is loss. The environmental future of Europe, like its past, will contain both signs of hope and times of tribulation.

## REFERENCES

ALAYEV, E. B. *et al.*, 'The Russian Plain', in B. L. Turner *et al.* (eds.), *The Earth as Transformed by Human Action*, (Cambridge, 1990), 543–60.

ARTOBOLEBSKIY, S., 'Environmental Problems in the USSR', *Geography Review*, 4 (1991), 12–16.

Bätzing, W., *Die Alpen—Entstehung und Gefährdung einer europäischen Kulturlandschaft*, (Munich, 1991).

Beach, H., 'Comparative Systems of Reindeer Herding', in J. G. Galaty and D. L. Johnson (eds.), *The World of Pastoralism*, (London, 1990), 255–98.

Berdnikov, K., 'Draft of the Povolje's Wood of the 17th Century as a Source for Historical Monitoring of the Environment', in I. G. Simmons and A. M. Mannion (eds.), *The Changing Nature of the People–Environment Relationship: Evidence from a Variety of Archives*, (Prague, 1995), 27–31.

Clapp, B. W., *An Environmental History of Britain since the Industrial Revolution*, (London and New York, 1994).

Douglas, I., *The Urban Environment*, (London, 1983).

Geddes, D. *et al.*, 'Postglacial Environments, Settlement and Subsistence in the Pyrenees: The Balma Margineda, Andorra', in C. Bonsall (ed.), *The Mesolithic in Europe*, (Edinburgh, 1989), 561–88.

Göransson, H., 'Man and the Forests of Nemoral Broad-Leafed Trees during the Stone Age', *Striae*, 24 (1986), 143–52.

Hächler, S., 'Überschwemmungen in Schweizer Alpenraum seit dem Spätmittelalter. Raum-zeitliche Rekonstruktion von Schadenmustern und gesellschaftliche Reaktionen', *Environmental History Newsletter*, 4 (1992), 12–18.

Herlihy, D., 'Ecological Conditions and Demographic Change', in R. L. DeMolen (ed.), *One Thousand Years: Western Europe in the Middle Ages*, (Boston, 1974), 3–43.

Hewitt, J., *European Environmental Almanac*, (London, 1995).

Jeleček, L., 'The Ecological Situation Changes in the Czech Republic 1948–1989: Some of their Historic-Geographical Causes and Connections', in I. G. Simmons and A. M. Mannion (eds.), *The Changing Nature of the People–Environment Relationship: Evidence from a Variety of Archives*, (Prague, 1995), 101–11.

Jochim, M. A., *Hunter-Gatherer Subsistence and Settlement: A Predictive Model*, (London and New York, 1976).

—— 'The Late Mesolithic in Southwestern Germany: Culture Change or Population Decline?', in P. M. Vermeersch and P. van Peer (eds.), *Contributions to the Mesolithic in Europe*, (Leuven, 1990), 183–91.

Jones, E. L., *The European Miracle: Environments, Economies and Geopolitics in the History of Europe and Asia*, (Cambridge, 1981).

—— 'Environment, Agriculture and Industrialization in Europe', *Agricultural History*, 51 (1977), 491–502.

Kont, A., Ratas, U., Puurman, E., Ainsaar, M., Pärtel, M., and Zobel, M., 'Eustatic fluctuations of the World Ocean and their Impact on the Environment and Social life of Estonia', in J.-M. Punning (ed.), *Estonia in the System of Global Climate Change*, Institute of Ecology Publication 4/96, (Tallinn, 1996), 104–22.

Lekakis, J. N., 'Social and Ecological Correlates of Rural Fires in Greece', *Journal of Environmental Management*, 43 (1995), 41–7.

Lepart, J., and Debussche, M., 'Human Impact on Landscape Patterning: Mediterranean Examples', in A. J. Hansen and F. di Castri (eds.), *Landscape Boundaries: Consequences for Biotic Diversity and Ecological Flows*, (New York and Heidelberg, 1993), 76–106.

Lourenco, L., 'Avaliacao do risco de incendio nas matas e florestas de Portugal Continental', *Finisterra*, 27 (1992), 115–40.

MANNION, A. M., *Agriculture and Environmental Change. Temporal and Spatial Dimensions*, (Chichester, 1995).

MICHELSEN, P., 'Irrigation in Norway and Elsewhere in Northern Europe', *Tools and Tillage*, 7 (1987), 243–59.

MORENO, D., 'Past Multiple Use of Tree Land in the Mediterranean Mountain: Experiments on the Sweet Chestnut Culture', *Environmental History Newsletter*, 2 (1990), 37–49.

NETTING, R. M., *Balancing on an Alp: Ecological Change and Continuity in a Swiss Mountain Community* (Cambridge, 1981).

OVIEDO, F. B., 'Estudio silvopastoral de la Dehesa Boyal de Alia (Caceres)', *Ecologia*, 3 (Madrid, 1989), 107–15.

PALMER, M., and NEAVERSON, P., *Industry in the Landscape, 1700–1900*, (London and New York, 1994).

PARK, C. C., *Chernobyl: The Long Shadow*, (London and New York, 1989).

PAULI, L., *The Alps: Archaeology and Early History*, (London, 1984).

PEARCE, D. W., and ATKINSON, G. D., *Are National Economies Sustainable? Measuring Sustainable Development* University College London: CSERGE Working Paper GEC 92–11, (London, 1992).

PEARCE, F., 'Dead in the Water', *New Scientist*, 145 (1995), 26–31.

PFISTER, C., 'The Early Loss of Ecological Stability in an Agrarian Region', in P. Brimblecombe and C. Pfister (eds.), *The Silent Countdown: Essays in European Environmental History*, (Berlin, 1990), 37–55.

PIUSSI, P., and STIAVELLI, S., 'Forest History of the Cerbaie Hills (Toscana, Italy)', in F. Salbitano (ed.), *Human Influence on Forest Ecosystems Development in Europe*, (Bologna, 1988), 109–20.

POUNDS, N. J. G., *An Historical Geography of Europe 1500–1840*, (Cambridge, 1979).

—— *An Historical Geography of Europe 1800–1914*, (Cambridge, 1985).

PRYDE, P., *Environmental Management in the Soviet Union*, (Cambridge, 1991).

RACKHAM, O., *Trees and Woodlands in the English Landscape*, (London, 1976).

—— *Ancient Woodland: Its History, Vegetation and Uses in England*, (London, 1980).

RASMUSSEN, P., 'Leaf-Foddering in the Earliest Neolithic Agriculture: Evidence from Switzerland and Denmark', *Acta Archaeologia*, 60 (1988), 71–86.

RICHARDS, J. F., 'World Environmental History and Economic Development', in W. C. Clarke and R. E. Munn (eds.), *Sustainable Development of the Biosphere*, (Laxenburg, Austria, 1986), 53–72.

ROBERTS, A., 'A Rising Cloud of Midges in the Scottish Highlands?', in R. A. Butlin and N. Roberts (eds.), *Ecological Relations in Historical Times: Human Impact and Adaptation*, (Oxford, 1995), 88–98.

ROWLEY-CONWY, P., 'The Laziness of the Short-Distance Hunter: The Origins of Agriculture in Western Denmark', *J. Anthropol. Archaeol.* 3 (1984), 300–24.

SALLARES, R., *The Ecology of the Ancient Greek World*, (Ithaca, NY, 1991).

SCHAMA, S., *Landscape and Memory*, (London, 1994).

SIIRIÄINEN, A., 'On the Cultural Ecology of the Finnish Stone Age', *Suomen Museo*, (1981), 5–40.

SIMMONS, I. G., *Environmental History*, (Oxford, 1993).

—— *The Environmental Impact of Later Mesolithic Cultures*, (Edinburgh, 1996).

STEVENSON, A. C., and HARRISON, R. J., 'Ancient Forests in Spain: A Model for

Land-Use and Dry Forest Management in South-west Spain from 4000 BC to 1900 AD', *Proceedings of the Prehistoric Society*, 58 (1992), 227–47.

STONE, P. (ed.), *The State of the World's Mountains: A Global Report*, (London and Atlantic Highlands, NJ, 1992).

SUSMEL, L., 'Tecnica dei Veneziani nei boschi di rovere', *Monti e Bosci*, 7 (1956), 212–22.

VASARI, Y., 'The Ecological Background to the Livelihood of Peasants in Kuusamo (NE Finland), during the Period 1670–1970', in P. Brimblecombe and C. Pfister (eds.), *The Silent Countdown: Essays in European Environmental History*, (Berlin, 1990), 125–34.

VIAZZO, P. P., 'An Anthropological Perspective of Environment, Population, and Social Structure in the Alps', in P. Brimblecombe and C. Pfister (eds.), *The Silent Countdown: Essays in European Environmental History*, (Berlin, 1990), 56–67.

WATSON, A. M., *Agricultural Innovation in the Early Islamic World: The Diffusion of Crops and Farming Techniques 700–1100*, (Cambridge, 1983).

WATSON, C., 'The Vegetational History of the Northern Apennines, Italy: Information from Three New Sequences and a Review of Regional Vegetational Change', *Journal of Biogeography*, 23 (1996), 805–41.

# Index

# Index